# A.D.A.M.® Interactive Anatomy Dissection Manual

Martha DePecol Sanner
Middlesex Community-Technical College

Harry Greer
Cayuga Community College

PRENTICE HALL
Upper Saddle River, NJ 07458

Executive Editor: David Kendric Brake
Senior Acquisitions Editor: Linda Schreiber
Assistant Vice President of Production and Manufacturing: David W. Riccardi
Special Projects Manager: Barbara A. Murray
A.D.A.M. Imagery and Content Courtesy of A.D.A.M. Software, Inc.
www.adam.com
Manufacturing Buyer: Benjamin Smith
Cover Design: Joseph Sengotta
Cover Imagery: Courtesy of A.D.A.M. Software, Inc.
www.adam.com

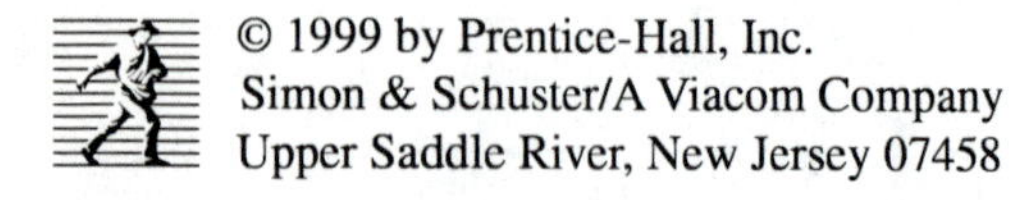



Printed in the United States of America

10 9 8 7 6 5 4 3 2 1

ISBN 0-13-082638-3

Prentice-Hall International (UK) Limited, *London*
Prentice-Hall of Australia Pty. Limited, *Sydney*
Prentice-Hall Canada, Inc., *Toronto*
Prentice-Hall Hispanoamericana, S.A., *Mexico*
Prentice-Hall of India Private Limited, *New Delhi*
Prentice-Hall of Japan, Inc., *Tokyo*
Simon & Schuster Asia Pte. Ltd., *Singapore*
Editora Prentice-Hall do Brasil, Ltda., *Rio de Janeiro*

The authors wish to thank Jennifer Davis
for her hard work in the development
of this manual.

The authors welcome suggestions for change.
Please send your comments to:

Martha DePecol Sanner
Middlesex Community-Technical College
100 Training Hill Road
Middletown, CT 06457
860-343-5780
MX_Sanner@Commnet.edu

Harry Greer
Cayuga Community College
197 Franklin Street
Auburn, NY 13021
315-255-1743 x314
ProfG@aol.com

# A.D.A.M.® Interactive Anatomy Dissection Manual

# Table of Contents

## User's Guide

# Tutorial: Major Organs

## Skeletal System

# Muscular System

## Nervous System

# Digestive System



# Urinary and Reproductive Systems

# Macintosh Guide to Using A.D.A.M. Interactive Anatomy

## Starting A.D.A.M. Interactive Anatomy

1. Turn on the computer and allow it to warm up. Place the A.D.A.M. Interactive CD-ROM in the CD tray.

2. The CD-ROM has been installed on your computer. Do not click on the CD-ROM icon. Find the folder (shown to the right) labeled "Interactive Anatomy 3.0" on your hard drive. Double click on that folder.

Interactive Anatomy 3.0

3. Double click on the icon shown to the right to open the program.

4. If you are asked if you want to switch to 256 colors, click on "switch".

## Content Types

There are four major content types within A.D.A.M. Interactive Anatomy: **Dissectible Anatomy**, **Atlas Anatomy**, **3D Anatomy**, and **Slide Shows**.

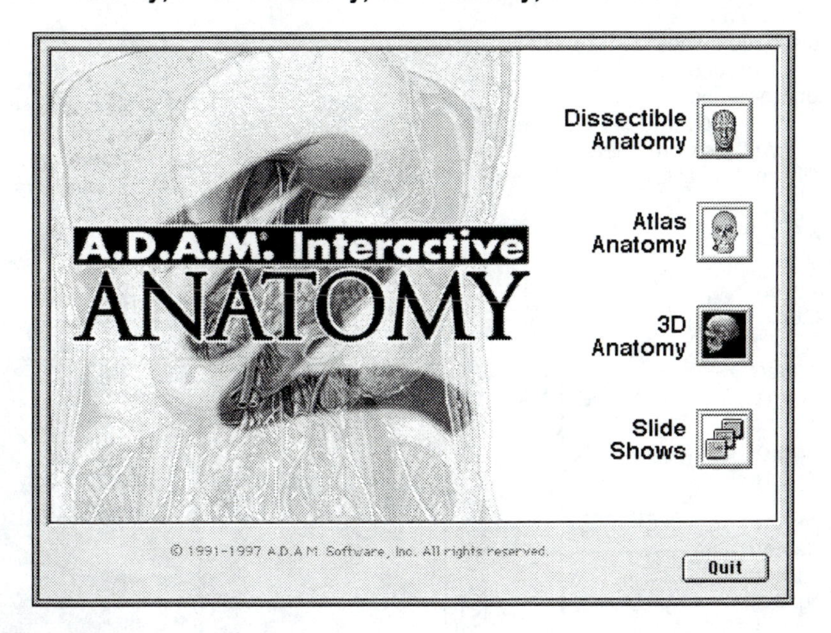


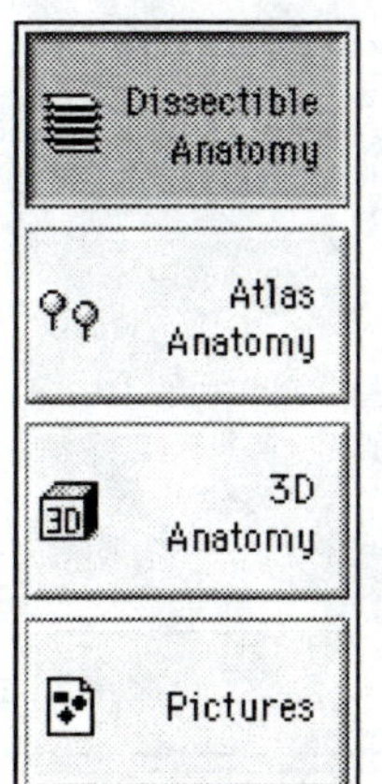


**Dissectible Anatomy:** Allows you to dissect layer by layer in anterior, posterior, lateral, or medial views and also male or female gender.

**Atlas Anatomy:** Allows you to choose pinned images, cadaver photographs, and radiographs.

**3D Anatomy:** Allows you to identify structures on 3D models.

**Pictures:** Allows you to add your own pictures or bit mapped images.

## Working With Dissectible Anatomy

Click on the **Dissectible Anatomy** button. You should see the following:

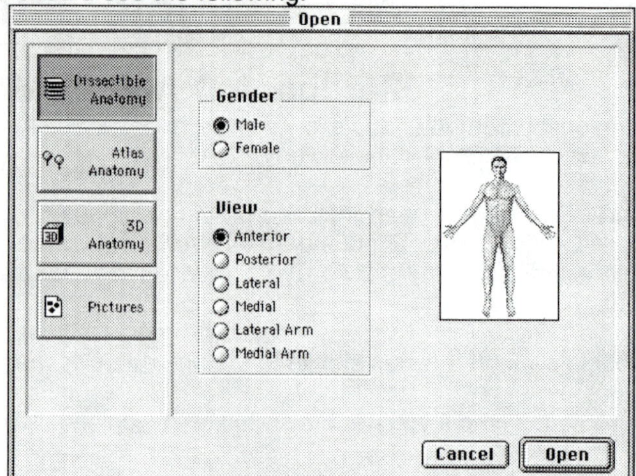


The opening screen allows you to choose:

- **Male** or **Female**
- **Anterior**, **Posterior**, **Lateral**, **Medial, Lateral Arm**, or **Medial Arm**
- By clicking on the buttons you may open **Dissectible Anatomy, Atlas Anatomy, and 3D Anatomy**.
- Choose a gender and view and click "open".

## Dissectible Anatomy Icons

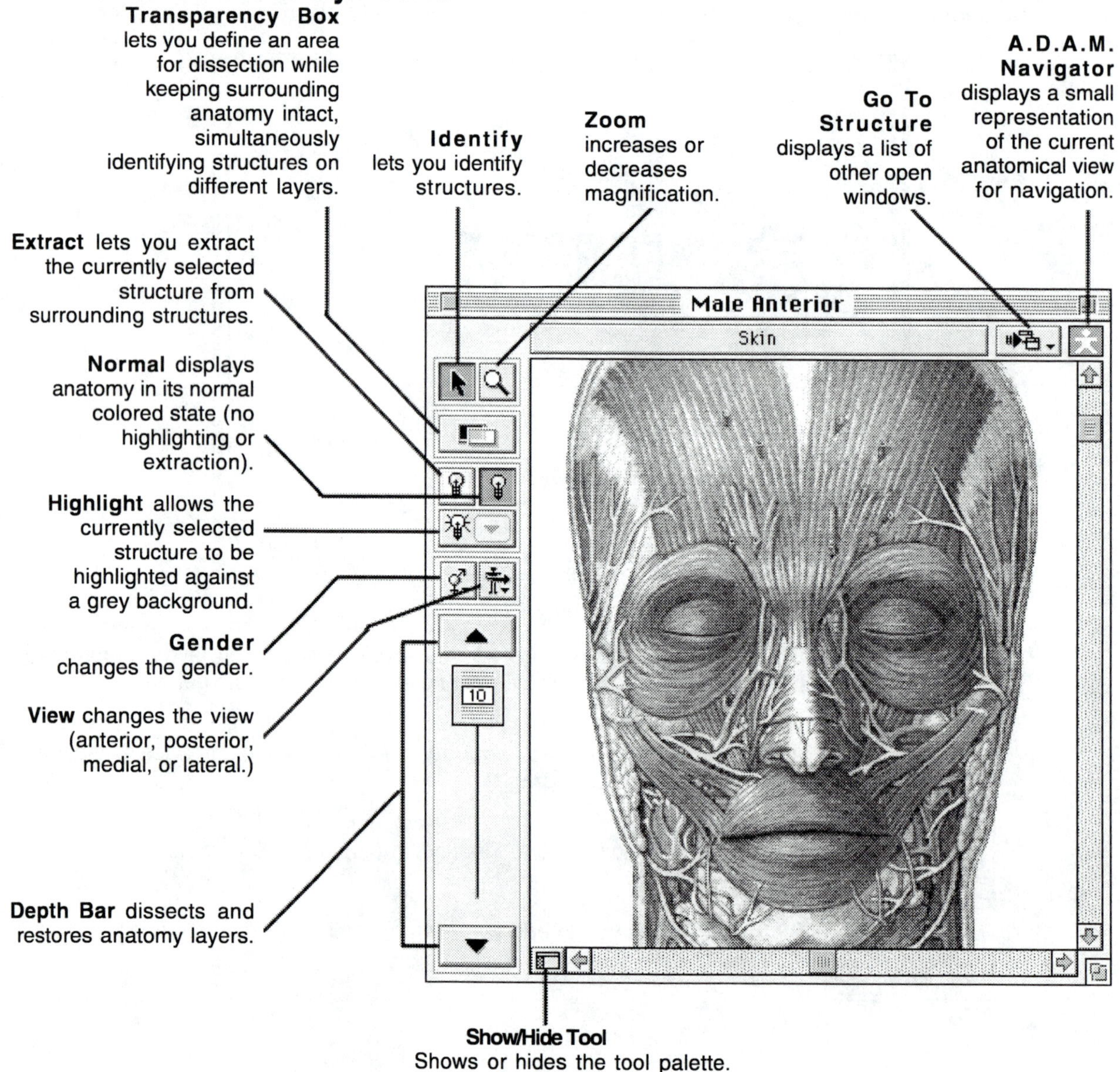

**Transparency Box** lets you define an area for dissection while keeping surrounding anatomy intact, simultaneously identifying structures on different layers.
**Identify** lets you identify structures.
**Zoom** increases or decreases magnification.
**Go To Structure** displays a list of other open windows.
**A.D.A.M. Navigator** displays a small representation of the current anatomical view for navigation.
**Extract** lets you extract the currently selected structure from surrounding structures.
**Normal** displays anatomy in its normal colored state (no highlighting or extraction).
**Highlight** allows the currently selected structure to be highlighted against a grey background.
**Gender** changes the gender.
**View** changes the view (anterior, posterior, medial, or lateral.)
**Depth Bar** dissects and restores anatomy layers.
Male Anterior
Skin
10
**Show/Hide Tool** Shows or hides the tool palette.

# Hints on Using Dissectible Anatomy

## 1. Opening Content: Dissectible Anatomy

**If A.D.A.M. Interactive Anatomy is launched but nothing appears on my screen:**

Go to the **File** on the menu and select **Open**.

A box will appear presenting the content types.

File Edit Options Win
New Slide Show... ⌘N
Open Slide Show...
Open... ⌘O
Close ⌘W
Save ⌘S
Save As...
Page Setup...
Print Preview...
Print... ⌘P
Quit ⌘Q

## 2. Enlarging the Image to Fit the Screen

There are two different ways to enlarge the image:

1. Click once on the **Zoom Box** to enlarge the screen. Click on it again to restore the screen to the original size.
2. Point to the **Size Box**, click and drag the box toward the lower right corner of the screen. Drag it back again to return the screen to its original position.

**Zoom Box** sizes the screen so that all of its contents are visible.

**Size Box** changes the shape or size of the screen.

Choose either option to enlarge the image.

## 3. Using Highlight & Extract

There are two ways to use the highlight:

- Click the **Highlight** button and select one of the systems listed in the drop down menu. Only the structures in that selected system will be displayed in color. Structures not in the selected system will be shown in gray.

- If you are trying to find different structures that are close together, such as the muscles of the leg, click the **Highlight** button and then the various muscles. As you click the muscles, they will be highlighted in color and their name will be displayed. This allows you to see the entire muscle highlighted while other structures are dimmed or gray.

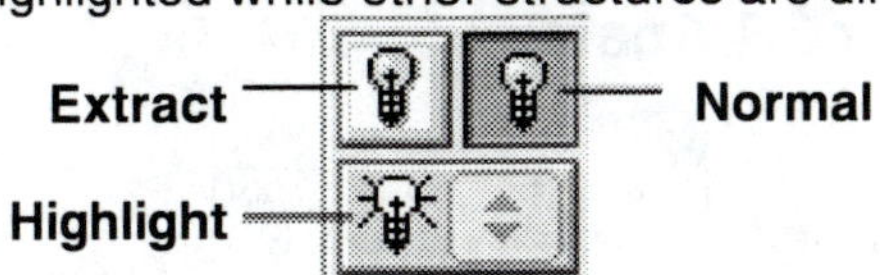


- After clicking on a particular structure, you can extract that structure from the surrounding tissue by clicking on the **Extract Tool**.

## 4. Making the Image Go From Black and White To Color

After using the **Highlight** button you can make the entire image colored again by clicking on the **Normal** button.

## 5. Centering the Image

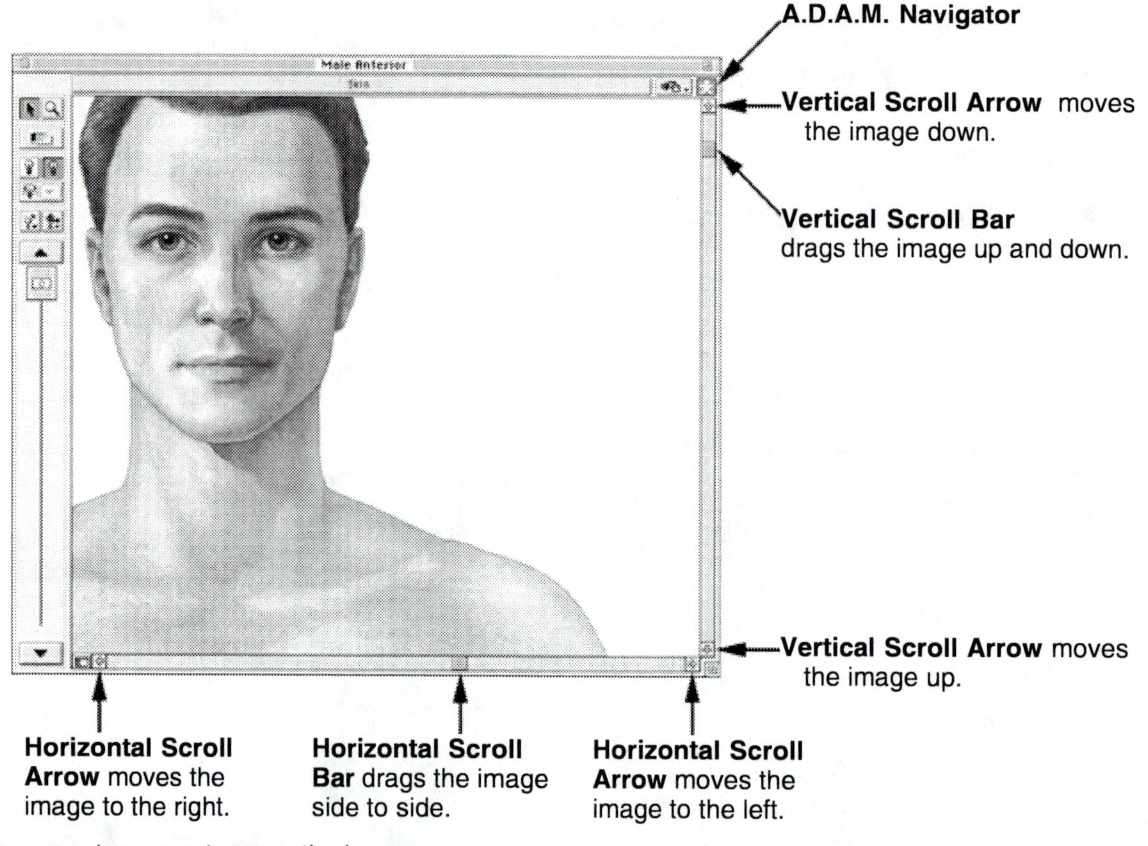

A.D.A.M. Navigator
Vertical Scroll Arrow moves the image down.
Vertical Scroll Bar drags the image up and down.
Vertical Scroll Arrow moves the image up.
Horizontal Scroll Arrow moves the image to the right.
Horizontal Scroll Bar drags the image side to side.
Horizontal Scroll Arrow moves the image to the left.

There are two ways to move the image:

1. Click on the **scroll arrows** to move the image to the desired location or drag on the **scroll bar** to move the image faster.

2. Use the **A.D.A.M. Navigator**
   1. Place your cursor over the **A.D.A.M. Navigator**.
   2. You will see an image of the entire body appear. The cursor will turn into a hand.
   3. Drag the hand (cursor) over the desired body part you want to go to.

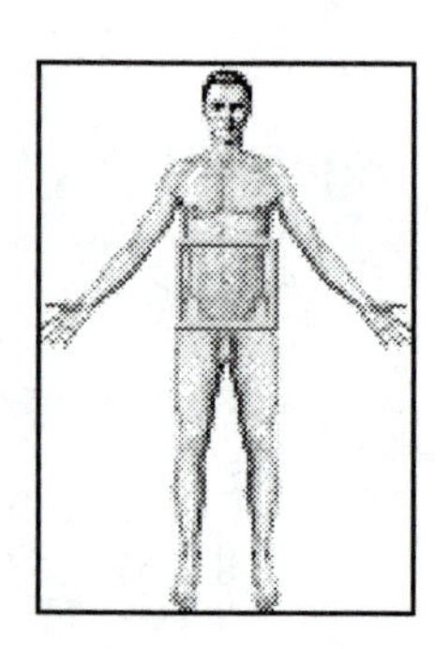

## 6. To Choose Atlas Anatomy or 3D Anatomy

Go to the **File** on the menu and select **Open**.

A dialog box will appear presenting the content types.

File Edit Options Win
New Slide Show... ⌘N
Open Slide Show...
Open... ⌘O
Close ⌘W
Save ⌘S
Save As...
Page Setup...
Print Preview...
Print... ⌘P
Quit ⌘Q

# Working with Atlas Anatomy

## Choosing Images in Atlas Anatomy

Go to the File menu and select open. Select Atlas Anatomy.

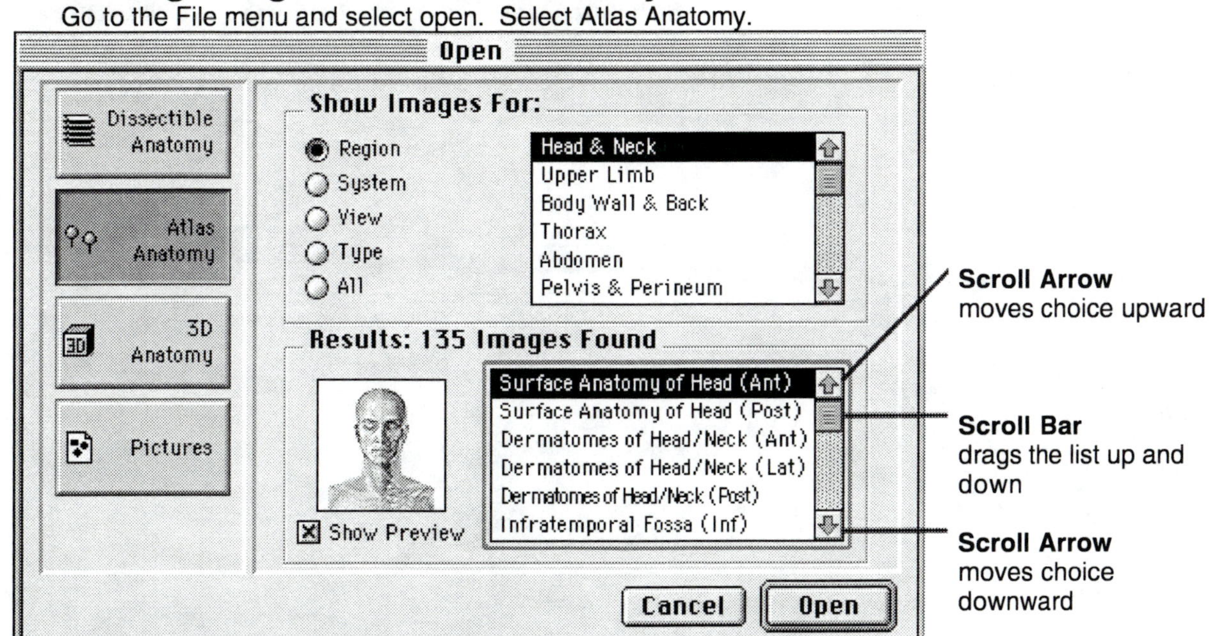


**1. Choose Region, System, View, Type, or All:**

You may need to scroll down to find the region, system, or view

**Region:** Allows you to search an area of the body. Choices are:

- *Head & Neck*
- *Upper Limb*
- *Body Wall & Back*
- *Thorax*
- *Abdomen*
- *Pelvis & Perineum*
- *Lower Limb*

**System:** Allows you to search a system of the body. Choices are:

- *Cardiovascular*
- *Digestive*
- *Endocrine*
- *Immune*
- *Integumentary*
- *Lymphatic*
- *Muscular*
- *Nervous*
- *Reproductive*
- *Respiratory*
- *Skeletal*
- *Urinary*

If you choose an image by system, only the pins corresponding to the system will be visible on the image.

**View:** Allows you to search a view of the body. Choices are:

- *Anterior*
- *Posterior*
- *Lateral*
- *Medial*
- *Superior*
- *Inferior (Most cross sections are inferior views.)*
- *Non-standard*

**Type:** There are three types of images. Choices are:

- *Illustrations*
- *Cadaver Photographs (including images of bones)*
- *Radiographs*

**All:** Allows you to review the entire list of images

**2. Use the scroll arrows and bar to find the image you are looking for.**

The images may or may not be in alphabetical order depending on the version of A.D.A.M. Interactive Anatomy you are using. If they are not in alphabetical order, follow directions to find the proper image by dragging the scroll button to the indicated depth. Click **System**, click **Nervous**. Scroll to **Spinal Cord, Vessels, & Meninges**. Click **Open.**

# Atlas Anatomy Icons

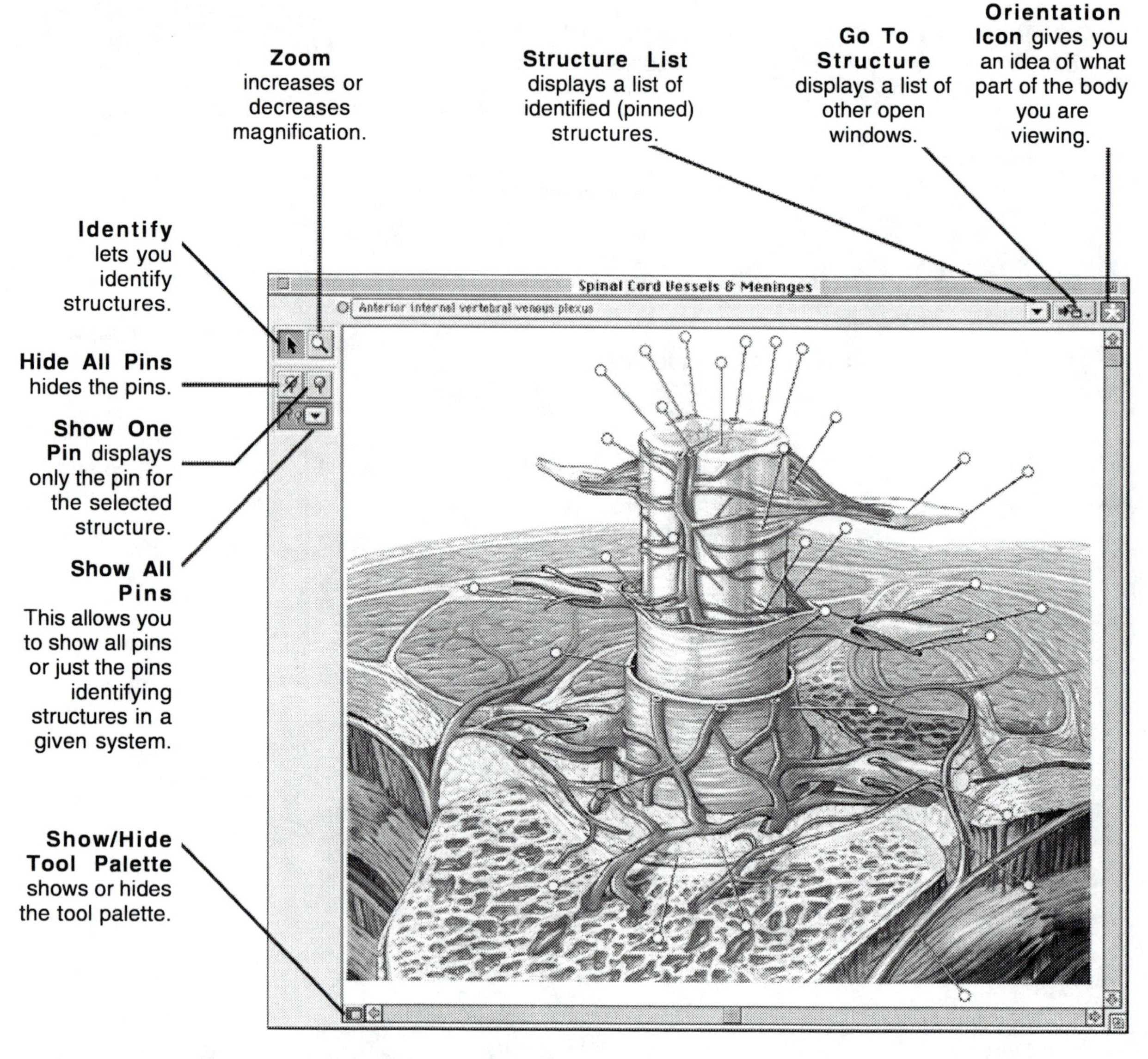

**Zoom** increases or decreases magnification.
**Structure List** displays a list of identified (pinned) structures.
**Go To Structure** displays a list of other open windows.
**Orientation Icon** gives you an idea of what part of the body you are viewing.
**Identify** lets you identify structures.
**Hide All Pins** hides the pins.
**Show One Pin** displays only the pin for the selected structure.
**Show All Pins** This allows you to show all pins or just the pins identifying structures in a given system.
**Show/Hide Tool Palette** shows or hides the tool palette.
Spinal Cord Vessels & Meninges
Anterior internal vertebral venous plexus


# Hints on Using Atlas Anatomy

## 1. Opening Content: Atlas Anatomy

**If A.D.A.M. Interactive Anatomy is launched but nothing appears on my screen:**

Go to the **File** on the menu and select **Open**.

A dialog box will appear presenting the content types.

File Edit Options Win

New Slide Show... ⌘N
Open Slide Show...
Open... ⌘O

Close ⌘W
Save ⌘S
Save As...

Page Setup...
Print Preview...
Print... ⌘P

Quit ⌘Q

## 2. To Show All Systems

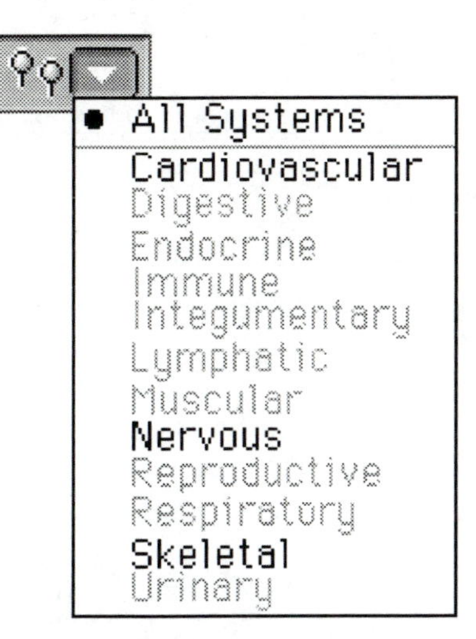


If you choose an image by system, only the pins corresponding to the system will be visible on the image. Click the **Show All Pins** button and select **All Systems** from the drop down menu to show all systems.

## 3. Atlas Anatomy Structure List

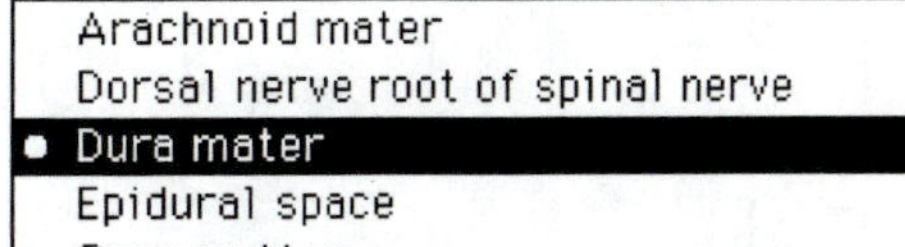


If you can't find a specific pinned structure, click and hold the **Atlas Anatomy Structure List** at the top of the image window. Drag down to the structure you are trying to find. (The structures are arranged alphabetically.) The head of the corresponding pin will turn green.

You can restrict the list to only structures within a certain system, such as the nervous system, by clicking on the **Show All Pins** button and drag to the system of interest.

## 4. Orientation Icon & Zoom Tool

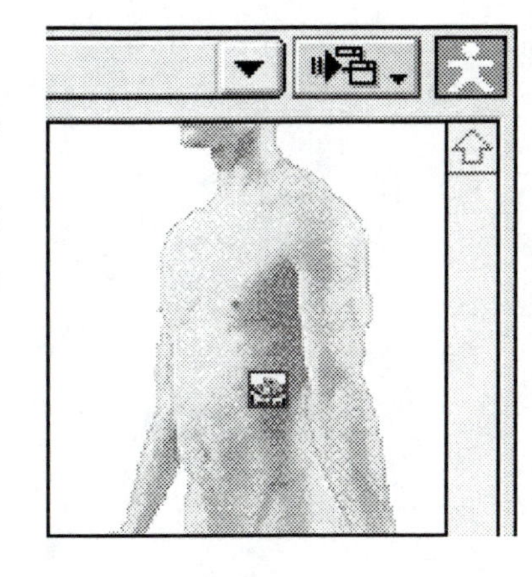

The orientation icon allows the user to determine what region of the body the image is from. To get an overall view of the image, place your cursor over the **Orientation Icon** in the upper right corner of the screen. This will show you the body location that is being observed.

Use the **Zoom Tool**  to fit the image within the window.

## 5. Scrolling

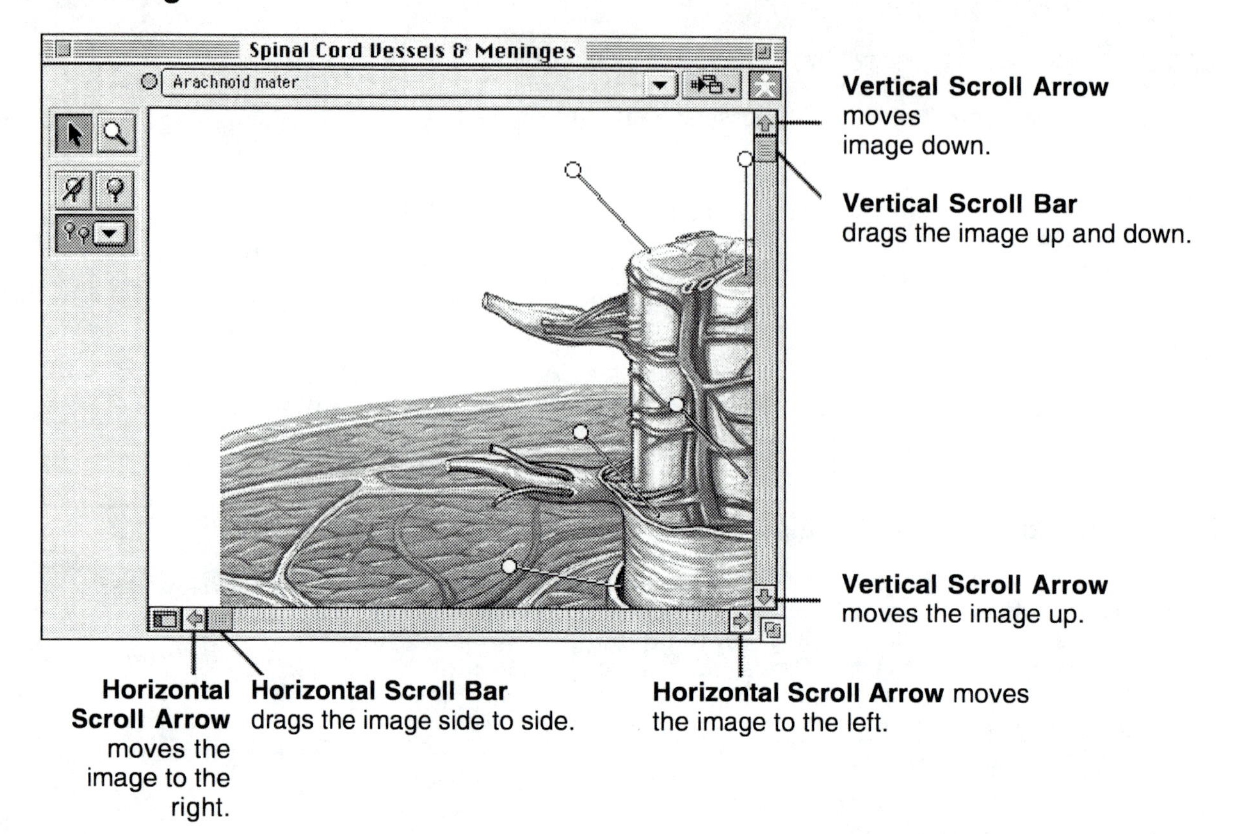

Spinal Cord Vessels & Meninges
Arachnoid mater
Vertical Scroll Arrow moves image down.
Vertical Scroll Bar drags the image up and down.
Vertical Scroll Arrow moves the image up.
Horizontal Scroll Arrow moves the image to the right.
Horizontal Scroll Bar drags the image side to side.
Horizontal Scroll Arrow moves the image to the left.

# Working with 3D Anatomy

## Choosing Images in 3D Anatomy

Go to the File menu and select open. Select 3D Anatomy. You may have additional 3D images installed on the computer you are using. Choose a 3D model you would like to view, then click Open.

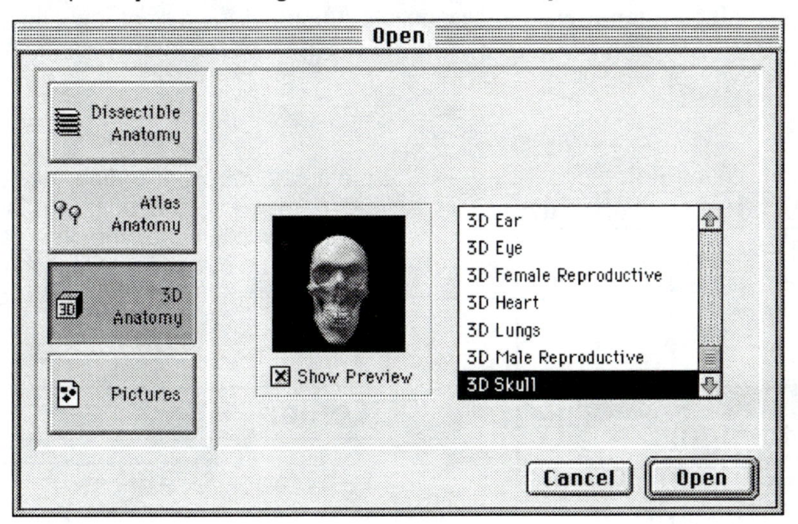


## 3D Anatomy Icons

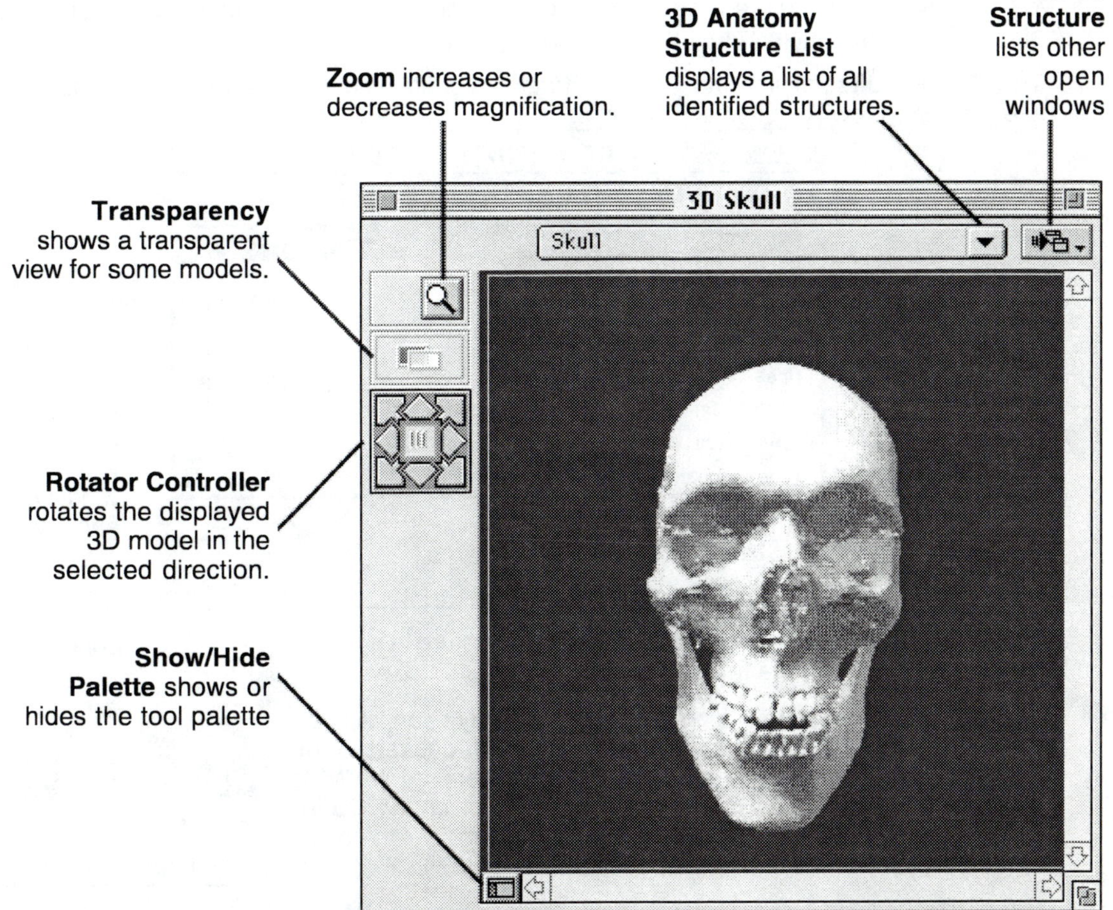

**Go To Structure** lists other open windows
**3D Anatomy Structure List** displays a list of all identified structures.
**Zoom** increases or decreases magnification.
**Transparency** shows a transparent view for some models.
**Rotator Controller** rotates the displayed 3D model in the selected direction.
**Show/Hide Palette** shows or hides the tool palette
3D Skull
Skull


Click and hold on the **3D Anatomy Structure List** button and locate or find a structure within the model. The selected structure will highlight in blue.

# Finding Structures

There are two ways to find structures:

1. **Find**
2. **List Manager**

Both the **Find** and **List Manager** work in Dissectible Anatomy, Atlas Anatomy and 3D Anatomy.

## Using Find

1. Choose **Find** from the **Edit** menu.

2. Lets say you want to find the femur. Type the word **Femur** into the box. It's not necessary to type in a whole word, the first part of a word will often do. You may want to limit the structure to the skeletal system. Click find when you are done.

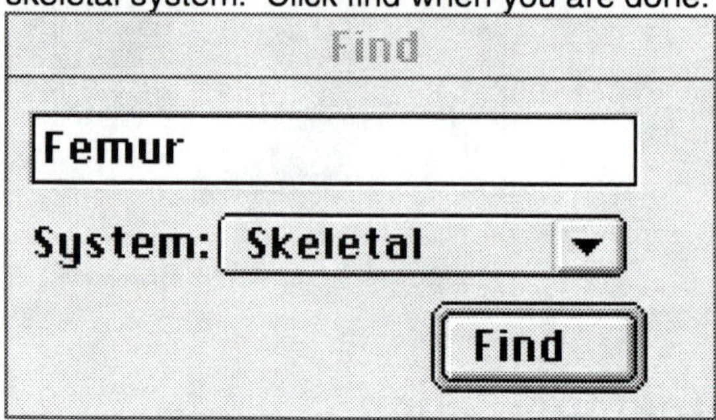


If you try to type in a word like **Large Intestine** it may not be found because the word is not specific enough. Try an alternative word such as **Colon**.

3. The **Find Results** box indicates that 25 structures containing the word femur were found. Scroll to locate the item you wish to find and select it.

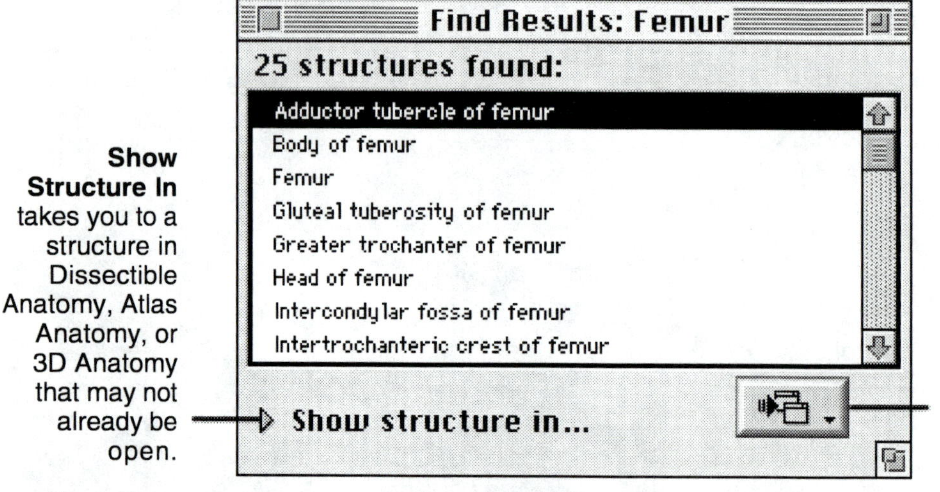


**Show Structure In** takes you to a structure in Dissectible Anatomy, Atlas Anatomy, or 3D Anatomy that may not already be open.

**Go To Structure Button** takes you to a structure in a window that is already open.

You now have two options. You can find the structure in a window that is already open or in a window that is not already open.

To display the structure in a window that is already open, click on the **Go To Structure** button and select the desired open window. For example, if you have been working in "Male Anterior" and you are trying to find the femur, click on the **Go To Structure** button and select "**Male Anterior**". Any open windows that do not contain the structure will be gray.

Male Anterior
All Windows

If you are working in Dissectible Anatomy you will be brought to the layer where the structure is best viewed. You may need to scroll back to the layer you were originally on to continue the dissection.

To return the image to full color, click on the **Normal** button.

To find a structure within a content type which is not open, click on "**Show Structure In**". The following box will appear:

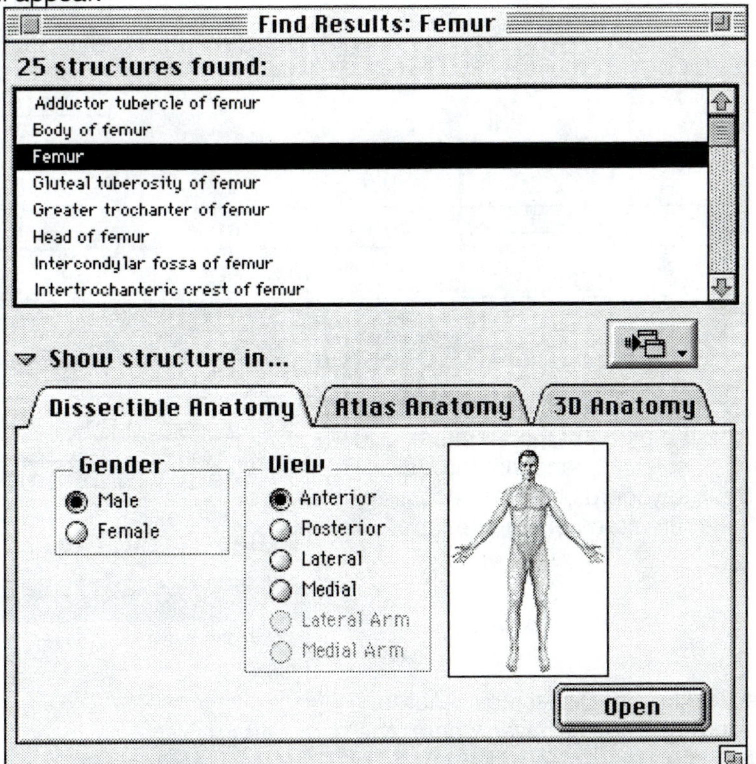


Choose the content type you want to view the structure in. Gray options are not available.

To return the image to full color, click on the **Normal** button.

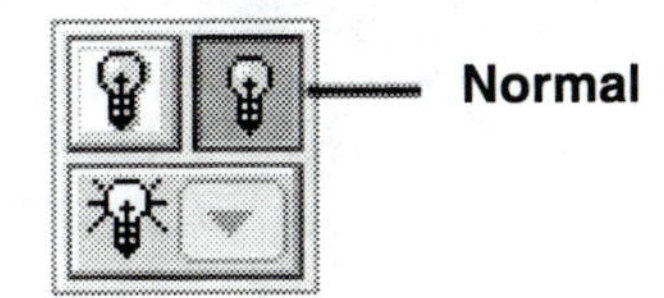

## Using the List Manager

1. **List Manager** works in Dissectible Anatomy, Atlas Anatomy, and 3D Anatomy. To open the **List Manager** choose **List Manager** from the **Edit** menu. If you are working in Dissectible Anatomy you can also access the **List Manager** by clicking on the **Structure List** button at the top of a Dissectible Anatomy Window.

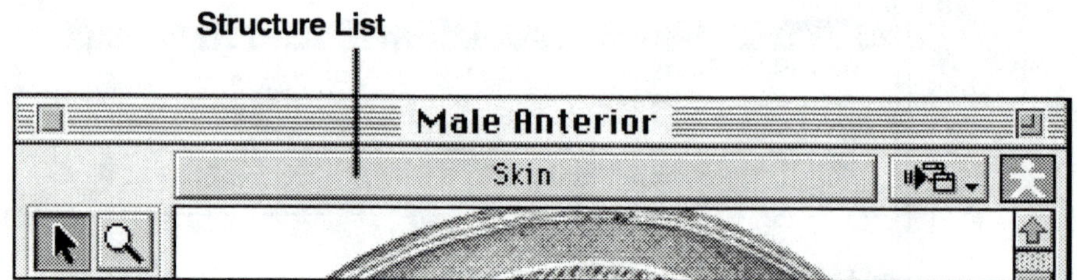


2. To quickly find a structure using **List Manager** type the name of the structure you wish to find , select it, and hit the return button on your keyboard. It's not necessary to type in the entire name of the structure.

3. If you are working in Dissectible Anatomy you will be brought to the layer where the structure is best viewed. You may need to scroll back to the layer you were originally on to continue the dissection.

4. To return the image to full color, click on the **Normal** button.

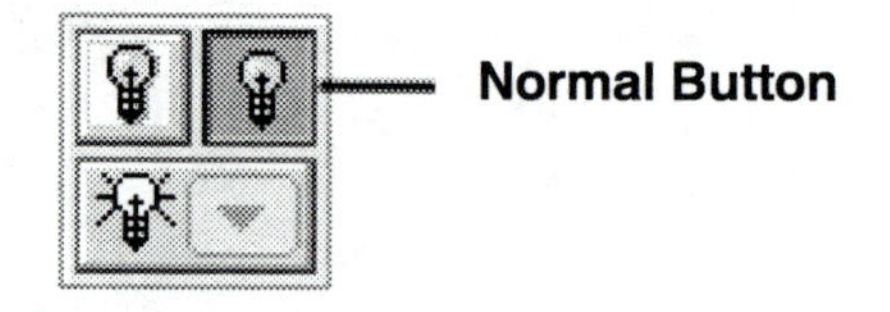

5. You may also choose to restrict the list of structures further. Click on **Restrict list to**. This will allow you to restrict the list to any system.

   If you are in Dissectible Anatomy, restricting the area to **Visible Area Only** will keep your image in the window from moving to a different part of the body, although you may go to a different layer. For example, if you are looking at Male Anterior head and you type in "stomach" you will not be shown the stomach because it's not in that area of the body.

## Changing the Language Lexicon

Go to the **Edit** menu and choose **Preferences....**

- **To remove a language lexicon:** Select the language under the primary lexicon or secondary lexicon box that you want to remove. Click this button [ < ].

- **To add a language lexicon:** Select the language under the available lexicon box. Click the [ > ] button next to either the primary lexicon or secondary lexicon box. The primary lexicon box must be empty to add a lexicon to it. Several languages may be added to the secondary lexicon box.

- For the purposes of this lab manual, use **English (undergraduate)** as a primary lexicon and **English** as a secondary lexicon.

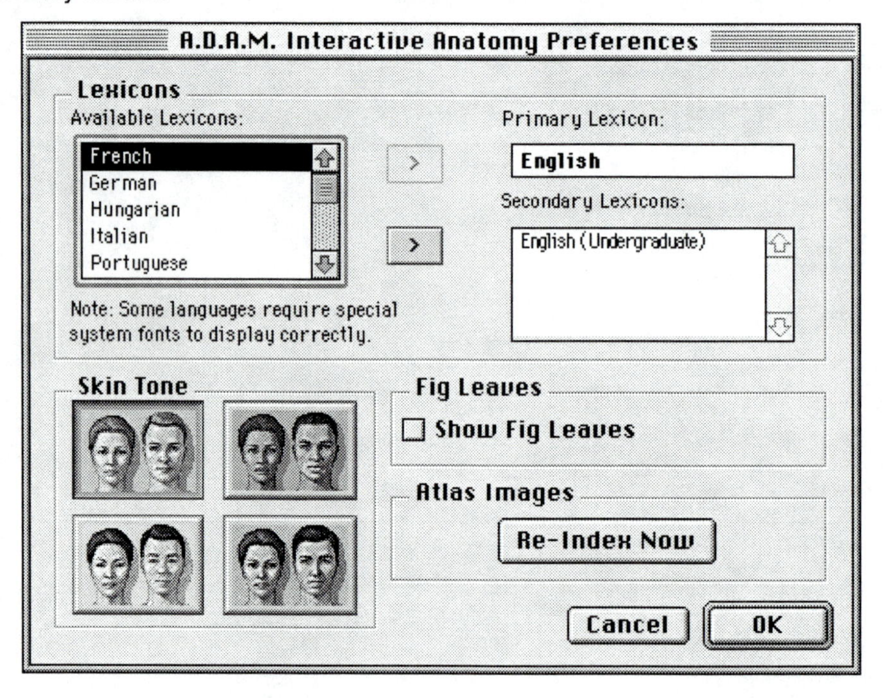

## Changing the Skin Tone

Go to the **Edit** menu and choose **Preferences....**

Edit Options Window
Can't Undo ⌘Z
Cut ⌘X
Copy ⌘C
Paste ⌘V
Clear
Find... ⌘F
List Manager... ⌘L
Internet Connection
Preferences...

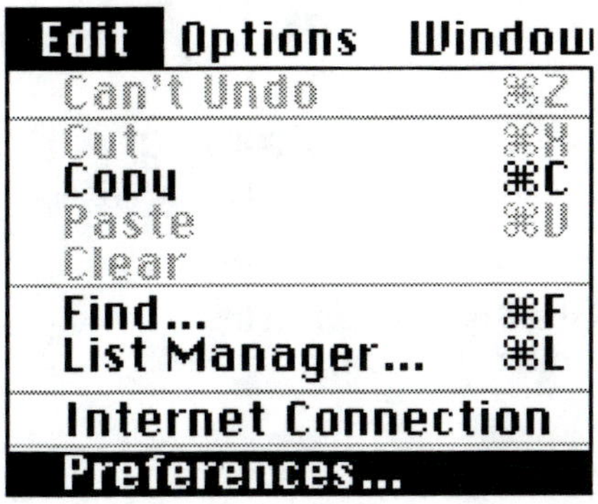

Select a skin tone from one of the available choices.

## Fig Leaf Option (Modesty Setting)

Go to the **Edit** menu and choose **Preferences....**

Edit Options Window
Can't Undo ⌘Z
Cut ⌘X
Copy ⌘C
Paste ⌘V
Clear
Find... ⌘F
List Manager... ⌘L
Internet Connection
Preferences...

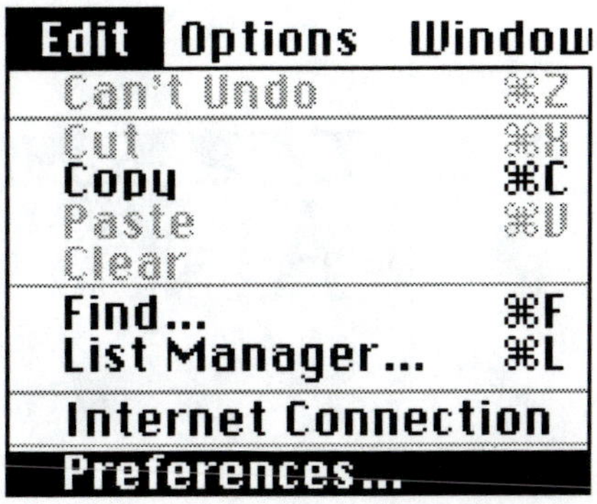

Clicking the **Show Fig Leaves** box covers the external genitalia of the male and female and the breasts of the female.

# Windows Guide to Using A.D.A.M. Interactive Anatomy

## Starting A.D.A.M. Interactive Anatomy

1. Turn on the computer and allow Windows to load. Place the A.D.A.M. Interactive CD-ROM in the C tray.

2. Click the **Start Button** in the lower left corner of the screen. Click **Programs**. Click **A.D.A.M. Interactive Anatomy**.

## Content Types

There are four major content types within A.D.A.M. Interactive Anatomy: **Dissectible Anatomy, Atlas Anatomy, 3D Anatomy,** and **Slide Shows**.

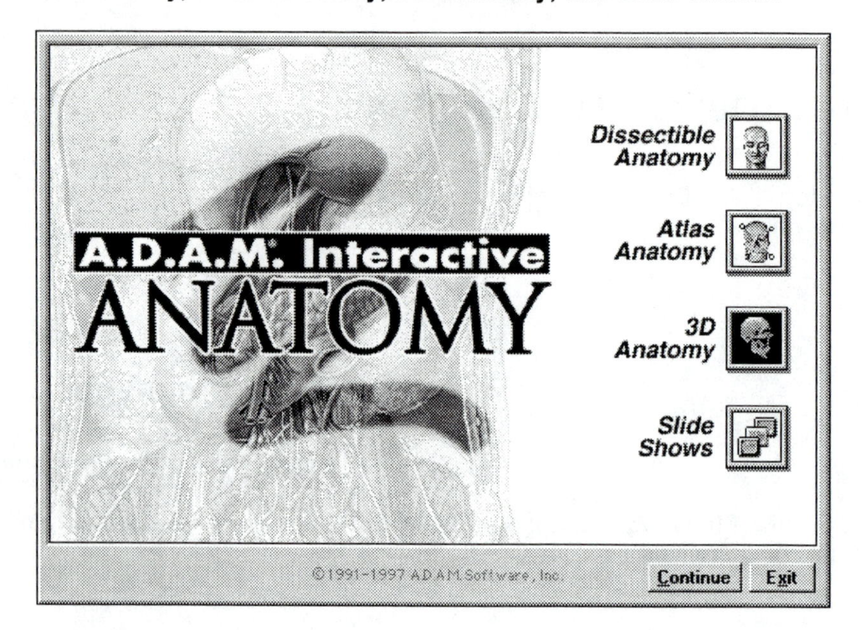


*Dissectible Anatomy* 

**Dissectible Anatomy:** Allows you to dissect layer by layer in ante posterior, lateral, or medial views and also male or female gender.

*Atlas Anatomy* 

**Atlas Anatomy:** Allows you to choose pinned images, cad photographs, and radiographs.

*3D Anatomy* 

**3D Anatomy:** Allows you to identify structures on 3D models.

*Slide Shows* 

**Pictures:** Allows you to add your own pictures or bit mapped images.

## Working With Dissectible Anatomy

Click on the **Dissectible Anatomy** button. You should see the following:

The opening screen allows you to choose:

- **Male** or **Female**
- **Anterior**, **Posterior**, **Lateral**, **Medial, Lateral Arm**, or **Medial Arm**
- By clicking on the tabs you may open **Dissectible Anatomy, Atlas Anatomy, and 3D Anatomy**.
- Choose a gender and view and click "open".

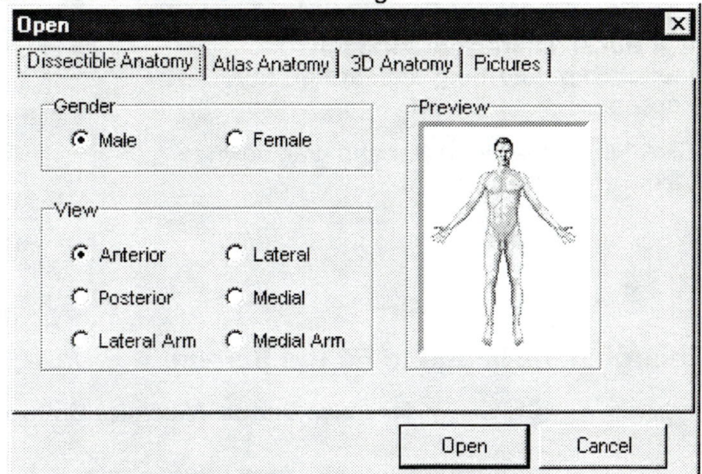

Open
Dissectible Anatomy
Atlas Anatomy
3D Anatomy
Pictures
Gender
Male
Female
Preview
View
Anterior
Lateral
Posterior
Medial
Lateral Arm
Medial Arm
Open
Cancel

Maximize window by clicking the maximize button and center image (see next page).

## Dissectible Anatomy Icons

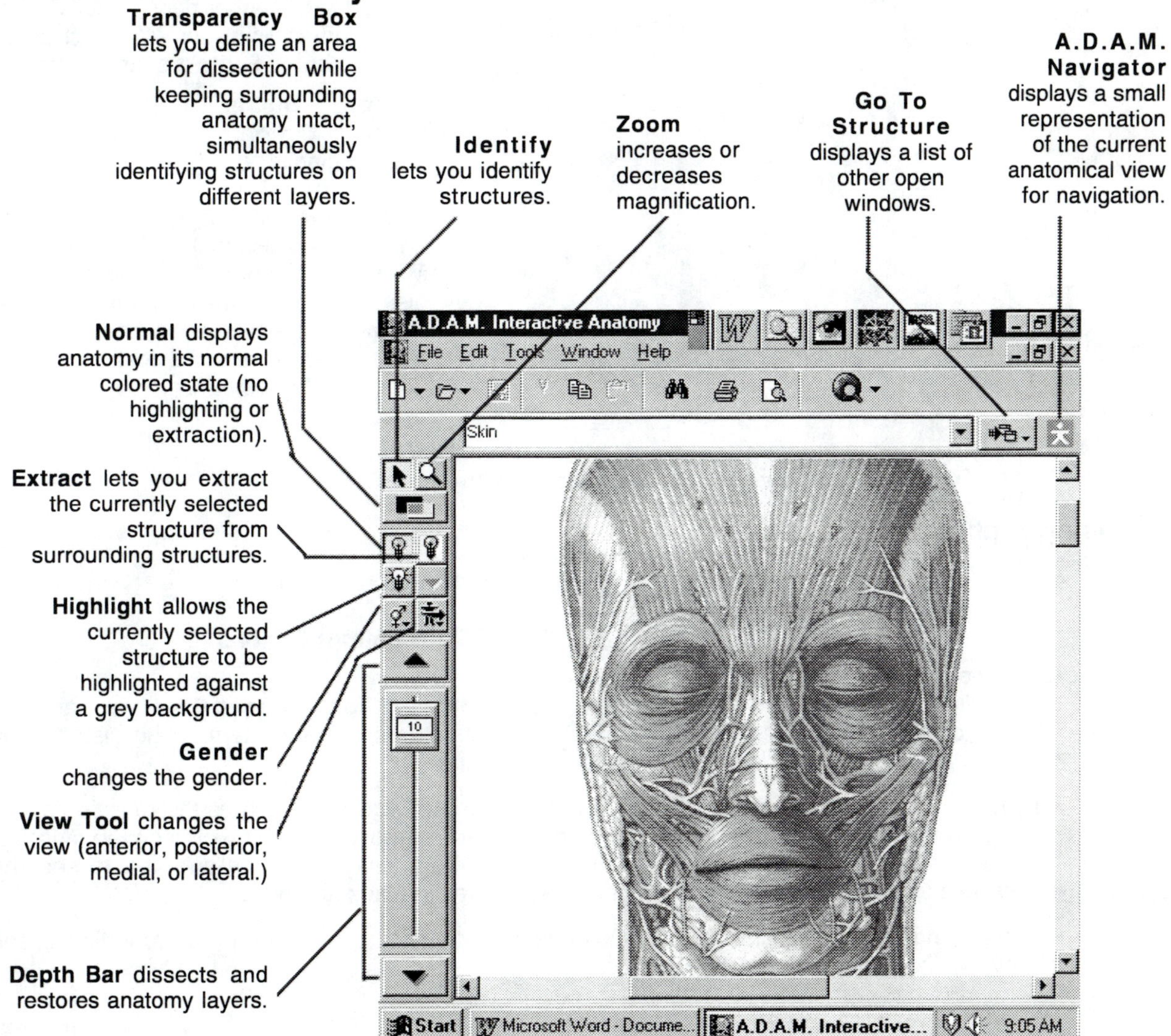

**Transparency Box** lets you define an area for dissection while keeping surrounding anatomy intact, simultaneously identifying structures on different layers.
**Identify** lets you identify structures.
**Zoom** increases or decreases magnification.
**Go To Structure** displays a list of other open windows.
**A.D.A.M. Navigator** displays a small representation of the current anatomical view for navigation.
**Normal** displays anatomy in its normal colored state (no highlighting or extraction).
**Extract** lets you extract the currently selected structure from surrounding structures.
**Highlight** allows the currently selected structure to be highlighted against a grey background.
**Gender** changes the gender.
**View Tool** changes the view (anterior, posterior, medial, or lateral.)
**Depth Bar** dissects and restores anatomy layers.
A.D.A.M. Interactive Anatomy
File
Edit
Tools
Window
Help
Skin
10
Start
Microsoft Word - Docume...
A.D.A.M. Interactive...
9:05 AM

# Hints on Using Dissectible Anatomy

## 1. Opening Content: Dissectible Anatomy

**If A.D.A.M. Interactive Anatomy is launched but nothing appears on my screen:**

Click **File** on the **Menu Bar** and drag down to **Open**. Then click **Content**.

A box will appear presenting the content types.

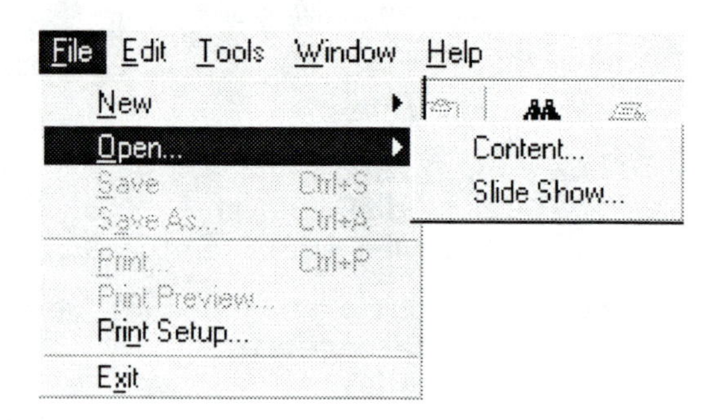


## 2. Enlarging the Image to Fit the Screen

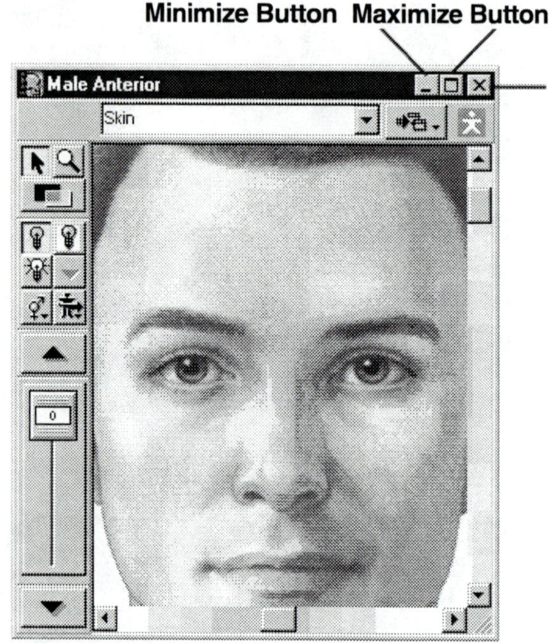


**Close Button**

To maximize the image on your screen, left click the **Maximize Button** located in the upper right corner of the **Application Window**. When clicked, the **Maximize Button** changes to the **Restore Button**. Clicking the **Restore Button** restores the screen to its original size.

To minimize the image, (to work within another application) left click the **Minimize Button**. The image will disappear from the screen but the program will remain open and its name will appear on the **Task Bar** at the bottom of the screen.

To open the program again, click the name on the **Task Bar**.
To close the window, click on the **Close Button**.

## 3. Using Highlight & Extract

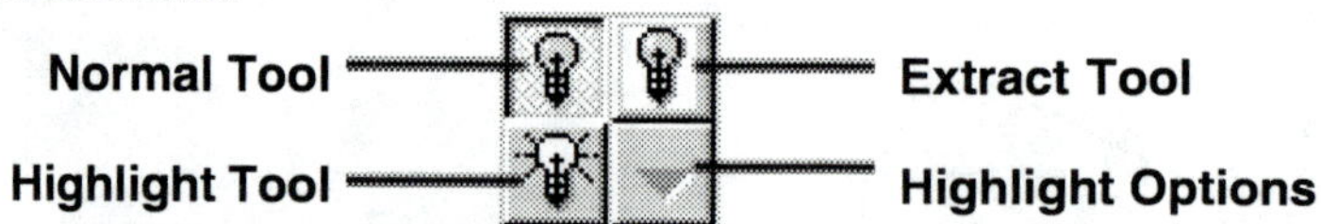


There are two useful ways to use the **Highlight Tool**:

- If you click on the **Highlight Tool** and then click on the **Highlight Options Tool**, you can then drag down to one of the systems listed. Only the structures in that system will be displayed in color. All other items will be shown in black and white.
- If you are trying to find different structures that are close together, such as the muscles of the leg, click on the **Highlight Tool** and then on the various muscles. As you click on the muscles, they will be highlighted in color and the name will be displayed. This allows you to see the entire muscle in color with the surrounding structures in black and white.
- After clicking on a particular structure, you can extract that structure from the surrounding tissue by clicking on the **Extract Tool**.

## 4. Making the Image Go From Black and White To Color

After using the **Highlight** button you can make the entire image colored again by clicking on the **Normal** button.

## 5. Centering the Image

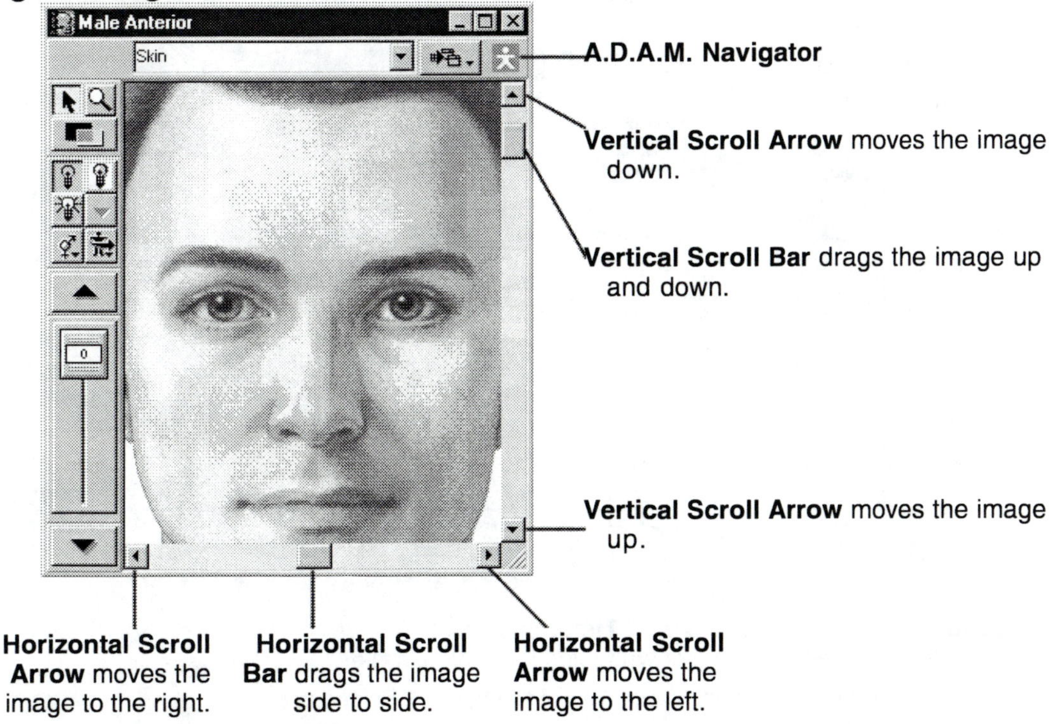

Male Anterior
Skin
**A.D.A.M. Navigator**
**Vertical Scroll Arrow** moves the image down.
**Vertical Scroll Bar** drags the image up and down.
**Vertical Scroll Arrow** moves the image up.
**Horizontal Scroll Arrow** moves the image to the right.
**Horizontal Scroll Bar** drags the image side to side.
**Horizontal Scroll Arrow** moves the image to the left.

There are two ways to move the image:

1. Click on the **scroll arrows** to move the image to the desired location or drag on the **scroll bar** to move the image faster.
2. Use the **A.D.A.M. Navigator**
   1. Place your cursor over the **A.D.A.M. Navigator**. You will see an image of the entire body with a rectangle superimposed over it.
   2. Place the cursor inside the rectangle, hold down the left mouse button, and drag until the rectangle is centered over the desired body part. Release the mouse button.
   3. Drag the cursor over the desired body part you want to go to.

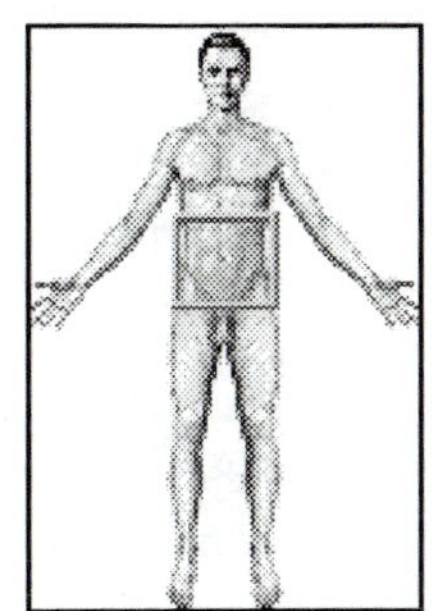

## 6. To Choose Atlas Anatomy or 3D Anatomy

Click **File** on the **Menu Bar** and drag down to **Open**. Then click **Content**.

A box will appear presenting the content types.

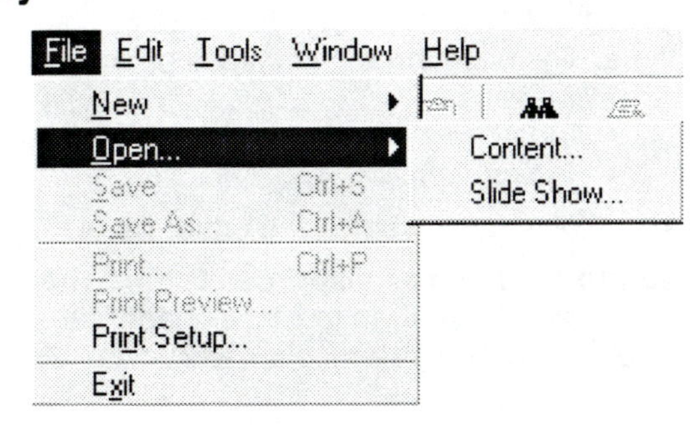

File
Edit
Tools
Window
Help
New
Open...
Save Ctrl+S
Save As... Ctrl+A
Print... Ctrl+P
Print Preview...
Print Setup...
Exit
Content...
Slide Show...

## Working with Atlas Anatomy

## Choosing Images in Atlas Anatomy

Go to the File menu and select open. Click content. Select Atlas Anatomy.

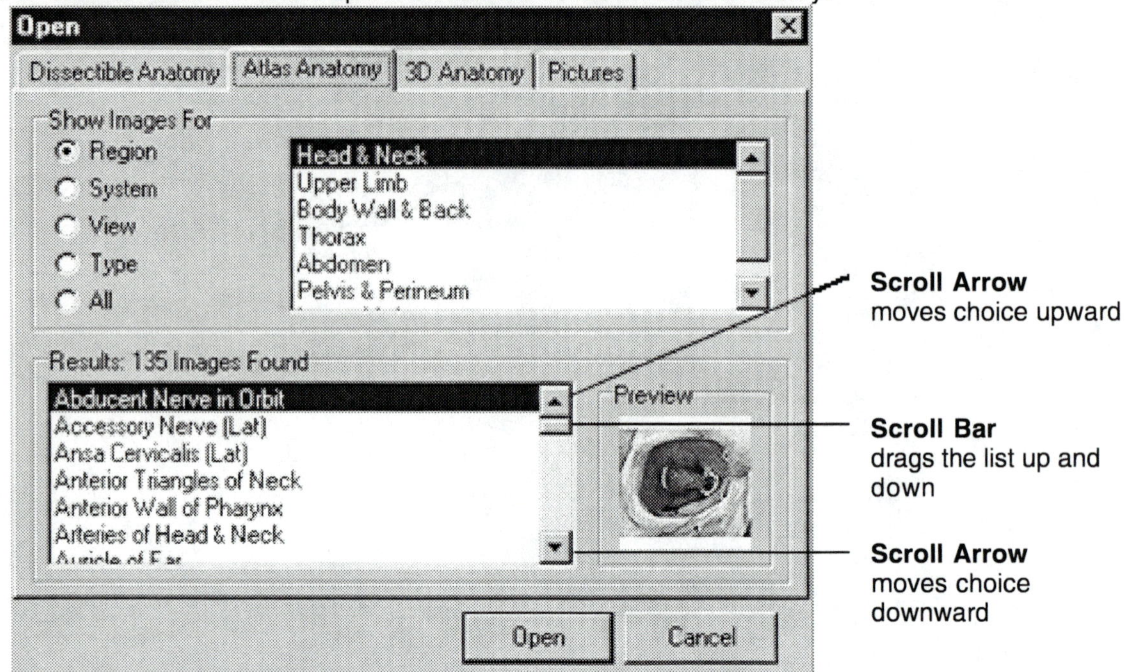


**1. Choose Region, System, View, Type, or All:**

You may need to scroll down to find the region, system, or view

**Region:** Allows you to search an area of the body. Choices are:

- *Head & Neck*
- *Upper Limb*
- *Body Wall & Back*
- *Thorax*
- *Abdomen*
- *Pelvis & Perineum*
- *Lower Limb*

**System:** Allows you to search a system of the body. Choices are:

- *Cardiovascular*
- *Digestive*
- *Endocrine*
- *Immune*
- *Integumentary*
- *Lymphatic*
- *Muscular*
- *Nervous*
- *Reproductive*
- *Respiratory*
- *Skeletal*
- *Urinary*

If you choose an image by system, only the pins corresponding to the system will be visible on the image.

**View:** Allows you to search a view of the body. Choices are:

- *Anterior*
- *Posterior*
- *Lateral*
- *Medial*
- *Superior*
- *Inferior (Most cross sections are inferior views.)*
- *Non-standard*

**Type:** There are three main types of images. Choices are:

- *Illustrations*
- *Cadaver Photographs (including images of bones)*
- *Radiographs*

**All:** Allows you to search the entire list of images

**2. Use the scroll arrows and bar to find the image you are looking for.**

The images are arranged in alphabetical order so it is quite easy to find a particular structure by scrolling. Click **System**, click **Nervous**. Scroll to **Spinal Cord, Vessels, & Meninges**. Click **Open.**

# Atlas Anatomy Icons

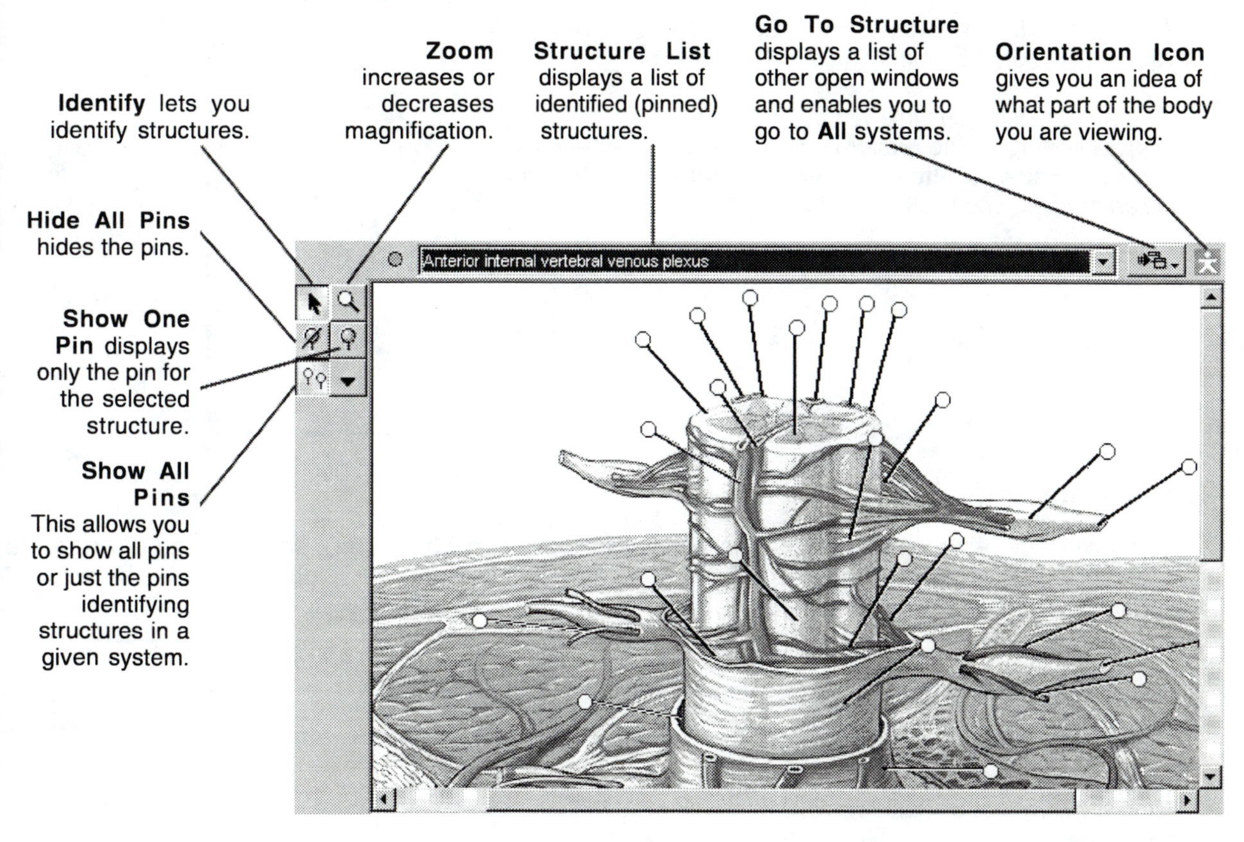

**Identify** lets you identify structures.
**Zoom** increases or decreases magnification.
**Structure List** displays a list of identified (pinned) structures.
**Go To Structure** displays a list of other open windows and enables you to go to **All** systems.
**Orientation Icon** gives you an idea of what part of the body you are viewing.
**Hide All Pins** hides the pins.
**Show One Pin** displays only the pin for the selected structure.
**Show All Pins** This allows you to show all pins or just the pins identifying structures in a given system.
Anterior internal vertebral venous plexus


The above commands can also be selected by clicking the **right** mouse button and selecting a particular command.

When you have finished, click the **Close** button.

# Hints on Using Atlas Anatomy

## 1. Opening Content: Atlas Anatomy

Click **File** on the **Menu Bar** and drag down to **Open**. Then click **Content**.

A box will appear presenting the content types.

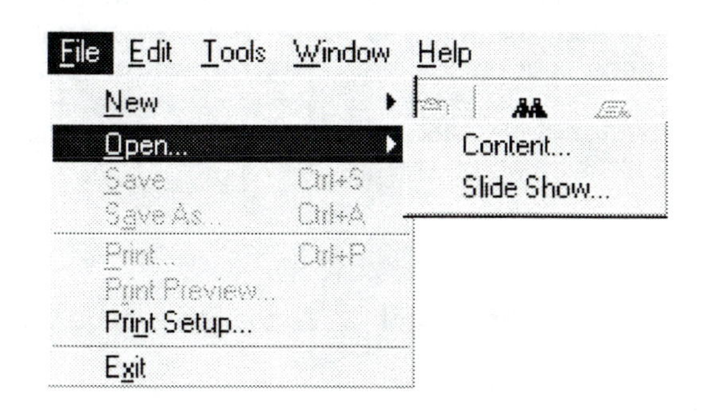

## 2. To Show All Systems

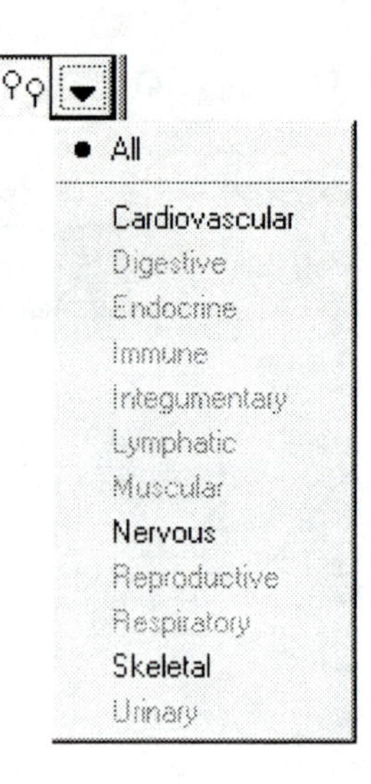


Click the **Atlas Anatomy** tab. Spinal Cord, Vessels, & Meninges should be highlighted. Click **Open**.

If you choose an image by system, only the pins corresponding to the system will be visible on the image. Click the **Show All Pins** button and select **All** from the drop down menu to show all systems.

## 3. Atlas Anatomy Structure List

If you can't find a structure, click on the **Atlas Anatomy Structure List**. Scroll or drag down to the name of the structure you are trying to find. (The structures are arranged alphabetically.) The head of the corresponding pin will turn green.

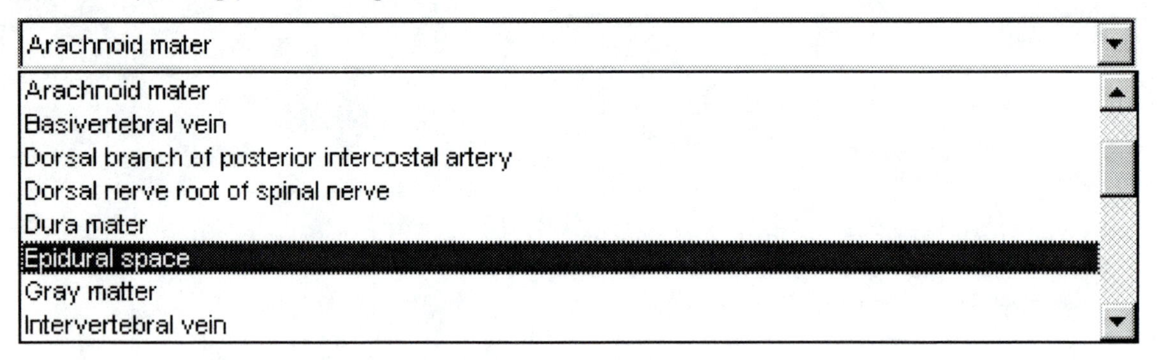


You can restrict the list to only structures within a certain system, such as the nervous system, by clicking on the **Show All Pins** button and drag to the system of interest.

## 4. Orientation Icon & Zoom Tool

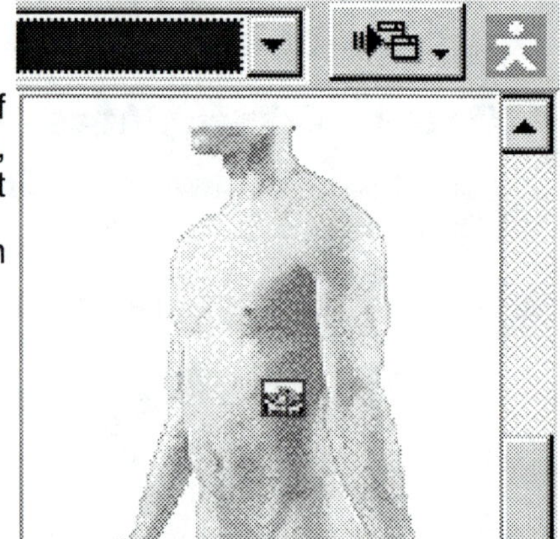

The orientation icon allows the user to determine what region of the body the image is from. To get an overall view of the image, place your cursor over the **Orientation Icon** in the upper right corner of the screen. This will show you the body location that is being observed.

Use the **Zoom Tool** to fit the image within the window.

## 5. Scrolling

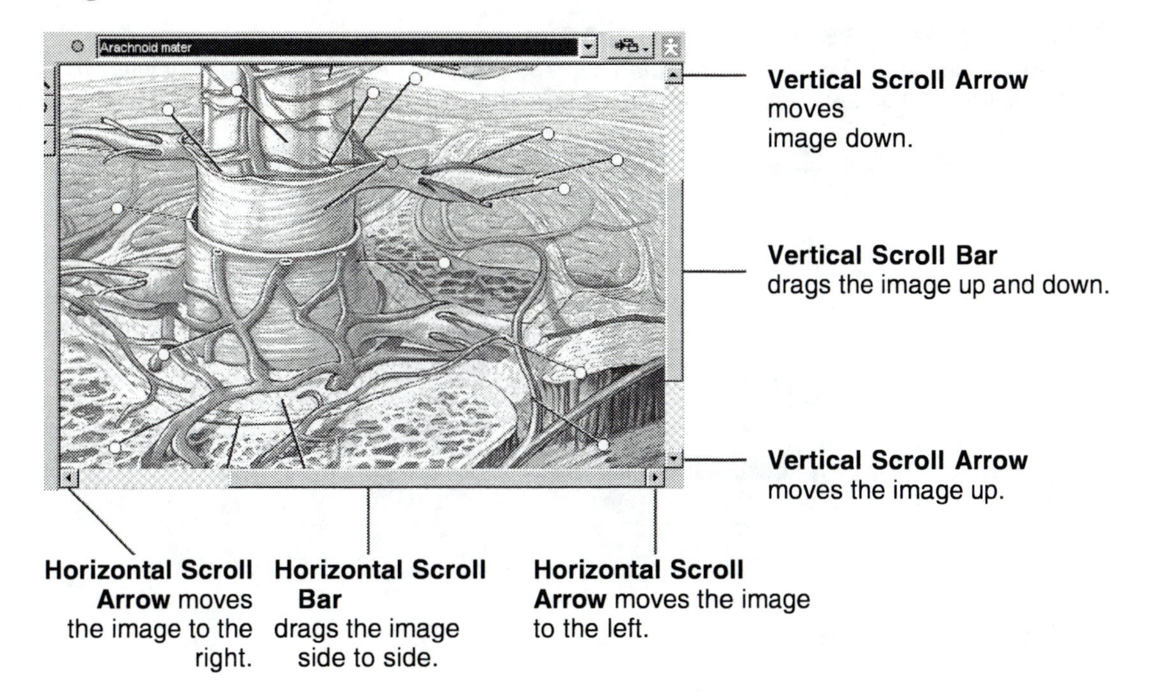
Arachnoid mater
Vertical Scroll Arrow moves image down.
Vertical Scroll Bar drags the image up and down.
Vertical Scroll Arrow moves the image up.
Horizontal Scroll Arrow moves the image to the right.
Horizontal Scroll Bar drags the image side to side.
Horizontal Scroll Arrow moves the image to the left.

## Working with 3D Anatomy

## Choosing Images in 3D Anatomy

Go to the file menu and select open, then content Select 3D Anatomy. You may have additional 3D images installed on the computer you are using. Choose a 3D model you would like to view, then click Open.

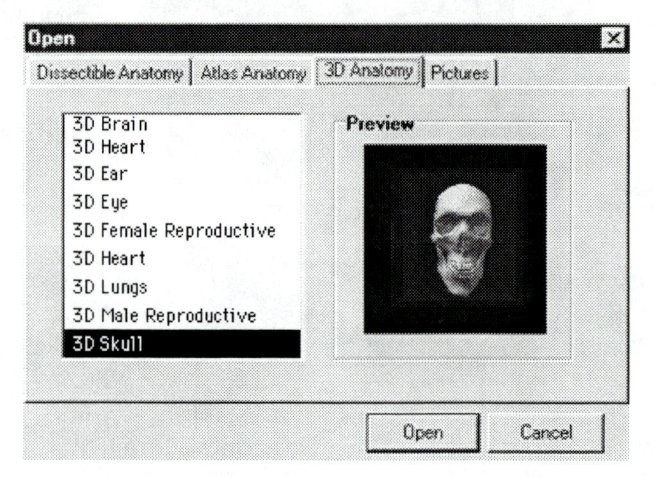

Open
Dissectible Anatomy
Atlas Anatomy
3D Anatomy
Pictures
3D Brain
3D Heart
3D Ear
3D Eye
3D Female Reproductive
3D Heart
3D Lungs
3D Male Reproductive
3D Skull
Preview
Open
Cancel

## 3D Anatomy Icon

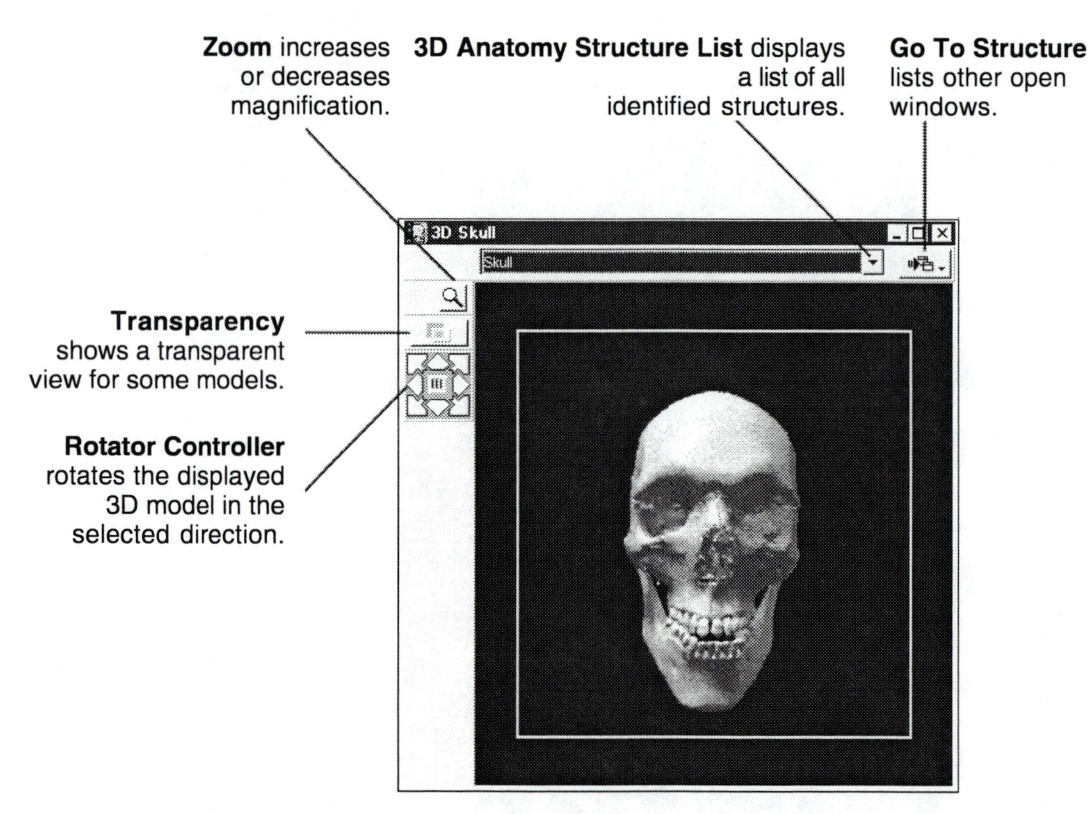

**Zoom** increases or decreases magnification.
**3D Anatomy Structure List** displays a list of all identified structures.
**Go To Structure** lists other open windows.
**Transparency** shows a transparent view for some models.
**Rotator Controller** rotates the displayed 3D model in the selected direction.
3D Skull
Skull

Click and hold on the **3D Anatomy Structure List** button and locate or find a structure within the model. The selected structure will highlight in blue.

# Finding Structures

There are two ways to find structures:

1. **Find**
2. **List Manager**

Both the **Find** and **List Manager** work in Dissectible Anatomy, Atlas Anatomy and 3D Anatomy.

## Using Find

1. Choose **Find** from the **Tools** menu.

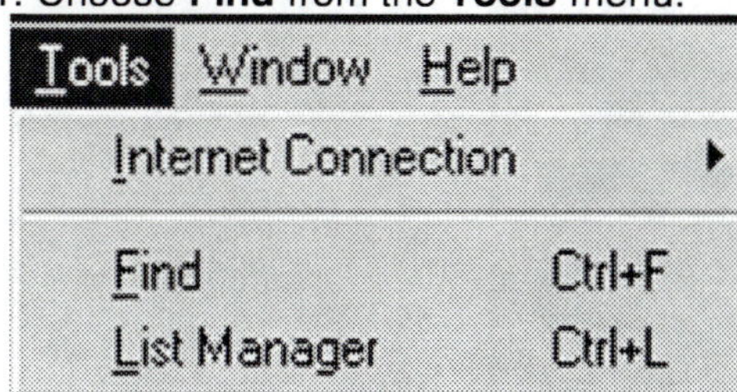


2. Lets say you want to find the femur. Type the word **Femur** into the box. Scroll to either a particular system (Skeletal) or choose **All.** Click find when you are done.

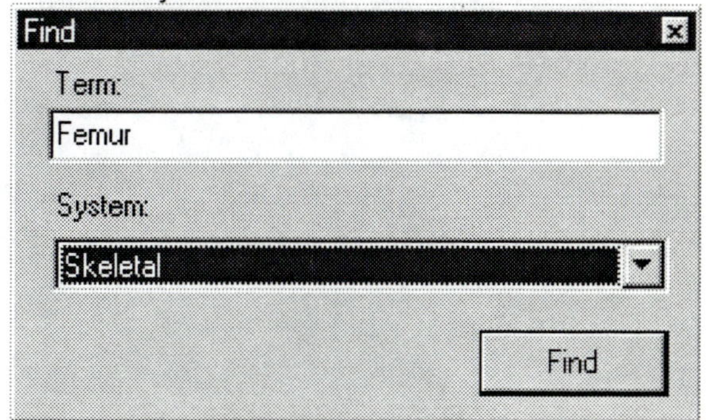


If you try to type in a word like **Large Intestine** it may not be found because the word is not specific enough. Try an alternative word such as **Colon**.

3. The **Find Results** box indicates that 25 structures containing the word femur were found. Scroll to locate the item you wish to find and select it.

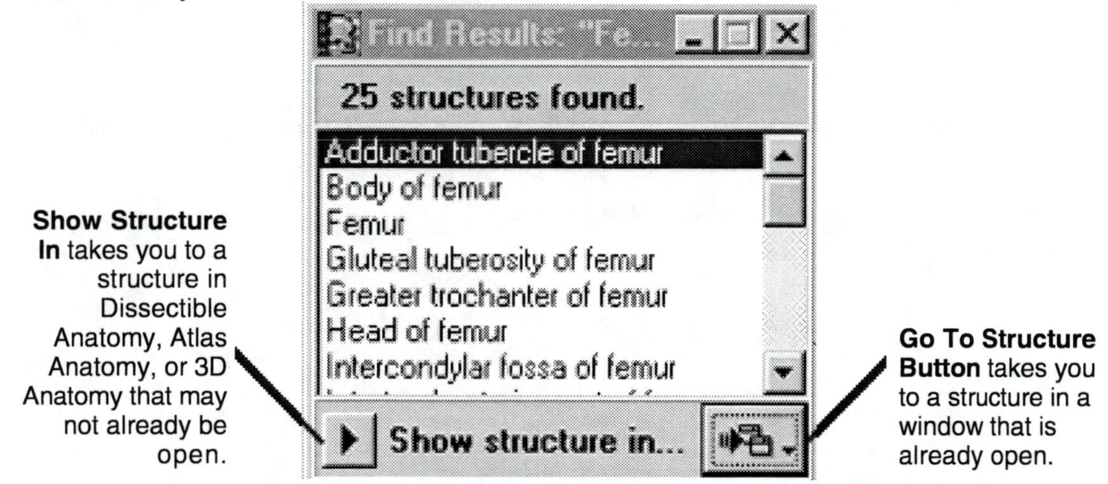


4. You now have two options. You can find the structure in a window that is already open or in a window that is not already open.

To display the structure in a window that is already open, click on the **Go To Structure** button and select the desired open window. For example, if you have been working in "Male Anterior" and you are trying to find the femur, click on the **Go To Structure** button and select "**Male Anterior**". Any open windows that do not contain the structure will be gray.

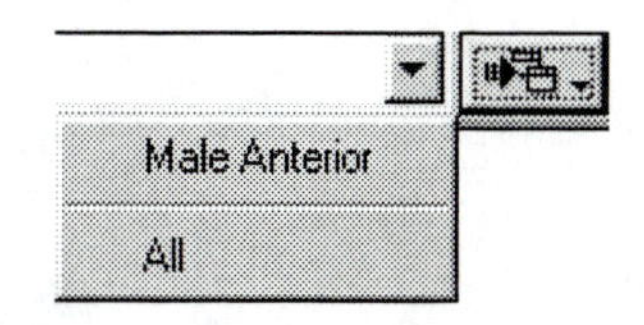


If you are working in Dissectible Anatomy you will be brought to the layer where the structure is best viewed. You may need to scroll back to the layer you were originally on to continue the dissection.

To return the image to full color, click on the **Normal** button.

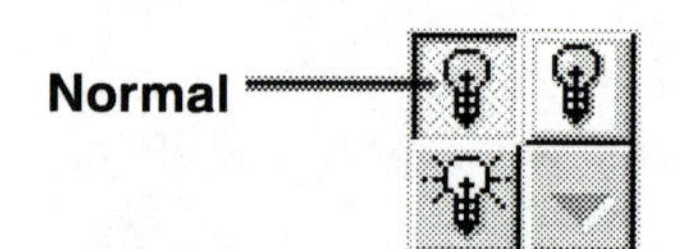


To find a structure within a content type which is not open, click on "**Show Structure In**". The following box will appear:

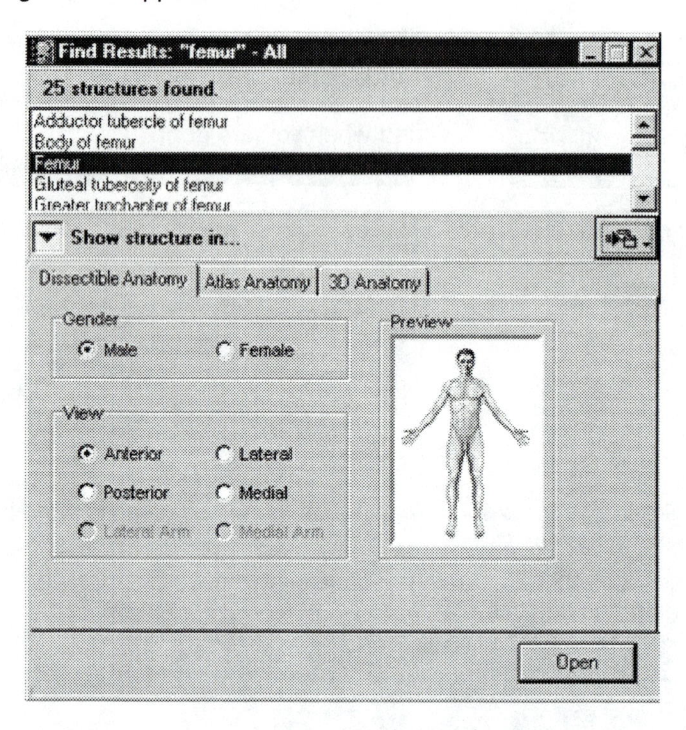


Choose the content type you want to view the structure in. Gray options are not available.

To return the image to full color, click on the **Normal** button.

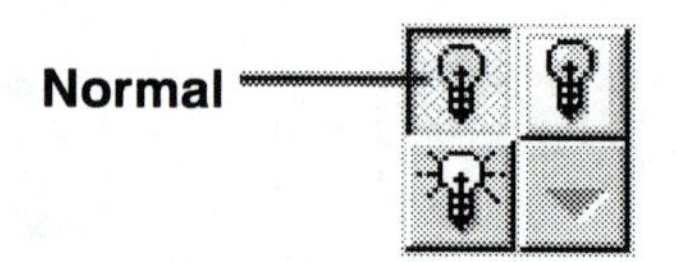

## Using the List Manager

1. **List Manager** works in Dissectible Anatomy, Atlas Anatomy, and 3D Anatomy. To open the **List Manager** click on the **Tools** menu and choose **List Manager** by clicking on the arrow located at the right of the **Structure List** at the top of the Dissectible Anatomy Window.

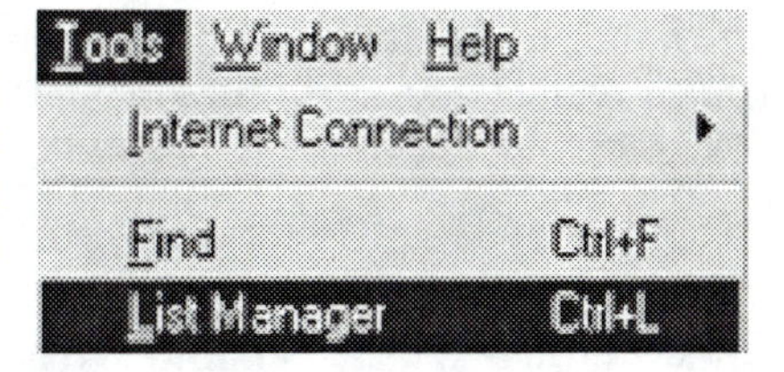


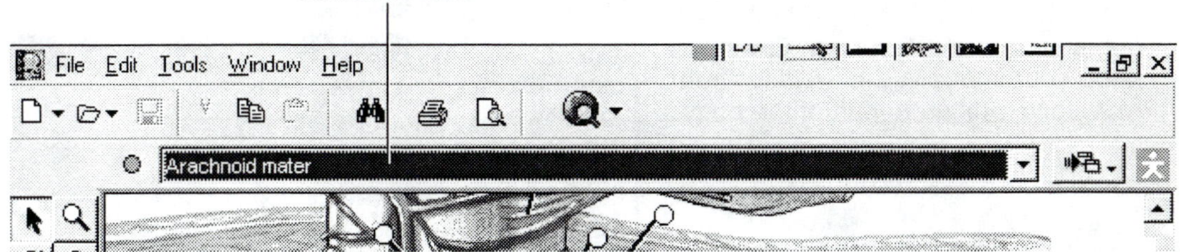


2. To quickly find a structure using **List Manager**, click the arrow located at the right of the **Structure List** at the top of the window. Type in the name of the structure you wish to find. As you type, you will notice that the **Structure List** will scroll so that the structure you are trying to find will appear below the **Structure List.** Click the name of the structure. In most instances, it will not be necessary to type in the entire name of the structure. Using the **Scroll Arrows** and **Bar**, scroll through the list (structures are arranged in alphabetical order) until you find the name of the structure you are searching for. Click the name.

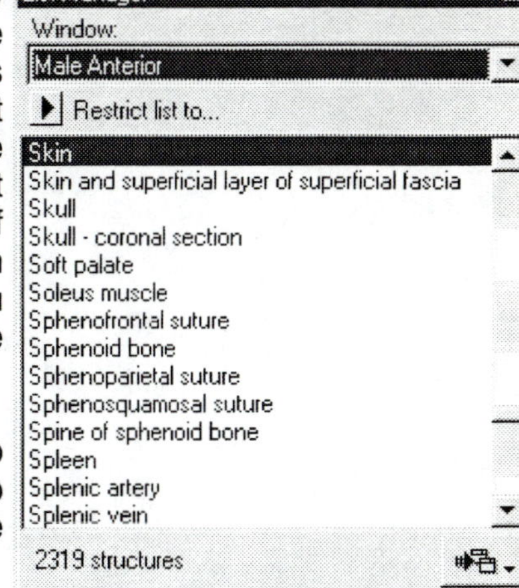


3. If you are working in Dissectible Anatomy, you will be brought to the layer where the structure is best viewed. You may need to scroll back to the layer you were originally on to continue the dissection.

4. To return the image to full color, click on the **Normal** button.

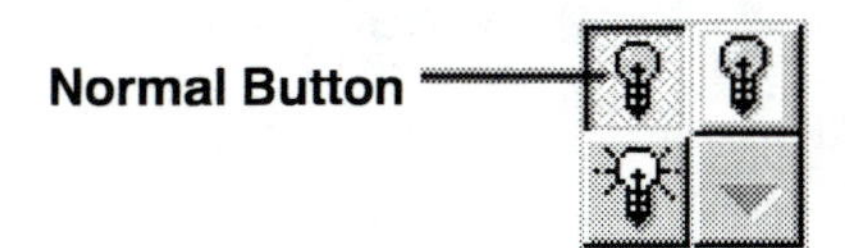

5. You may also choose to restrict the list of structures further. Click on the arrow next to **System**. This will allow you to restrict the **Structure List** to any particular system or you can choose **All**.

   Click the arrow next to **Area**. You can select **Entire View, Visible Area Only,** or **Transparency Box.** If you are in Dissectible Anatomy, restricting the area to **Visible Area Only** will keep your image in the window from moving to a different part of the body.

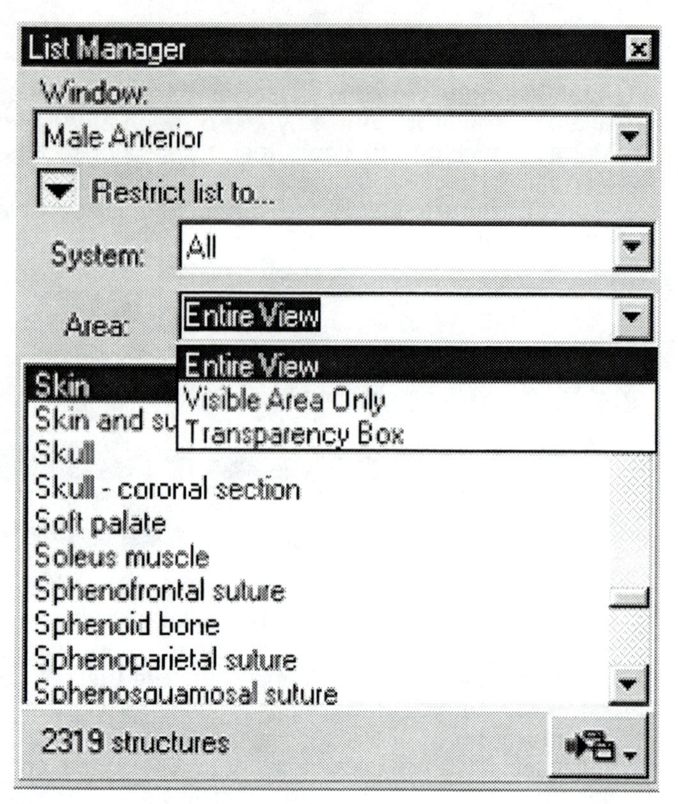

## Changing the Language Lexicon

To add or remove a **language lexicon**, click on the **Edit** menu, drag down to **Options**, and click on **Lexicons.**

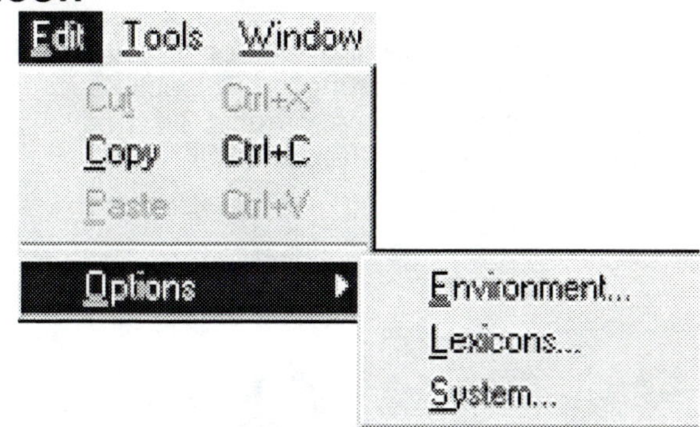


This image will appear:

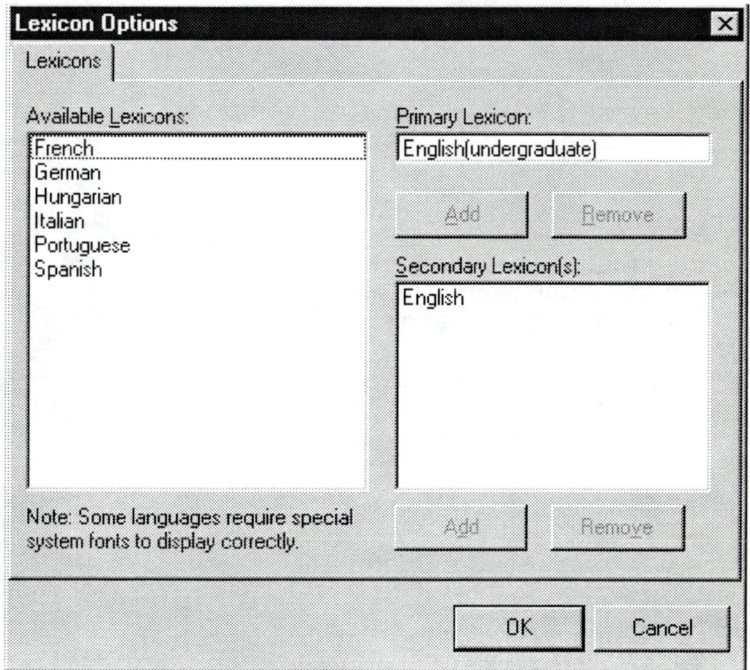


- **To remove a language lexicon:** Highlight the language in the **Primary Lexicon** or **Secondary Lexicon** box and click **Remove**.
- **To add a language lexicon:** Highlight the language in the **available lexicon** box and click **add**. The primary lexicon box must be empty to add a lexicon to it. Several languages may be added to the **secondary lexicon** box.
- For the purposes of this lab manual, it's best to use **English (undergraduate)** as a primary lexicon and **English** as a secondary lexicon.

## Changing the Skin Tone or Fig Leaf Option (Modesty Setting)

To change the **skin tone** or to add or remove **Fig Leaves (Modesty Setting)** in Dissectible Anatomy, right click and the graphic on the right will appear. Left click **Properties**.

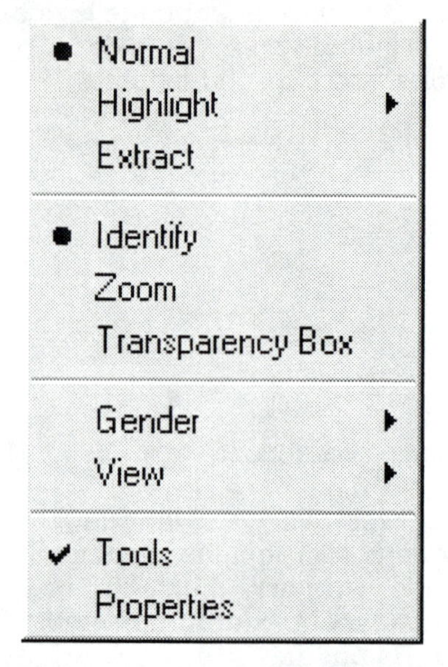


The following graphic will appear.

To change the **Skin Tone**, click on any of the four choices and click OK.

Clicking the **Show Fig Leaves** box covers the genitalia of both the male and the female and the breasts of the female.

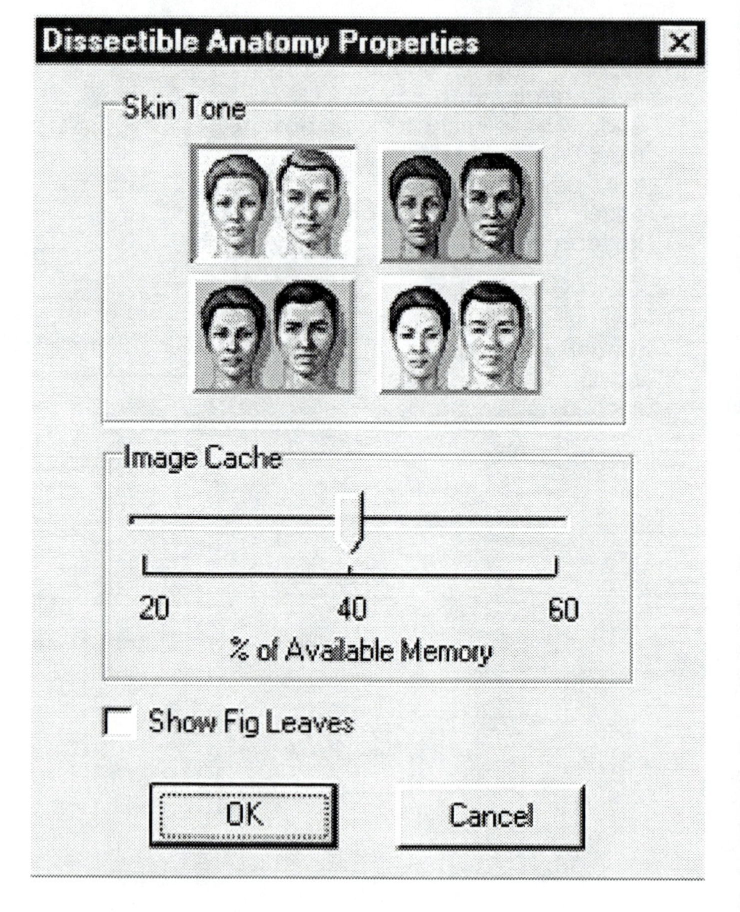

# A.D.A.M. Interactive Anatomy Tutorial for the Macintosh: A.D.A.M.'s Major Organs

## Opening A.D.A.M. Interactive Anatomy

- To open A.D.A.M. Interactive Anatomy, double click on the file that is labeled "Interactive Anatomy 3.0", which has been installed on the hard drive.

Interactive Anatomy 3.0

- Double click on the "Interactive Anatomy 3.0" icon shown below which is on your hard drive:

Interactive Anatomy 3.0

- If you are asked if you want to switch to 256 colors, click on "switch".
- Click on the icon corresponding to "Dissectible Anatomy".

- Select **male** and **anterior**. Click **Open**.

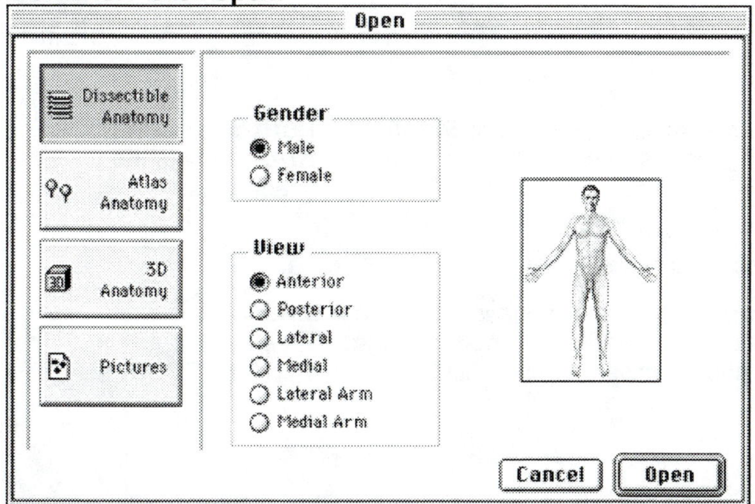


## Enlarging the Image

You will see the following image:

There are two different ways to enlarge the image:

1. Click once on the **Zoom Box** to enlarge the screen. Click on it again to restore the screen to the original size.
2. Point to the **Size Box**, click and drag the box toward the lower right corner of the screen. Drag it back again to return the screen to its original position.

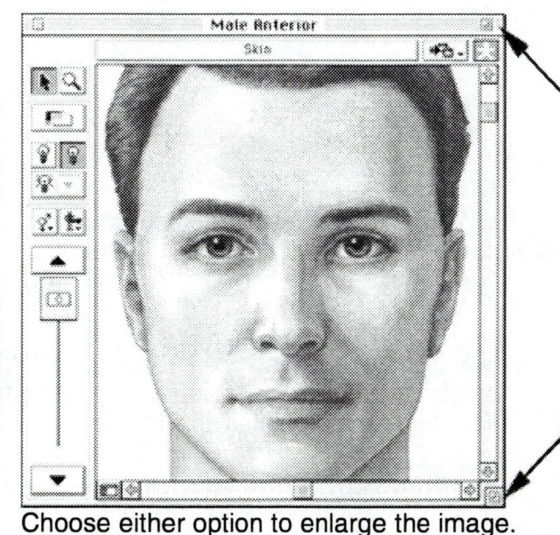


**Zoom Box** sizes the screen so that all of its contents are visible.

**Size Box** changes the shape or size of the screen.

Choose either option to enlarge the image.

## Centering the Image Over the Desired Body Part

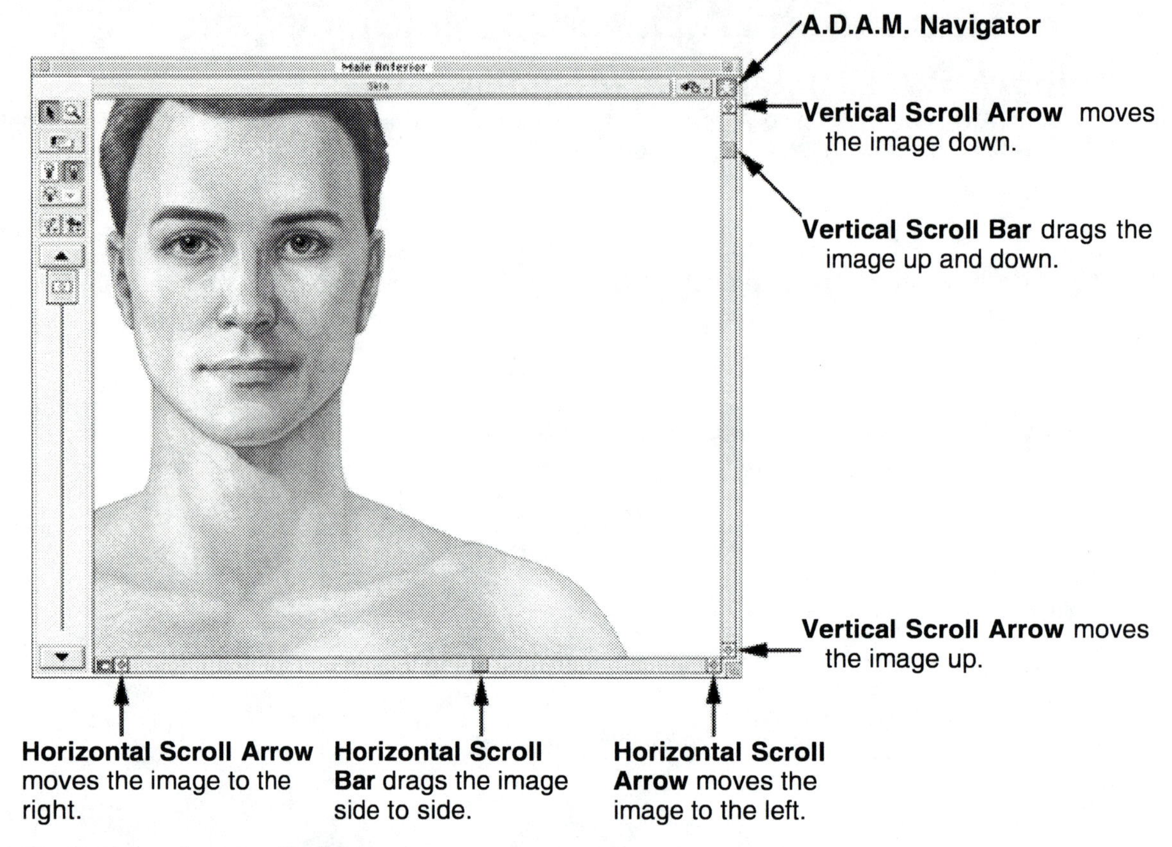
A.D.A.M. Navigator
Vertical Scroll Arrow moves the image down.
Vertical Scroll Bar drags the image up and down.
Vertical Scroll Arrow moves the image up.
Horizontal Scroll Arrow moves the image to the right.
Horizontal Scroll Bar drags the image side to side.
Horizontal Scroll Arrow moves the image to the left.

There are two ways to move the image:

1. Click on the **scroll arrows** or drag the **scroll buttons**.
2. The fastest way to move the image is to use the **A.D.A.M. Navigator:**
   - Place your cursor over the **A.D.A.M. Navigator**, then move it to the left and downward (do not click).

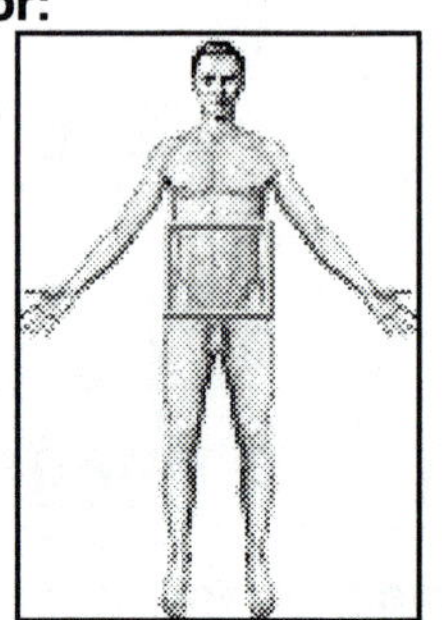

   - You will see an image of the entire body appear. The cursor will turn into a hand.
   - Drag the hand (cursor) so that the box is over the abdomen.

## Reducing the Image Size

- Click on the **Zoom Button**. Notice that the cursor turns into a minus (-) sign indicating that you will reduce the size of the image if you click. Using the mouse, move the cursor (minus sign) to A.D.A.M.'s navel. Click the mouse button. Notice that you reduced the size of the image so that now you are able to see more of his body.
- Click the **Zoom Button** again and move the tool back to A.D.A.M.'s navel. Notice that there is now a plus (+) sign indicating that you will increase the size of the magnification if you click. Click the mouse button and you will return to the original size image.

# Anterior Dissection of Abdominopelvic Cavity

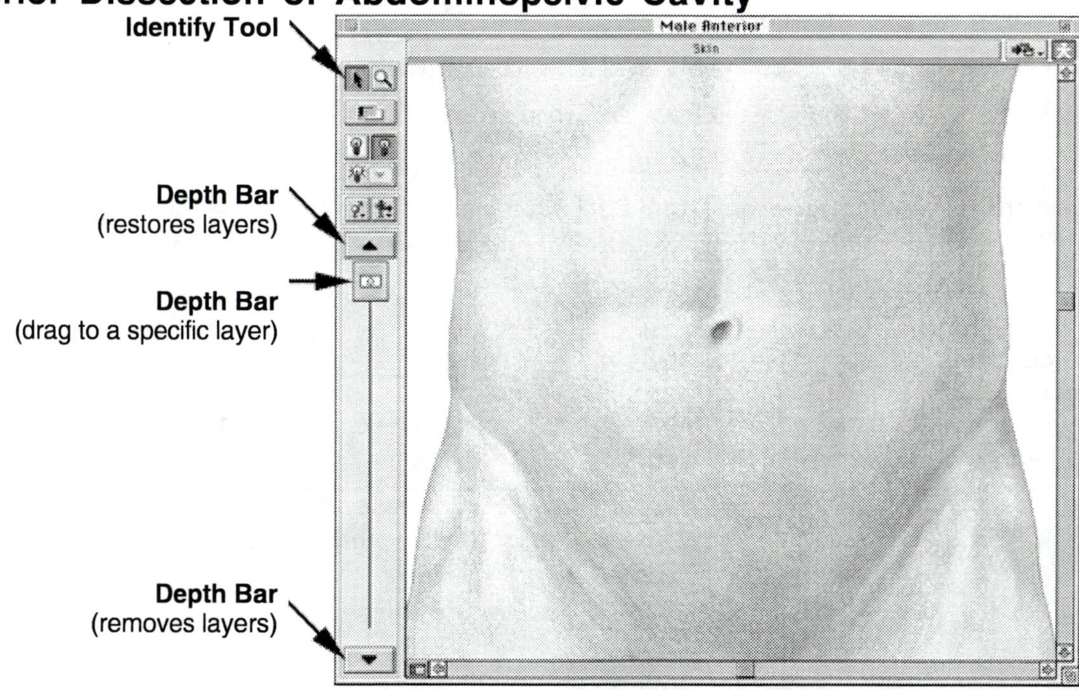


Click once on the downward arrow on the **Depth Bar** [▼] to remove the skin. You are now at **layer 1**. The first layer below the skin is fat. Identify the **Fat** (superficial fascia) by holding down the mouse button over the fat. (Make sure the **Identify Tool** [↖] is highlighted before you do this.)

Try to resist scrolling down to the level indicated since this will not give you a "dissection experience".

The structures you will need to know for the laboratory practical exam are shown in bold face print. If the program calls the structure something else, the structure name will be in parentheses.

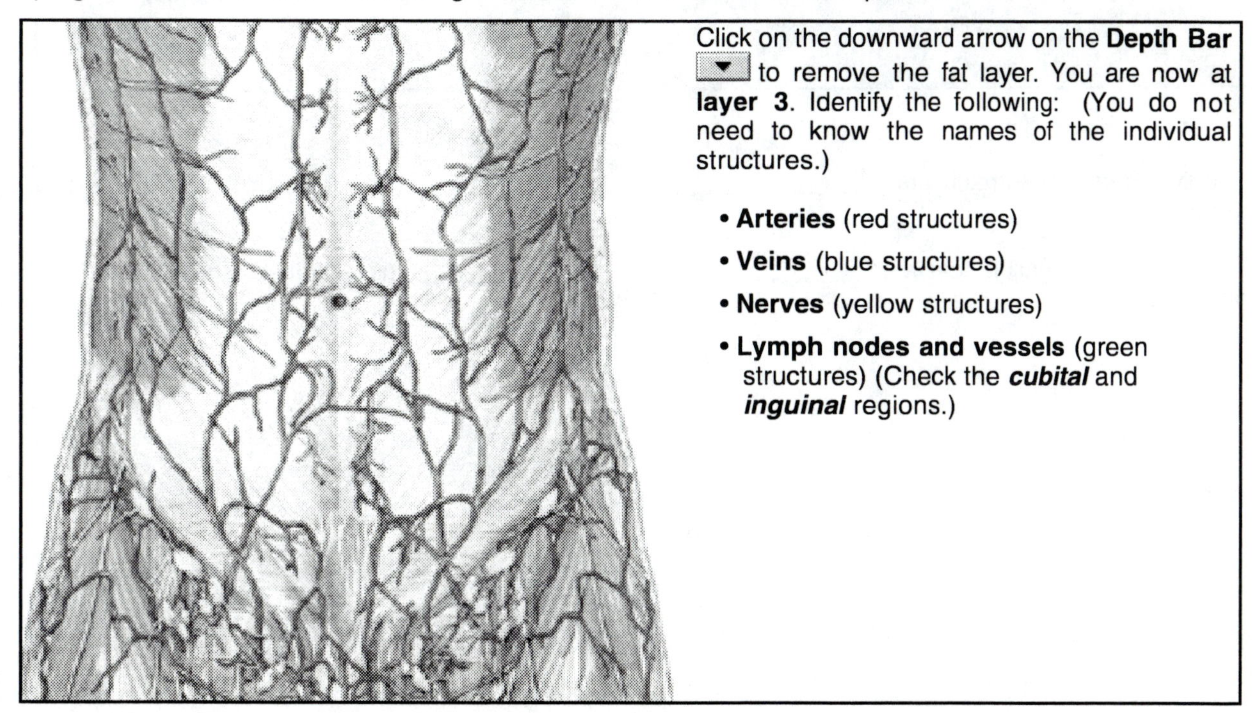

Click on the downward arrow on the **Depth Bar** [▼] to remove the fat layer. You are now at **layer 3**. Identify the following: (You do not need to know the names of the individual structures.)

- **Arteries** (red structures)
- **Veins** (blue structures)
- **Nerves** (yellow structures)
- **Lymph nodes and vessels** (green structures) (Check the ***cubital*** and ***inguinal*** regions.)

As you move through the laboratory, connect the names of the structures with the corresponding diagrams by drawing lines or arrows.

You may have to scroll up or down to find all of the structures listed.

Do not try to click when you see the black and white circle. This means that the computer is busy doing something.

Click several times on the downward **Depth Bar** [▼] to remove the blood vessels, lymphatic structures, nerves and fascia, until you reach **layer 18**. Note the various **Superficial muscles**.

Drag the **Depth Bar** [18] down to **layer 194** [194]. Click and identify the **Parietal peritoneum**, which is the serous membrane against the wall of the abdominal cavity.

Click the downward arrow on the **Depth Bar** [▼] once to reach **layer 195**. Identify the **Greater Omentum** which covers the abdominal organs.

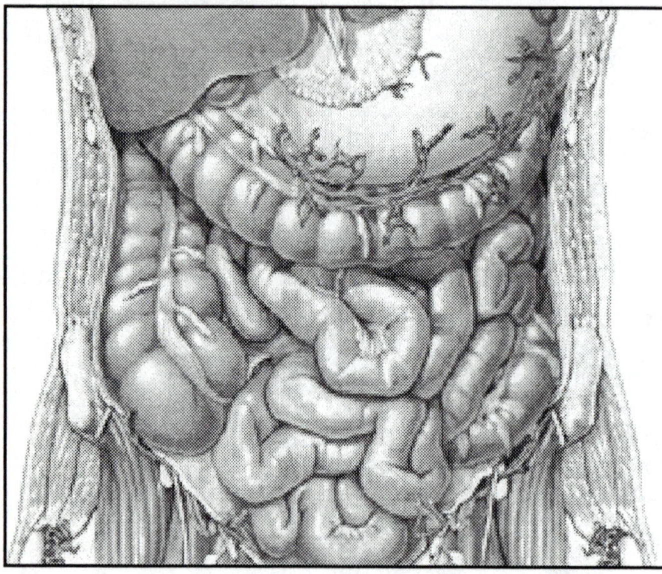

Click several times on the downward arrow on the **Depth Bar** [▼] until you reach **layer 198**. Identify:

- **Liver**
- **Stomach**
- **Gallbladder**
- **Large intestine** (ascending colon, transverse colon, sigmoid colon)
- **Small intestine** (jejunum & ileum)

You will be responsible for items listed in bold font. Items in parentheses are used to find particular structures. To find the organs above follow the directions on the next page.

The liver and the stomach are ______________________ (ipsilateral, contralateral) to one another.

The liver and the gallbladder are ______________________ (ipsilateral, contralateral) to one another.

# Using the List Manager to Identify Structures in Dissectible Anatomy

We are going to find the liver using the **List Manager.** Go to the **Edit Menu** and select **List Manager.**

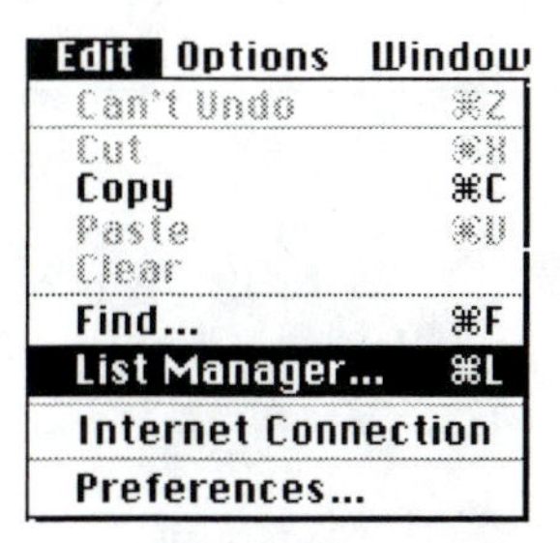


You will see the image shown on the right:

Type **Liver** and press **Return** on your keyboard. You will see the liver highlighted in color.

Click the **Normal** button to return the screen to color.

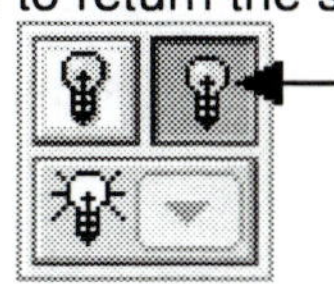

Now let's find the stomach using the **List Manager.** An easier way to access the List Manager is to click on the **Structure List** at the top of the window.

You will see the following image:

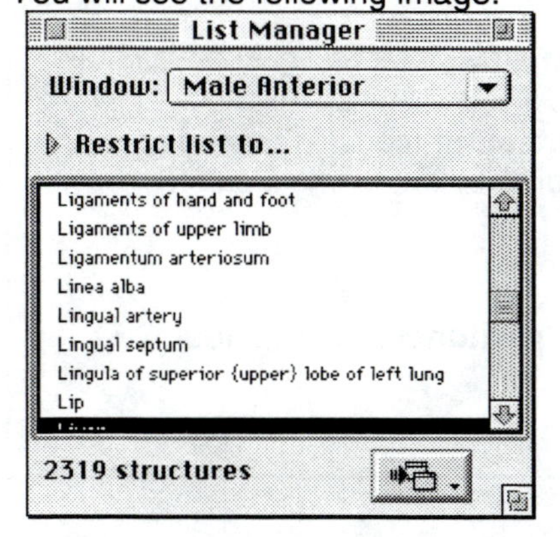


Type **Stomach** and press **Return** on the keyboard. You will see A.D.A.M.'s stomach highlighted in color.

Notice you have been taken from **layer 198** to **layer 204** because this gives you the best view of the stomach. Drag the **Depth Bar** back up to **layer 198** and you will still see the stomach highlighted, but now there are other structures covering it.

Click the **Normal** button to return the screen to color.

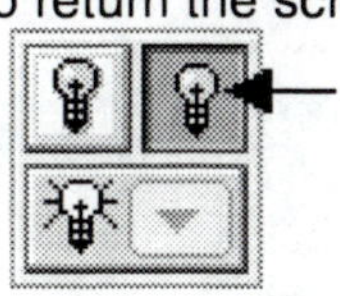

Identify the **Large intestine**, **Small intestine**, and **Gall bladder** by either using the **List Manager** or by pointing and clicking on the image.

When you identify the intestines using the **List Manager**, be sure to type in the words listed in parentheses above because large or small intestine is not specific enough.

- **Large intestine** (ascending colon, transverse colon, sigmoid colon)
- **Small intestine** (jejunum & ileum)

Click several times on the downward **Depth Bar** ▼ until you reach **layer 202**. Click and identify: **Gallbladder**. (Use the List Manager if necessary to find it.) No graphic has been provided for this layer.

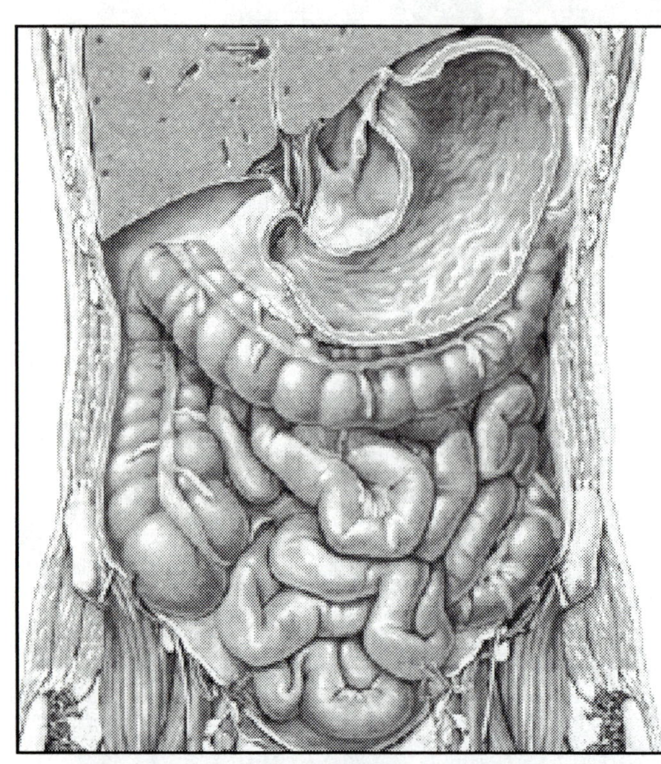

Click several times on the downward **Depth Bar** ▼ until you reach **layer 205**. Identify:

- **Inside of the stomach** (rugae)
- **Opening to the small intestine** (pyloric sphincter).
- **Liver**
- **Stomach**
- **Large intestine** (ascending colon, transverse colon, sigmoid colon)
- **Small intestine** (jejunum & ileum)

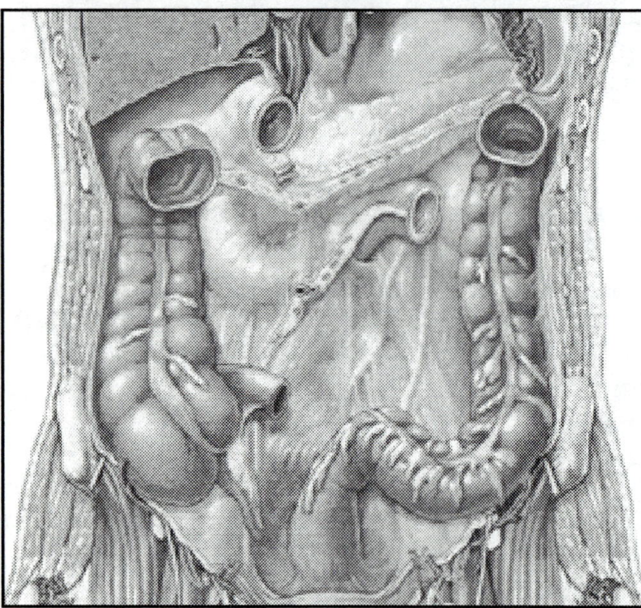

Click a few times on the **Depth Bar** ▼ to remove the stomach and small intestine. You are now at **layer 209**. Identify:

- **Liver**
- **Opening to the small intestine** (pyloric sphincter)
- **Small intestine** (cut off jejunum & ileum)
- **Large intestine** (cecum, ascending colon, descending colon, sigmoid colon)
- **Rectum**
- **Parietal peritoneum** (peritoneum)

## Using Find to Identify Structures in Dissectible Anatomy

At this level let's identify the appendix using **Find**. Go to the **Edit Menu** and select **Find.**

Edit Options Window
Can't Undo ⌘Z
Cut ⌘X
Copy ⌘C
Paste ⌘V
Clear
Find... ⌘F
List Manager... ⌘L
Internet Connection
Preferences...

You will see the following image:

Find
Peritoneum
System: Any System
Find

Type **Appendix**, click on **Find**.

You will then see the following image:

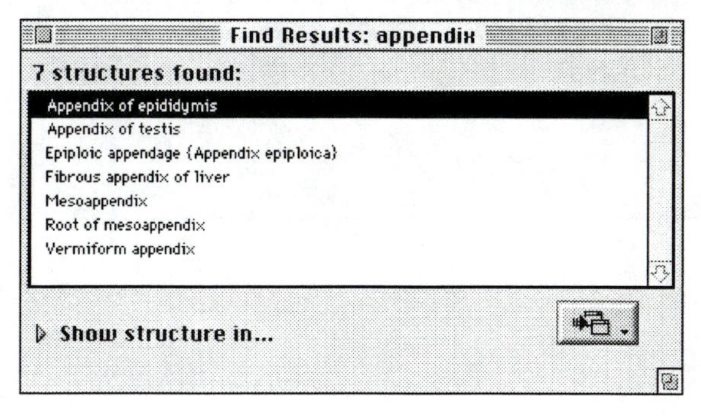


Select **Vermiform Appendix**.

Click on the **Go To Structure** button and choose **Male Anterior**. You will see the appendix highlighted in color.

Click on the **Normal** button and note the appendix on the color image.

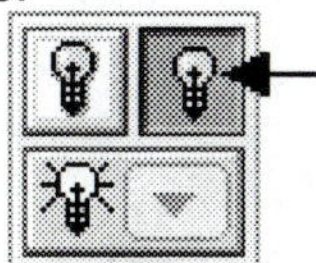

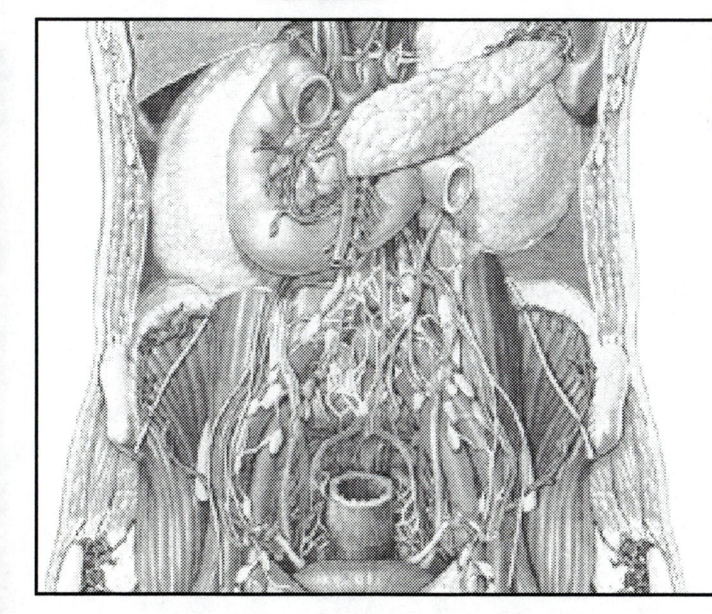

Click a few times on the downward **Depth Bar** to remove most of the intestines and peritoneum until you reach **layer 211**. Identify:

- **Liver**
- **Opening to the small intestine** (pyloric sphincter)
- **Small intestine** (jejunum & duodenum)
- **Pancreas** (body, tail, and head)
- **Spleen**
- **Urinary bladder**

List several retroperitoneal organs (found posterior to the peritoneum).

______________________________________________________________________

______________________________________________________________________

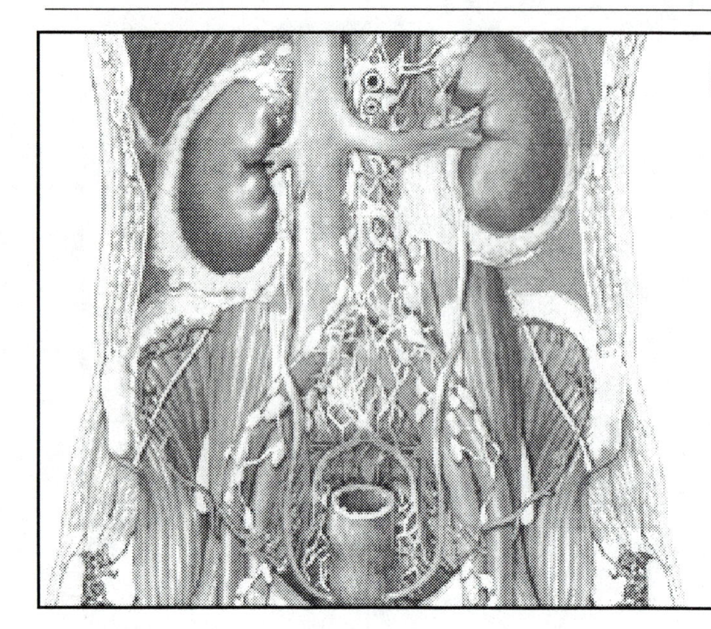

Click many times on the downward **Depth Bar** ▼ until you reach **layer 230**. Identify:

- **Kidneys**
- **Adrenal gland**
- **Ureter**
- **Rectum**

## Changing the View, Changing the Gender, and Using the Highlight Tool

Use the **Vertical Scroll Arrows** or **A.D.A.M. Navigator** to move the image down over the hip and buttock area. Click on the **View** button and choose **Medial**. Click on the **Normal** button. Scroll to **layer 0**.

You are now looking at a midsagittal view.

Click on the **Highlight** button and select **Digestive** from the drop down menu. This feature allows you to see a highlighted view of all structures associated with the Digestive System.

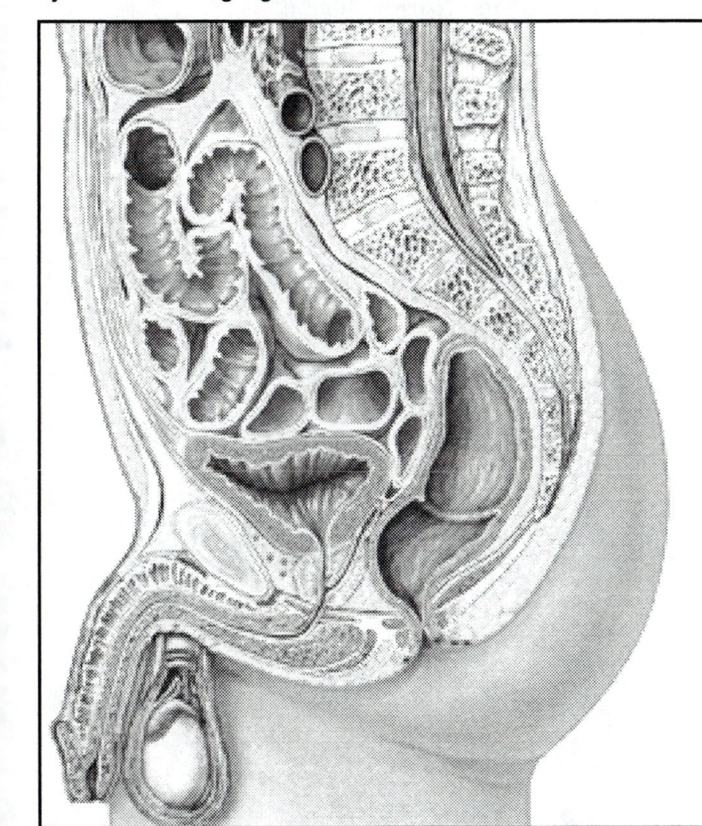

Identify:
- **Rectum**

Click on the **Highlight** button again and select **Urinary** from the drop down menu. Identify:
- **Urinary bladder**

Click on the **Highlight** button again and choose **Reproductive** from the drop down menu. Identify:

- **Prostate Gland**
- **Testis**

Click on the **Gender** button and select **Female**.

Click on the **Normal** button:

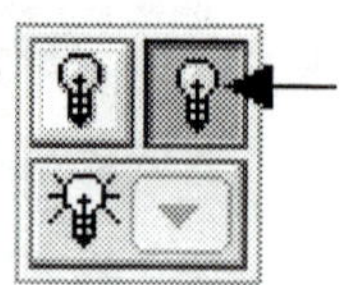

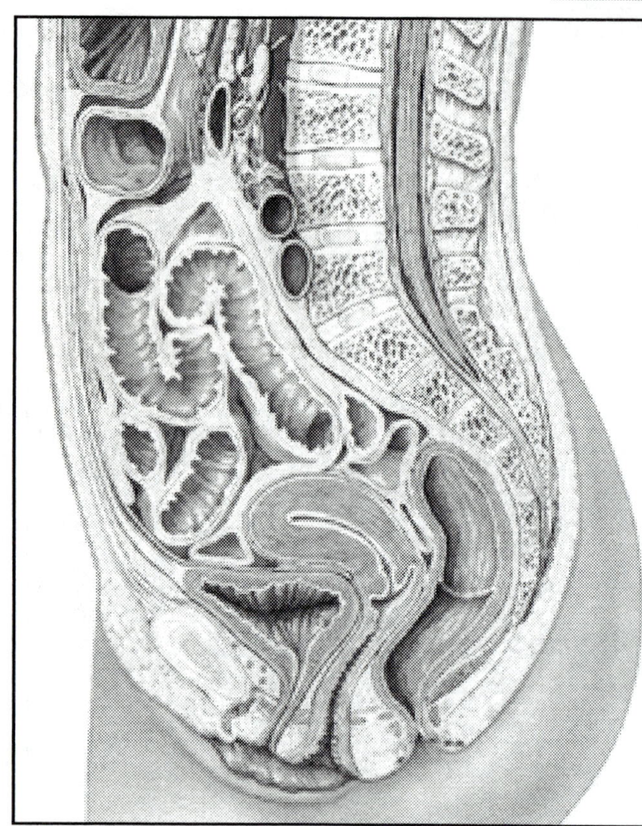

Click on the **Highlight** button and select **Reproductive**. Identify:

- **Uterus** (body of uterus)
- **Vagina**

Click on the **Normal** button.

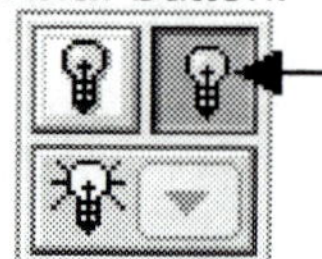

Identify the following on the female image:

- **Rectum**
- **Urinary Bladder**

## Anterior Dissection of Thoracic Cavity

Click on the **View** button and choose **Anterior**.

Click on the **Gender** button and choose **Male**.

Use the **Vertical Scroll Arrows** or **A.D.A.M. Navigator** to move the image over the chest area.

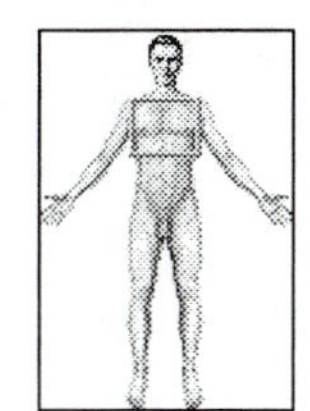

Click on the **Normal Tool.**

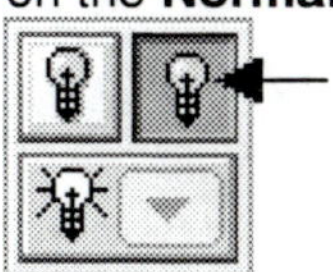

Scroll up to **layer 0**.

Click once on the downward **Depth Bar** to remove the skin. You are now at **layer 1**. Identify:

- **Fat** (superficial fascia)

Click on the downward **Depth Bar** to remove the fat. You are now at **layer 5**. Identify the following (you do not need to know the names of the individual structures):

- **Arteries** (red structures)
- **Veins** (blue structures)
- **Nerves** (yellow structures)

Click several times on the downward **Depth Bar** to remove the blood vessels, nerves and fascia, until you reach **layer 18**. Note the various **Superficial muscles**. List the name of a superficial muscle of the chest.

______________________________________________________________

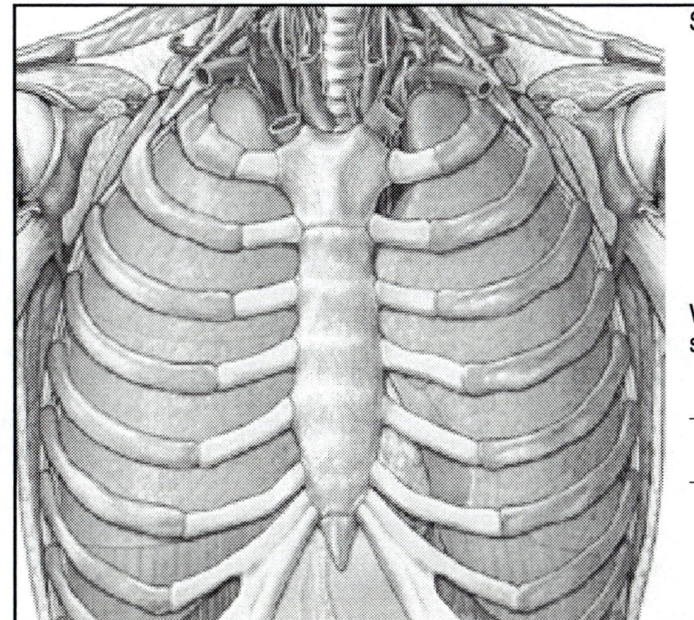

Scroll down to **layer 158**. Identify:

- **Ribs**
- **Sternum**
- **Trachea**
- **Lungs** (under parietal pleura)

What is the protective role of the ribs and sternum?

______________________________________

______________________________________

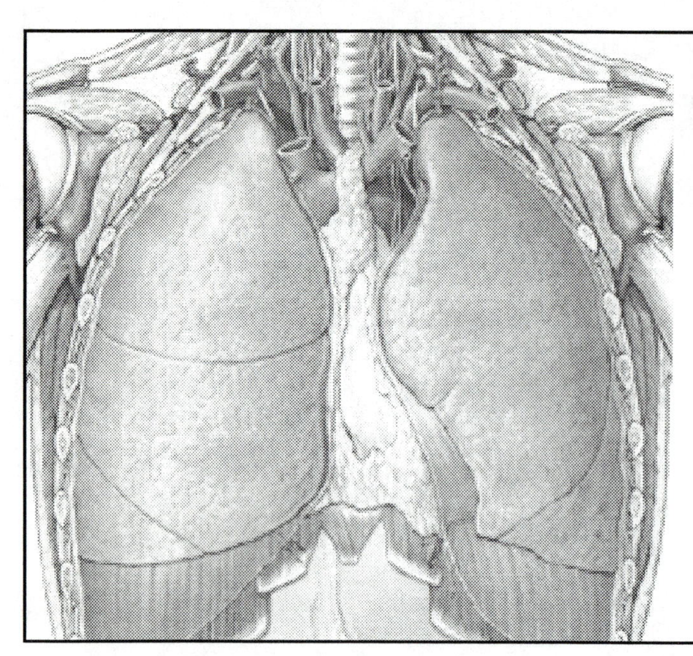

Click several times on the downward **Depth Bar** [▼] to remove the ribs, sternum, and parietal pleura until you reach **layer 162**. Identify:

- **Lungs**
- **Thymus gland**
- **Pericardial sac**
- **Diaphragm**
- **Trachea**

What organ is found within the pericardial sac? (Hint: Drag down to **layer 173** to find out.)

______________________________________________________________________

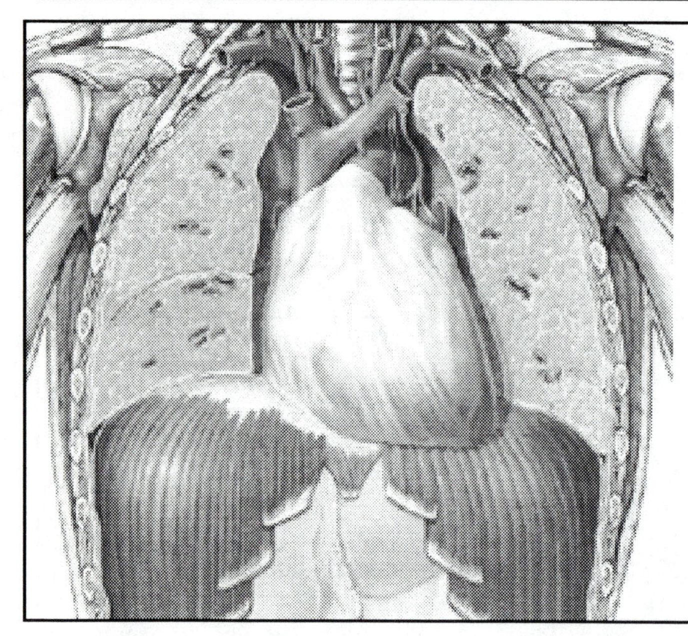

Click several times on the downward **Depth Bar** [▼] until you reach **layer 169**. Identify:

- **Lungs**
- **Pericardial sac**
- **Diaphragm**
- **Trachea**

What muscle separates the thoracic cavity from the abdominal cavity?

______________________________________________________________________

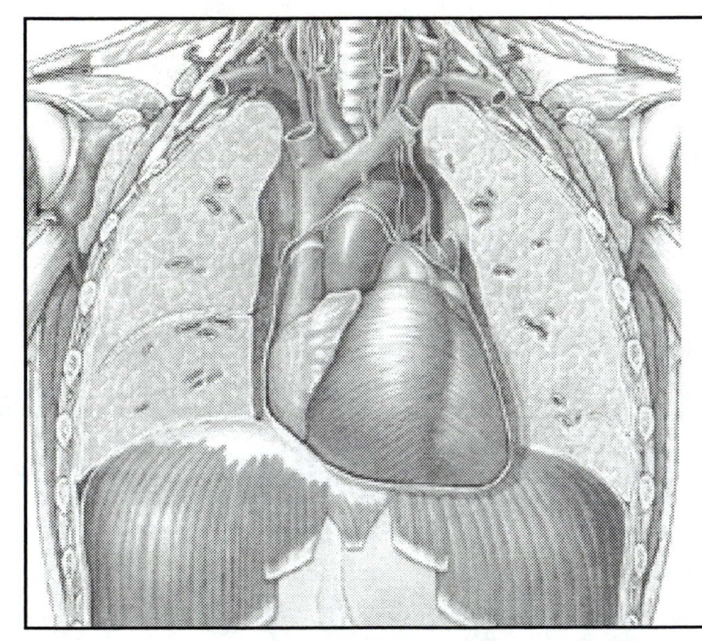

Click several times on the downward **Depth Bar** [▼] until you reach **layer 173**. Identify:

- **Lungs**
- **Heart** (right ventricle, left ventricle, right atrium)
- **Diaphragm**
- **Trachea**

Which of the organs above are found in the mediastinum?

---

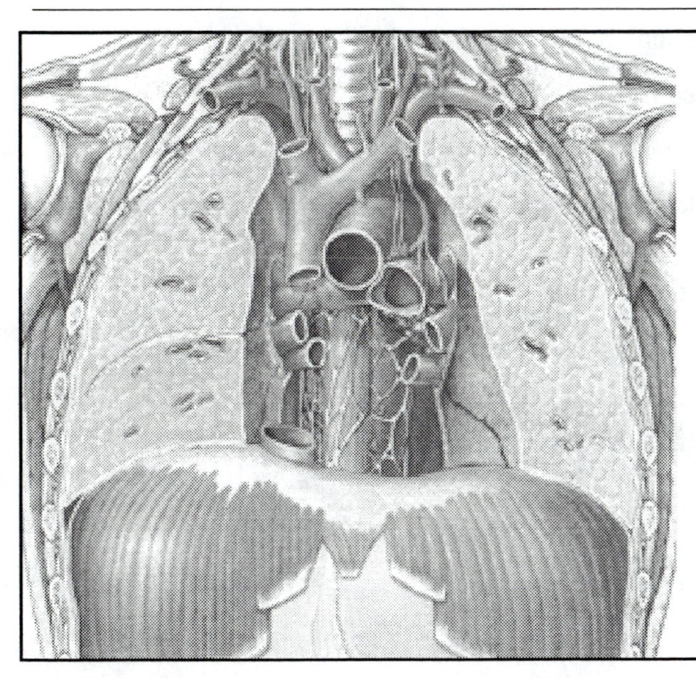

Click several times on the downward **Depth Bar** [▼] to remove the heart until you reach **layer 176**. Identify:

- **Lungs**
- **Esophagus**
- **Diaphragm**
- **Trachea**
- **Parietal peritoneum** (around liver)

Note the location of the esophagus behind the heart. Why is the trachea (windpipe) not found behind the heart as well? (To find the answer, click several times to remove the blood vessels found behind the heart.)

---

When you are done, click the **Close Window Box** in the upper left corner of the window.

## 3D Anatomy

Under the **File** Menu, select **Open**.

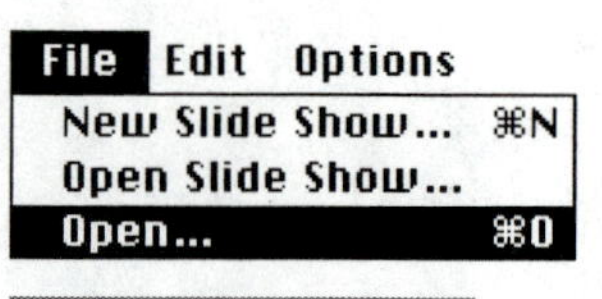


Choose **3D Anatomy**.

Choose **3D Lungs**. Click **Open**.

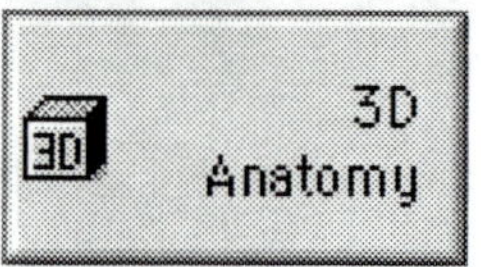


In the upper right corner of the window, click on the **Zoom Box** to allow the image to fill the screen.

Move the 3D image of the lungs around by holding down the arrows on the **Rotator Controller**.

Click on the **Transparency Tool** to show the structures within the lung.

Click on the **Transparency Tool** again to restore the outside of the lungs.

Click on the **Zoom Tool** to increase the magnification.

Click on the **Zoom Tool** again to decrease the magnification.

Hold down the **3D Anatomy Structure List** and select **Trachea**. The trachea should highlight in blue.

Hold down the **3D Anatomy Structure List** and drag to **Right main (primary) bronchus - Anterior**, which should highlight in blue.

Hold down the **3D Anatomy Structure List** and drag to **Carina**.

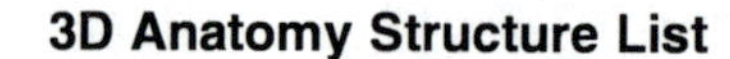

**3D Anatomy Structure List**

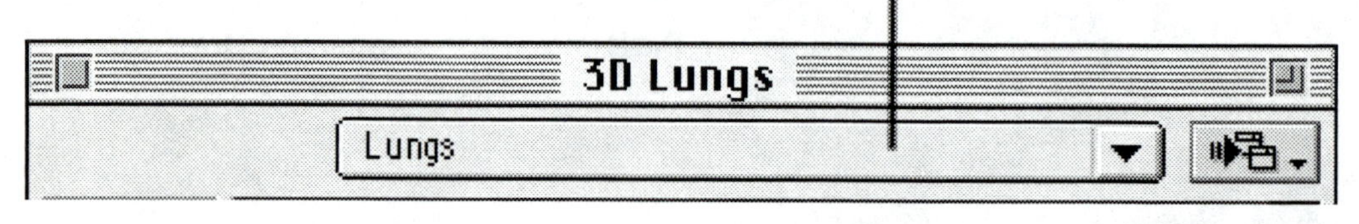


When you are done exploring 3D Anatomy, click the **Close Window Box** in the upper left corner of the window.

# Atlas Anatomy: Anatomical Landmarks

Under the **File** Menu, select **Open**.

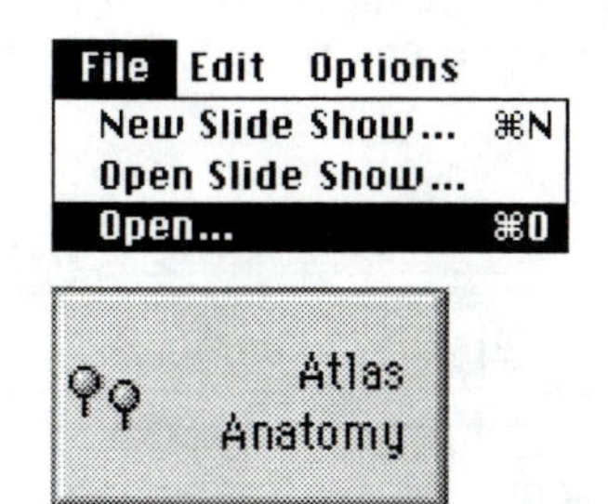


Select **Atlas Anatomy**.

Select **System** and **Digestive**.

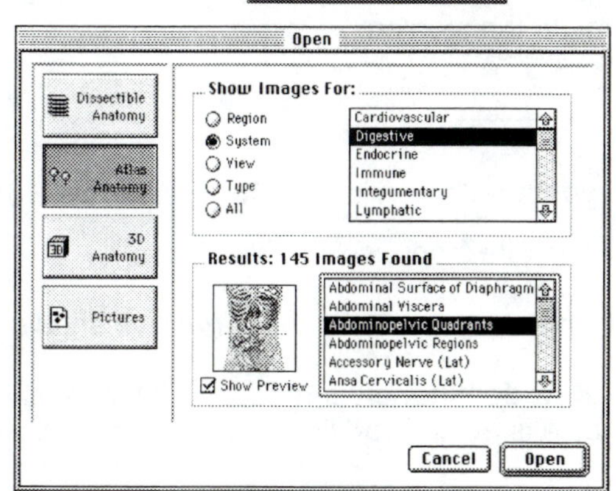


Scroll to **Abdominopelvic Quadrants**. If the list is not in alphabetical order, it will be about two-fifths of the way down.

Click **Open**.

Click on the **Show All Pins** button and scroll down to **All Systems**. This will allow you to view structures in all systems, not just the digestive system.

You may not be able to see the entire image on screen so you may need to scroll to find the structures. Identify the following by clicking on the heads of the pins.

- **Diaphragm**
- **Small intestine** (duodenum)
- **Gallbladder**
- **Left inferior quadrant**
- **Left superior quadrant**
- **Kidney**
- **Liver**
- **Median plane**
- **Right inferior quadrant**
- **Right superior quadrant**
- **Spleen**
- **Stomach**
- **Umbilicus**
- **Appendix** (vermiform appendix)

If you need help finding a structure, follow the directions on the next page.

To find a structure in **Atlas Anatomy**, pull down the **Atlas Anatomy Structure List**. Select the structure you are trying to find. The head of the corresponding pin will turn green.

**Atlas Anatomy Structure List**

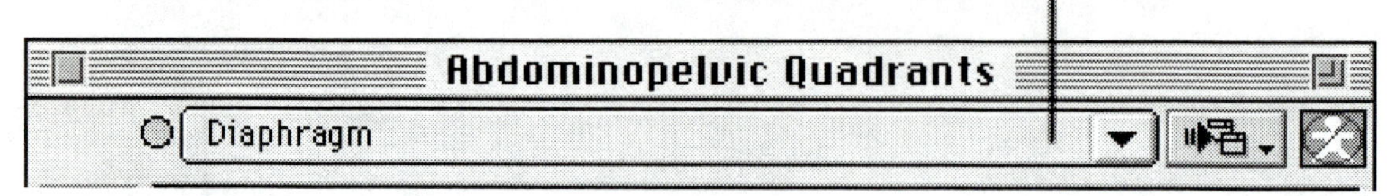


If a patient arrives at the emergency room of the hospital with a severe pain in the right inferior quadrant, what might the problem be?

---

When you are done, click the **Close Window Box** in the upper left corner of the window.

Under the **File** Menu, select **Open**.

**Atlas Anatomy**, **System**, and **Digestive** should still be selected.

Scroll to **Abdominopelvic Regions**. If the list is not in alphabetical order the choice is about two-fifths down the list.

Click **Open**.

Click on the **Show All Pins** button and scroll down to **All Systems**.

You may not be able to see the entire image on screen so you may need to scroll to find the structures. Identify:

- **Diaphragm**
- **Small intestine** (duodenum)
- **Gallbladder**
- **Epigastric area**
- **Hypogastric area**
- **Kidney**
- **Liver**
- **Left inguinal area**
- **Left hypochondriac area**
- **Right lumbar area**
- **Spleen**
- **Stomach**
- **Umbilicus**
- **Appendix** (vermiform appendix)
- **Right inguinal area**
- **Right hypochondriac area**
- **Umbilical area**
- **Ribs**

When you are done, click the **Close Window Box** in the upper left corner of the window.

# Atlas Anatomy: Cross Sectional Anatomy

Under the **File** Menu, select **Open**.

File Edit Options
New Slide Show... ⌘N
Open Slide Show...
Open... ⌘O

**Atlas Anatomy**, **System**, and **Digestive** should still be selected.

Scroll to **Thoracic Surface of Diaphragm**. If the list is not in alphabetical order the choice is about two-fifths down the list.

Click **Open**.

Click on the **Show All Pins** button and scroll down to **All Systems**.

This is a transverse section of the thorax right above the diaphragm. To get a better idea of what you are looking at place your cursor over the **Orientation Icon** in the upper right corner of the screen.

You may not be able to see the entire image on screen. For that reason, you may click on the **Zoom Tool** then click anywhere on the image to make it fit onto your screen. Click on the image again to make it larger.

You may need to scroll to find the all the structures below. Identify the following by clicking on the heads of the pins or by using the **Atlas Anatomy Structure List.**

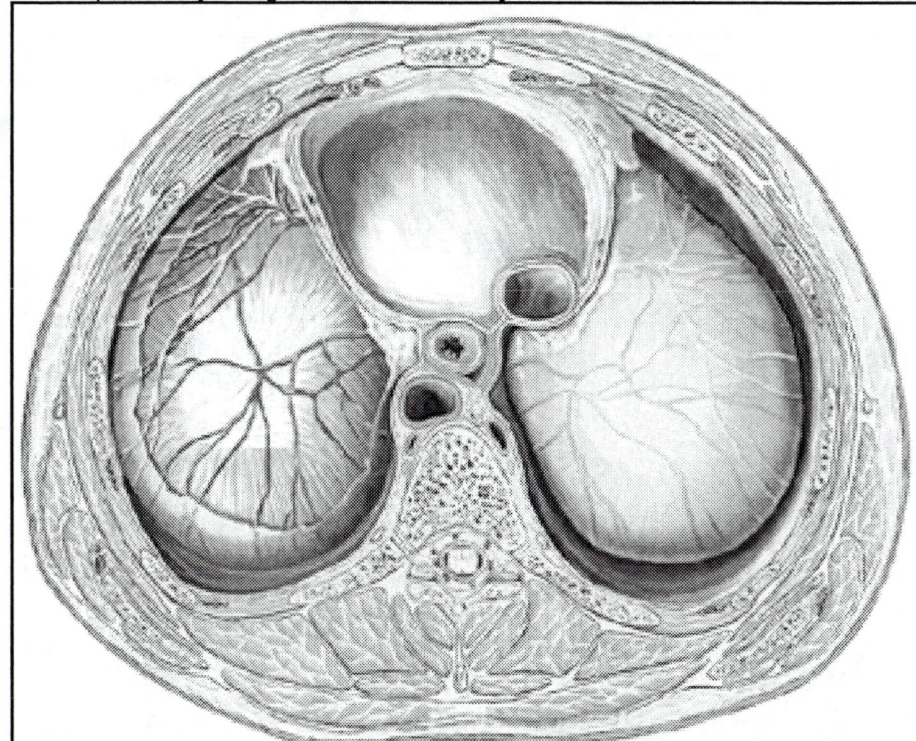

- **Parietal pleura**
- **Costal parietal pleura**
- **Diaphragmatic parietal pleura**
- **Esophagus**
- **Pericardial sac** (parietal pericardium)
- **Pericardium** (parietal pericardium)
- **Sternum**

What organs would be found superior to the diaphragm in this image?

______________________________________________

When you are done, click the **Close Window Box**  in the upper left corner of the window.

Under the **File** Menu, select **Open**.

File Edit Options
New Slide Show... ⌘N
Open Slide Show...
Open... ⌘O

**Atlas Anatomy**, **System**, and **Digestive** should also still be selected.

Scroll to **T8 Vertebra (Inf)**. If the list is not in alphabetical order the choice is about one-third of the way down the list.

Click **Open**.

Click on the **Show All Pins** button and scroll down to **All Systems**.

This is an inferior view of a transverse section of the thorax at the level of the lungs and heart. To get a better idea of what you are looking at place your cursor over the **Orientation Icon** in the upper right corner of the screen.

You may not be able to see the entire image on screen. For that reason, you may click on the **Zoom Tool** then click anywhere on the image to make it fit onto your screen. Click on the image again to make it larger.

You may need to scroll to find the all the structures below. Identify the following by clicking on the heads of the pins or by using the **Atlas Anatomy Structure List.**

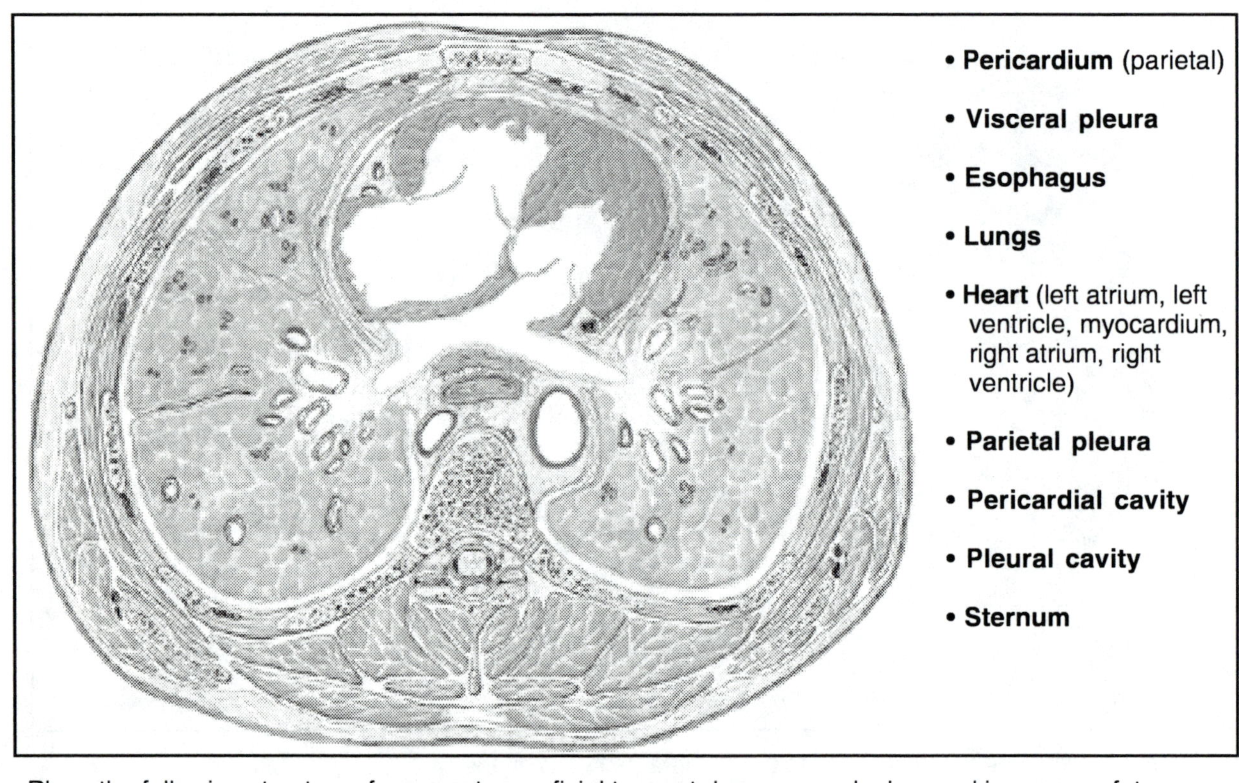

- **Pericardium** (parietal)
- **Visceral pleura**
- **Esophagus**
- **Lungs**
- **Heart** (left atrium, left ventricle, myocardium, right atrium, right ventricle)
- **Parietal pleura**
- **Pericardial cavity**
- **Pleural cavity**
- **Sternum**

Place the following structures from most superficial to most deep: muscle, bone, skin, organs, fat

_______________________________________________

When you are done, click the **Close Window Box** in the upper left corner of the window.

## Cadaver Images

Under the **File** Menu, select **Open**.

File Edit Options
New Slide Show... ⌘N
Open Slide Show...
Open... ⌘O

**Atlas Anatomy**, **System**, and **Digestive** should still be selected.

Scroll to **Abdominal Viscera**. If the list is not in alphabetical order the choice is about half way down the list.

Click **Open**.

Click on the **Show All Pins** button and scroll down to **All Systems**.

This is a cadaver image of the abdomen. To get a better idea of what you are looking at place your cursor over the **Orientation Icon** in the upper right corner of the screen.

You may not be able to see the entire image on screen. For that reason, you may click on the **Zoom Tool** then click anywhere on the image to make it fit onto your screen. Click on the image again to make it larger.

You may need to scroll to find the all the structures below. Identify the following by clicking on the heads of the pins or by using the **Atlas Anatomy Structure List:**

- **Large intestine** (ascending colon, cecum, sigmoid colon, transverse colon)
- **Stomach** (body, fundus)
- **Diaphragm**
- **Esophagus**
- **Gall bladder**
- **Small intestine** (ileum, jejunum)
- **Liver**
- **Spleen**

Based on this image, how much fluid would you expect to find in the peritoneal cavity?

---

When you are done, click the **Close Window Box** in the upper left corner of the window.

Under the **File** Menu, select **Open**.

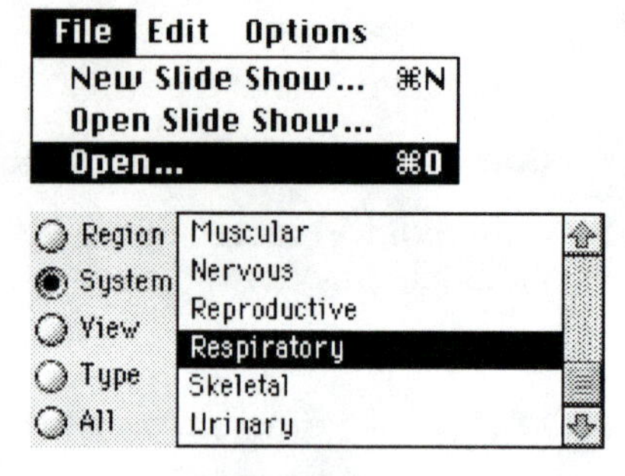


**Atlas Anatomy** should already be selected.

Choose **System** and **Respiratory**.

Scroll to **Thoracic Viscera (Ant)**. If the list is not in alphabetical order the choice is about four-fifths of the way down the list.

Click **Open**.

Click on the **Show All Pins** button and scroll down to **All Systems**.

This is an anterior image of the thorax. To get a better idea of what you are looking at place your cursor over the **Orientation Icon** in the upper right corner of the screen.

You may not be able to see the entire image on screen. For that reason, you may click on the **Zoom Tool** then click anywhere on the image to make it fit onto your screen. Click on the image again to make it larger.

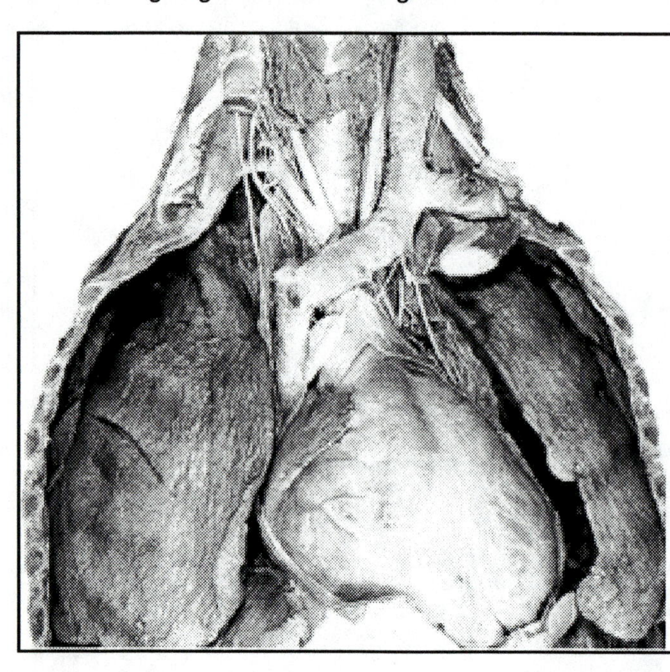

You may need to scroll to find the all the structures below. Identify the following by clicking on the heads of the pins or by using the **Atlas Anatomy Structure List:**

- **Heart** (apex, auricle of left atrium, anterior interventricular sulcus, left ventricle, right ventricle)
- **Lungs**
- **Pericardium**
- **Trachea**

When you are done, click the **Close Window Box** in the upper left corner of the window.

# A.D.A.M. Interactive Anatomy Tutorial for Windows: A.D.A.M.'s Major Organs

## Opening A.D.A.M. Interactive Anatomy

- Turn on the computer and allow Windows to load.
- Place the A.D.A.M. Interactive Anatomy CD-ROM disk in the CD tray.
- Click the **Start Button** in the lower left corner of the screen, click **Programs**, and click **A.D.A.M. Interactive Anatomy**.
- You should now see the A.D.A.M. Interactive Anatomy Introduction screen with four "hot buttons". Click on the icon corresponding to "Dissectible Anatomy".
- Select **Male** and **Anterior**. Click **Open**.

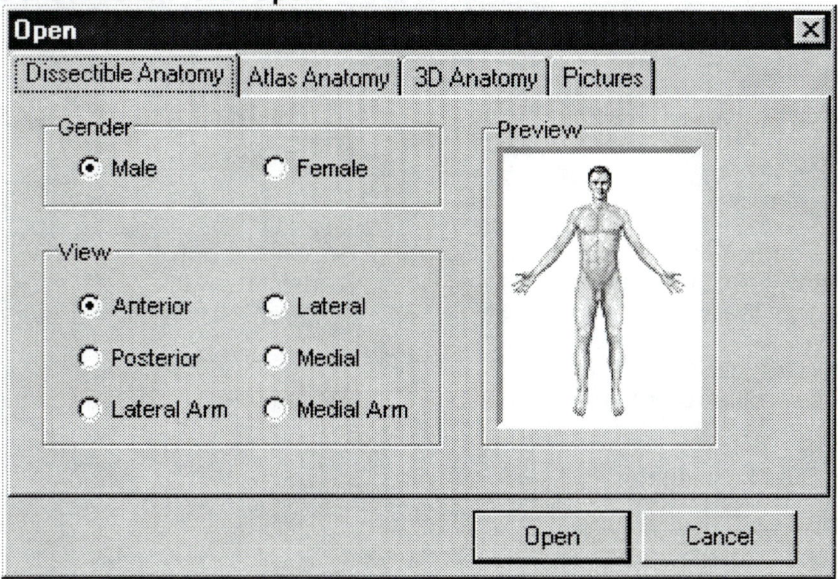


## Enlarging the Image

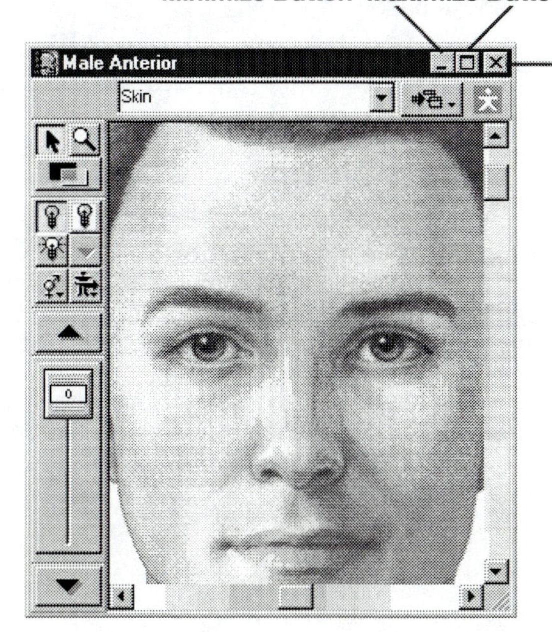


**Close Button**

To maximize the image on your screen, left click the **Maximize Button** located in the upper right corner of the **Application Window**. When clicked, the **Maximize Button** changes to the **Restore Button**. Clicking the **Restore Button** restores the screen to its original size.

To minimize the image, (to work within another application) left click the **Minimize Button**. The image will disappear from the screen but the program will remain open and its name will appear on the **Task Bar** at the bottom of the screen.

## Centering the Image Over the Desired Body Part

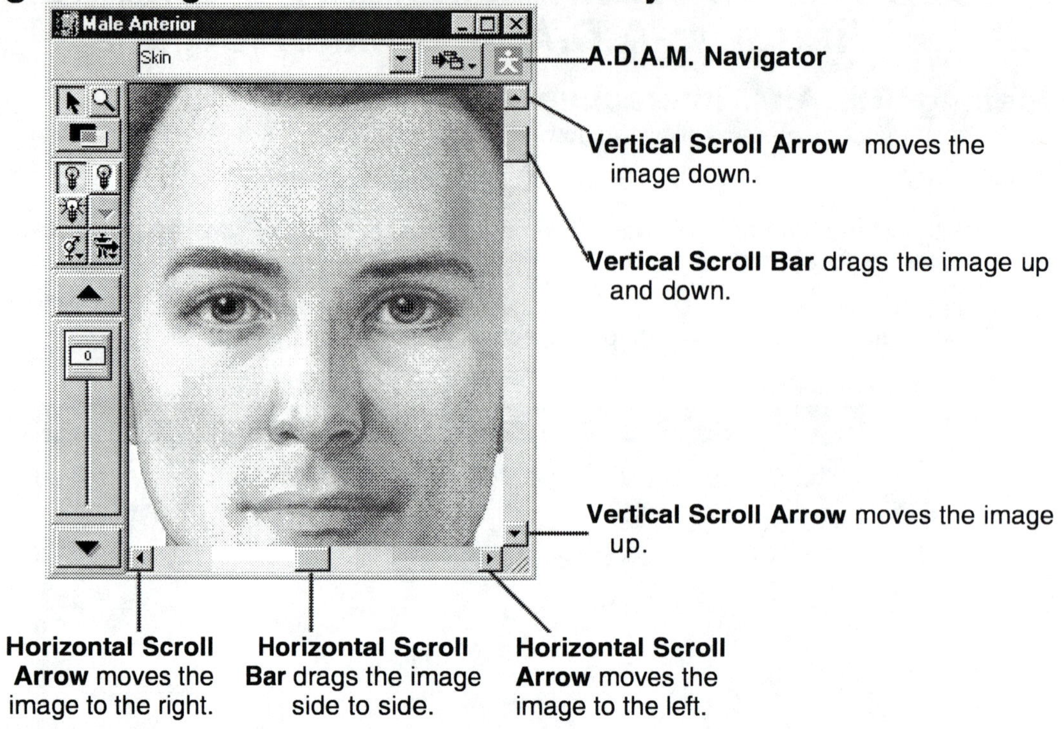
Male Anterior
Skin
A.D.A.M. Navigator
Vertical Scroll Arrow moves the image down.
Vertical Scroll Bar drags the image up and down.
Vertical Scroll Arrow moves the image up.
Horizontal Scroll Arrow moves the image to the right.
Horizontal Scroll Bar drags the image side to side.
Horizontal Scroll Arrow moves the image to the left.

There are two ways to move the image:

1. Click on the **scroll arrows** to move the image to the desired location or drag on the **scroll buttons** to move the image faster.

2. Use the **A.D.A.M. Navigator**:

   • Place the cursor over the **A.D.A.M. Navigator**. You will see an image of the entire body with a rectangle superimposed over it.

   • Place the cursor inside the rectangle, hold down the left mouse button, and drag until the rectangle is centered over the desired body part. Release the mouse button.

   • Drag the cursor so that the box is over the abdomen.

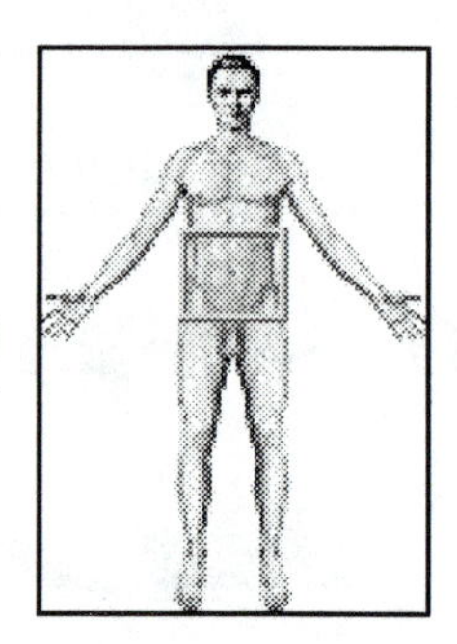

## Reducing the Image Size

• Click on the **Zoom Tool**. Using the mouse, move the **Zoom Tool** to A.D.A.M.'s abdomen. Notice that a minus (-) sign appears in the center of the tool. This means that you have reached the maximum magnification of the **Zoom** function. Click the left mouse button. Notice that you now have a smaller image of A.D.A.M. but you are able to see more of his body.

• Click the **Zoom Tool** again and move the tool to A.D.A.M.'s abdomen. Notice that there is now a (+) in the center of the tool. Click the left mouse button. Notice that you are now back at the original image.

## Anterior Dissection of Abdominopelvic Cavity

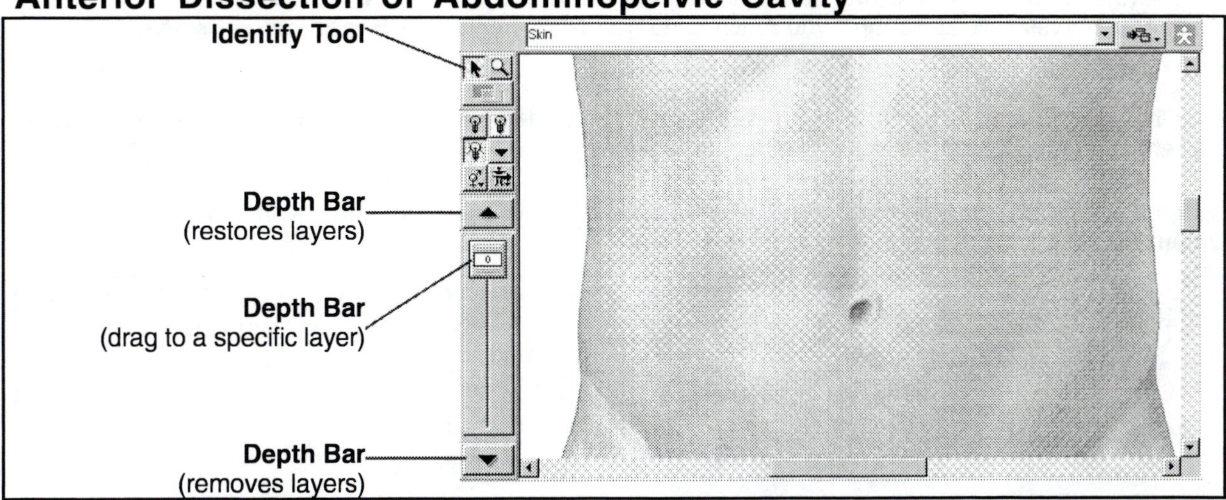


Click once on the downward arrow on the **Depth Bar** [▼] to remove the skin. You are now at **layer 1**. The first layer below the skin is fat. Identify:

**Fat** (superficial fascia)

To identify structures, click on the **Identify Tool** [arrow], move the tool over the structure, and hold down the left mouse button.

Try to resist scrolling down to the layer indicated since this will not give you a "dissection experience".

The structures you will need to know for the laboratory practical exam are shown in bold face print. If the program calls the structure something else, the structure name will be in parentheses.

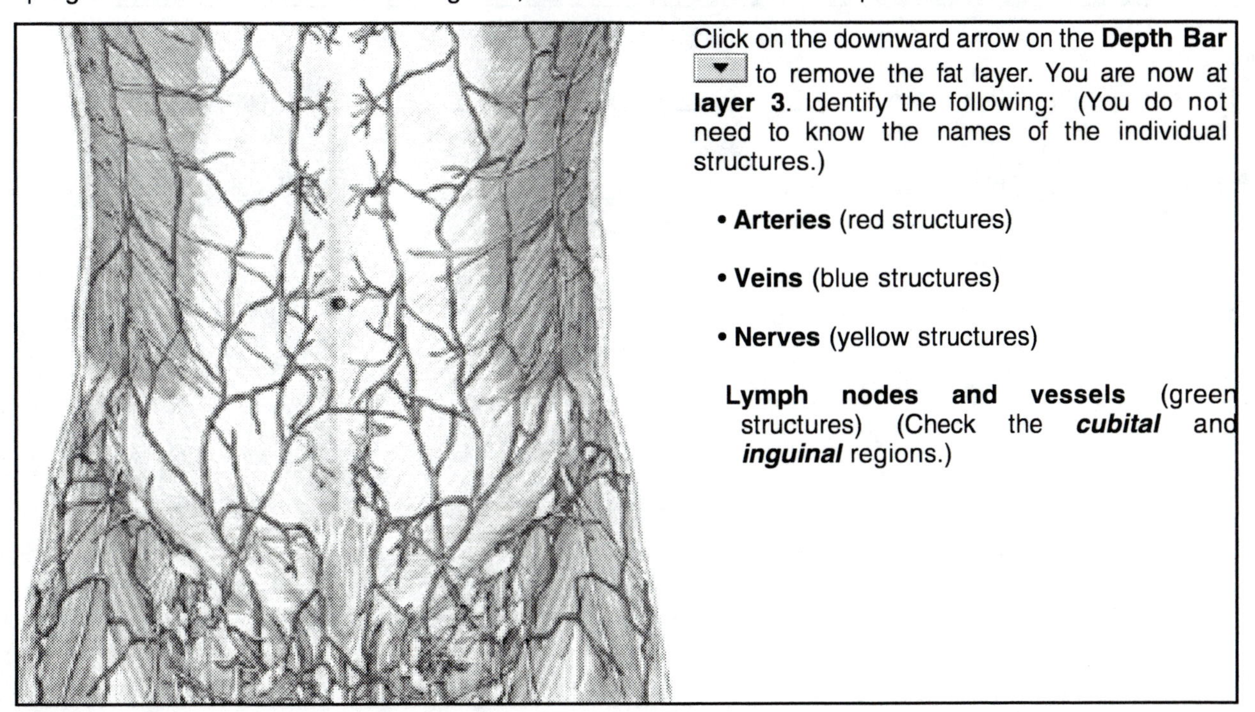

Click on the downward arrow on the **Depth Bar** [▼] to remove the fat layer. You are now at **layer 3**. Identify the following: (You do not need to know the names of the individual structures.)

- **Arteries** (red structures)
- **Veins** (blue structures)
- **Nerves** (yellow structures)

**Lymph nodes and vessels** (green structures) (Check the ***cubital*** and ***inguinal*** regions.)

As you move through the laboratory, connect the names of the structures with the corresponding diagrams by drawing lines or arrows.

You may have to scroll up or down to find all of the structures listed.

Do not try to click when you see the hour-glass. This means that the computer is busy doing something.

Click several times on the downward arrow on the **Depth Bar** to remove the blood vessels, lymphatic structures, nerves and fascia, until you reach **layer 18**. Note the various **Superficial muscles**.

Drag the **Depth Bar** 18 down to **layer 194** 194. Click and identify the **Parietal peritoneum**, which is the serous membrane against the wall of the abdominal cavity.

Click the downward arrow on the **Depth Bar** once to reach **layer 195**. Identify the **Greater Omentum** which covers the abdominal organs.

Click several times on the downward arrow on the **Depth Bar** until you reach **layer 198**. Identify:

- **Liver**
- **Stomach**
- **Gallbladder**
- **Large intestine** (ascending colon, transverse colon, sigmoid colon)
- **Small intestine** (jejunum & ileum)

You will be responsible for items listed in bold font. Items in parentheses are used to find particular structures. To find the organs above follow the directions below.

The liver and the stomach are ______________________ (ipsilateral, contralateral) to one another.

The liver and the gallbladder are ______________________ (ipsilateral, contralateral) to one another.

# Using the List Manager to Identify Structures in Dissectible Anatomy

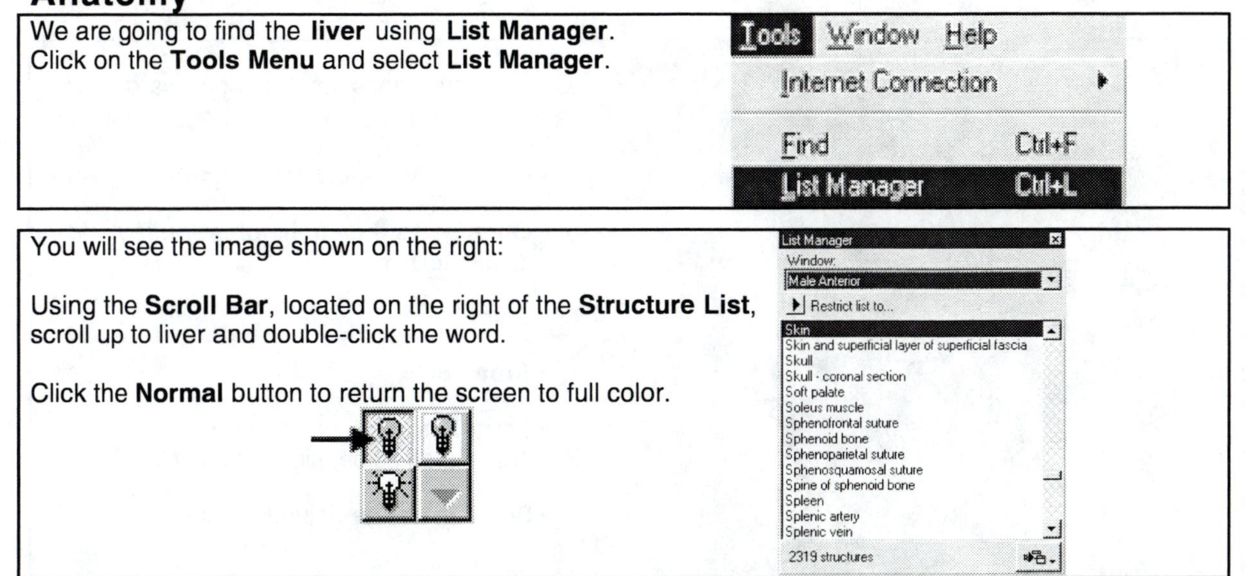


We are going to find the **liver** using **List Manager**.
Click on the **Tools Menu** and select **List Manager**.

You will see the image shown on the right:

Using the **Scroll Bar**, located on the right of the **Structure List**, scroll up to liver and double-click the word.

Click the **Normal** button to return the screen to full color.

Now let us find the **stomach** using the **List Manager** another way.

Click anywhere in the **Structure List** to highlight a structure. Type stomach and left click. You will see A.D.A.M.'s stomach highlighted in color.

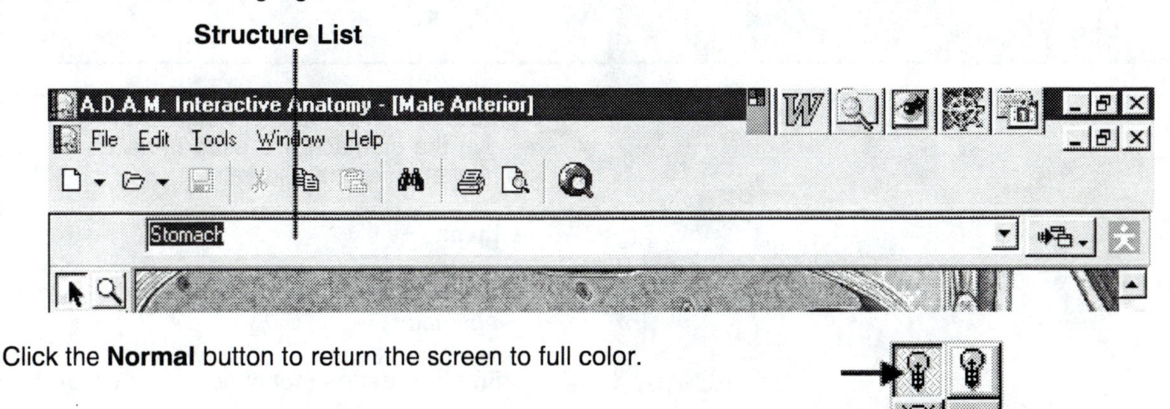


Click the **Normal** button to return the screen to full color.

Identify the **Large intestine**, **Small intestine**, and **Gall bladder** by either using the **List Manager** or by pointing and clicking on the image.

When you identify the intestines using the **List Manager**, be sure to type in the words listed in parentheses above because large or small intestine is not specific enough.

- **Large intestine** (ascending colon, transverse colon, sigmoid colon)
- **Small intestine** (jejunum & ileum)

 

Click several times on the downward **Depth Bar** ▼ until you reach **Layer 202**. Click and identify: **Gall bladder**. No graphic has been provided for this layer.

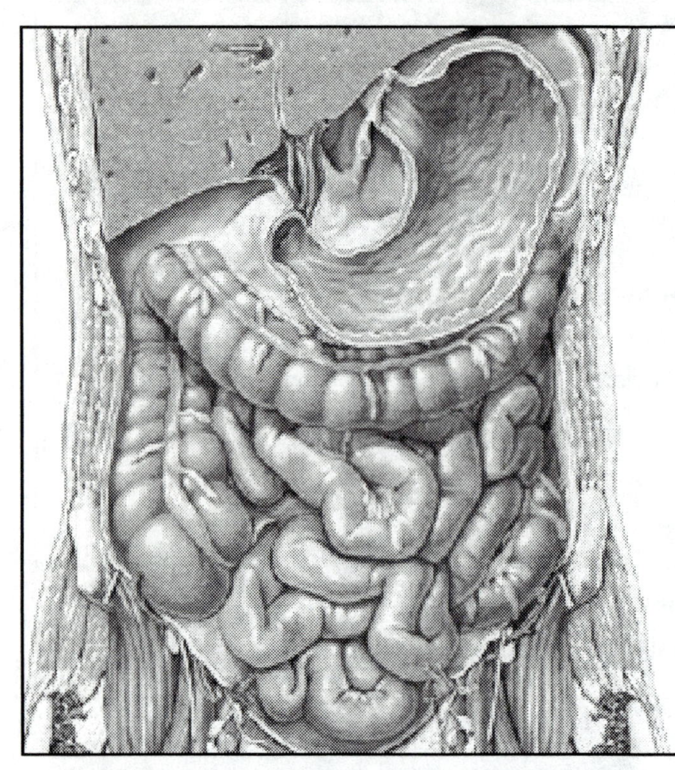

Click several times on the downward **Depth Bar** ▼, until you reach **layer 205**. Identify:

- **Inside of the stomach** (rugae)
- **Opening to the small intestine** (pyloric sphincter).
- **Liver**
- **Stomach**
- **Large intestine** (ascending colon, transverse colon, sigmoid colon)
- **Small intestine** (jejunum & ileum)

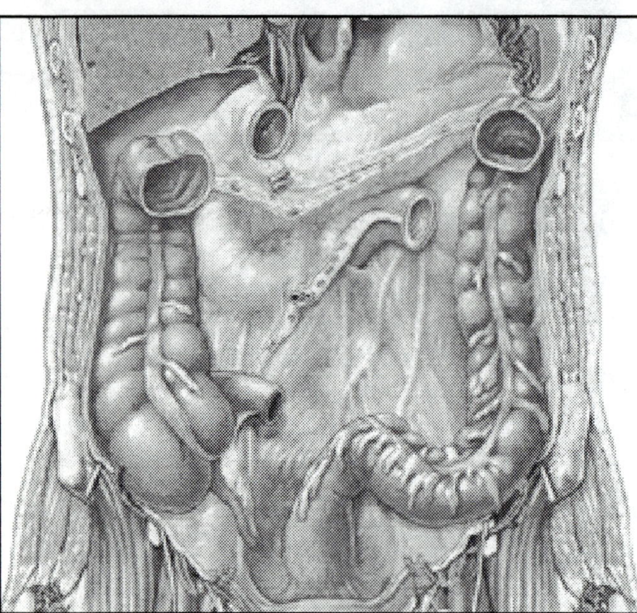

Click several times on the **Depth Bar** ▼ to remove the stomach and small intestine. You are now at **layer 209**. Identify:

- **Liver**
- **Opening to the small intestine** (pyloric sphincter)
- **Small intestine** (cut off jejunum & ileum)
- **Large intestine** (cecum, ascending colon, descending colon, sigmoid colon)
- **Rectum**
- **Parietal peritoneum** (peritoneum)

## Using Find to Identify Structures in Dissectible Anatomy

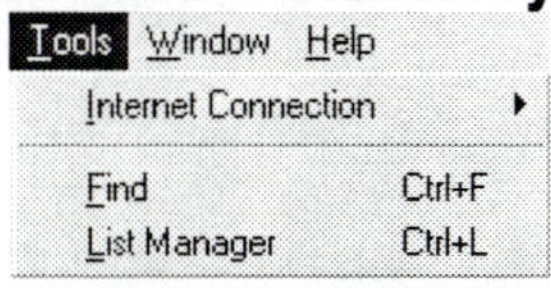


At this level let's identify the appendix using **Find**. Go to the **Tools Menu** and select **Find**.

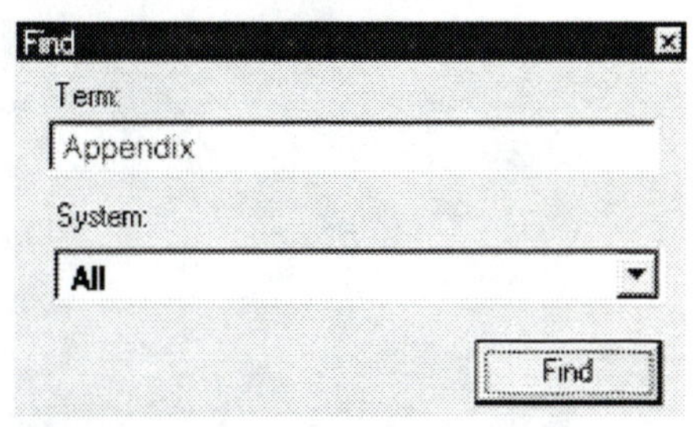


Click on the word(s) in the **Term Box** and type **Appendix**. Click **Find**.

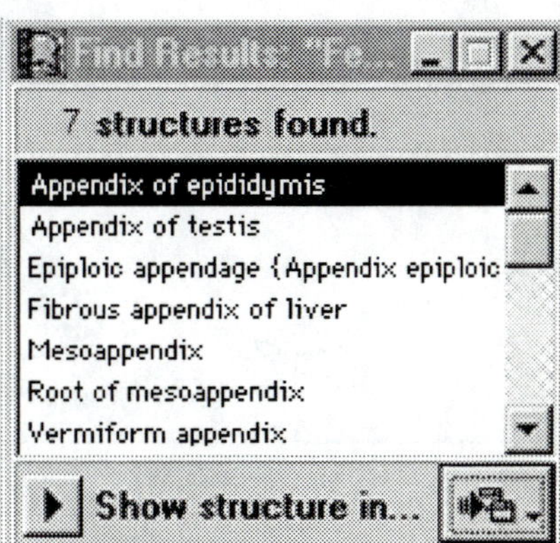


You will then see the following image:

Select **Vermiform appendix**.

Click on the **Show structure in** ...button and choose **Male Anterior.** You will see the appendix highlighted in color.

Click the **Normal** button and note the appendix on the color image.

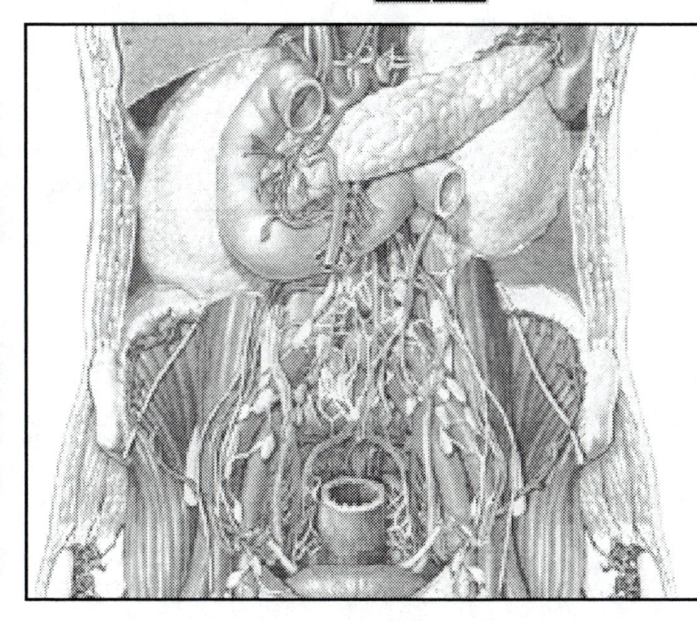

Click a few times on the downward **Depth Bar** to remove most of the intestines and peritoneum until you reach **layer 211**. Identify:

- **Liver**
- **Opening to the small intestine** (pyloric sphincter)
- **Small intestine** (jejunum & duodenum)
- **Pancreas** (body, tail, and head)
- **Spleen**
- **Urinary bladder**

List several retroperitoneal organs (found posterior to the peritoneum).

______________________________________________________________________

______________________________________________________________________

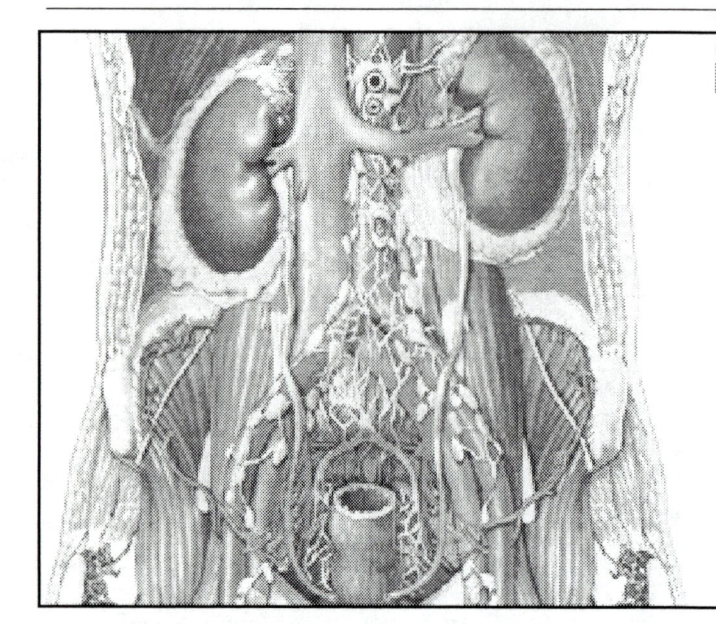

Click many times on the downward **Depth Bar** ▼ until you reach **layer 230**. Identify:

- **Kidneys**
- **Adrenal gland**
- **Ureter**
- **Rectum**

## Changing the View, Changing the Gender, and Using the Highlight Tool

Use the **Vertical Scroll Arrows** or **A.D.A.M. Navigator** to move the image down over the hip and buttock area. Click on the **View** button and choose **Medial**.

Click on the **Normal** button. Scroll to **layer 0**.

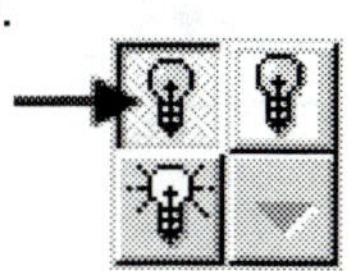

You are now looking at a midsagittal view.

Click on the **Highlight** button and select **Digestive** from the drop down menu. This feature allows you to see a highlighted view of all structures associated with the Digestive System.

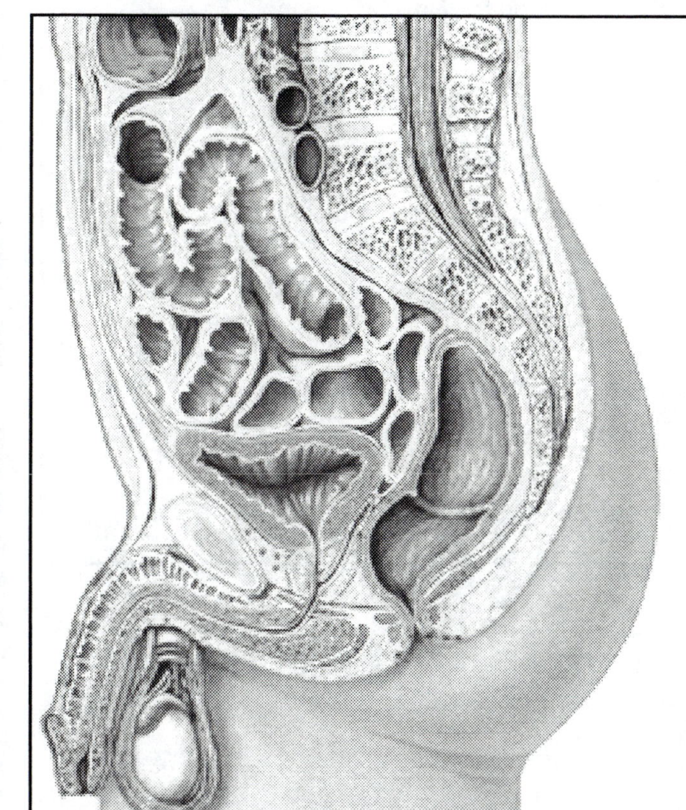

Identify:

- **Rectum**

Click on the **Highlight** button again and select **Urinary** from the drop down menu. Identify:

- **Urinary bladder**

Click on the **Highlight** button again and choose **Reproductive** from the drop down menu. Identify:

- **Prostate Gland**

- **Testis**

Click on the **Gender** button and select **Female**.

Click on the **Normal** button:

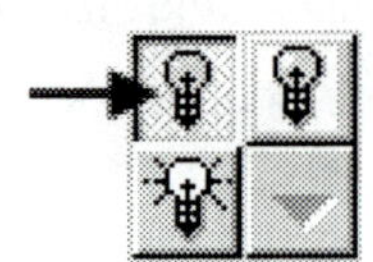

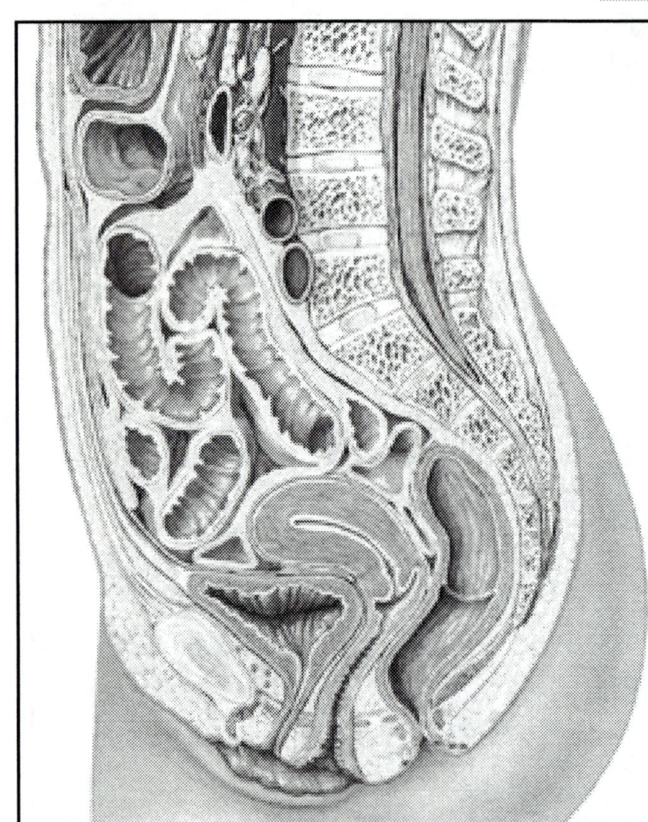

Click on the **Highlight** button  and select **Reproductive**. Identify:

- **Uterus** (body of uterus)
- **Vagina**

Click on the **Normal** button.

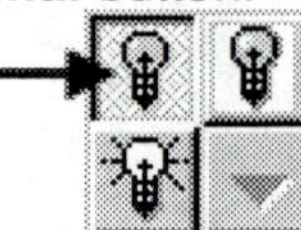

Identify the following on the female image:

- **Rectum**
- **Urinary Bladder**

## Anterior Dissection of Thoracic Cavity

Click on the **View** button and choose **Anterior**.

Click on the **Gender** button and choose **Male**.

Use the **Vertical Scroll Arrows** or **A.D.A.M. Navigator** to move the image over the chest area.

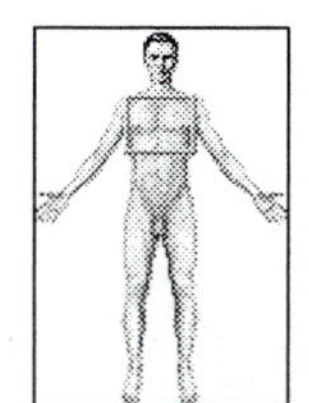

Click on the **Normal Tool.**

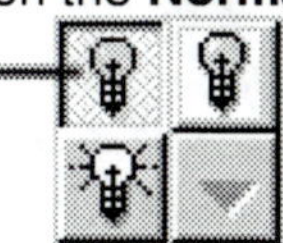

Scroll up to **layer 0**.

Click once on the downward **Depth Bar** to remove the skin. You are now at **layer 1**. Identify:
- **Fat** (superficial fascia)

Click on the downward **Depth Bar** to remove the fat. You are now at **layer 5**. Identify the following (you do not need to know the names of the individual structures):
- **Arteries** (red structures)
- **Veins** (blue structures)
- **Nerves** (yellow structures)

Click several times on the downward **Depth Bar** to remove the blood vessels, nerves and fascia, until you reach **layer 19**. Note the various **Superficial muscles**. List the name of a superficial muscle of the chest.

___________________________________________

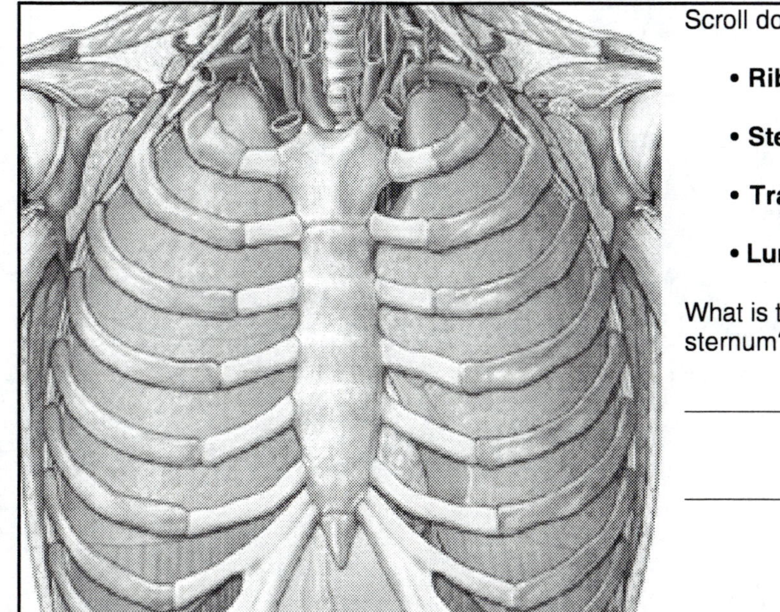

Scroll down to **layer 158**. Identify:

- **Ribs**
- **Sternum**
- **Trachea**
- **Lungs** (under parietal pleura)

What is the protective role of the ribs and sternum?

___________________________________________

___________________________________________

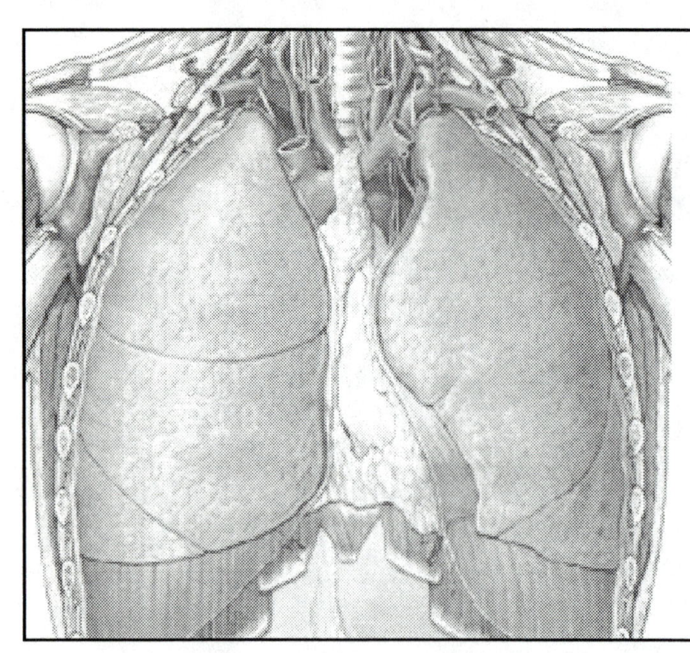

Click several times on the downward **Depth Bar** ▼ to remove the ribs, sternum, and parietal pleura until you reach **layer 162**. Identify:

- **Lungs**
- **Thymus gland**
- **Pericardial sac**
- **Diaphragm**
- **Trachea**

What organ is found within the pericardial sac? (Hint: Drag down to **layer 173** to find out.)

---

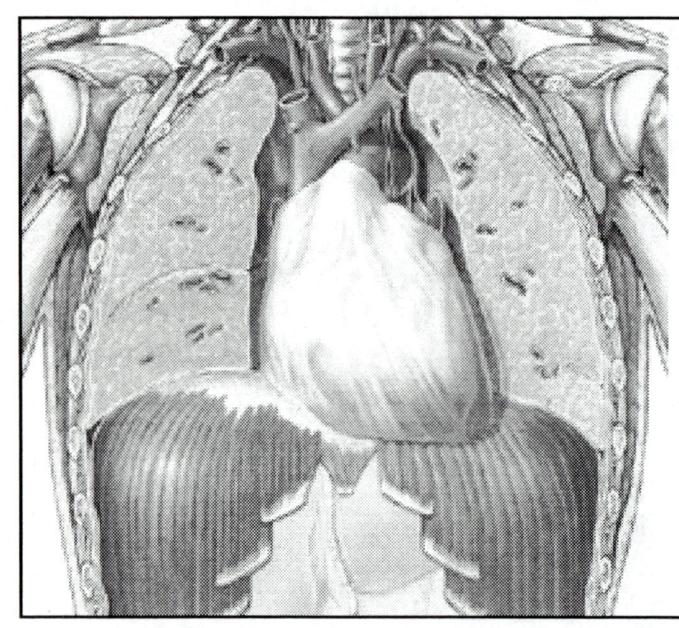

Click several times on the downward **Depth Bar** ▼ until you reach **layer 169**. Identify:

- **Lungs**
- **Pericardial sac**
- **Diaphragm**
- **Trachea**

What muscle separates the thoracic cavity from the abdominal cavity?

---

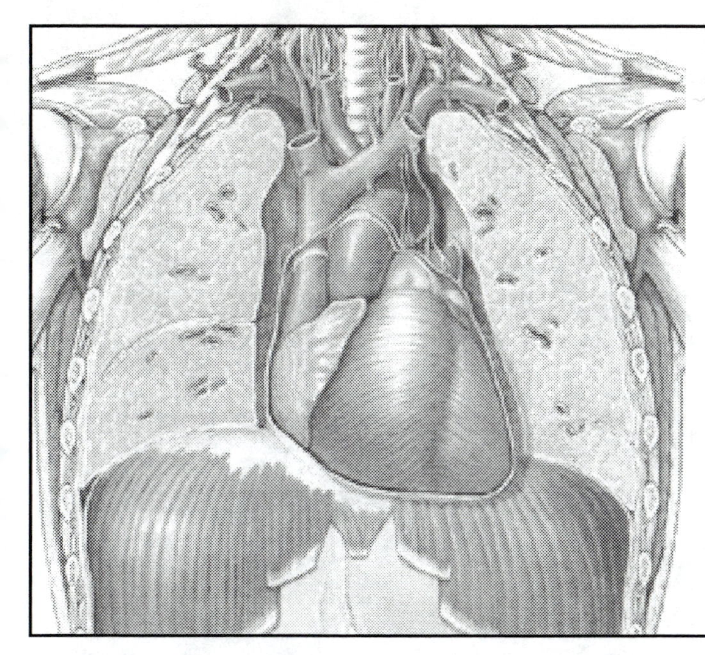

Click several times on the downward **Depth Bar** ▼ until you reach **layer 173**. Identify:

- **Lungs**
- **Heart** (right ventricle, left ventricle, right atrium)
- **Diaphragm**
- **Trachea**

Which of the organs above are found in the mediastinum?

---

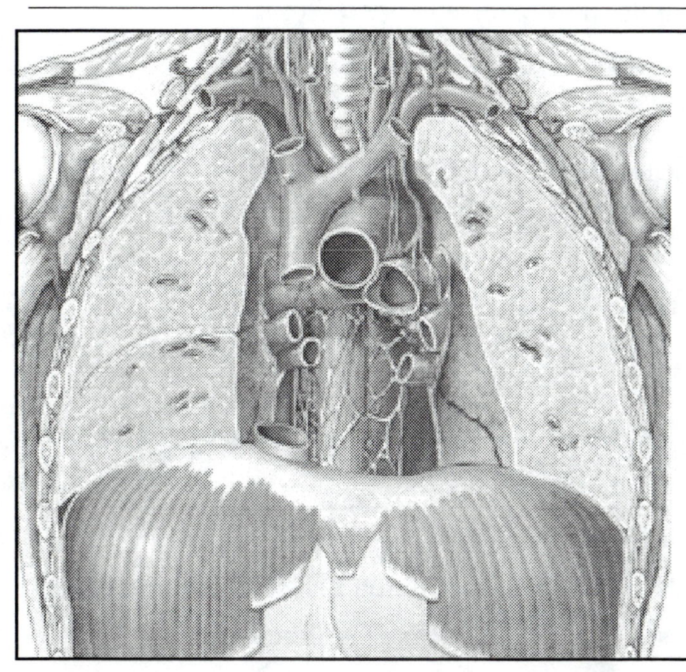

Click several times on the downward **Depth Bar** ▼ to remove the heart until you reach **layer 176**. Identify:

- **Lungs**
- **Esophagus**
- **Diaphragm**
- **Trachea**
- **Parietal peritoneum** (around liver)

Note the location of the esophagus behind the heart. Why is the trachea (windpipe) not found behind the heart as well? (To find the answer, click several times to remove the blood vessels found behind the heart.)

---

When you are done, click the **Close Button** ✕ in the upper right corner of the window.

## 3D Anatomy

Click **File** on the **Menu Bar**, drag down to **Open**, and click **Content**.

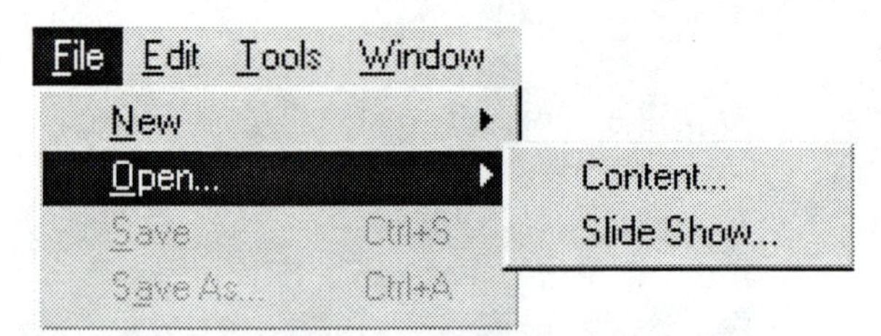


Click the **3D Anatomy Tab**, choose **Lungs**, click **Open**.

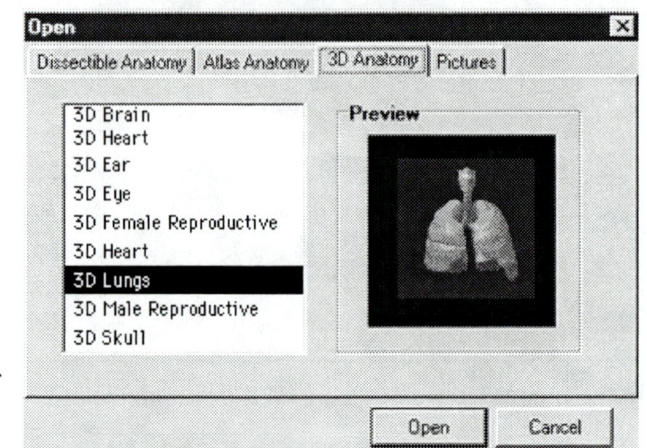


Click the **Maximize Button** in the upper right corner of the screen.

Move the 3D image of the lungs around by holding down the arrows on the **Rotator Controller**.

Click on the **Transparency Tool** to show the structures within the lung.

Click on the **Transparency Tool** again to restore the outside of the lungs.

Click on the **Zoom Tool** to increase the magnification.

Click on the **Zoom Tool** again to decrease the magnification.

Click on the **3D Anatomy Structure List Scroll Arrow** and, using the **Scroll Bar**, scroll to **Trachea**. The trachea should highlight in blue.

Click on the **3D Anatomy Structure List Scroll Arrow** and, using the **Scroll Bar**, scroll to **Right main (primary) bronchus - Anterior** which should highlight in blue.

Click on the **3D Anatomy Structure List Scroll Arrow** and, using the **Scroll Bar**, scroll to **Carina**.

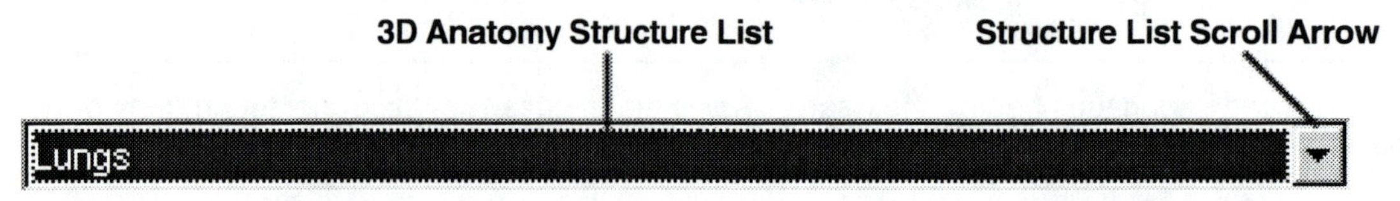


When you have finished exploring 3D Anatomy, click the **Close Button** in the upper right corner of the window.

# Atlas Anatomy: Anatomical Landmarks

Click **File** on the **Menu Bar**, drag down to **Open**, and click **Content**.

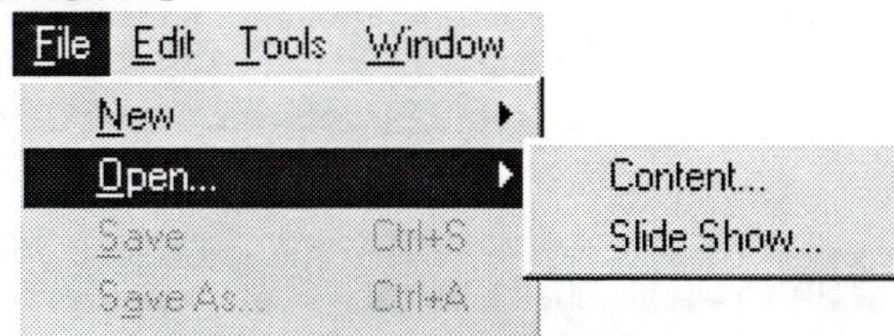


Click the **Atlas Anatomy Tab**.

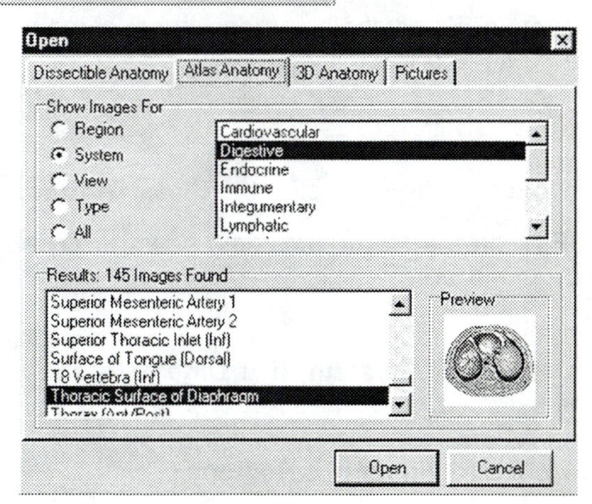


Select **System** and **Digestive**.

Using the **Structure List Scroll Bar**, scroll to **Abdominopelvic Quadrants**.

Click **Open**.

Maximize the image.

You may not be able to see the entire image on screen so you may need to scroll to find the structures. Identify the following by clicking on the heads of the pins.

- **Diaphragm**
- **Small intestine** (duodenum)
- **Gallbladder**
- **Left inferior quadrant**
- **Left superior quadrant**
- **Kidney**
- **Liver**
- **Median plane**
- **Right inferior quadrant**
- **Right superior quadrant**
- **Spleen**
- **Stomach**
- **Umbilicus**
- **Appendix** (vermiform appendix)

If you need help finding a structure, follow the directions on the next page.

To find a structure in **Atlas Anatomy**, click on the **Atlas Anatomy Structure List Scroll Arrow**. Select the name of the structure you are trying to find. The head of the corresponding pin will turn green.

If a patient arrives at the emergency room of the hospital with a severe pain in the **right inferior quadrant**, what might the problem be?

_______________________________________________

When you have finished, click the **Close Button** in the upper right corner of the screen.

Click **File** on the **Menu Bar**, drag down to **Open**, and click **Content**.

**Atlas Anatomy**, **System**, and **Digestive** should still be selected.

Scroll to **Abdominopelvic Regions**.

Click **Open**.

Maximize the image.

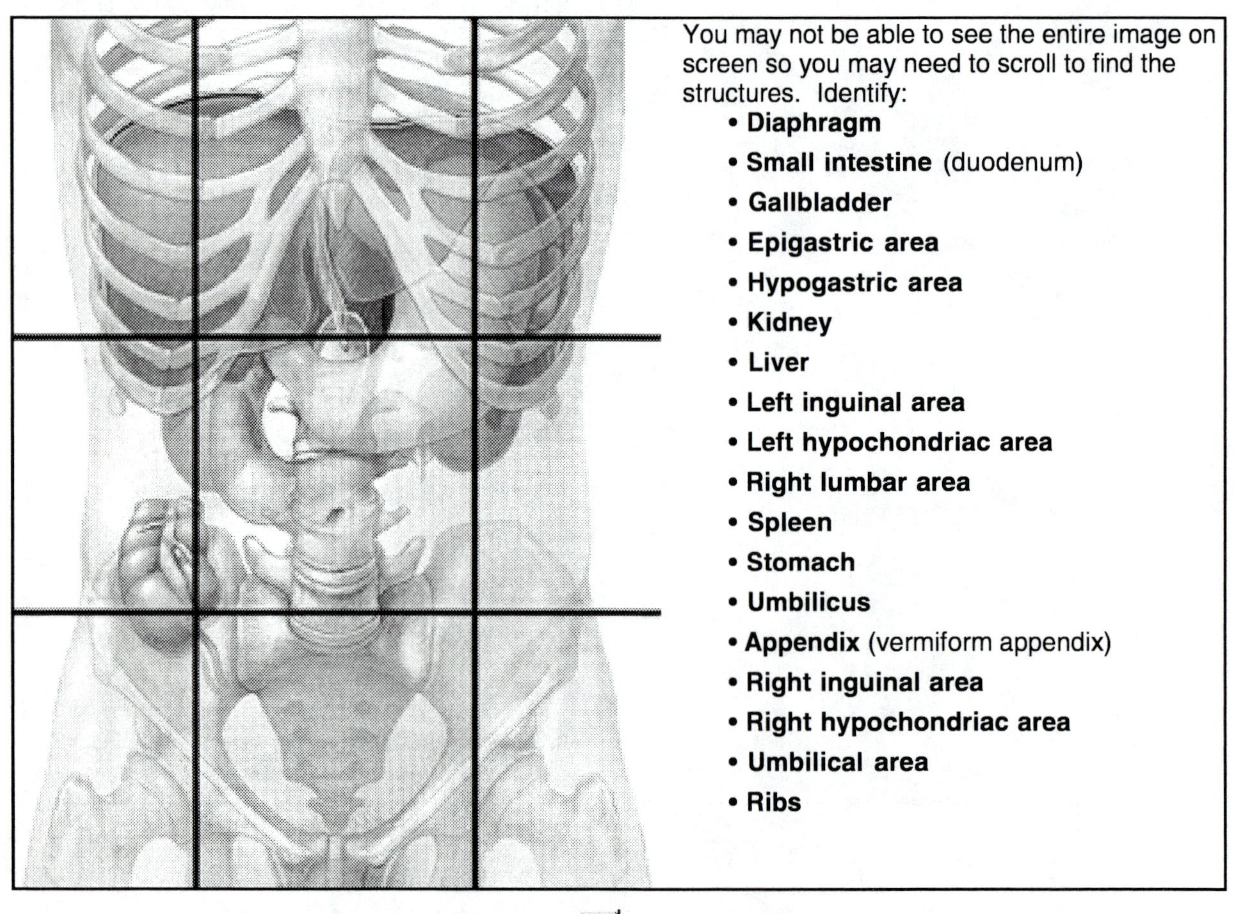

You may not be able to see the entire image on screen so you may need to scroll to find the structures. Identify:

- **Diaphragm**
- **Small intestine** (duodenum)
- **Gallbladder**
- **Epigastric area**
- **Hypogastric area**
- **Kidney**
- **Liver**
- **Left inguinal area**
- **Left hypochondriac area**
- **Right lumbar area**
- **Spleen**
- **Stomach**
- **Umbilicus**
- **Appendix** (vermiform appendix)
- **Right inguinal area**
- **Right hypochondriac area**
- **Umbilical area**
- **Ribs**

When you have finished, click the **Close Button** in the upper right corner of the window.

# Atlas Anatomy: Cross Sectional Anatomy

Click **File** on the **Menu Bar**, drag down to **Open**, and click **Content**.

File Edit Tools Window
New
Open... | Content... | Slide Show...
Save Ctrl+S
Save As... Ctrl+A

**Atlas Anatomy**, **System**, and **Digestive** should still be selected.

Scroll to **Thoracic Surface of Diaphragm**.
Click **Open**.
Maximize the image.

This is a transverse section of the thorax right above the diaphragm. To get a better idea of what you are looking at place your cursor over the **Orientation Icon** in the upper right corner of the screen.

You may not be able to see the entire image on screen. For that reason, you may click on the **Zoom Tool** then place your cursor any place on the image and click. Click on the image again to make it larger.

You may need to scroll to find the all the structures below. Identify the following by clicking on the heads of the pins or by using the **Atlas Anatomy Structure List**.

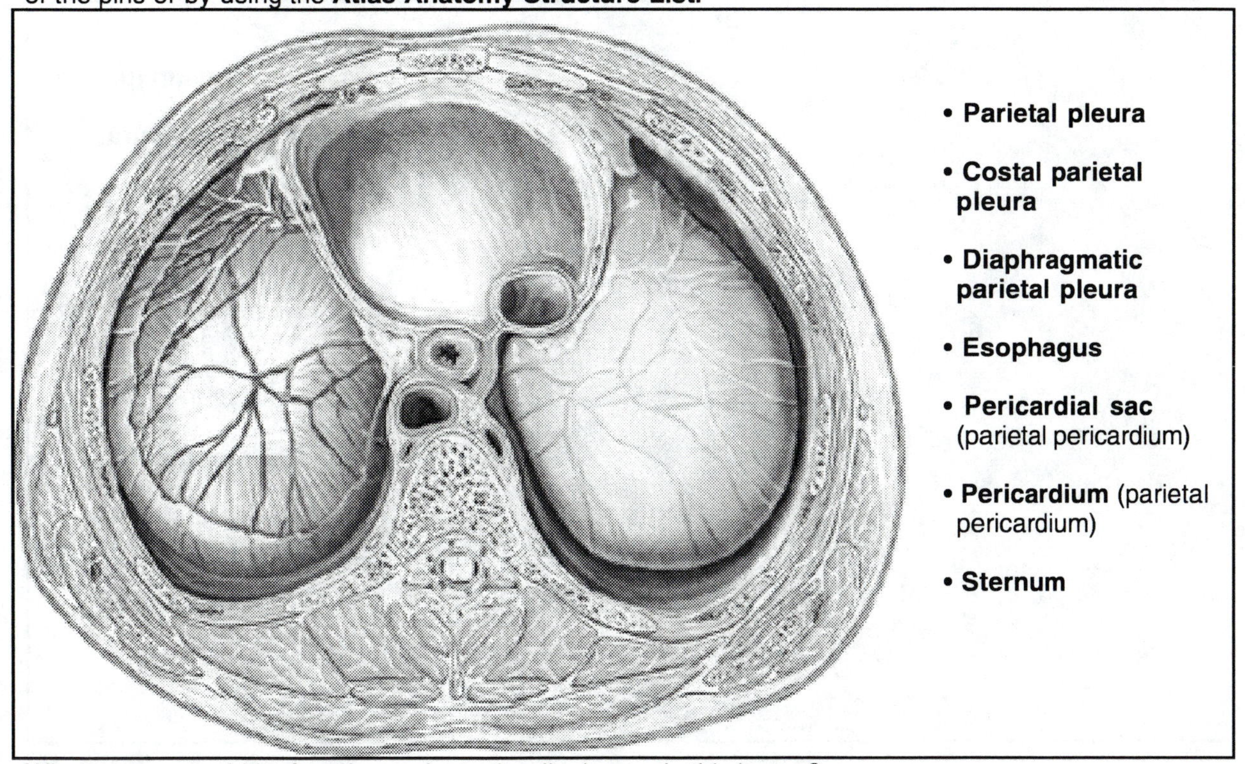


What organs would be found superior to the diaphragm in this image?

---

When you have finished, click the **Close Button** in the upper right corner of the window.

Click **File** on the **Menu Bar**, drag down to **Open**, and click **Content**.

File Edit Tools Window
New
Open... Content...
Save Ctrl+S Slide Show...
Save As... Ctrl+A

**Atlas Anatomy**, **System**, and **Digestive** should still be selected.

Scroll to **T8 Vertebra (Inf)**. Click **Open**.

Maximize the image.

This is an inferior view of a transverse section of the thorax at the level of the lungs and heart. To get a better idea of what you are looking at place your cursor over the **Orientation Icon** in the upper right corner of the screen.

You may not be able to see the entire image on the screen. To reduce the size of the image to fit it on the screen, click the **Zoom Tool** and then place your cursor any place on the image and click. The image will become smaller. Click on the image again to make it larger.

You may need to scroll to find all of the structures listed below. Identify the following by either clicking on the heads of the pins or by using the **Atlas Anatomy Structure List**.

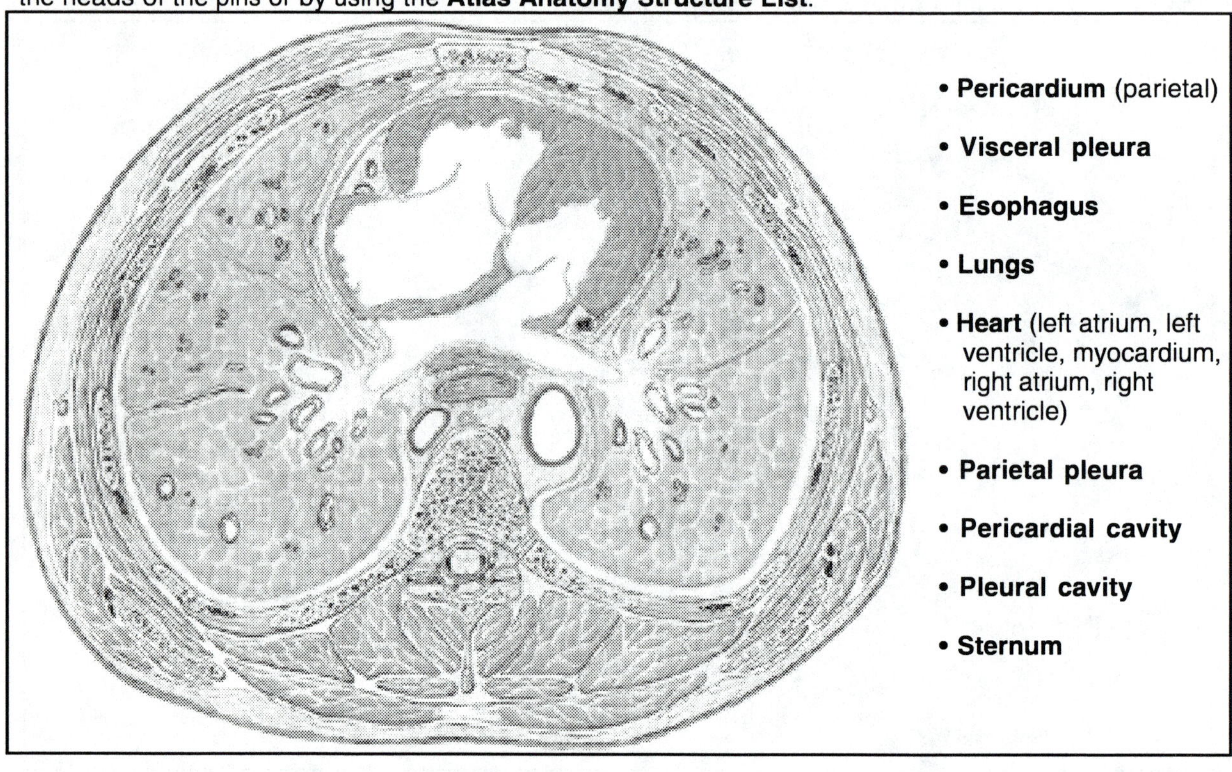

- **Pericardium** (parietal)
- **Visceral pleura**
- **Esophagus**
- **Lungs**
- **Heart** (left atrium, left ventricle, myocardium, right atrium, right ventricle)
- **Parietal pleura**
- **Pericardial cavity**
- **Pleural cavity**
- **Sternum**

Place the following structures from most superficial to most deep: muscle, bone, skin, organs, fat

________________________________________________________

When you have finished, click the **Close Button** ☒ in the upper right corner of the window.

## Cadaver Images

Click **File** on the **Menu Bar**, drag down to **Open**, and select **Content**.

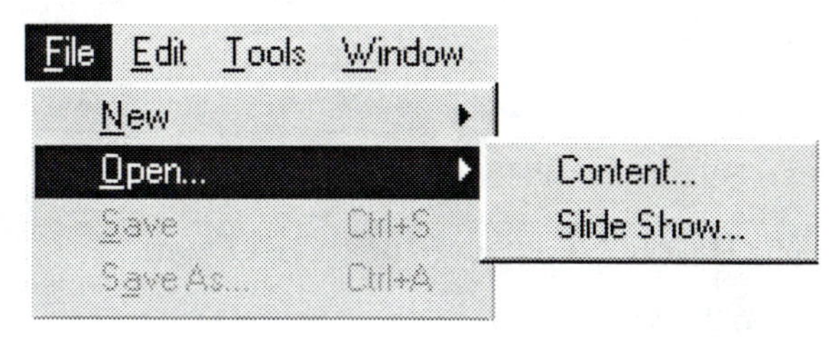


**Atlas Anatomy**, **System**, and **Digestive** should also still be selected.

Scroll to **Abdominal Viscera**. Click **Open**.

Maximize the image.

This is a cadaver image of the abdomen. To get a better idea of where in the body you are looking, place your cursor on the **Orientation Icon** in the upper right corner of the screen.

You may not be able to see the entire image on the screen. To reduce the size of the image to fit it on the screen, click the **Zoom Tool** and then place your cursor any place on the image and click. The image will become smaller. Click on the image again to make it larger.

You may need to scroll to find the all the structures below. Identify the following by clicking on the heads of the pins or by using the **Atlas Anatomy Structure List:**

- **Large intestine** (ascending colon, cecum, sigmoid colon, transverse colon)
- **Stomach** (body, fundus)
- **Diaphragm**
- **Esophagus**
- **Gall bladder**
- **Small intestine** (ileum, jejunum)
- **Liver**
- **Spleen**

Based on this image, how much fluid would you expect to find in the peritoneal cavity?

---

When you have finished, click the **Close Button** [X] in the upper right corner of the window.

Click **File** on the **Menu Bar**, drag down to **Open**, and select **Content**.

**Atlas Anatomy** should still be selected.

Choose **System** and **Respiratory**.

Scroll to **Thoracic Viscera (Ant)**. Click **Open**.

Maximize the image.

This is an anterior image of the thorax. To get a better idea of what you are looking at place your cursor over the **Orientation Icon** in the upper right corner of the screen.

You may not be able to see the entire image on screen. For that reason, you may click on the **Zoom Tool** then click anywhere on the image to make it fit onto your screen. Click on the image again to make it larger.

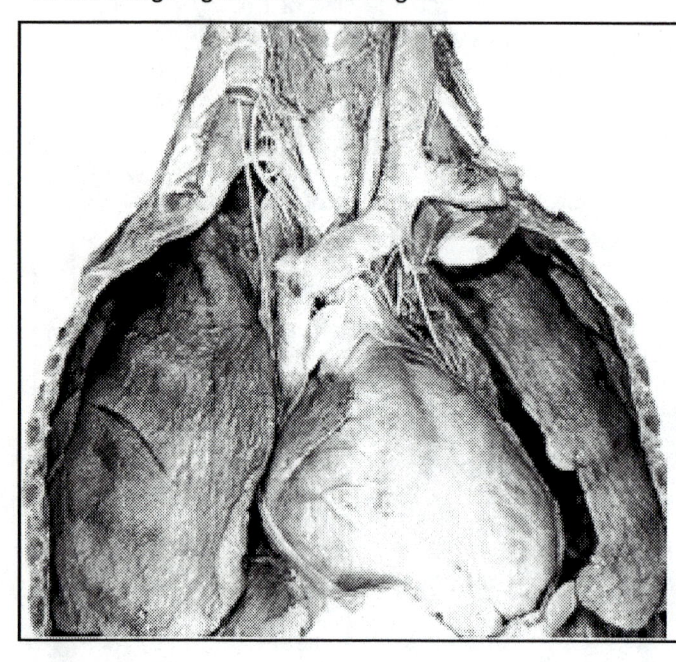

You may need to scroll to find the all the structures below. Identify the following by clicking on the heads of the pins or by using the **Atlas Anatomy Structure List:**

- **Heart** (apex, auricle of right atrium, anterior interventricular sulcus, left ventricle, right ventricle)
- **Lungs**
- **Pericardium**
- **Trachea**

When you have finished, click the **Close Button** in the upper right corner of the window.

# Skull

## Opening A.D.A.M. Interactive Anatomy

Open A.D.A.M. Interactive Anatomy according to the directions in the Tutorial.

## 3D Anatomy of the Skull Bones

Choose **3D Anatomy**. Choose **Skull.** Identify the following bones by choosing bones from the 3D Anatomy Structure List. If a skull is available, it is highly suggested you hold the skull in your hand and identify the items below on the skull as you proceed.

- **Sphenoid bone** - isolated
- **Ethmoid bone** - isolated
- **Sinuses of skull**
- **Maxillary Sinus** - anterior and lateral
- **Frontal sinus** - anterior and lateral
- **Ethmoidal cells** - anterior and lateral
- **Sphenoidal sinus** - anterior and lateral

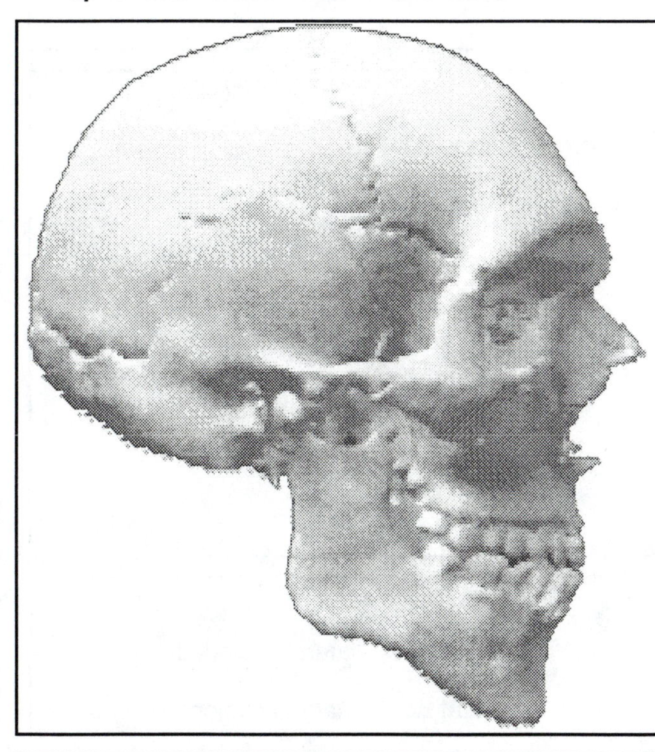

Identify:

- **Frontal bone** - default
- **Parietal bone** - default
- **Temporal bone** - default
- **Coronal suture**
- **Zygomatic process** of the temporal bone
- **External auditory** (acoustic) **meatus**
- **Zygomatic bone** - default
- **Condyle** of mandible
- **Mental foramen** of the mandible
- **Lambdoid suture**
- **Squamosal suture**

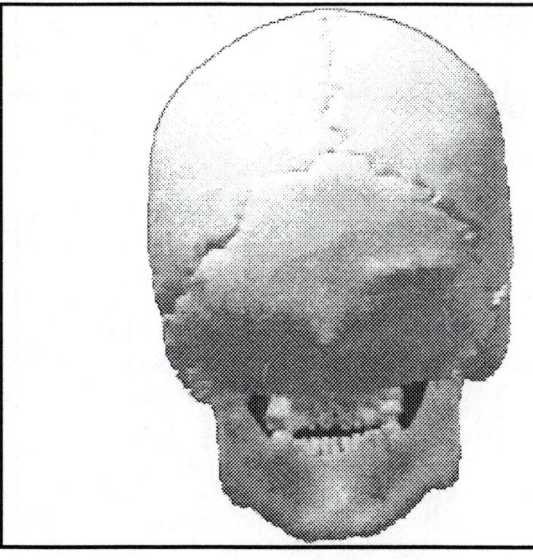

Identify:

- **Occipital bone** - default
- **Sagittal suture**

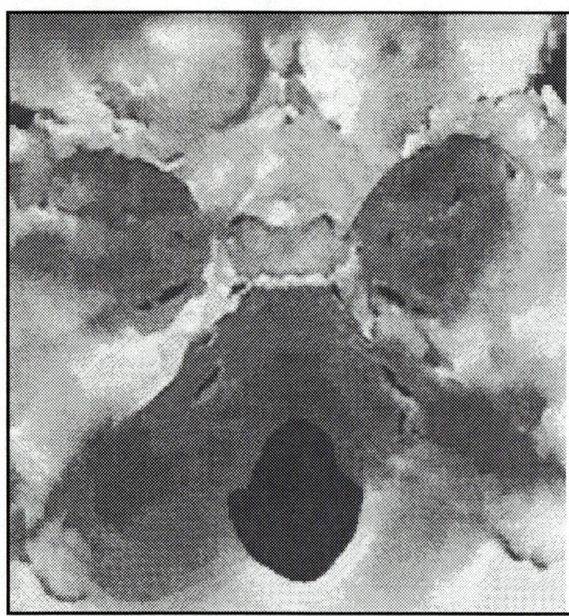

Identify:

- **Greater wing of sphenoid bone**
- **Lesser wing of sphenoid bone**
- **Jugular foramen** of the temporal bone-superior
- **Foramen ovale** of the sphenoid bone - superior
- **Internal auditory** (acoustic) **meatus** - superior
- **Optic canal** - superior
- **Sella turcica** - superior
- **Cribriform plate** of ethmoid bone
- **Crista galli** of ethmoid bone
- **Foramen lacerum** - superior
- **Foramen rotundum** - superior

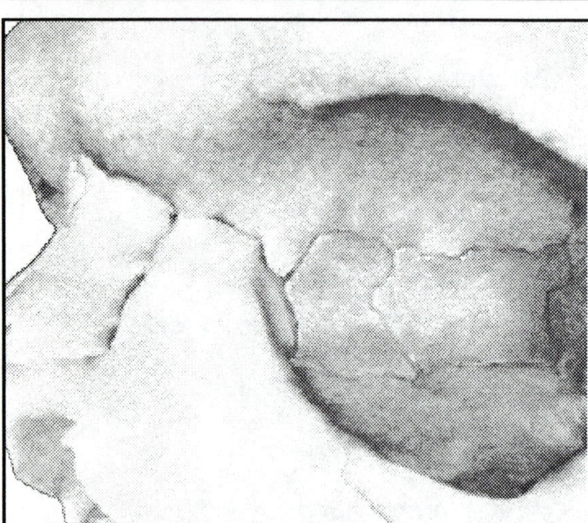

Identify:

- **Ethmoid bone** - default
- **Lacrimal Bone**

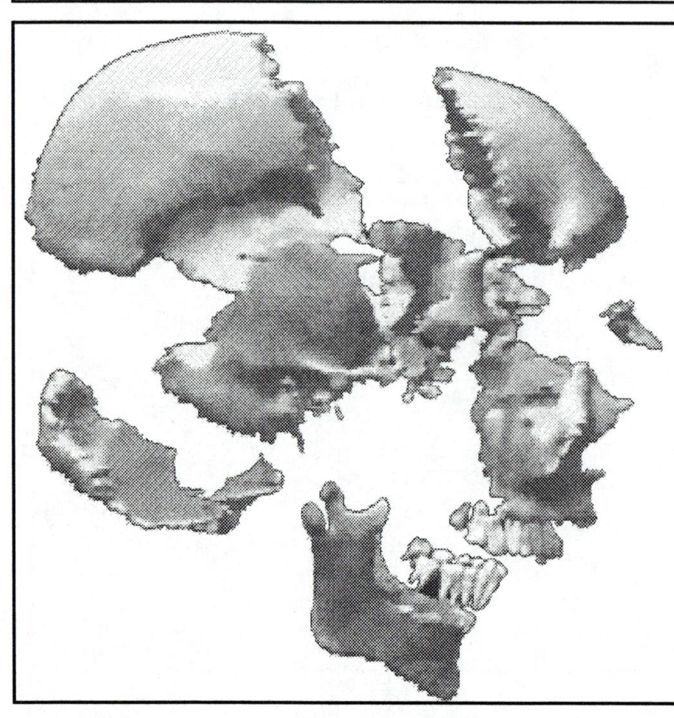

Identify:

- **Frontal bone** - lateral exploded
- **Parietal bone** - lateral exploded
- **Temporal bone** - lateral exploded
- **Occipital bone** - lateral exploded
- **Mandible** - lateral exploded
- **Maxilla** - lateral exploded
- **Nasal bone** - lateral exploded
- **Zygomatic bone** - lateral exploded

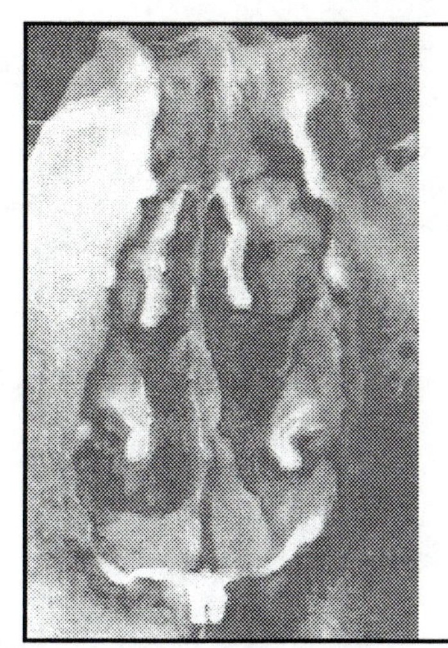

Identify:

- **Inferior Nasal Conchae**
- **Middle Nasal Conchae** of the ethmoid bone

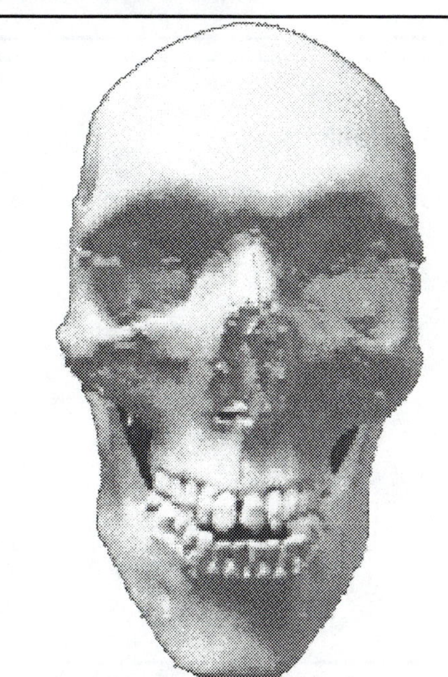

Identify:

- **Mandible** - default
- **Maxilla** - default
- **Nasal bone** - default

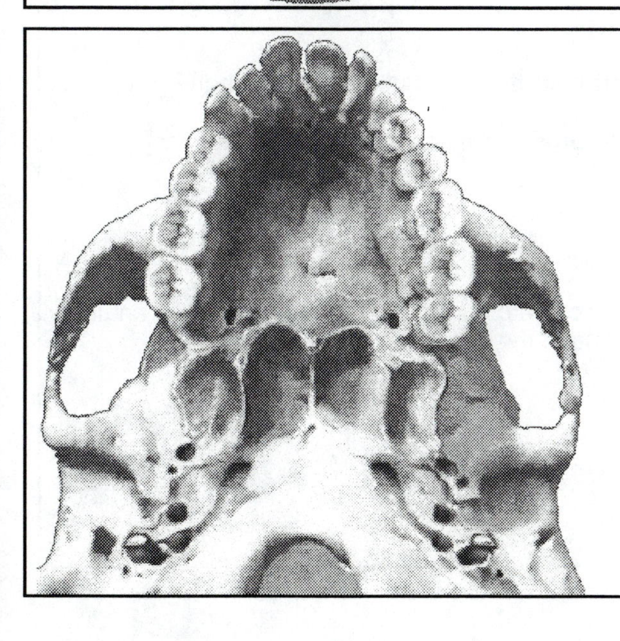

Identify:

- **Palatine bone**
- **Vomer** - inferior
- **Carotid canal** of the temporal bone
- **Jugular foramen** of the temporal bone - inferior
- **Foramen ovale** of the sphenoid bone - inferior
- **Foramen lacerum** - inferior
- **Foramen rotundum** - inferior

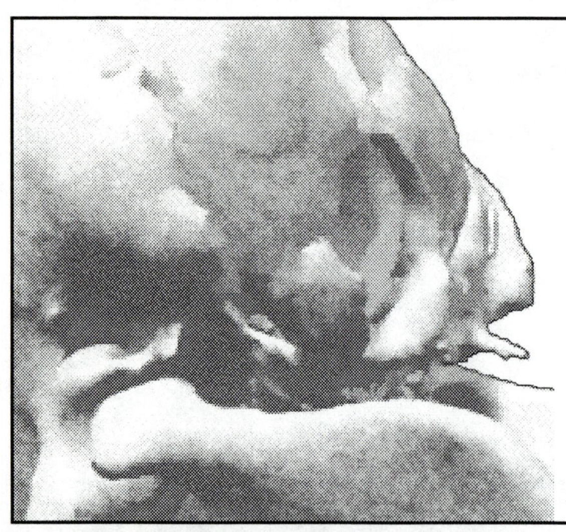

Identify:

- **Mastoid process** of the temporal bone
- **Styloid process** of the temporal bone

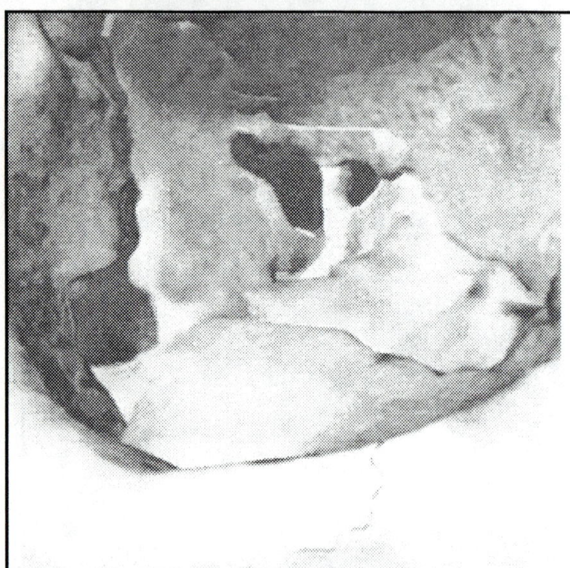

Identify:

- **Optic canal** - default
- **Superior orbital fissure** of the sphenoid bone

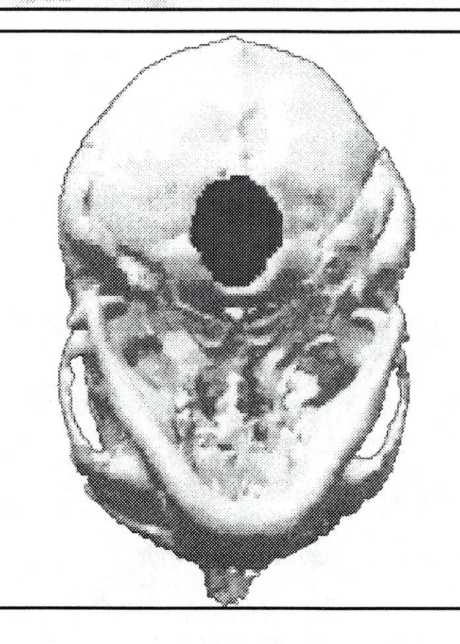

Identify:

- **Foramen magnum** of the occipital bone
- **Occipital condyles** of the occipital bone

When you are finished, **close** the 3D anatomy according to the directions in the Tutorial.

# Lateral Dissection of the Skull Bones

Under the **File** menu, drag down to **Open** (then to **Content** if you are working in Windows).

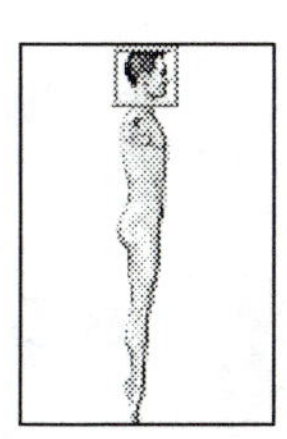

Click on **Dissectible Anatomy**. Choose either male or female. Select **Lateral**. Click **Open**.

Position the navigator rectangle over the head. Center and maximize the image. You may need to scroll to find all the structures.

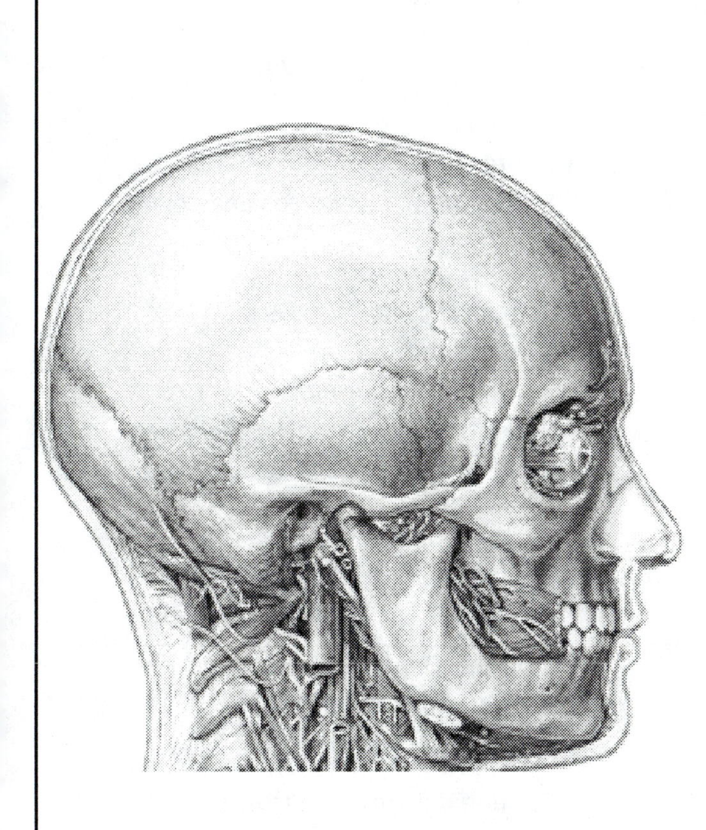

Click on the **Depth Bar** to remove skin, fat, blood vessels, and muscles until you reach **layer 118**. Identify:

- **Frontal bone**
- **Parietal bone**
- **Temporal bone**
- **Occipital bone**
- **Greater wing of sphenoid bone**
- **Coronal suture**
- **Lambdoid suture**
- **Squamosal suture**
- **Zygomatic process of temporal bone**
- **External auditory (acoustic) meatus**
- **Mastoid process of temporal bone**
- **Zygomatic bone**
- **Temporal process of zygomatic bone**
- **Maxilla**
- **Nasal bone**
- **Nasal cartilage**
- **Condylar process of mandible**
- **Coronoid process of mandible**
- **Ramus of mandible**
- **Body of mandible**
- **Hyoid bone**

What is the only movable joint of the skull?

___

What bone markings form the only movable joint of the skull?

___

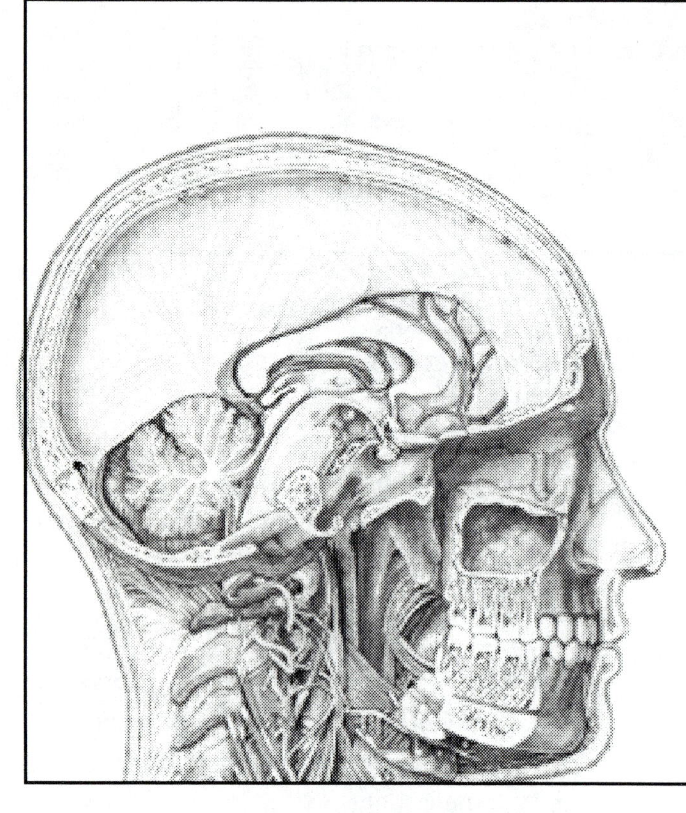

Click on the **Depth Bar** [▼] until you reach **layer 208**. Carefully observe the various structures as they are removed. Identify:

- **Frontal sinus**
- **Maxillary sinus**
- **Sphenoid bone**
- **Lacrimal bone**
- **Maxilla**
- **Ethmoid bone**
- **Optic canal**
- **Mandible**
- **Nasal bone**
- **Hyoid Bone**
- **Frontal bone** (cut portion and portion in orbit)
- **Parietal bone** (cut)
- **Temporal bone** (petrous part)
- **Occipital bone** (cut)
- **Foramen Magnum** (margin of)
- **Sella turcica** (hypophyseal fossa of sphenoid bone)

What bones of the orbit are visible at this level?

---

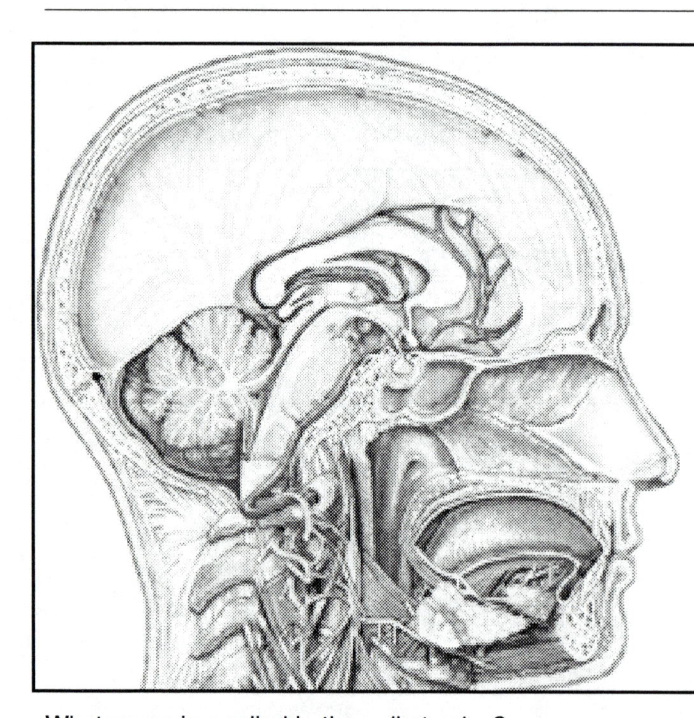

Click on the **Depth Bar** [▼] until you reach **layer 209**. Note the removal of the bones of the orbit. Identify:

- **Frontal sinus** (mucosa)
- **Sphenoidal sinus** (mucosa)
- **Sphenoid bone**
- **Palatine process of maxilla**
- **Palatine bone**
- **Perpendicular plate of ethmoid bone**
- **Cribriform plate of ethmoid bone**
- **Crista galli of ethmoid bone**
- **Mandible**
- **Vomer**
- **Sella turcica**

What organ is cradled in the sella turcica?

---

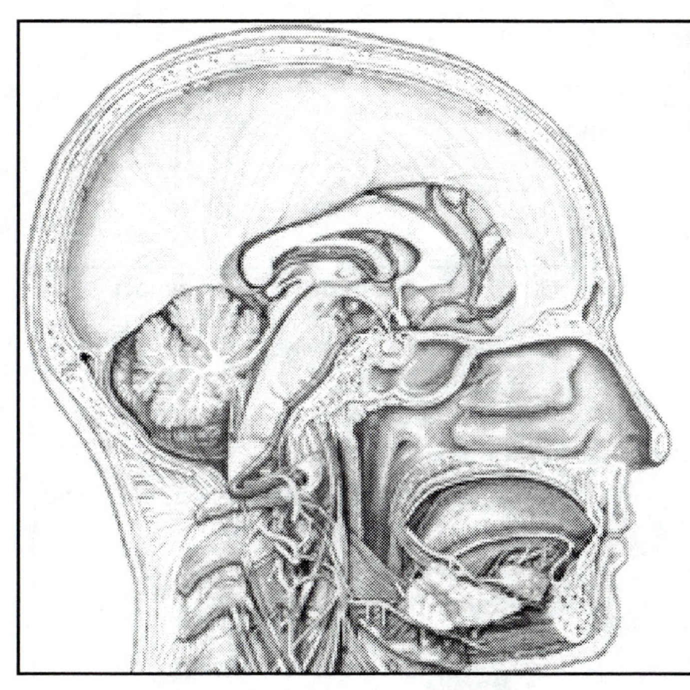

Click on the **Depth Bar** [▼] until you reach **layer 225**. Note the removal of the nasal septum. Identify:

- **Inferior nasal conchae**
- **Middle nasal conchae** of the ethmoid bone
- **Superior nasal conchae** of the ethmoid bone

Contrast the appearance of the superior nasal conchae to the middle and inferior nasal conchae.

________________________________________________________________________

________________________________________________________________________

## Anterior Dissection of the Skull Bones

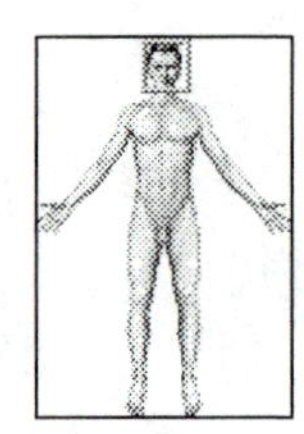

Click on the **View Button** and choose **Anterior**.

Click on the **Normal** button. Windows— —Mac

Drag the **Depth Bar** to **layer 0.**

Position the navigator rectangle over the head.

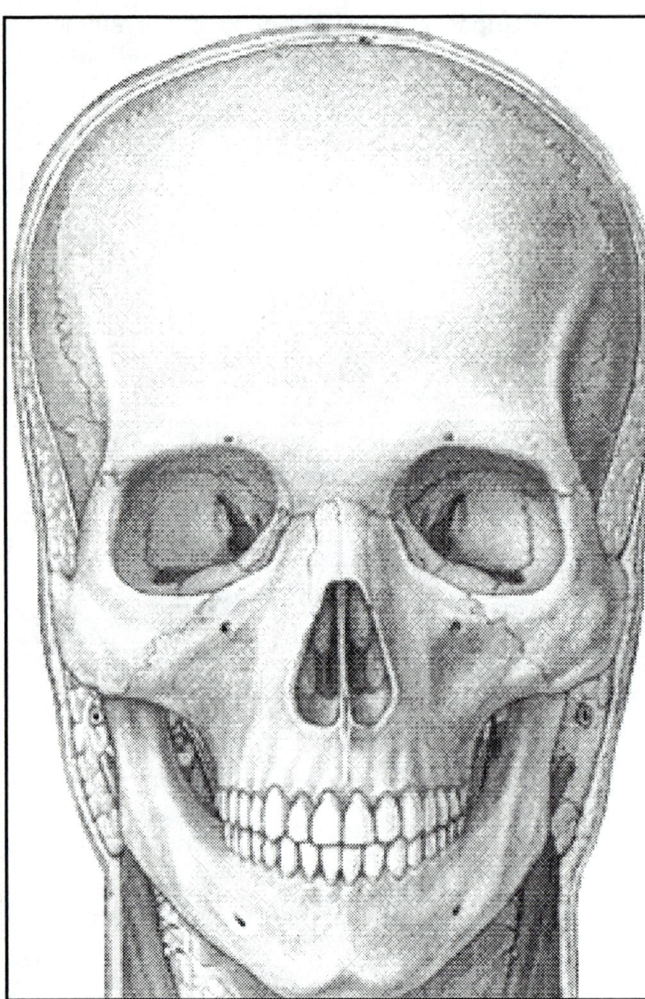

Click on the **Depth Bar** to remove skin, fat, blood vessels, and muscles until you reach **layer 48**. Identify:

- **Frontal bone**
- **Parietal bone**
- **Temporal bone**
- **Greater wing of sphenoid bone** (including the orbital part)
- **Coronal suture**
- **Supraorbital foramen**
- **Zygomatic Bone**
- **Maxilla**
- **Lacrimal Bone**
- **Ethmoid bone** (in orbit)
- **Perpendicular plate of ethmoid bone**
- **Optic canal**
- **Superior orbital fissure**
- **Mandible**
- **Vomer**
- **Inferior nasal concha**
- **Middle nasal concha (Ethmoid bone)**
- **Nasal bone**
- **Mental foramen**
- **Infraorbital foramen**

List three bones that form the bony structure of the nose.

___

___

___

Click on the **Depth Bar** ▼ until you reach **level 49**. Identify:

- **Frontal bone** (orbital part)
- **Greater wing of sphenoid bone** (orbital part)
- **Zygomatic Bone** (cut)
- **Maxilla** (cut)
- **Inferior meatus**
- **Middle meatus**
- **Superior meatus**
- **Lacrimal Bone** (cut)
- **Ethmoid bone** (orbital part)
- **Perpendicular plate of ethmoid bone**
- **Crista galli of ethmoid bone**
- **Vomer**
- **Inferior nasal concha** (cut)
- **Middle nasal concha (Ethmoid bone)** (cut)
- **Palatine process of maxilla** (cut)
- **Superior orbital fissure**
- **Ethmoidal cells** (sinuses)
- **Optic canal**

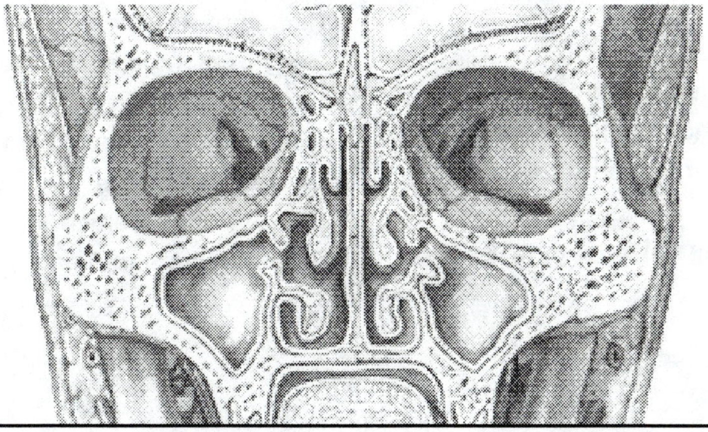

What bones make up the bony portion of the perpendicular plate of the nose?

________________________________________

________________________________________

List three parts of the ethmoid bone.

________________________________________

________________________________________

________________________________________

When you are finished, **close** dissectible anatomy.

## Atlas Anatomy of the Skull Bones

If a skull is available, it is highly suggested you hold the skull in your hand and identify the items below on the skull as you proceed.

Under the **File** menu, drag down to **Open** (then to **Content** if you are working in Windows). Click on **Atlas Anatomy**. Select **System** and **Skeletal.** Choose **Skull (Lat) 1**. If list is not in alphabetical order, the image is close to the top of the list. You may need to scroll to find all the structures.

Identify:

- **Frontal bone**
- **Parietal bone**
- **Temporal bone**
- **Occipital bone**
- **Greater wing of sphenoid bone**
- **Coronal suture**
- **Lambdoid suture**
- **Squamosal suture**
- **Zygomatic process of temporal bone**
- **External auditory (acoustic) meatus**
- **Mastoid process of temporal bone**
- **Zygomatic bone**
  - **Temporal process of zygomatic bone**
- **Maxilla**
- **Nasal bone**
- **Condylar process of mandible** (condyle of mandible)
- **Mental foramen**
- **Temporomandibular joint**
- **Coronoid process of mandible**
- **Ramus of mandible**
- **Body of mandible**
- **Lacrimal bone**
  - **Mandibular fossa of temporal bone**
- **Occipital condyle**
- **Supraorbital foramen**
- **Sutural (wormian) bone**

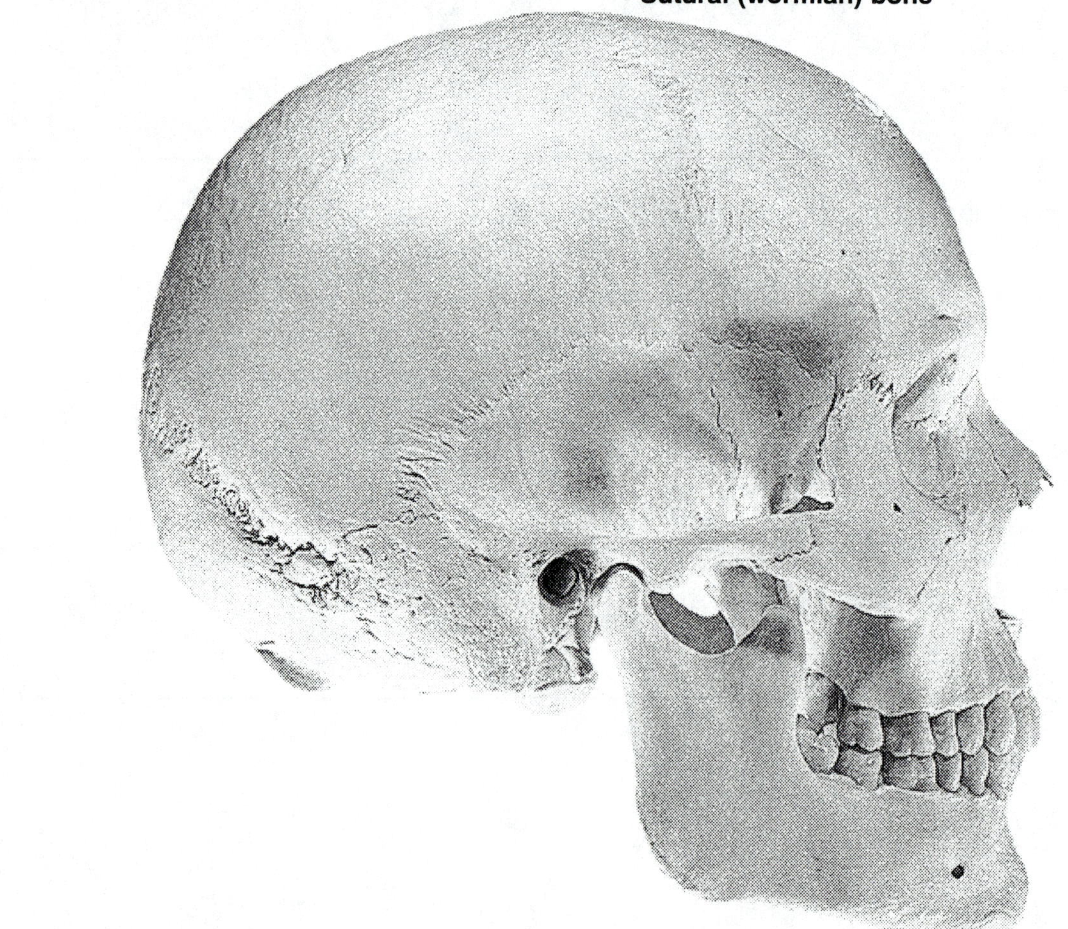

When you are finished, **close** the window.

Under the **File** menu, drag down to **Open** (then to **Content** if you are working in Windows). **Atlas Anatomy** should still be selected. **System** and **Skeletal** should still be selected. Choose **Muscle Atts - Skull (Ant)**. If list is not in alphabetical order, the image is close to the top of the list.

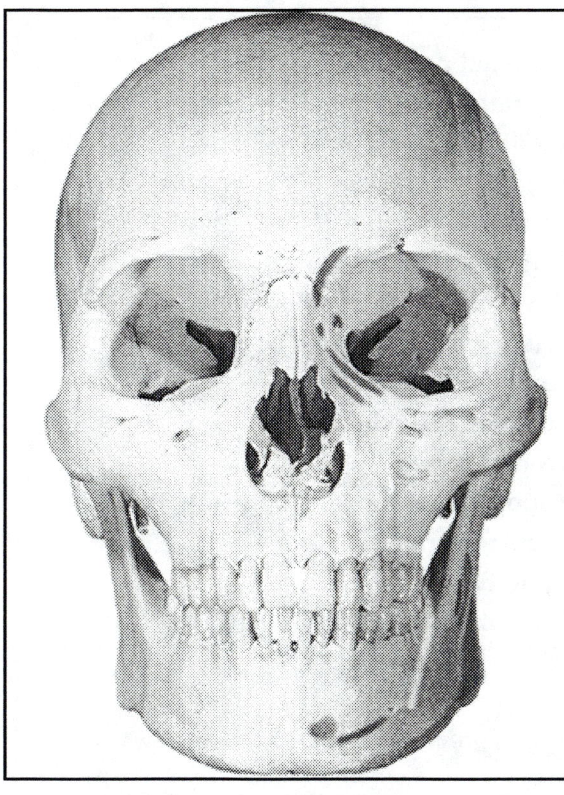

Identify:

- **Frontal bone**
- **Greater wing of sphenoid bone** (orbital part)
- **Supraorbital foramen**
- **Zygomatic Bone**
- **Maxilla**
- **Lacrimal Bone**
- **Perpendicular plate of ethmoid bone**
- **Superior orbital fissure**
- **Body of mandible**
- **Vomer**
- **Inferior nasal concha**
- **Middle nasal concha (Ethmoid bone)**
- **Nasal bone**
- **Infraorbital foramen**
- **Mental foramen**

When you are finished, **close** the window.

Under the **File** menu, drag down to **Open** (then to **Content** if you are working in Windows). **Atlas Anatomy** should still be selected. **System** and **Skeletal** should still be selected. Choose **Muscle Atts - Skull (Inf)**. If list is not in alphabetical order, the image is close to the top of the list.

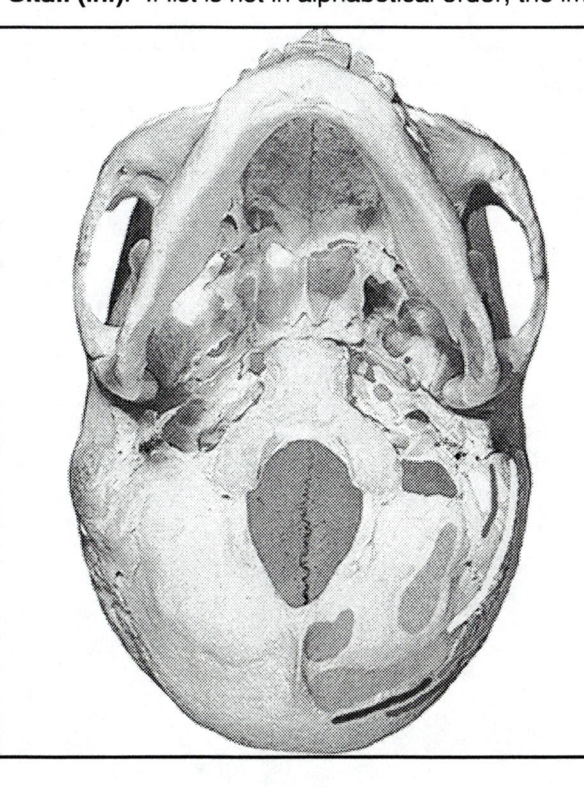

Identify :

- **Occipital bone**
- **Sphenoid bone** (various parts)
- **Body of mandible**
- **Maxilla**
- **Horizontal plate of palatine bone**
- **Vomer**
- **Zygomatic bone**
- **Zygomatic process** of the temporal bone
- **Foramen magnum** of the occipital bone
- **Foramen ovale** of the sphenoid bone
- **Mastoid process** of the temporal bone
- **Occipital condyles** of the occipital bone
- **Sagittal suture**
- **Palatine process of maxilla**
- **Temporal process of zygomatic bone**

When you are finished, **close** the window.

Under the **File** menu, drag down to **Open** (then to **Content** if you are working in Windows). **Atlas Anatomy** should still be selected. **System** and **Skeletal** should still be selected. Choose **Cranial Cavities (Sup)1**. If list is not in alphabetical order, the image is close to the top of the listing.

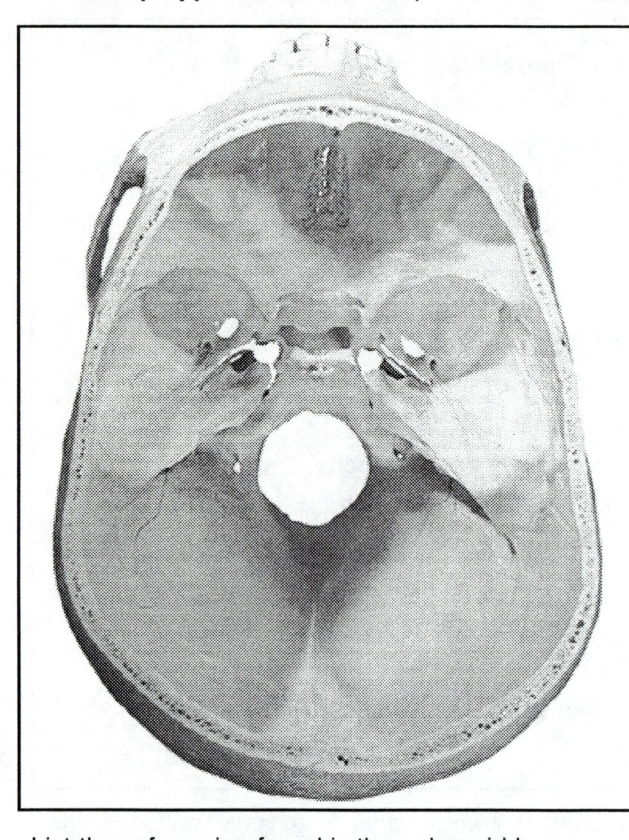

Identify :

- **Occipital bone**
- **Greater wing of sphenoid bone**
- **Maxilla**
- **Zygomatic process** of the temporal bone
- **Foramen magnum** of the occipital bone
- **Foramen ovale** of the sphenoid bone
- **Cribriform plate of ethmoid bone**
- **Crista galli of ethmoid bone**
- **Foramen lacerum**
- **Foramen rotundum**
- **Frontal bone**
- **Hypophyseal fossa of sphenoid bone**
- **Jugular Foramen**
- **Optic canal**
- **Temporal bone** (petrous part)

When you are finished, **close** the window.

List three foramina found in the sphenoid bone.

______________________________

______________________________

______________________________

List two foramina found in the temporal bone.

______________________________

______________________________

List one foramen found in the occipital bone.

______________________________

Under the **File** menu, drag down to **Open** (then to **Content** if you are working in Windows). **Atlas Anatomy** should still be selected. **System** and **Skeletal** should still be selected. Choose **Mus Atts - Skull (Med).** If list is not in alphabetical order, the image is close to the top of the list.

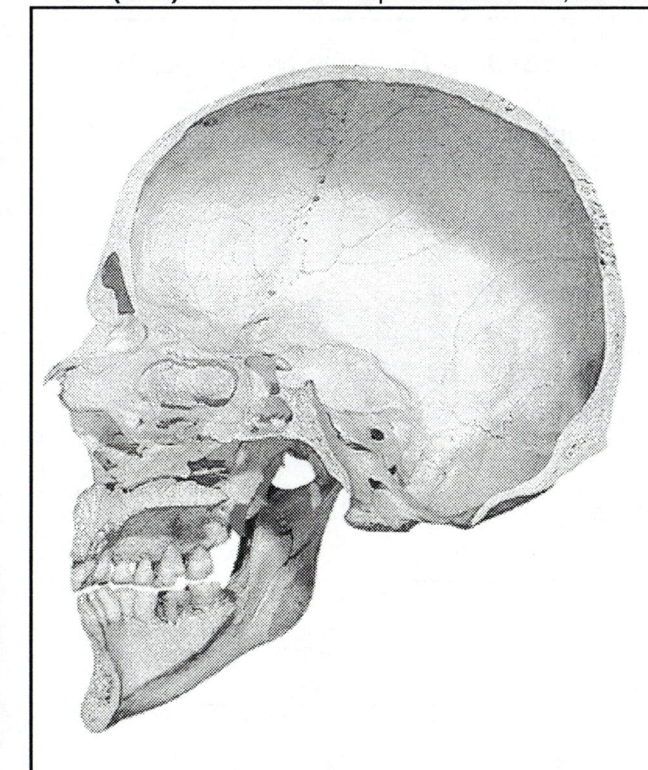

Identify :

- **Frontal bone**
- **Parietal bone**
- **Occipital bone**
- **Temporal bone** (squamous part)
- **Hypophyseal fossa of sphenoid bone**
- **Lambdoid suture**
- **Occipital condyle**
- **Coronal suture**
- **Foramen magnum**
- **Internal auditory (acoustic) meatus**
- **Frontal sinus**
- **Nasal Bone**
- **Perpendicular plate of ethmoid bone**
- **Vomer**
- **Palatine process of maxilla**
- **Palatine bone** (horizontal plate)
- **Cribriform plate of ethmoid bone**
- **Body of mandible**
- **Inferior nasal conchae**
- **Superior nasal conchae**

When you are finished, **close** the window.

Under the **File** menu, drag down to **Open** (then to **Content** if you are working in windows). **Atlas Anatomy** should still be selected. **System** and **Skeletal** should still be selected. Choose **Mus Atts - Infratemporal fossa.** If list is not in alphabetical order, the image is close to the top of the list.

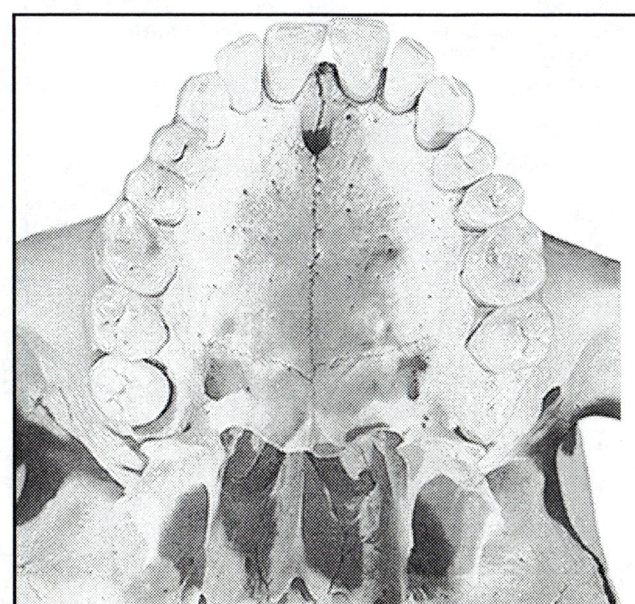

Identify :

- **Maxilla**
- **Palatine process of maxilla**
- **Horizontal plate of palatine bone**
- **Vomer**

The hard palate is composed of what two bones?

____________________________________

____________________________________

When you are finished, **close** the window.

Under the **File** menu, drag down to **Open** (then to **Content** if you are working in windows). **Atlas Anatomy** should still be selected. **System** and **Skeletal** should still be selected. Choose **Walls of Orbit (Ant).** If list is not in alphabetical order, the image is one-fifth from the top of the list.

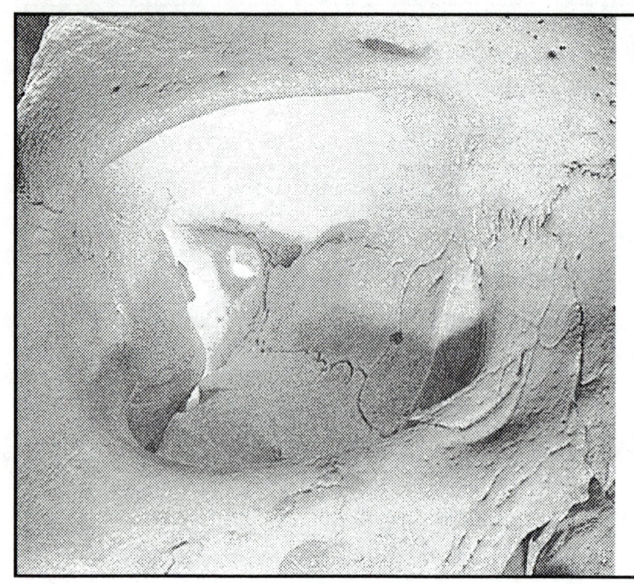

Identify :

- **Frontal Bone** (zygomatic process, supraorbital margin)
- **Supraorbital foramen of frontal bone**
- **Nasal Bone**
- **Maxilla**
- **Lacrimal Bone**
- **Ethmoid Bone**
- **Sphenoid Bone** (lesser wing and body)
- **Zygomatic Bone**
- **Superior Orbital fissure**
- **Optic Canal**

When you are finished, **close** the window.

Name six bones found in the orbit of the eye.

_______________________________________________

_______________________________________________

Under the **File** menu, drag down to **Open** (then to **Content** if you are working in Windows). **Atlas Anatomy** should still be selected. **System** and **Skeletal** should still be selected. Choose **Bony Wall of Nasal Cavity.** If list is not in alphabetical order, the image is one-fourth from the top of the list.

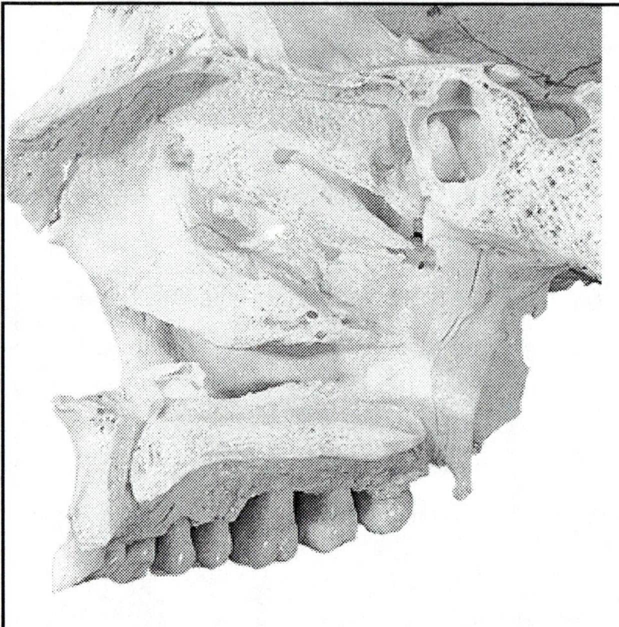

If you are working in Mac, click on the **Show All Pins** button and scroll to "All Systems".

Identify :

- **Frontal bone**
- **Body of sphenoid bone**
- **Hypophyseal fossa of sphenoid bone**
- **Sphenoidal sinus**
- **Nasal Bone**
- **Maxilla** (frontal process of)
- **Palatine process of maxilla**
- **Palatine bone** (horizontal plate)
- **Cribriform plate of ethmoid bone**
- **Inferior nasal concha**
- **Inferior nasal meatus**
- **Superior nasal concha (ethmoid bone)**
- **Superior nasal meatus**
- **Middle nasal concha (ethmoid bone)**
- **Middle nasal meatus**

When you are finished, **close** the window.

## Cadaver Images of the Skull

Under the **File** menu, drag down to **Open** (then to **Content** if you are working in Windows). **Atlas Anatomy** should still be selected. **System** and **Skeletal** should still be selected. Choose **Sagittal Section of Head and Neck.** If list is not in alphabetical order, the image is close to the top of the list.

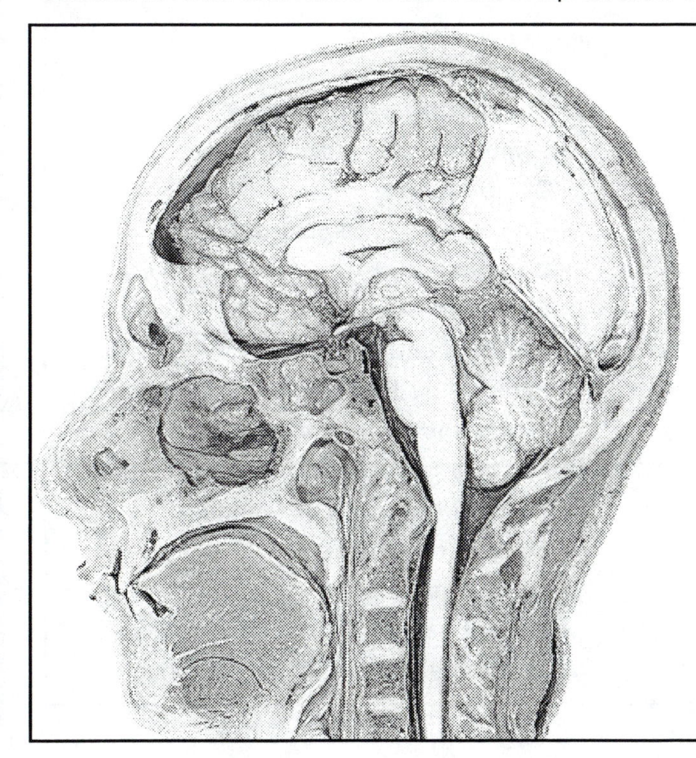

Identify :

- **Nasal Bone**
- **Nasal septum**
- **Palatine process of maxilla**
- **Palatine bone** (horizontal plate)
- **Hyoid bone**
- **Inferior nasal concha**
- **Superior nasal concha (ethmoid bone)**
- **Middle nasal concha (ethmoid bone)**
- **Frontal bone**
- **Parietal bone**
- **Occipital bone**
- **Sphenoid bone** (dorsum sellae of)
- **Lambdoid suture**
- **Coronal suture**
- **Foramen magnum**

When you are finished, **close** the window.

What goes through the foramen magnum?

---

Under the **File** menu, drag down to **Open** (then to **Content** if you are working in Windows). **Atlas Anatomy** should still be selected. **System** and **Skeletal** should still be selected. Choose **Lateral Wall of Palate.** If list is not in alphabetical order, the image is one-fifth of the way from the top of the list.

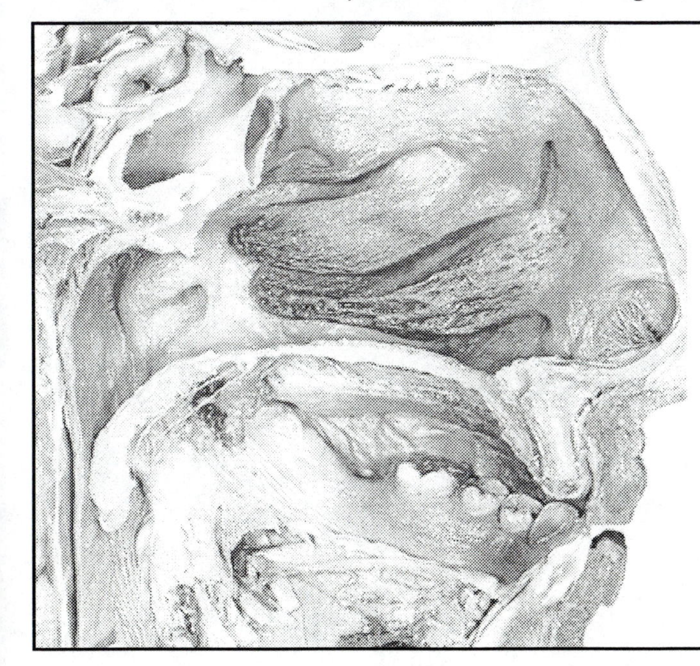

Identify :

- **Inferior nasal concha**
- **Inferior nasal meatus**
- **Superior nasal concha (ethmoid bone)**
- **Superior nasal meatus**
- **Middle nasal concha (ethmoid bone)**
- **Middle nasal meatus**
- **Hard palate**

When you are finished, **close** the window.

 

## X-Ray of Skull

Under the **File** menu, drag down to **Open** (then to **Content** if you are working in Windows). **Atlas Anatomy** should still be selected. **System** and **Skeletal** should still be selected. Choose **Skull (Lat) 2.** If list is not in alphabetical order, the image is close to the top of the list.

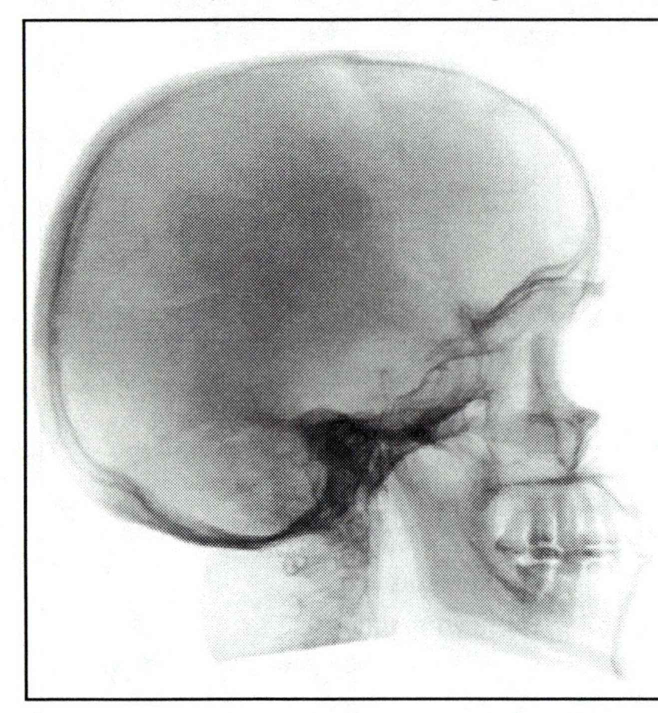

Identify :

- **Frontal bone**
- **Parietal bone**
- **Temporal bone** (articular tubercle)
- **Occipital bone**
- **Greater wing of sphenoid bone**
- **Coronal suture**
- **Sphenoid bone**
- **Lambdoid suture**
- **External auditory (acoustic) meatus**
- **Zygomatic bone** (frontal process)
- **Palatine process of maxilla**
- **Coronoid process of mandible**
- **Body of mandible**
- **Frontal sinus**
- **Hypophyseal fossa of sphenoid bone**

When you are finished, **close** the window.

## Objectives: Skull

Student will be able to identify the following bones and bone markings of the skull:

**Skull bones:**

- Frontal bone
- Parietal bone
- Temporal bone
- Occipital bone
- Sphenoid bone
- Ethmoid bone

**Bone Markings on Skull Bones:**

On frontal bone:
- Supraorbital foramen

On temporal bone:
- Zygomatic process
- Carotid canal
- Jugular foramen
- Internal auditory (acoustic) meatus
- External auditory (acoustic) meatus
- Mastoid process
- Styloid process
- Mandibular fossa

On occipital bone:
- Foramen magnum
- Occipital condyles

On sphenoid bone
- Optic canal
- Foramen ovale
- Superior orbital fissure
- Sella turcica
- Foramen lacerum
- Foramen rotundum
- Greater wing
- Hypophyseal fossa
- Body

On ethmoid bone:
- Cribriform plate
- Middle Nasal Conchae
- Crista galli
- Perpendicular plate
- Middle nasal conchae
- Superior nasal conchae

**Facial bones:**

- Inferior nasal conchae
- Lacrimal Bone
- Mandible
- Maxilla
- Nasal bone
- Palatine bone
- Vomer
- Zygomatic bone

**Sutures:**

- Coronal suture
- Sagittal suture
- Lambdoid suture
- Squamosal suture

**Sinuses:**

- Maxillary sinus
- Frontal sinus
- Ethmoidal sinus or cells
- Sphenoidal sinus

**Bone Markings on Facial Bones:**

On mandible
- Condyle
- Condylar process
- Mental foramen
- Coronoid process
- Ramus
- Body

On maxilla:
- Infraorbital foramen
- Palatine process of maxilla

On palatine bone:
- Horizontal plate

On zygomatic bone:
- Temporal process

**Other:**

- Nasal cartilage
- Hyoid bone
- Inferior meatus
- Middle meatus
- Superior meatus
- Temporomandibular joint
- Sutural (wormian) bone
- Hard palate

# Vertebrae, Ribs, and Sternum

## Opening A.D.A.M Interactive Anatomy

Open A.D.A.M. Interactive Anatomy according to the directions in the Tutorial.

## Anterior Dissection of Vertebrae, Ribs, and Sternum

Choose **Dissectible Anatomy**.

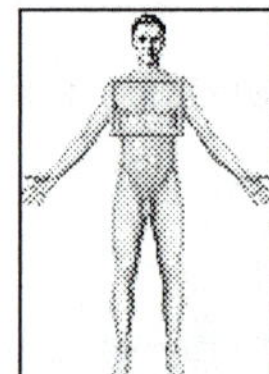

Choose either male or female. Select **Anterior.** Click **Open**.

Enlarge the window and position the navigation rectangle over the chest.

Click on the **Depth Bar** ▼ to remove skin, fat, blood vessels, and muscles until you reach **layer 19**. Slowly click on the **Depth Bar** ▼ until you reach **layer 56**. Note the removal of the muscles and other structures.

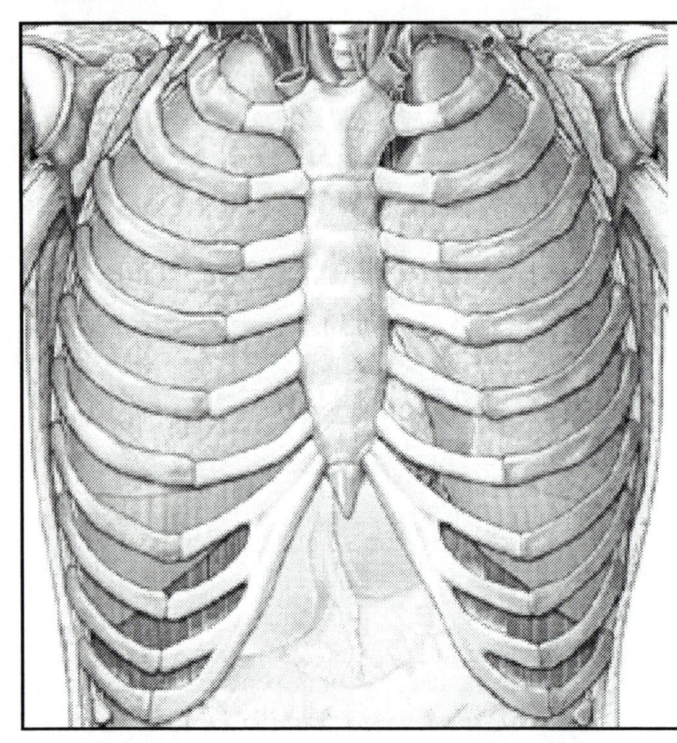

Scroll to **layer 148**, then slowly click on the **Depth Bar** ▼ until you reach **layer 158**. You may need to scroll to see the entire image. Identify:

- **Manubrium**
- **Body of Sternum**
- **Xiphoid process**
- **Costal Cartilage**
- **Ribs 1-8**
- **Lungs** (costal parietal pleura)
- **Clavicular notch of sternum**
- **Jugular notch of sternum**

Indicate the **sternocostal joints** in this diagram.

Which ribs are called "true ribs" because they connect directly to the sternum?

______________________________________________

Which ribs are the "false ribs"?

______________________________________________

Which ribs are the "floating ribs"?

______________________________________________

What are the three parts to the sternum?

______________________________________________

Click several times on the **Depth Bar** [▼] to expose the lungs and the heart that are protected by the ribs and sternum, then scroll to **layer 268**.

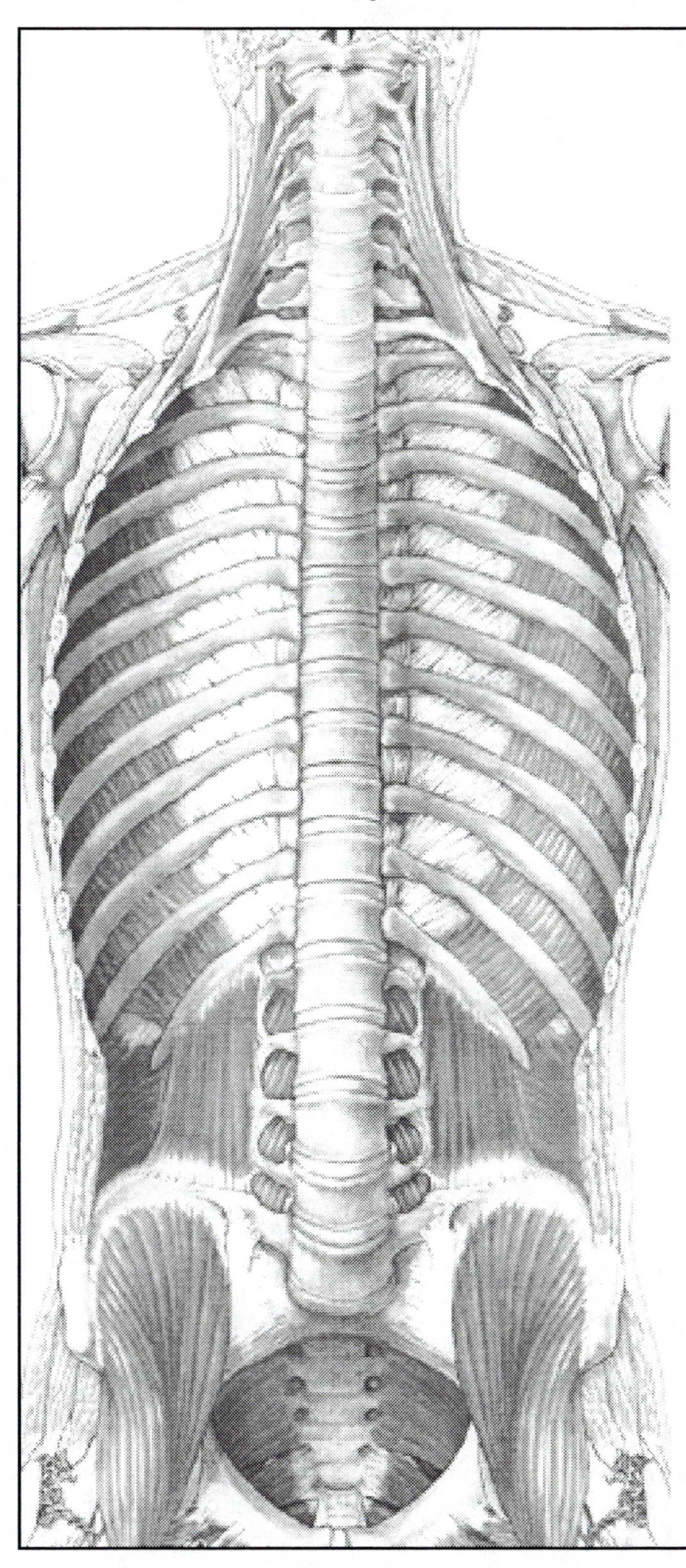

Click several times on the **Depth Bar** [▼] until you reach **layer 277**. You may need to scroll to see the entire image. Identify:

- **C1 vertebrae (atlas)**
- **Body of C2 - C7**
- **Intervertebral Disks**
- **Ribs 1-12**
- **Transverse process of C1 - C7**
- **Dens of C2 (axis)** (above and below C1)
- **Occipital condyles**
- **Superior articular process of C1**
- **Body of T1- T12**
- **Body of L1 - L5**
- **Transverse process of L1 - L5**
- **Sacrum** (including the wing)
- **Anterior sacral foramen**
- **Coccyx**

Indicate the location of these joints in this diagram:

- **Atlanto-occipital joint**
- **Atlanto-axial joint**
- **Intervertebral joints**
- **Vertebrocostal joints**

What do you notice about the size of the vertebrae from superior to inferior?

_______________________________________________

_______________________________________________

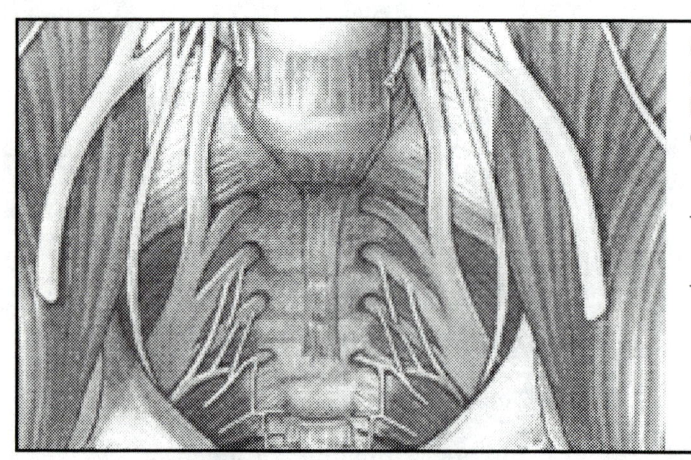

Click several times on the **Depth Bar** until you reach **layer 271**. (No graphic has been provided for this layer.) What structures emerge from the **anterior sacral foramen**?

_______________________________________________

_______________________________________________

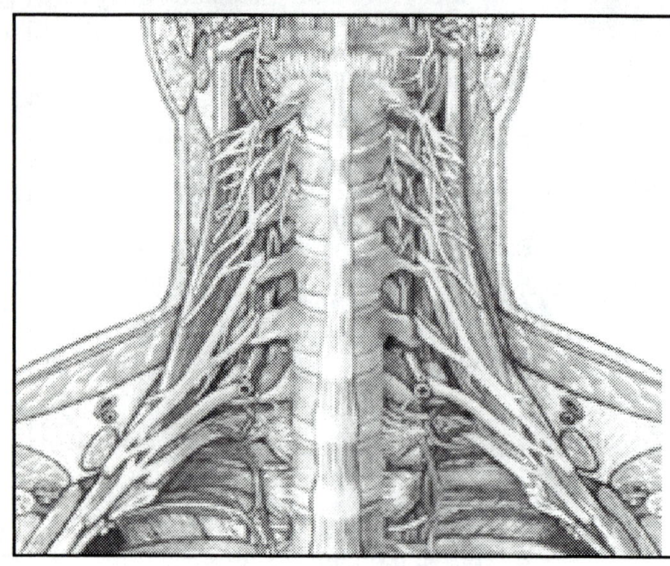

While you are at **layer 271**, move the navigator up to the neck area. What structures are found in the transverse foramina of the cervical vertebrae?

_______________________________________________

_______________________________________________

## Posterior Dissection of the of Vertebrae, Ribs, and Sternum

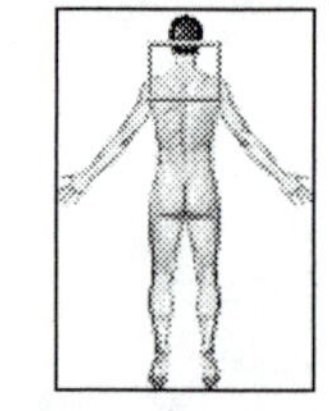

Click on the **View** button and choose **Posterior**.

Click on the **Normal** button. Windows— —Mac

Drag the **Depth Bar** to **layer 0.**

Position the navigator rectangle over the posterior neck.

Click several times on the **Depth Bar** to remove the skin, and fascia until you reach **layer 10**. Scroll down to **layer 34**, then to **74**, **89**, **101**, and finally to **175**. Note the layers of muscles that cover the spinal column.

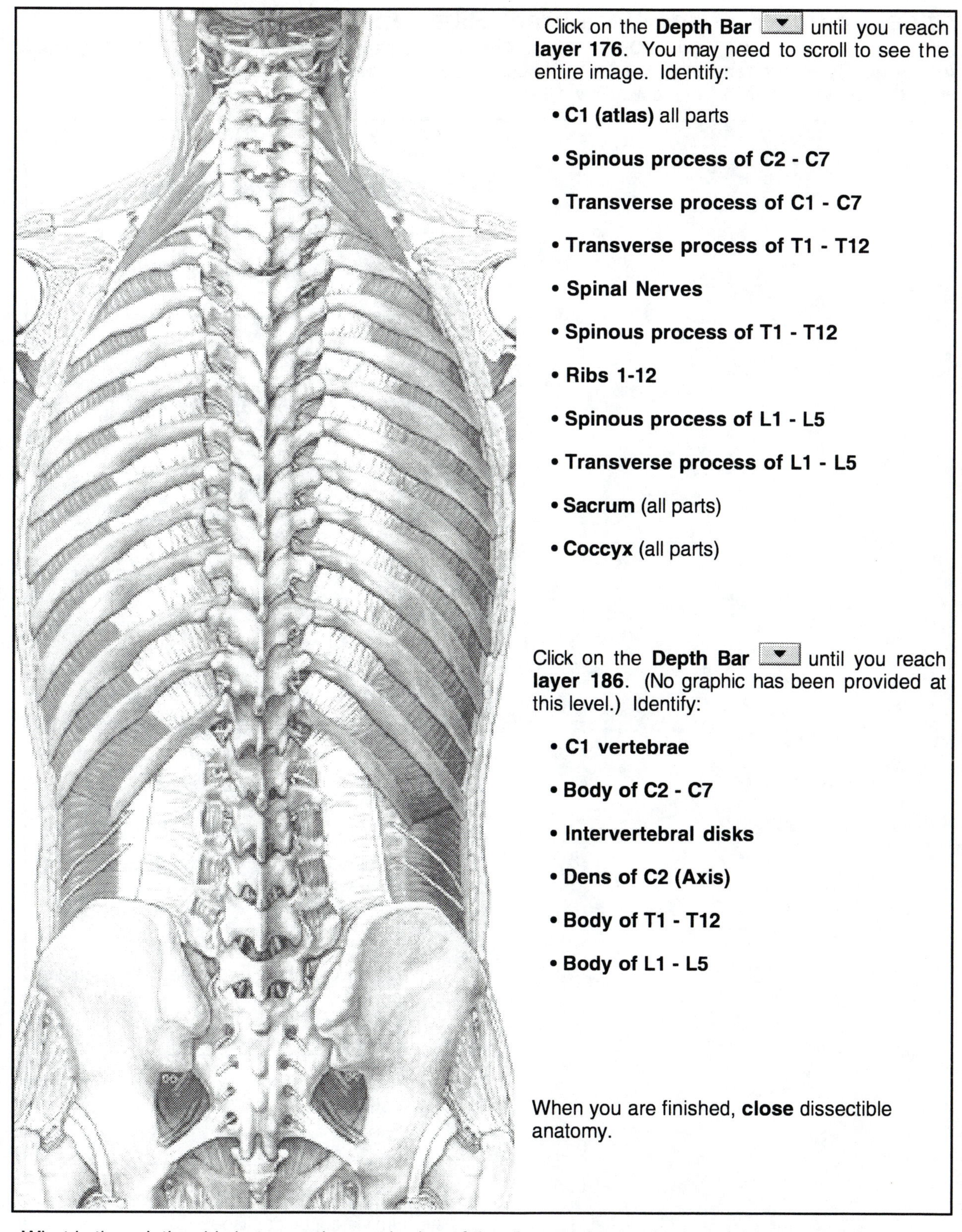

Click on the **Depth Bar** until you reach **layer 176**. You may need to scroll to see the entire image. Identify:

- **C1 (atlas)** all parts
- **Spinous process of C2 - C7**
- **Transverse process of C1 - C7**
- **Transverse process of T1 - T12**
- **Spinal Nerves**
- **Spinous process of T1 - T12**
- **Ribs 1-12**
- **Spinous process of L1 - L5**
- **Transverse process of L1 - L5**
- **Sacrum** (all parts)
- **Coccyx** (all parts)

Click on the **Depth Bar** until you reach **layer 186**. (No graphic has been provided at this level.) Identify:

- **C1 vertebrae**
- **Body of C2 - C7**
- **Intervertebral disks**
- **Dens of C2 (Axis)**
- **Body of T1 - T12**
- **Body of L1 - L5**

When you are finished, **close** dissectible anatomy.

What is the relationship between the numbering of the ribs and the numbering of the vertebrae?

______________________________________________________________________

## Atlas Anatomy of the Vertebrae, Ribs, and Sternum

Under the **File** menu, drag down to **Open** (then to **Content** if you are working in windows). Click on **Atlas Anatomy**. Select **System** and **Skeletal.** Choose **Bones of Trunk (Med)**. If list is not in alphabetical order the image is about half of the way down the list.

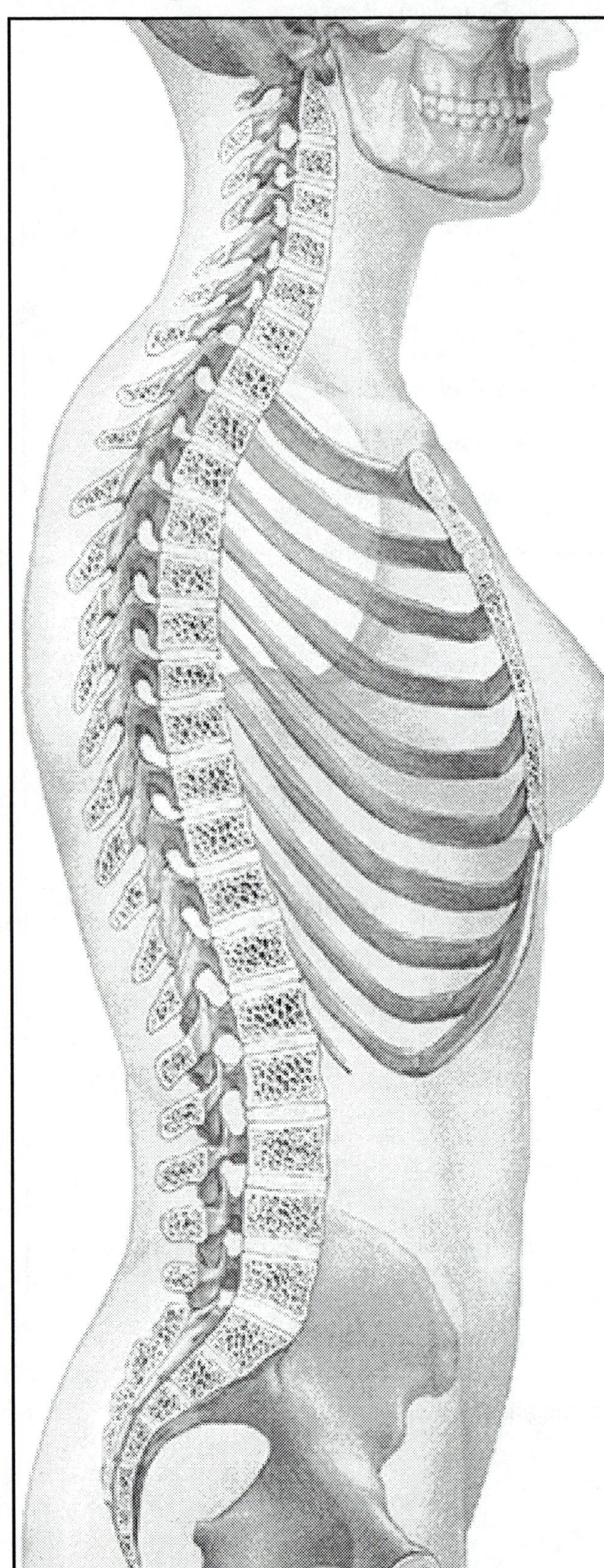

You may need to scroll to see the entire image. Identify:

- **C2 vertebrae (axis)** (body of)
- **Body of C2, C3, C4, C7 Vertebrae**
- **Body of L1 to L5 Vertebrae**
- **Body of Sternum**
- **Coccyx**
- **Ribs 1,2,3,4,5,6,8,7,9,10,11**
- **Intervertebral disks**
- **Intervertebral foramen**
- **Spinous process of C7, L1, T1, T12 vertebrae**
- **Manubrium**
- **Xiphoid process**
- **Sacrum**

Note the curvatures of the spine. Is the curvature of the cervical vertebrae primary or secondary?

______________________________

Is the curvature of the thoracic vertebrae primary or secondary?

______________________________

Is the curvature of the lumbar vertebrae primary or secondary?

______________________________

Is the curvature of the sacral vertebrae primary or secondary?

______________________________

Which ribs do not connect anteriorly to the sternum via costal cartilage?

______________________________

When you are finished, **close** the window.

Under the **File** menu, drag down to **Open** (then to **Content** if you are working in Windows). **Atlas Anatomy** should still be selected. **System** and **Skeletal** should still be selected. Choose **Median Section of Lumbar Spine**. If list is not in alphabetical order the image is about half way down the list.

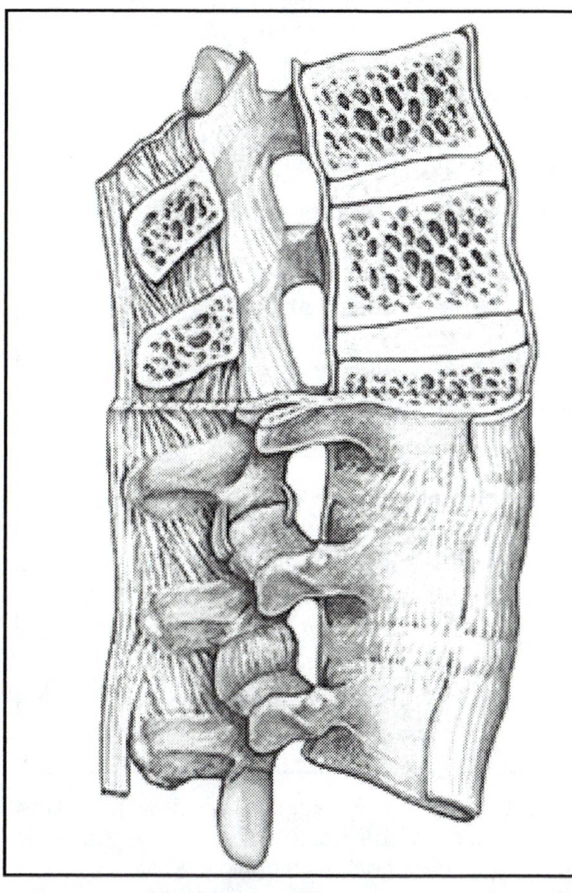

Identify:

- **Body of L1 - L5 Vertebrae**
- **Intervertebral disks** (nucleus pulposus & anulus fibrosus)
- **Intervertebral foramen**
- **Spinous process of L1 - L5 vertebrae**
- **Vertebral canal**

What is the function of the intervertebral disks?

_______________________________________________

Based on previous images you have seen, what is present in the vertebral canal?

_______________________________________________

What is present in the intervertebral foramen?

_______________________________________________

When you are finished, **close** the window.

Under the **File** menu, drag down to **Open** (then to **Content** if you are working in Windows). **Atlas Anatomy** should still be selected. **System** and **Skeletal** should still be selected. Choose **T12 Vertebrae (Sup)**. If list is not in alphabetical order, the image is about half way down the list.

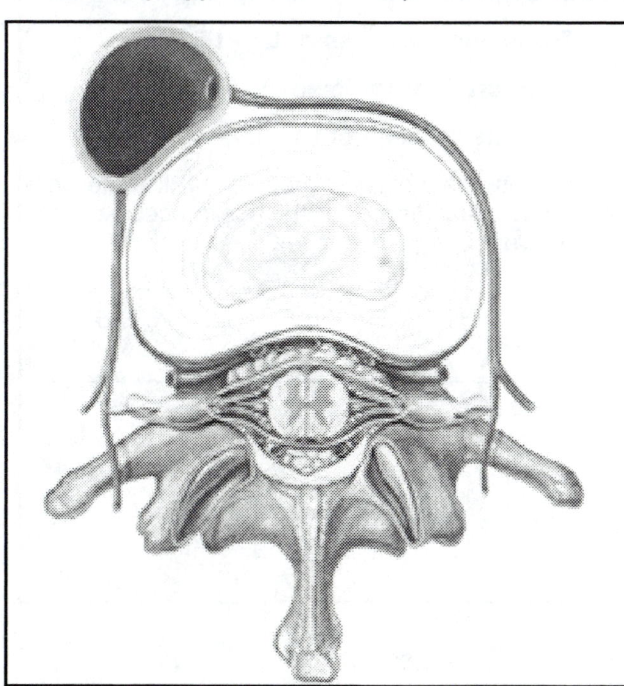

If you are working in Mac, click on the **Show All Pins** button and scroll to "All Systems".

Identify :

- **Intervertebral disk** (Anulus fibrosus & Nucleus pulposus)
- **Spinous process**
- **Spinal nerve**
- **Transverse process**
- **Spinal cord** (gray matter & white matter)

This image nicely shows the spinal nerves emerging from the intervertebral foramen. It also shows that the spinal cord is not the only structure within the vertebral foramen. List one additional structure within the vertebral foramen.

_______________________________________________

When you are finished, **close** the window.

If these bones are available, it is suggested you hold them in your hand and identify the items below as you proceed.

Under the **File** menu, drag down to **Open** (then to **Content** if you are working in Windows). **Atlas Anatomy** should still be selected. **System** and **Skeletal** should still be selected. Choose **Cervical Vertebrae (Post/Lat).** If list is not in alphabetical order, the image is close to the top of the list.

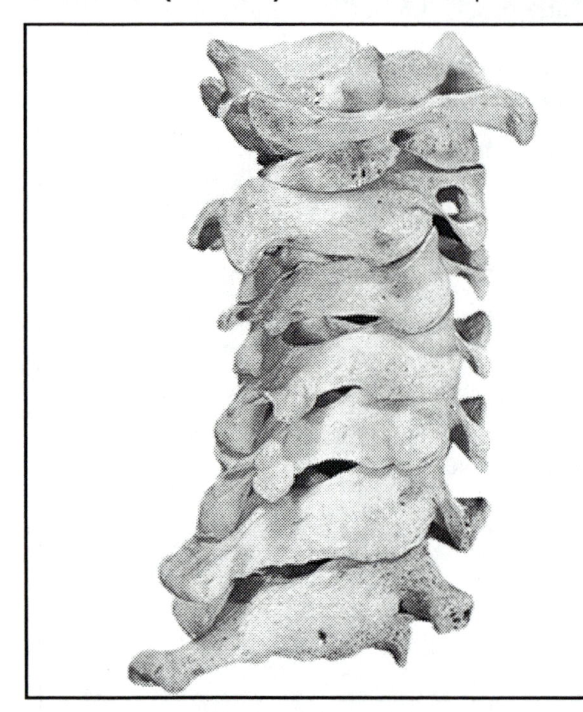

Identify :

- **Atlanto-occipital joint**
- **Dens of C2 vertebrae(axis)**
- **C1 (atlas)** (all parts)
- **Transverse process of C1 - C7**
- **Transverse foramen**
- **C2 - C7 Vertebrae** (lamina of)
- **Spinous process of C2 - C7**

Indicate the atlantoaxial joint. What type of movement does this joint allow?

_______________________________________________

_______________________________________________

When you are finished, **close** the window.

Under the **File** menu, drag down to **Open** (then to **Content** if you are working in Windows). **Atlas Anatomy** should still be selected. **System** and **Skeletal** should still be selected. Choose **Lumbar Vertebrae (Post/Lat).** If list is not in alphabetical order, the image is about half way down the list.

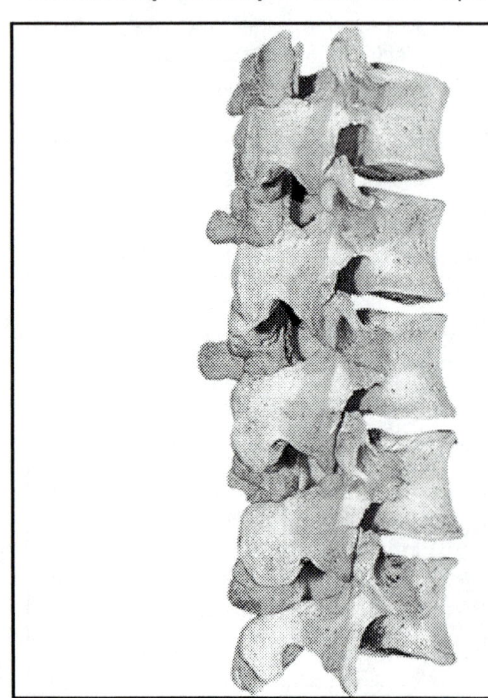

Identify :

- **Body of L1 - L5**
- **Transverse process of L1 - L5**
- **Intervertebral foramen**
- **Spinous process of L1 - L5**

How does the spinous process of the cervical vertebrae differ from the spinous process of the lumbar vertebrae (see above)?

_______________________________________________

_______________________________________________

When you are finished, **close** the window.

Under the **File** menu, drag down to **Open** (then to **Content** if you are working in Windows). **Atlas Anatomy** should still be selected. **System** and **Skeletal** should still be selected. Choose **Isolated Vertebrae (Post/Lat)**. If list is not in alphabetical order, the image is about half way down the list.

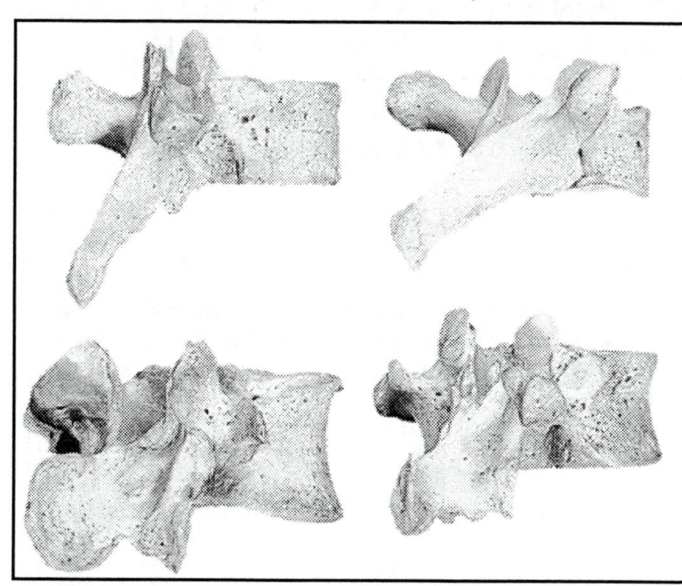

Before you identify any features of these vertebrae, predict which one will be T2, T8, T11, & L2.

______________________________

Identify:

- **Body of T2, T8, T11, L2**
- **Transverse process**
- **Spinous process**

When you are finished, **close** the window.

Under the **File** menu, drag down to **Open** (then to **Content** if you are working in Windows). **Atlas Anatomy** should still be selected. **System** and **Skeletal** should still be selected. Choose **1st, 3rd, & 8th Ribs.** If list is not in alphabetical order, the image is about half way down the list.

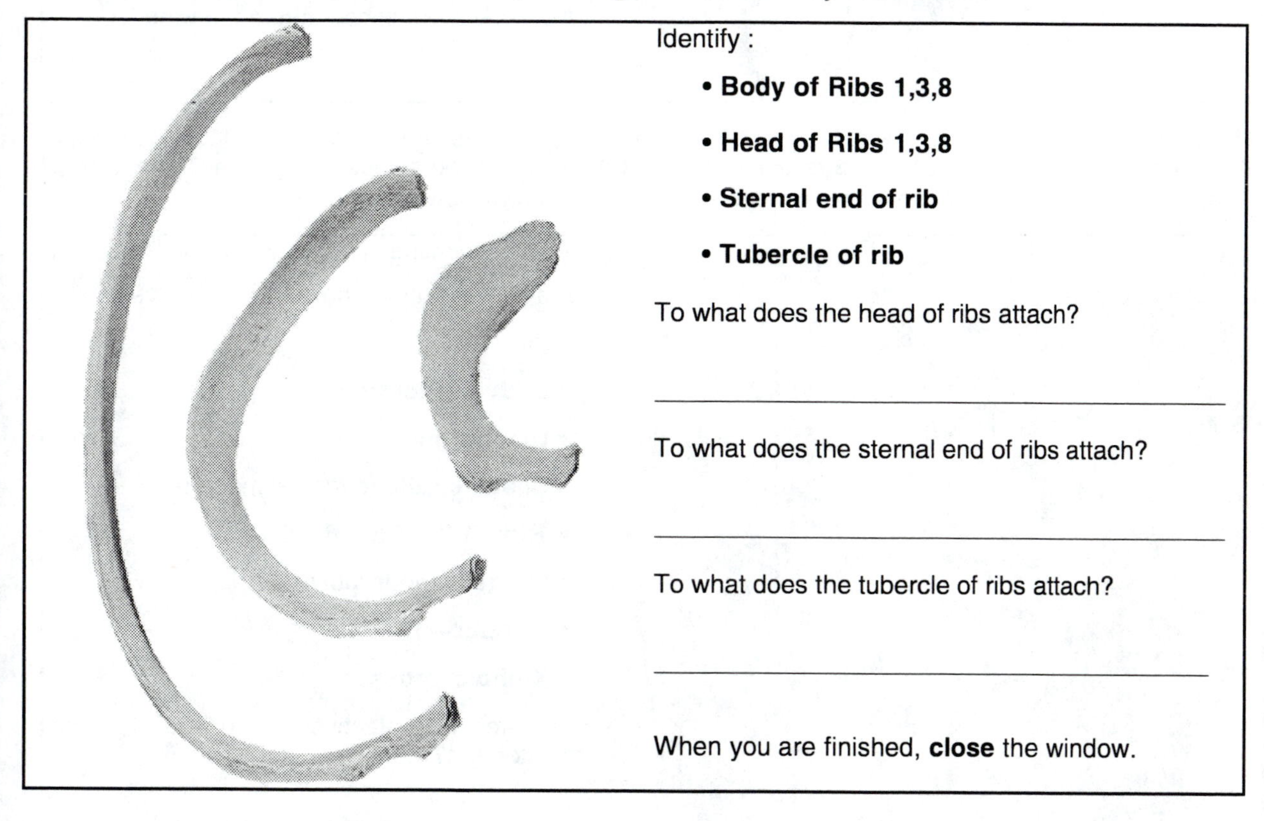

Identify :

- **Body of Ribs 1,3,8**
- **Head of Ribs 1,3,8**
- **Sternal end of rib**
- **Tubercle of rib**

To what does the head of ribs attach?

______________________________

To what does the sternal end of ribs attach?

______________________________

To what does the tubercle of ribs attach?

______________________________

When you are finished, **close** the window.

## Cadaver Images of the Vertebrae, Ribs, and Sternum

Under the **File** menu, drag down to **Open** (then to **Content** if you are working in Windows). **Atlas Anatomy** should still be selected. **System** and **Skeletal** should still be selected. Choose **Sagittal Section of Head and Neck.** If list is not in alphabetical order, the image is close to the top of the list.

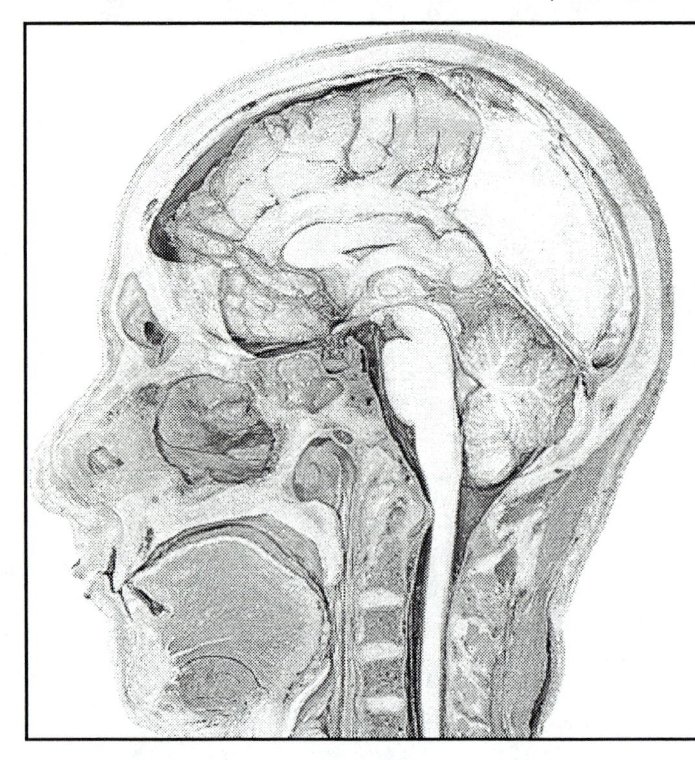

Identify :

- **Body of C6 vertebrae**
- **Dens of C2 vertebrae**
- **C1 vertebrae (atlas)**

Note the appearance of the intervertebral disks which are the light areas between the body of the vertebrae.

The spinal cord passes through what foramen?

______________________________

When you are finished, **close** the window.

Under the **File** menu, drag down to **Open** (then to **Content** if you are working in Windows). **Atlas Anatomy** should still be selected. **System** and **Skeletal** should still be selected. Choose **Dissection of Thorax (Ant).** If list is not in alphabetical order, the image is half way down the list.

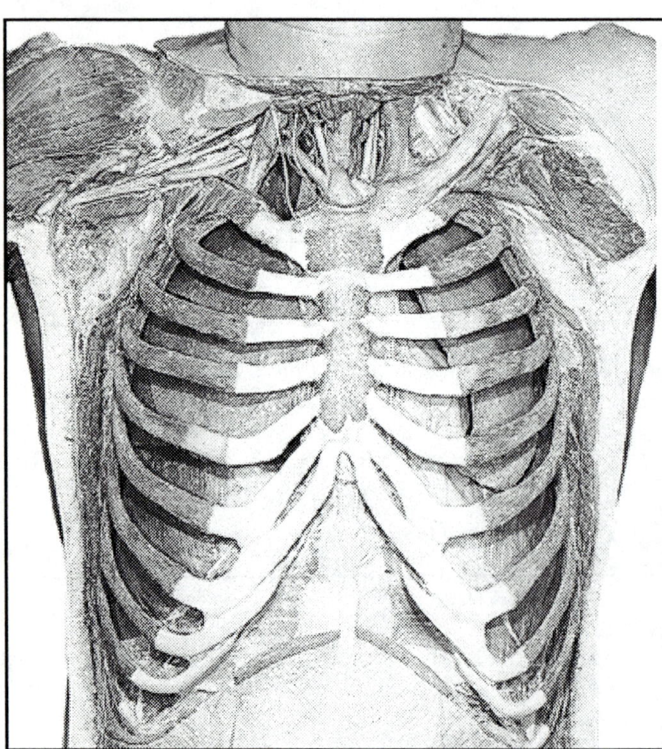

If you are working in Mac, click on the **Show All Pins** button and scroll to "All Systems".

Identify :

- **Body of Sternum**
- **Diaphragm**
- **Costal cartilage** (costal margin)
- **Ribs 1,2,3,4,5,6,8,7,9,10**
- **Lungs** (various lobes)
- **Manubrium**
- **Xiphoid process**

What are the protective functions of the ribs and sternum?

______________________________

When you are finished, **close** the window.

## X-Rays of the Vertebrae:

Under the **File** menu, drag down to **Open** (then to **Content** if you are working in Windows). **Atlas Anatomy** should still be selected. **System** and **Skeletal** should still be selected. Choose **Flexed Cervical Spine (Lat)**. If list is not in alphabetical order, the image is close to the top of the list.

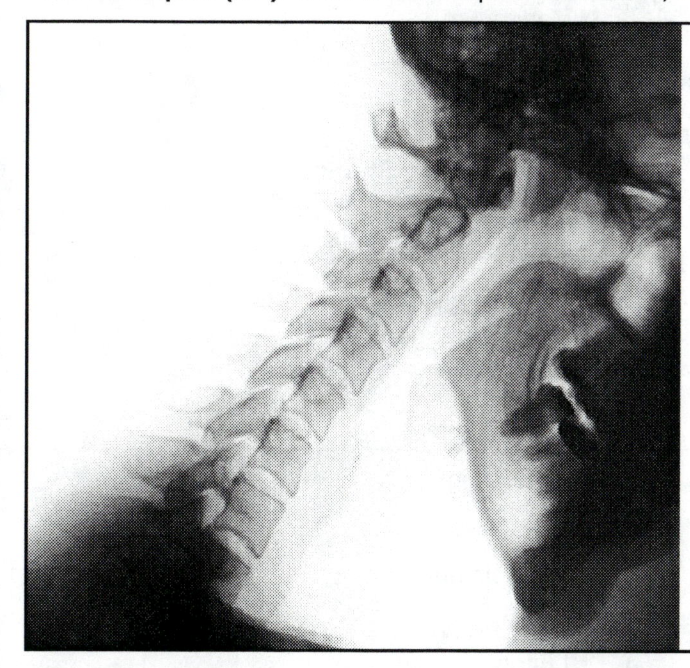

Identify:

- **Anterior arch of C1 vertebrae (atlas)**
- **Posterior arch of C1 vertebrae (atlas)**
- **Body of C3 - C7**
- **C2 vertebra (axis)**
- **Dens of C2 vertebrae (axis)**
- **Intervertebral foramen**
- **Spinous process of C2, C3, C4, C7**
- **Transverse process of C2, C3**
- **Mandible**

When you are finished, **close** the window.

Under the **File** menu, drag down to **Open** (then to **Content** if you are working in Windows). **Atlas Anatomy** should still be selected. **System** and **Skeletal** should still be selected. Choose **Lumbar Spine (Lat)**. If list is not in alphabetical order, the image is about half way from the top of the list.

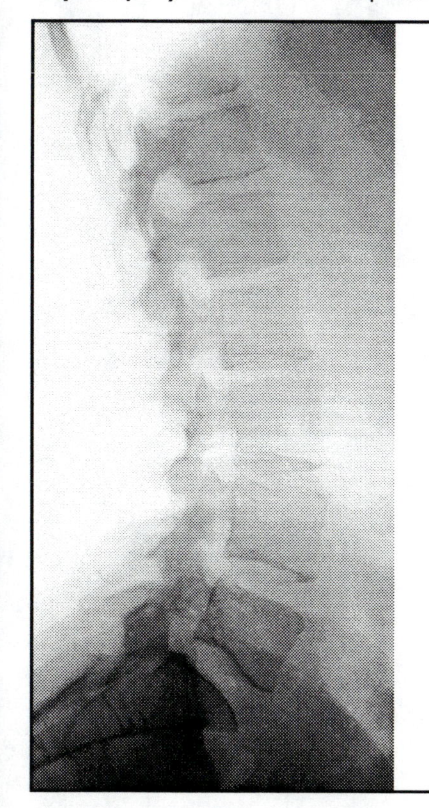

Identify:

- **Ribs 10, 11, 12**
- **Body of T11, T12**
- **Body of L1-L5**
- **Body of S1, S2**
- **Sacrum**
- **Intervertebral foramen**
- **Spinous process of L4, L5**

Where do you think the intervertebral disks will be found in this X-Ray?

________________________________________

When you are finished, **close** the window.

# Objectives: Vertebrae, Ribs, and Sternum

Student will be able to identify the following bones, bone markings and other structures associated with the vertebrae, ribs, and sternum:

**Vertebrae**

- C1 vertebra (atlas)
- C2 vertebra (axis)
- Cervical vertebrae
- Thoracic vertebrae
- Lumbar vertebrae
- Sacrum
- Coccyx

**Bone Markings on Vertebrae**

- Spinous process
- Body of vertebrae
- Intervertebral disks
- Dens of C2 (Axis)
- Superior articular process of C1
- Transverse process
- Anterior sacral foramen
- Intervertebral foramen
- Vertebral canal
- Anterior arch of C1 vertebrae
- Posterior arch of C1 vertebrae

**Joints**

- Atlanto-occipital joint
- Atlantoaxial joint
- Intervertebral joints
- Vertebrocostal joints
- Sternocostal joints

**Sternum**

- Manubrium
- Body of sternum
- Xiphoid process

**Bone Markings on Sternum**

- Clavicular notch of sternum
- Jugular notch of sternum

**Ribs & Bone Markings**

- Costal cartilage
- Ribs
- Body of ribs
- Head of ribs
- Sternal end of rib
- Tubercle of rib

**Other**

- Lungs
- Spinal nerves
- Occipital condyles
- Spinal cord
- Diaphragm

# Upper Appendicular Skeleton

## Opening A.D.A.M Interactive Anatomy

Open A.D.A.M. Interactive Anatomy according to the directions in the Tutorial.

## Anterior Dissection of Upper Appendicular Skeleton

Choose **Dissectible Anatomy**.

Choose either male or female. Select **Anterior.** Click **Open**.

Enlarge the window and position the navigation rectangle over the upper right shoulder.

Click on the **Depth Bar** ▼ to remove skin, fat, blood vessels, and muscles until you reach **layer 19**. Note the **clavicle.**

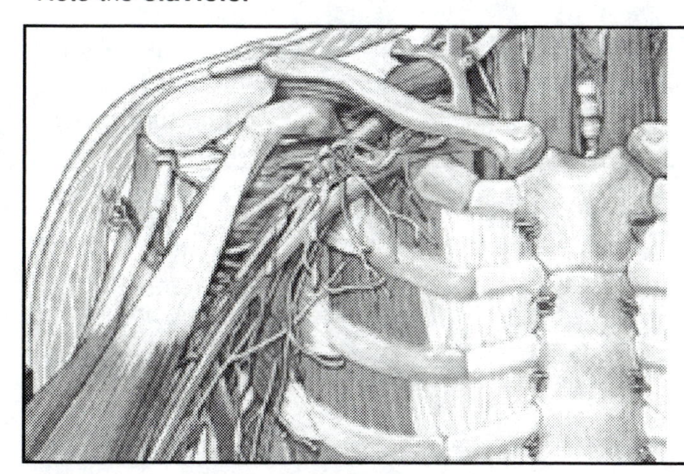

Click on the **Depth Bar** ▼ until you reach **layer 56**, noting the removal of the muscles and other structures. Identify:

- **Clavicle**
- **Acromion process of scapula**
- **Humerus**
- **Coracoid process of scapula**
- **Clavicular notch of sternum**
- **Jugular notch of sternum**

Locate these joints in this diagram:

- **Acromioclavicular joint**
- **Sternoclavicular joint**

What structure attaches to the coracoid process of the scapula?

---

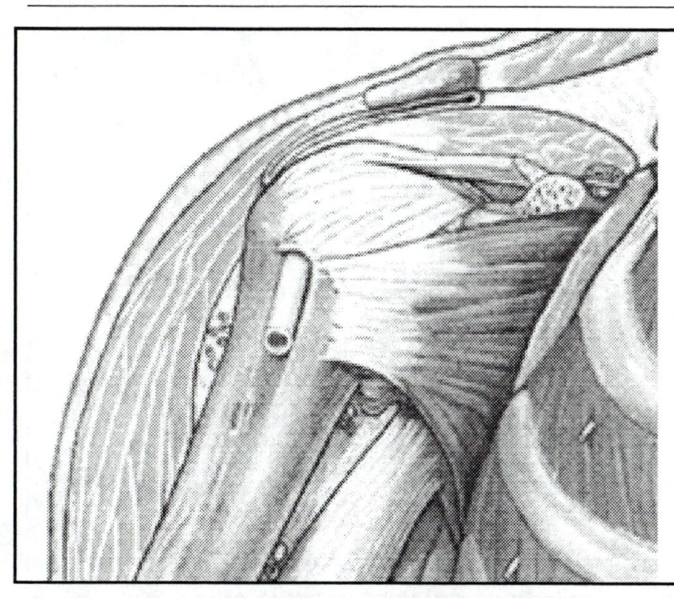

Click on the **Depth Bar** ▼ to remove muscles and other structures until you reach **layer 95**. Identify:

- **Scapula**
- **Humerus**
- **Acromion process of scapula**
- **Ligament** (superior glenohumeral)
- **Bursa**
- **Tendon sheath**

Scroll to **layer 143**. Note the removal of the ligaments and the underlying **synovial capsule of the shoulder joint**. No graphic has been provided for this layer.

Slowly click on the **Depth Bar** ▼ until you reach **layer 145**. Note the removal of the synovial capsule of the shoulder joint. No graphic has been provided for this layer. Identify:

- **Articular cartilage** of the shoulder joint

Scroll down to **layer 328**, then click once on the **Depth Bar** until you reach **layer 329**. Position the image over the upper arm. You may need to scroll to see the entire image. Identify:

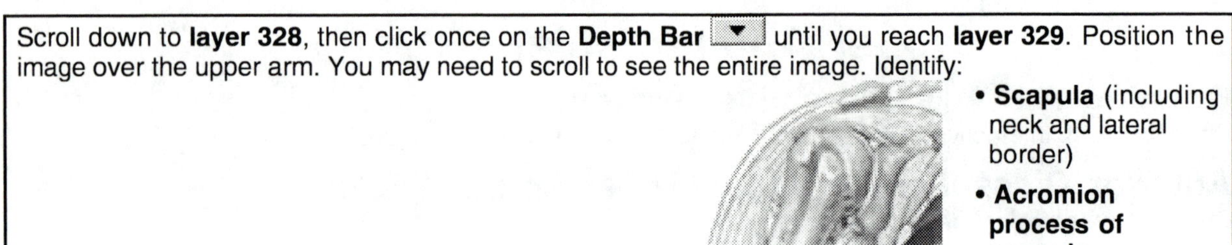

- **Scapula** (including neck and lateral border)
- **Acromion process of scapula**
- **Humerus**
- **Head of humerus**
- **Articular cartilage of humerus**
- **Greater tubercle of humerus**
- **Lesser tubercle of humerus**
- **Deltoid tuberosity of humerus**
- **Medial epicondyle of humerus**
- **Trochlea of humerus**
- **Coronoid fossa of humerus**
- **Capitulum of humerus**
- **Lateral epicondyle of humerus**
- **Head of radius**
- **Radial tuberosity**

- **Carpal Bones:**
  - **trapezium**
  - **scaphoid**
  - **trapezoid**
  - **lunate**
  - **triquetrum**
  - **pisiform**
  - **capitate**
  - **hamate**
- **Metacarpals**
- **Proximal, middle & distal phalanges** (phalanx)
- **Interosseous membrane of forearm**
- **Coronoid process of ulna**
- **Styloid process of ulna**
- **Styloid process of radius**
- **Ulnar tuberosity**
- **Ulnar notch**
- **Ulna** (anterior & lateral surfaces)
- **Radius**
- **Head of ulna** of radius

Locate these joints in this diagram:

- **Glenohumeral joint (shoulder joint)**
- **Elbow joint**
- **Proximal radioulnar joint**
- **Distal radioulnar joint**
- **Radiocarpal joint (wrist joint)**
- **Intercarpal joints**
- **Carpometacarpal joints**
- **Metacarpophalangeal joints (knuckle joints)**
- **Interphalangeal joints**

## Lateral Dissection of the Pectoral Girdle

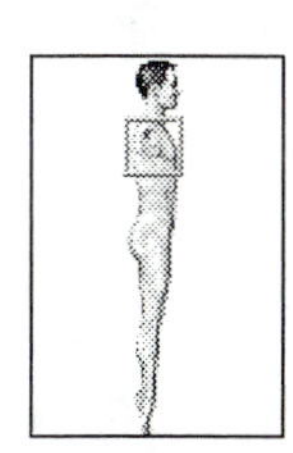

Click on the **View** button and choose **Lateral**.

Click on the **Normal** button.

Drag the **Depth Bar** to **layer 0.**

Position the navigator rectangle over the right shoulder.

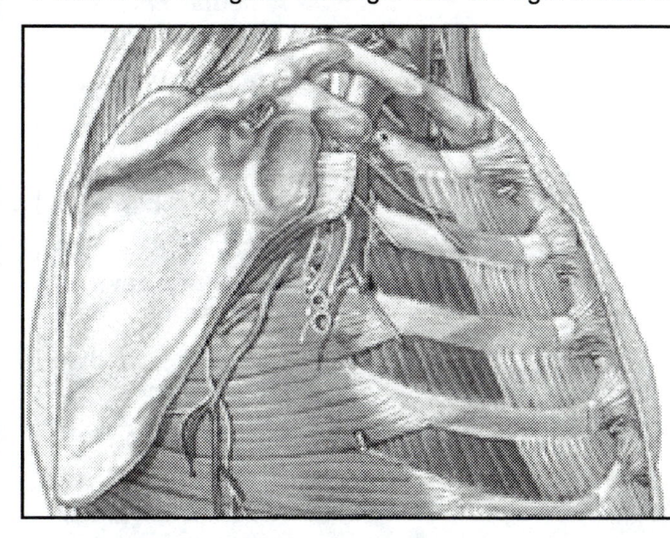

Click several times on the **Depth Bar** to remove the skin, fascia, and muscles until you reach **layer 41**. Identify:

- **Clavicle**
- **Scapula**
- **Acromion process of the scapula**.
- **Spine of the Scapula**
- **Coracoid process of scapula**
- **Glenoid cavity of scapula**
- **First rib**

Locate this joint in this diagram:

- **Acromioclavicular joint**

## Posterior Dissection of the Bones of the Mid Arm

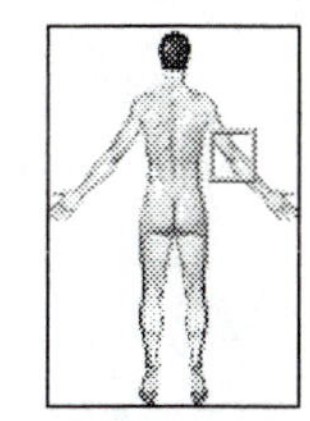

Click on the **View** button and choose **Posterior**.

Click on the **Normal** button. Windows— —Mac

Drag the **Depth Bar** to **layer 0.**

Position the navigator rectangle over the right elbow area.

Click several times on the **Depth Bar** to remove the skin, and fascia until you reach **layer 27**. No graphic has been provided for this layer. Identify:

- **Olecranon bursa**.

Scroll down to **layer 66**. No graphic has been provided for this layer. Identify:

- **Fibrous capsule of the elbow joint**

Click several times on the **Depth Bar** until you reach **layer 70**. No graphic has been provided for this layer. Identify:

- **Synovial capsule of the elbow joint.**

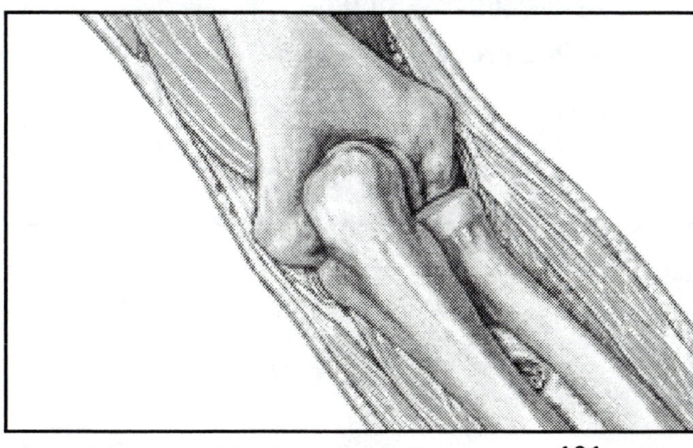

Scroll down to **level 185.** Identify:

- **Humerus**
- **Ulna** (all parts)
- **Olecranon of ulna**
- **Head of radius**
- **Radius**
- **Tuberosity of radius**
- **Interosseous membrane of forearm**

Click once on the **Depth Bar** [▼] until you reach **layer 186**. Identify:

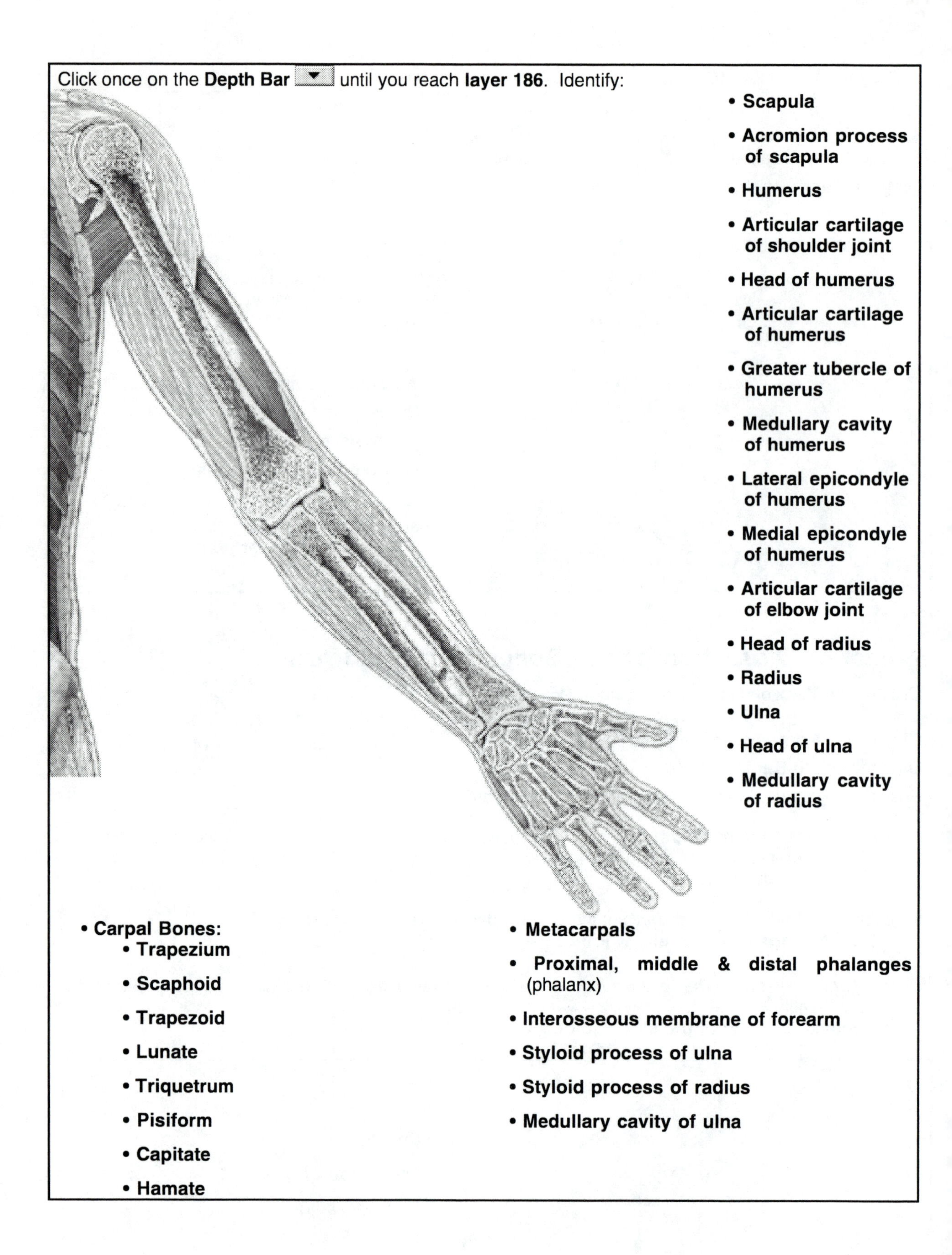

- **Scapula**
- **Acromion process of scapula**
- **Humerus**
- **Articular cartilage of shoulder joint**
- **Head of humerus**
- **Articular cartilage of humerus**
- **Greater tubercle of humerus**
- **Medullary cavity of humerus**
- **Lateral epicondyle of humerus**
- **Medial epicondyle of humerus**
- **Articular cartilage of elbow joint**
- **Head of radius**
- **Radius**
- **Ulna**
- **Head of ulna**
- **Medullary cavity of radius**

- **Carpal Bones:**
  - **Trapezium**
  - **Scaphoid**
  - **Trapezoid**
  - **Lunate**
  - **Triquetrum**
  - **Pisiform**
  - **Capitate**
  - **Hamate**

- **Metacarpals**
- **Proximal, middle & distal phalanges** (phalanx)
- **Interosseous membrane of forearm**
- **Styloid process of ulna**
- **Styloid process of radius**
- **Medullary cavity of ulna**

## Medial Dissection of the Upper Appendicular Skeleton

Click on the **View** button and choose **Medial Arm**.

Click on the **Normal** button. Windows— —Mac

Drag the **Depth Bar** to **layer 0.**

Position the navigator rectangle over the upper arm & elbow area.

Click several times on the **Depth Bar** to remove the skin, fascia and muscles until you reach **layer 5**. Scroll down to **layer 56** then click on the **Depth Bar** until you reach **layer 70**. No graphic has been provided for this layer. Identify:

- **Articular cartilage of shoulder joint**.

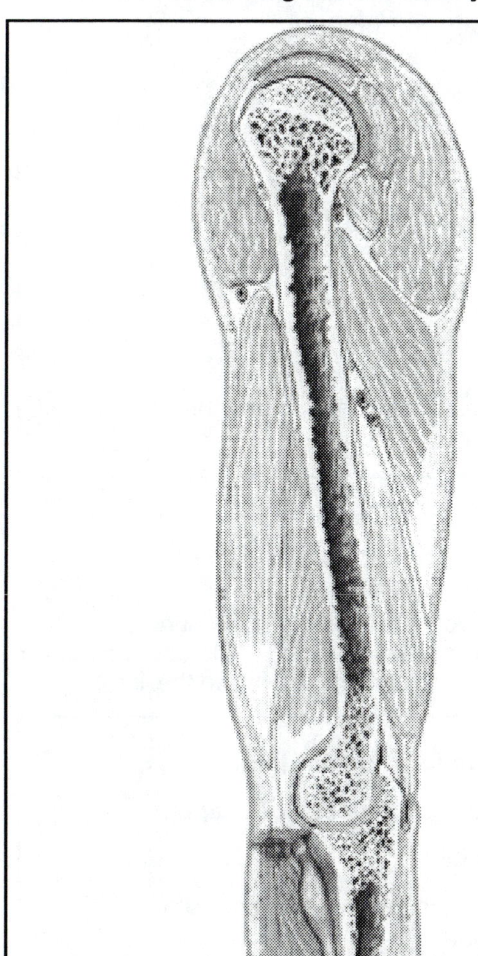

Click a few times on the **Depth Bar** until you reach **level 72**. You may need to scroll to see the entire image. Identify:

- **Head of humerus**
- **Humerus**
- **Medullary cavity of humerus**
- **Epiphyseal line of humerus**
- **Trochlea of humerus**
- **Articular cartilage of elbow joint**
- **Ulna**
- **Olecranon of ulna**
- **Coronoid process of ulna**
- **Radius**
- **Tuberosity of radius**
- **Head of radius**
- **Medullary cavity of ulna**

Locate this joint in this diagram:

- **Elbow joint**

What is the difference between the **elbow joint** and the **proximal radioulnar joint**?

_______________________________________________

When you are finished, **close** dissectible anatomy.

## Cadaver Image of the Upper Appendicular Skeleton

Under the **File** menu, drag down to **Open** (then to **Content** if you are working in Windows). Click on **Atlas Anatomy**. Select **System** and **Skeletal.** Choose **Axillary Fossa**. If the list is not in alphabetical order, the image is about a third of the way down the list.

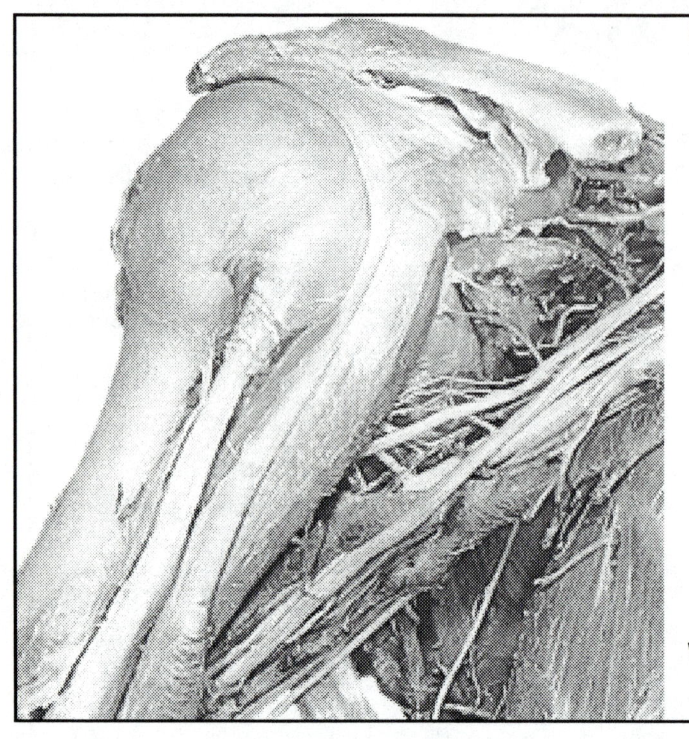

Identify:

- **Clavicle**
- **Coracoid process of scapula**
- **Acromion process of scapula**
- **Head of humerus**
- **Body of humerus**

When you are finished, **close** the window.

## X-Rays of the Upper Appendicular Skeleton

Under the **File** menu, drag down to **Open** (then to **Content** if you are working in Windows). **Atlas Anatomy** should still be selected. **System** and **Skeletal** should still be selected. Choose **Shoulder (Ant/Post)**. If the list is not in alphabetical order, the image is about a third of the way down the list.

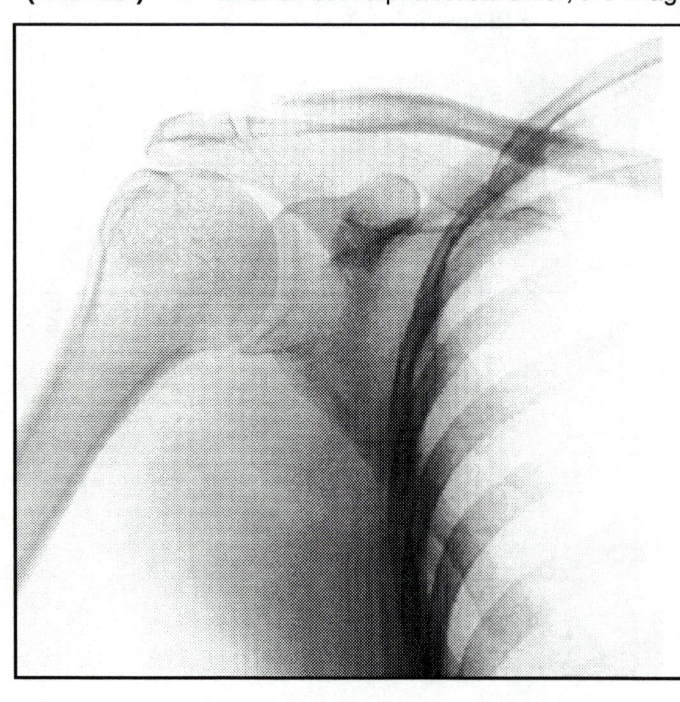

Identify:

- **Acromioclavicular joint**
- **Acromion process of scapula**
- **Clavicle**
- **Coracoid process of scapula**
- **Ribs**
- **Glenohumeral joint**
- **Glenoid cavity**
- **Humerus**
- **Head of humerus**
- **Scapula**
- **Greater tubercle of humerus**
- **Lesser tubercle of humerus**

When you are finished, **close** the window.

Under the **File** menu, drag down to **Open** (then to **Content** if you are working in Windows). **Atlas Anatomy** should still be selected. **System** and **Skeletal** should still be selected. Choose **Flexed Elbow (lat)**. If the list is not in alphabetical order, the image is about a third of the way down the list.

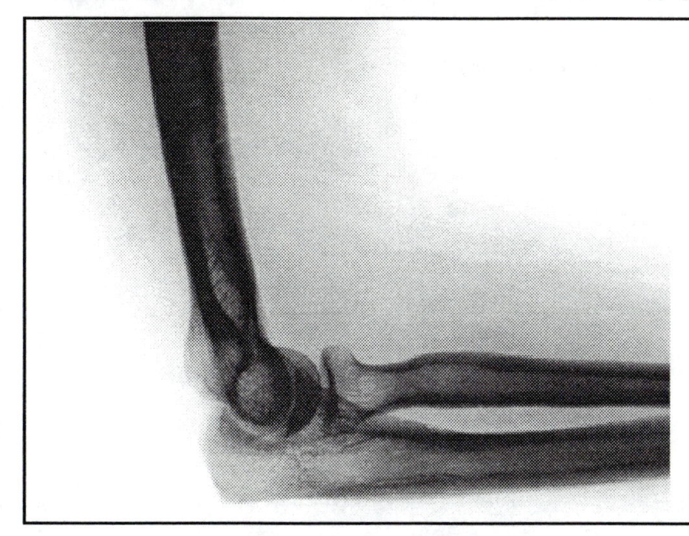

Identify:

- **Capitulum of humerus**
- **Coronoid process of ulna**
- **Head of radius**
- **Humerus**
- **Lateral epicondyle of humerus**
- **Olecranon of ulna**
- **Ulna**
- **Radius**
- **Tuberosity of radius**

When you are finished, **close** the window.

Under the **File** menu, drag down to **Open** (then to **Content** if you are working in Windows). **Atlas Anatomy** should still be selected. **System** and **Skeletal** should still be selected. Choose **Hand (Post/Ant)**. If the list is not in alphabetical order, the image is about half way down the list.

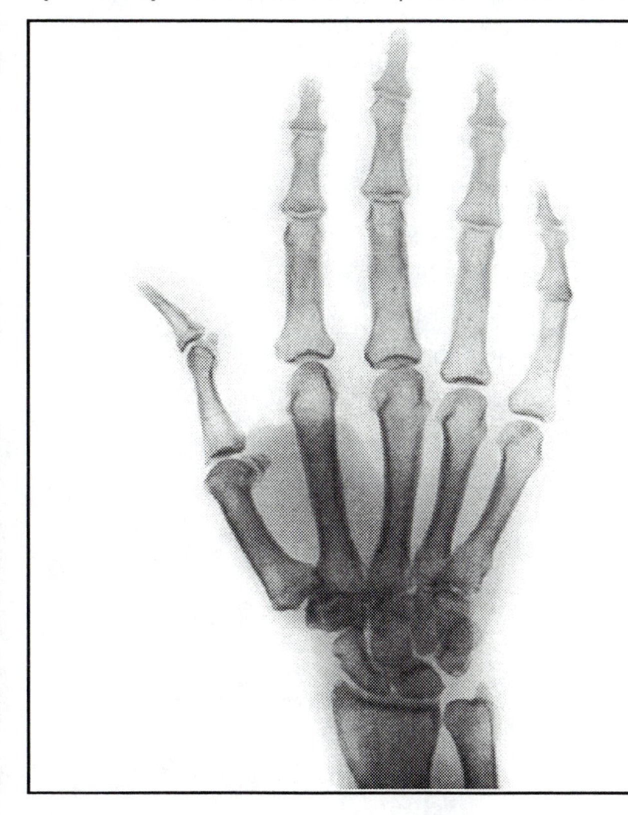

You may need to scroll to see the entire image.
Identify:

- **Styloid process of ulna**
- **Styloid process of radius**
- **Ulna**
- **Radius**
- **Carpal Bones:**
    - **trapezium**
    - **scaphoid**
    - **trapezoid**
    - **lunate**
    - **triquetrum**
    - **pisiform**
    - **capitate**
    - **hamate**
- **Metacarpals**
- **Proximal, middle & distal phalanges** (phalanx)
- **Carpometacarpal joints**
- **Distal interphalangeal joints**
- **Metacarpophalangeal joint**

When you are finished, **close** the window.

## Objectives: Upper Appendicular Skeleton

Student will be able to identify the following the bones and bone markings of the upper appendicular skeleton.

**Bones**

- Clavicle
- Scapula
- Humerus
- Radius
- Ulna
- Carpals
  - Trapezium
  - Scaphoid
  - Trapezoid
  - Lunate
  - Triquetrum
  - Pisiform
  - Capitate
  - Hamate
- Metacarpals
- Proximal, Middle & Distal Phalanges (Phalanx)

**Joints**

- Sternoclavicular joint
- Acromioclavicular joint
- Glenohumeral joint (shoulder joint)
- Elbow joint
- Proximal radioulnar joint
- Distal radioulnar joint
- Radiocarpal joint (wrist joint)
- Intercarpal joints
- Carpometacarpal joints
- Metacarpophalangeal joints (knuckle joints)
- Interphalangeal joints

**Other**

- Clavicular notch of sternum
- Jugular notch of sternum
- Ligaments (various)
- Bursa
- Tendon sheath
- Synovial capsules (various)
- Articular cartilage
- Interosseous membrane of forearm
- Olecranon bursa
- Fibrous capsules (various)
- Medullary cavities (various)

**Bone Markings:**

On scapula:

- Acromion process
- Coracoid process
- Spine
- Glenoid cavity

On humerus:

- Head
- Greater tubercle
- Lesser tubercle
- Deltoid tuberosity
- Medial epicondyle
- Trochlea
- Coronoid fossa
- Capitulum
- Lateral epicondyle
- Epiphyseal line

On radius:

- Head
- Radial tuberosity
- Ulnar notch
- Styloid process

On ulna:

- Ulnar tuberosity
- Coronoid process
- Head
- Styloid process
- Olecranon

# Lower Appendicular Skeleton

## Opening A.D.A.M Interactive Anatomy

Open A.D.A.M. Interactive Anatomy according to the directions in the Tutorial.

## Anterior Dissection of Lower Appendicular Skeleton

Choose **Dissectible Anatomy**.

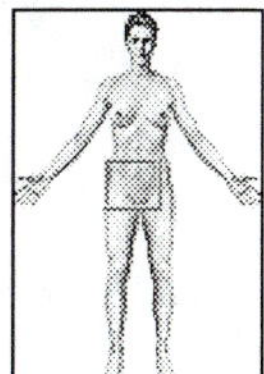

Choose **female**. Select **Anterior.** Click **Open**.

Enlarge the window and position the navigation rectangle over the right hip.

Click on the **Depth Bar** ▼ to remove skin, fat, blood vessels, and muscles until you reach **layer 18**. No graphic has been provided for this layer. Identify:

- **Iliac crest**
- **Pubis**
- **Pubic symphysis**

Scroll down to **layer 208**, then click on the **Depth Bar** ▼ until you reach **layer 266**. No graphic has been provided for this layer. Identify:

- **Iliac crest**
- **Pubis**
- **Pubic symphysis**

Click on the **Depth Bar** ▼ until you reach **layer 300**. No graphic has been provided for this layer. Identify:

- **Sacroiliac ligament** (ventral)
- **Iliofemoral ligament**

Click on the **Depth Bar** ▼ until you reach **layer 308**. No graphic has been provided for this layer. Identify:

- **Synovial capsule of the hip joint**

Click on the **Depth Bar** ▼ until you reach **layer 327.** No graphic has been provided for this layer. Identify:

- **Articular cartilage of the hip joint**

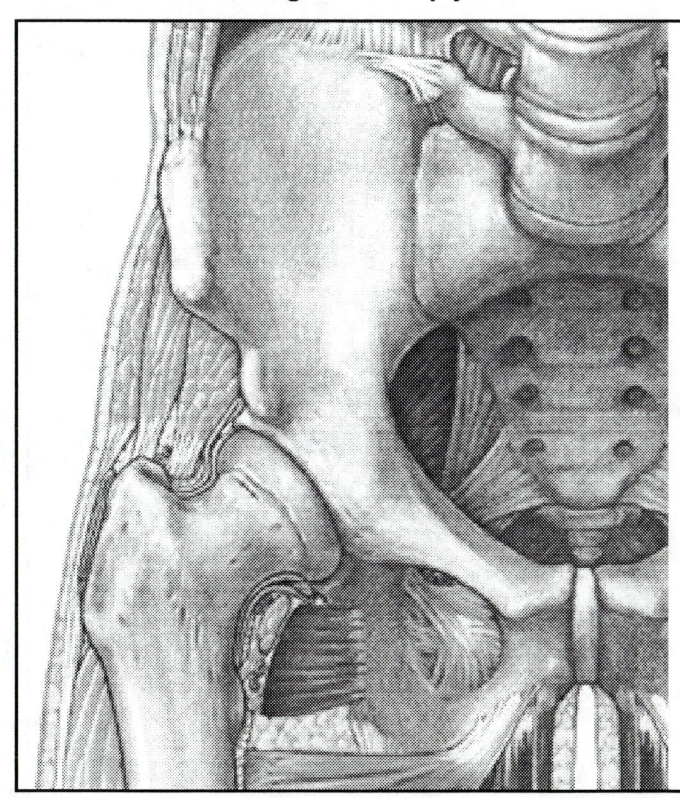

Click on the **Depth Bar** ▼ until you reach **layer 328**. Identify:

- **Body of L5 vertebra**
- **Transverse process of L5 vertebra**
- **Intervertebral disc**
- **Sacrum** (including wing)
- **Ilium**
- **Iliac crest**
- **Anterior superior iliac spine**
- **Pubis**
- **Pubic symphysis**
- **Obturator foramen** (obturator membrane is in the foramen)
- **Acetabulum**
- **Ischium**
- **Head of femur**
- **Neck of femur**
- **Greater trochanter of femur**
- **Lesser trochanter of femur**
- **Femur**

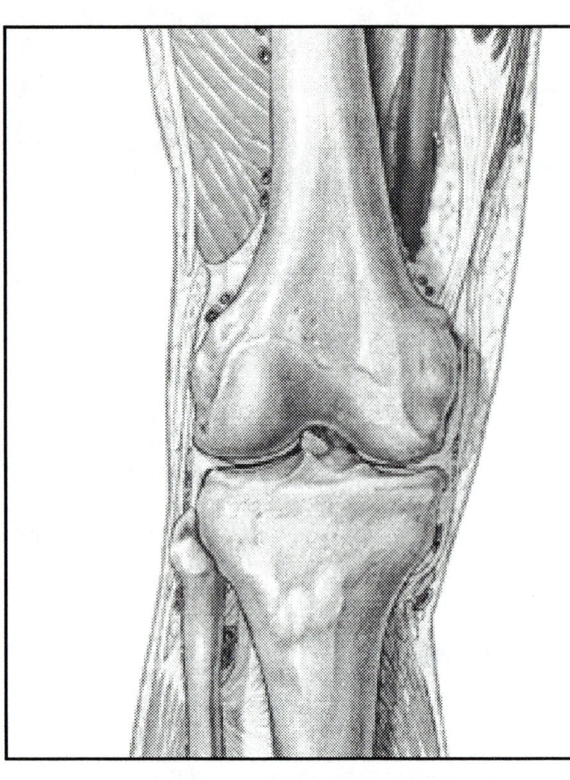

Move down to the right knee area. You are still at **layer 329**. Identify:

- **Femur**
- **Lateral condyle of femur**
- **Medial condyle of femur**
- **Lateral epicondyle of femur**
- **Medial epicondyle of femur**
- **Head of fibula**
- **Fibula** (neck, lateral and medial surface)
- **Tibia** (lateral and medial surface)
- **Tibial tuberosity**
- **Medial condyle of tibia**
- **Lateral condyle of tibia**
- **Interosseous membrane of the leg**
- **Articular cartilage** of the condyles

Identify these joints:

- **Knee (tibiofemoral) joint**
- **Proximal tibiofibular joint**

You will now do a "reverse dissection" of the knee. Click on the **Depth Bar** [▲] to go to **layer 327**. No graphic has been provided for this layer. Identify:

- **Articular cartilage of the knee joint**.

Click on the **Depth Bar** [▲] to go to **layer 319**. No graphic has been provided for this layer. Identify:

- **Synovial capsule of the knee joint**.

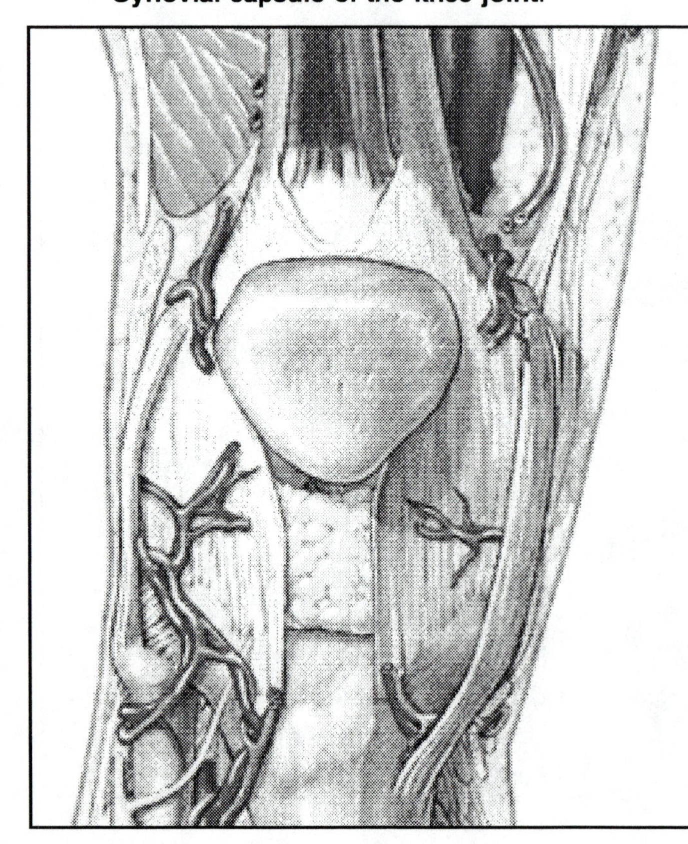

Click on the **Depth Bar** [▲] to go to **layer 292**. Identify:

- **Patella**
- **Femur**
- **Head of fibula**
- **Fibula** (neck, lateral and medial surface)
- **Tibia** (tuberosity, lateral and anterior surface)
- **Medial condyle of tibia**
- **Lateral condyle of tibia**
- **Interosseous membrane of the leg**

Click on the **Depth Bar** [▲] to go to **layer 290**.

What type of bone is the patella (long, short, flat, irregular, sutural, sesamoid)?

---

Position the image over the right foot. Click several times on the **Depth Bar** [▼] to remove muscles and other structures until you reach **layer 323**. No graphic has been provided for this layer. Note the various **ligaments**.

Click several times on the **Depth Bar** [▼] until you reach **layer 327**. No graphic has been provided for this layer. Note the **articular cartilage** around the various joints.

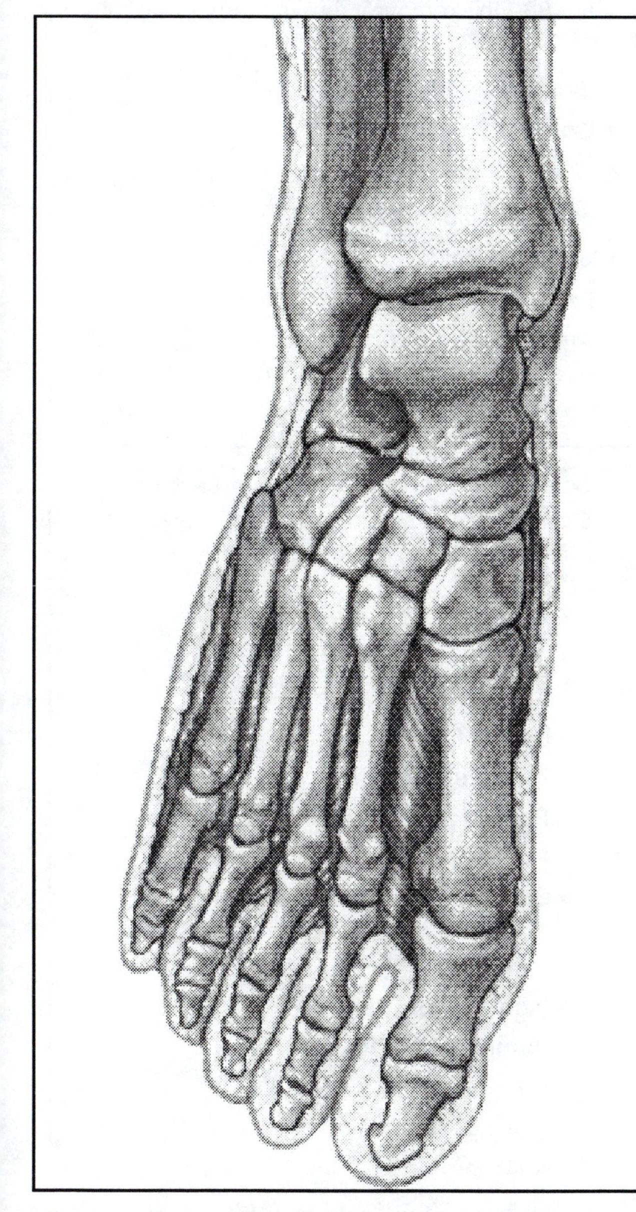

Click on the **Depth Bar** [▼] until you reach **level 329**. Identify:

- **Fibula** (all parts)
- **Tibia** (all parts)
- **Interosseous membrane of the leg**
- **Medial malleolus of tibia**
- **Lateral malleolus of fibula**
- **Tarsals:**
  - **Talus bone**
  - **Calcaneus bone**
  - **Navicular bone**
  - **Medial cuneiform bone**
  - **Intermediate cuneiform bone**
  - **Lateral cuneiform bone**
  - **Cuboid bone**
- **Metatarsals**
- **Proximal, middle, and distal phalanges** (phalanx)

On this diagram, indicate the following joints:

- **Distal tibiofibular joint**
- **Ankle joint**
- **Intertarsal joints**
- **Tarsometatarsal joints**
- **Metatarsophalangeal joint**
- **Interphalangeal joints**

Contrast the number of phalanges in the big toe compared to the other toes.

---

## Posterior Dissection of the Lower Appendicular Skeleton

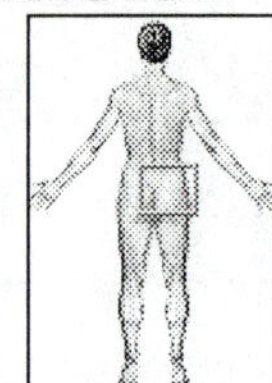

Click on the **View** button and choose **Posterior**.

Click on the **Normal** button. Windows— —Mac

Drag the **Depth Bar** to **layer 0.**

Position the navigator rectangle over the right buttock area.

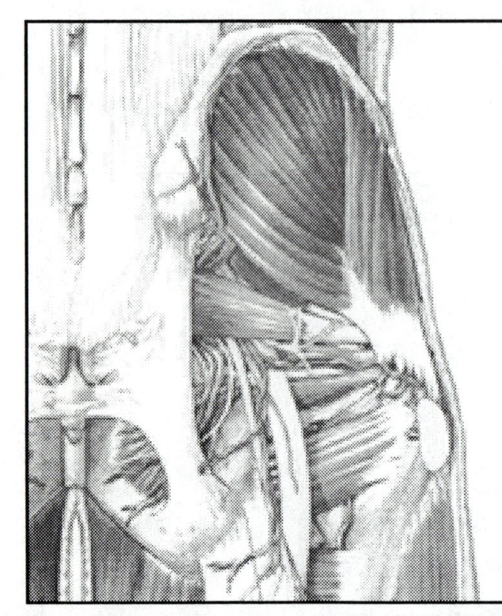

Click several times on the **Depth Bar** to remove the skin, fascia, and blood vessels until you reach **layer 50**. Scroll down to **layer 82**, then click on the **Depth Bar** until you reach **layer 83** to remove the gluteus maximus muscle. Identify:

- **Sacrum**
- **Coccyx**
- **Ilium** & **Iliac crest**
- **Pubis** (inferior ramus, body)
- **Ischium** (including tuberosity)
- **Femur**
- **Pubic Symphysis**

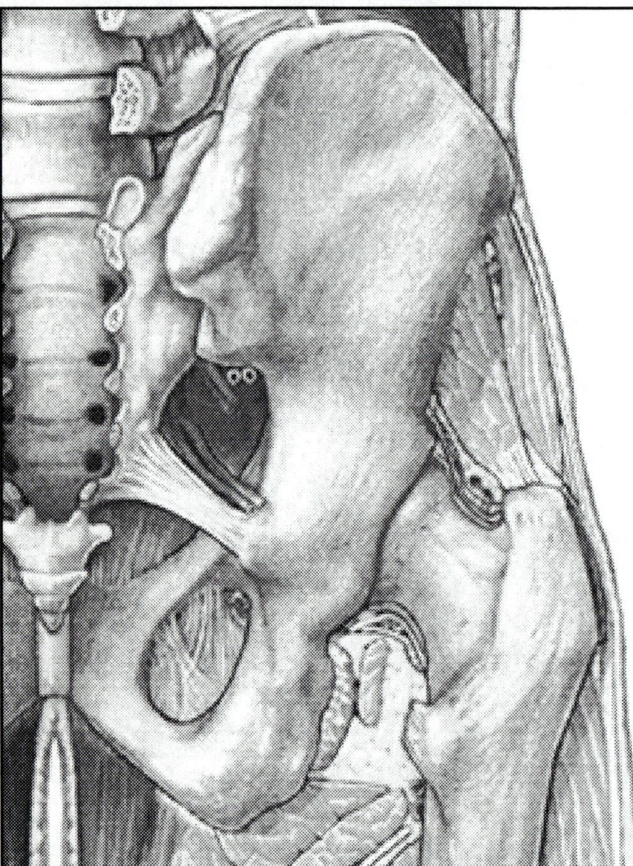

Scroll down to **layer 128**, then click on the **Depth Bar** until you reach **layer 185**. Identify:

- **Sacrum**
- **Coccyx**
- **Ilium**
- **Iliac crest**
- **Body of pubis**
- **Pubic symphysis**
- **Posterior superior iliac spine**
- **Posterior inferior iliac spine**
- **Obturator foramen** (obturator membrane is in the foramen)
- **Ischium**
- **Ischial tuberosity**
- **Ramus of ischium**
- **Superior ramus of pubis**
- **Head of femur**
- **Neck of femur**
- **Greater trochanter of femur**
- **Lesser trochanter of femur**
- **Gluteal tuberosity of femur**

Identify the following joints on the diagram above:

- **Sacroiliac joint**
- **Hip (coxal) joint**

When you sit down, what bone do you "sit on"?

---

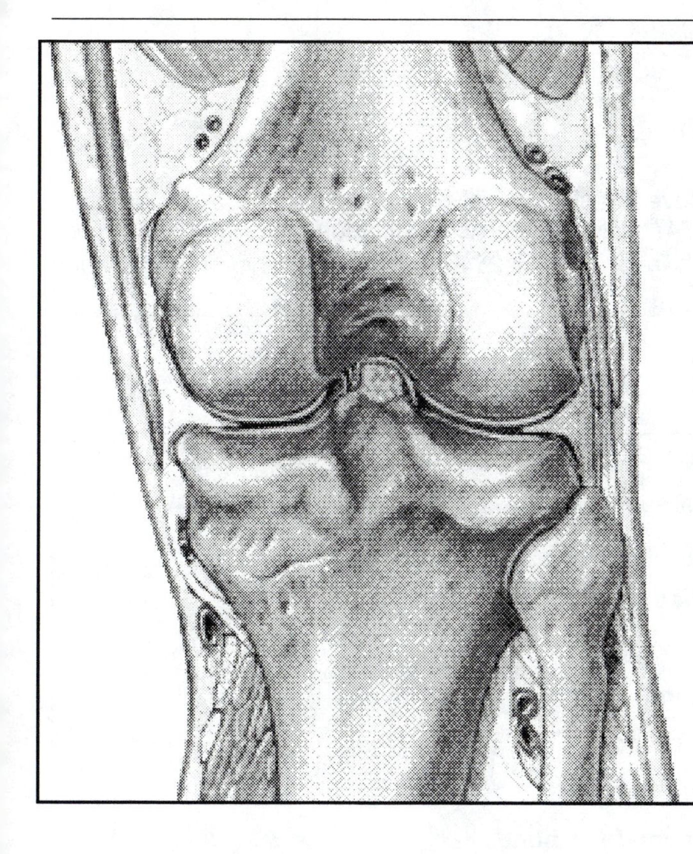

Move down to the right knee area. You are still at **layer 185**. Identify:

- **Femur**
- **Lateral condyle of femur**
- **Medial condyle of femur**
- **Lateral epicondyle of femur**
- **Medial epicondyle of femur**
- **Head of fibula**
- **Fibula**
- **Tibia**
- **Medial condyle of tibia**
- **Lateral condyle of tibia**
- **Interosseous membrane of the leg**
- **Articular cartilage** of the condyles

Identify these joints:

- **Knee (tibiofemoral) joint**
- **Proximal tibiofibular joint**

## Lateral Dissection of the Bones of the Foot

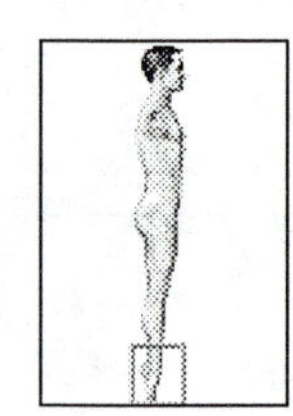

Click on the **View** button and choose **Lateral**.

Click on the **Normal** button. Windows— —Mac

Drag the **Depth Bar** to **layer 0.**

Position the navigator rectangle over the foot area.

Click several times on the **Depth Bar** to remove the skin, fascia, blood vessels, muscles, and other connective tissue until you reach **layer 273 or 274**. Note **ligaments** and fibrous capsule of the various joints. Click on the **Depth Bar** until you reach **layer 283 or 284**. No graphics have been provided for these layers. Identify:

- **Articular cartilage**
- **Interosseous membrane of the leg**.

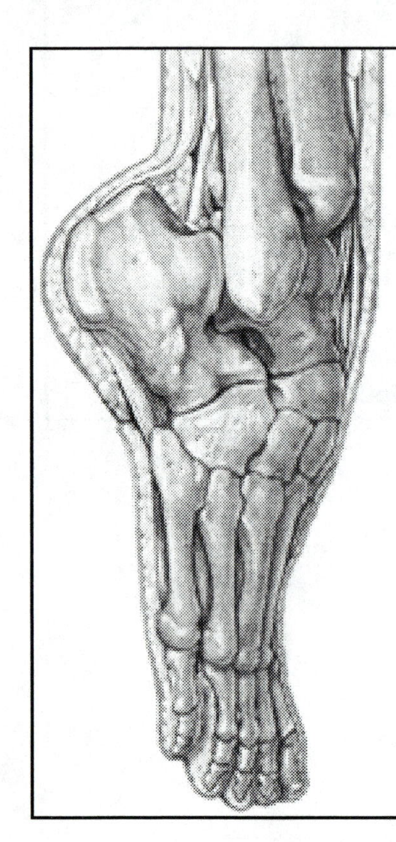

Click on the **Depth Bar** until you reach **layer 287**. Identify:

- **Fibula** (lateral surface)
- **Tibia** (lateral surface)
- **Lateral malleolus of fibula**
- **Tarsals:**
  - **Talus bone**
  - **Calcaneus bone**
  - **Navicular bone**
  - **Medial cuneiform bone**
  - **Intermediate cuneiform bone**
  - **Lateral cuneiform bone**
  - **Cuboid bone**
- **Metatarsals**
- **Proximal, middle, and distal phalanges** (phalanx)

When you are finished, **close** dissectible anatomy.

## Comparison of Male & Female Pelvis

In this section you will open images of male and female pelvises at the same time and compare them to one another.

Under the **File** menu, drag down to **Open** (then to **Content** if you are working in Windows). Click on **Atlas Anatomy**. Select **System** and **Skeletal.** Choose **Female Bony Pelvis (Ant)**. If list is not in alphabetical order, the image is about a two-thirds of the way down the list. Click on the **Zoom** button then click anywhere on the image to make it fit onto your screen.

Under the **File** menu, drag down to **Open** (then to **Content** if you are working in Windows). Click on **Atlas Anatomy**. Select **System** and **Skeletal.** Choose **Male Bony Pelvis (Ant)**. If list is not in alphabetical order, the image is about three-fourths of the way down the list.

Click on the **Zoom** button to reduce the size of the image, then click anywhere on the image to make it fit onto your screen.

Position the female and male bony pelvis images so they can both be seen at the same time on the screen.

- If you are working with a Macintosh, do this by dragging the gray or blue bar above the images:
- If you are working with Windows, click **Window** on the Tool Bar. Click **Tile**. Adjust each image so they are side by side.

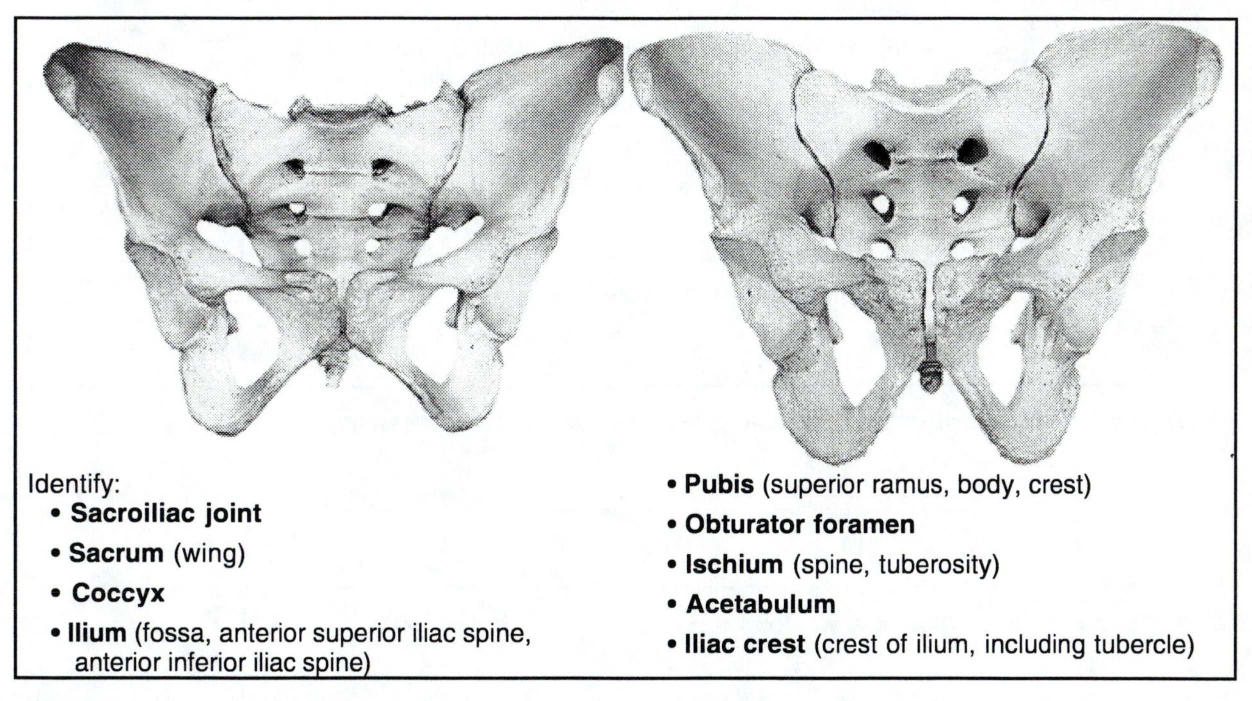

Identify:

- **Sacroiliac joint**
- **Sacrum** (wing)
- **Coccyx**
- **Ilium** (fossa, anterior superior iliac spine, anterior inferior iliac spine)
- **Pubis** (superior ramus, body, crest)
- **Obturator foramen**
- **Ischium** (spine, tuberosity)
- **Acetabulum**
- **Iliac crest** (crest of ilium, including tubercle)

Which is male and which is female? Indicate at least one difference between the two.

____________________________________________________________

____________________________________________________________

When you are finished, **close** both windows.

Under the **File** menu, drag down to **Open** (then to **Content** if you are working in Windows). Click on **Atlas Anatomy**. Select **System** and **Skeletal.** Choose **Female Bony Pelvis (Lat)**. If list is not in alphabetical order, the image is about three-fourths of the way down the list.

Click on the **Zoom** button to reduce the size of the image, then click anywhere on the image to make it fit onto your screen.

Under the **File** menu, drag down to **Open** (then to **Content** if you are working in Windows). Click on **Atlas Anatomy**. Select **System** and **Skeletal.** Choose **Male Bony Pelvis (Lat)**. If list is not in alphabetical order, the image is about three-fourths of the way down the listing.

Click on the **Zoom** button to reduce the size of the image, then click anywhere on the image to make it fit onto your screen.

Position the female and male bony pelvis images so they can both be seen at the same time on the screen.

- If you are working with a Macintosh, do this by dragging the gray or blue bar above the images:

- If you are working with Windows, click **Window** on the Tool Bar. Click **Tile**. Adjust each image so they are side by side.

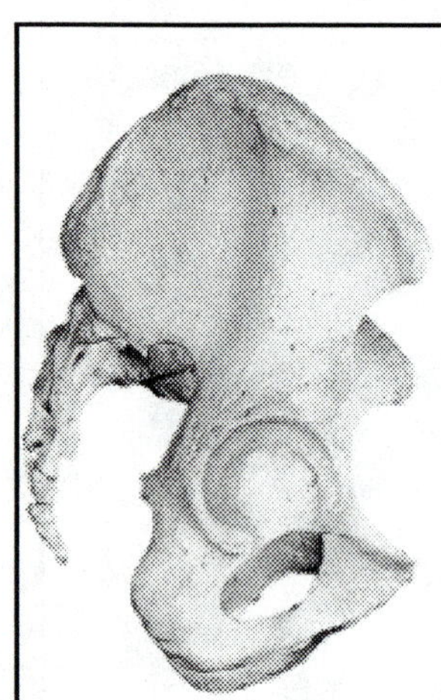

Identify:

- **Sacrum**
- **Coccyx**
- **Ilium** (posterior superior iliac spine, posterior inferior iliac spine, anterior superior iliac spine, anterior inferior iliac spine, body)
- **Iliac crest** (crest of ilium)
- **Pubis** (tubercle, superior ramus, body, inferior ramus)
- **Obturator foramen**
- **Ischium** ( body, tuberosity)
- **Acetabulum**

Which is male and which is female? Indicate at least one difference between the two.

____________________________________________________________

____________________________________________________________

When you are finished, **close** both windows.

Under the **File** menu, drag down to **Open** (then to **Content** if you are working in Windows). Click on **Atlas Anatomy**. Select **System** and **Skeletal.** Choose **Female Bony Pelvis (Sup)**. If list is not in alphabetical order , the image is about three-fourths of the way down the list.

Click on the **Zoom** button to reduce the size of the image, then click anywhere on the image to make it fit onto your screen.

Under the **File** menu, drag down to **Open** (then to **Content** if you are working in Windows). Click on **Atlas Anatomy**. Select **System** and **Skeletal.** Choose **Male Bony Pelvis (Sup)**. If list is not in alphabetical order, the image is about three-fourths of the way down the listing.

Click on the **Zoom** button to reduce the size of the image, then click anywhere on the image to make it fit onto your screen.

Position the female and male bony pelvis images so they can both be seen at the same time on the screen.

- If you are working with a Macintosh, do this by dragging the gray or blue bar above the images:

- If you are working with Windows, click **Window** on the Tool Bar. Click **Tile**. Adjust each image so they are side by side.

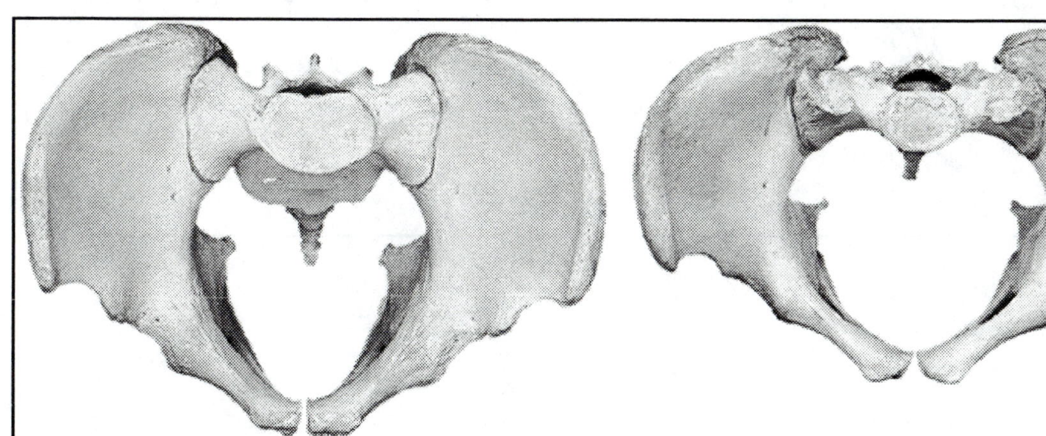

Identify:

- **Sacroiliac joint**
- **Sacrum** (intermediate crest, median crest, canal)
- **Ilium** (fossa, anterior superior iliac spine, anterior inferior iliac spine, posterior superior iliac spine)
- **Iliac crest** (crest of ilium)
- **Pubis** (inferior ramus, superior ramus, crest)
- **Ischium** (spine)
- **Pubic Symphysis**
- **Coccyx**

When you are finished, **close** both windows.

Indicate the **greater (false) pelvis** and the **lesser (true) pelvis** on these diagrams.

Which is male and which is female? Indicate at least one difference between the two.

______________________________________________________________________

## Cadaver Image of Lower Appendicular Skeleton

Under the **File** menu, drag down to **Open** (then to **Content** if you are working in Windows). **Atlas Anatomy** should still be selected. **System** and **Skeletal** should still be selected. Choose **Dissection of Medial Leg and Foot**. If the list is not in alphabetical order, the image is almost at the bottom of the list.

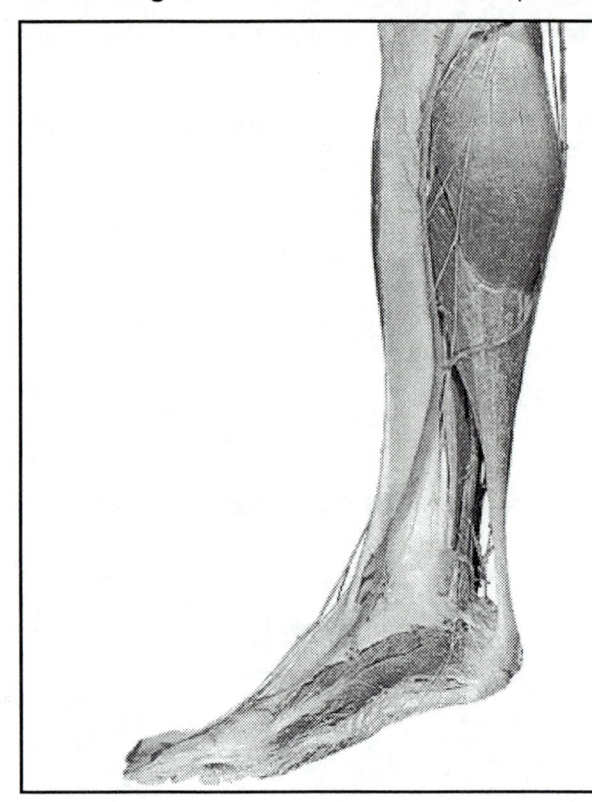

Identify:

- **Tibia**
- **Medial malleolus**
- **Calcaneous bone** (tuberosity of)

What's the name of the tendon that inserts into the calcaneous bone? (If you are working in Mac, click on the **Show All Pins** button and scroll to **All Systems**.)

_______________________________________________

When you are finished, **close** the window.

## X-Rays of Bones of the Lower Appendicular Skeleton

Under the **File** menu, drag down to **Open** (then to **Content** if you are working in Windows). **Atlas Anatomy** should still be selected. **System** and **Skeletal** should still be selected. Choose **Pelvis (Ant/Post)**. If the list is not in alphabetical order, the image is about three fourths of the way down the list.

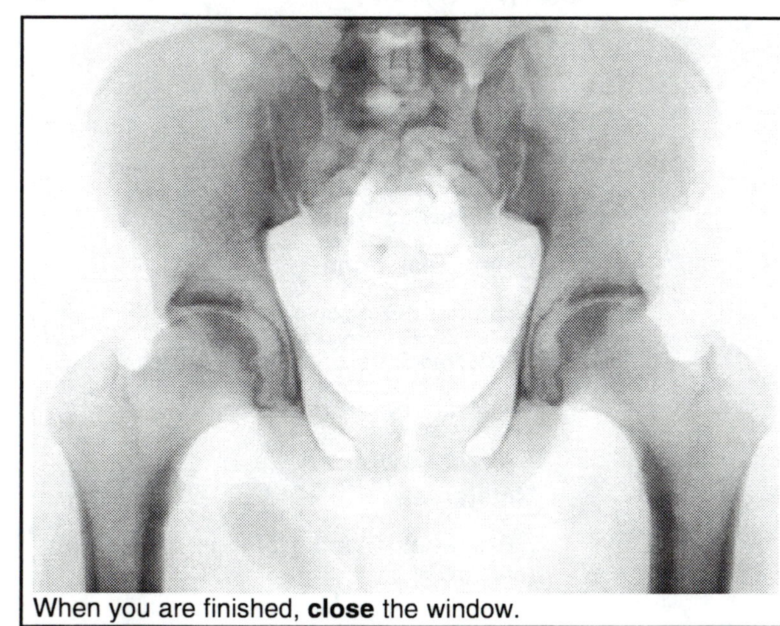

Identify:

- **Sacrum** (including wing)
- **Ilium** (including wing)
- **Iliac crest** (crest of ilium)
- **Anterior superior iliac spine**
- **Pubis** (body, inferior ramus, superior ramus)
- **Pubic symphysis**
- **Acetabulum**
- **Ischium** (including tuberosity)
- **Sacroiliac joint**
- **Head of femur**
- **Neck of femur**
- **Greater trochanter of femur**
- **Femur**
- **Lesser trochanter of femur**

When you are finished, **close** the window.

Under the **File** menu, drag down to **Open** (then to **Content** if you are working in Windows). **Atlas Anatomy** should still be selected. **System** and **Skeletal** should still be selected. Choose **Flexed Knee (lat)**. If the list is not in alphabetical order, the image is almost to the bottom of the list.

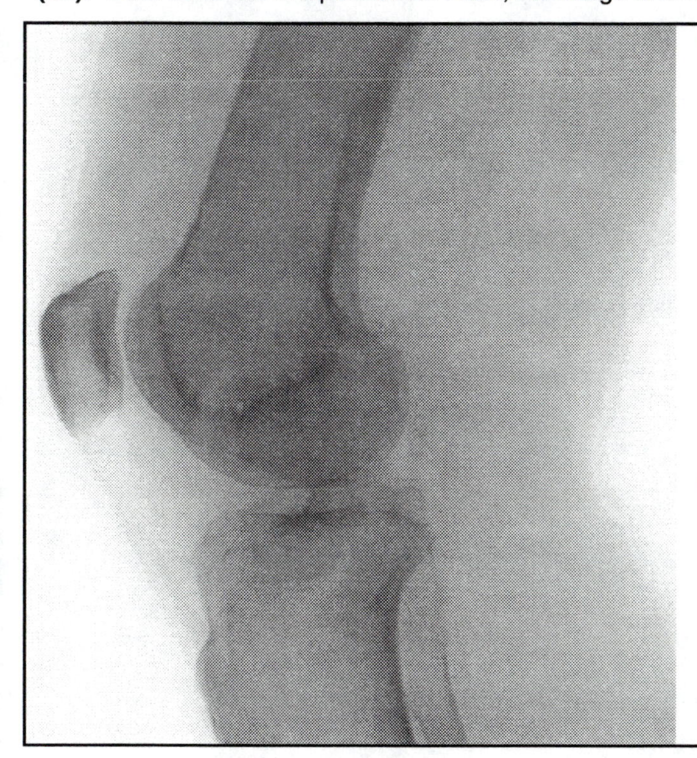

Identify:

- **Patella**
- **Femur**
- **Lateral condyle of femur**
- **Lateral epicondyle of femur**
- **Fibula**
- **Tibia**
- **Tibial tuberosity**
- **Lateral condyle of tibia**

Identify these joints on this diagram:

- **Knee (tibiofemoral) joint**
- **Femoropatellar joint**

When you are finished, **close** the window.

Under the **File** menu, drag down to **Open** (then to **Content** if you are working in windows). **Atlas Anatomy** should still be selected. **System** and **Skeletal** should still be selected. Choose **Foot (Lat)**. If the list is not in alphabetical order, the image is all the way down the list.

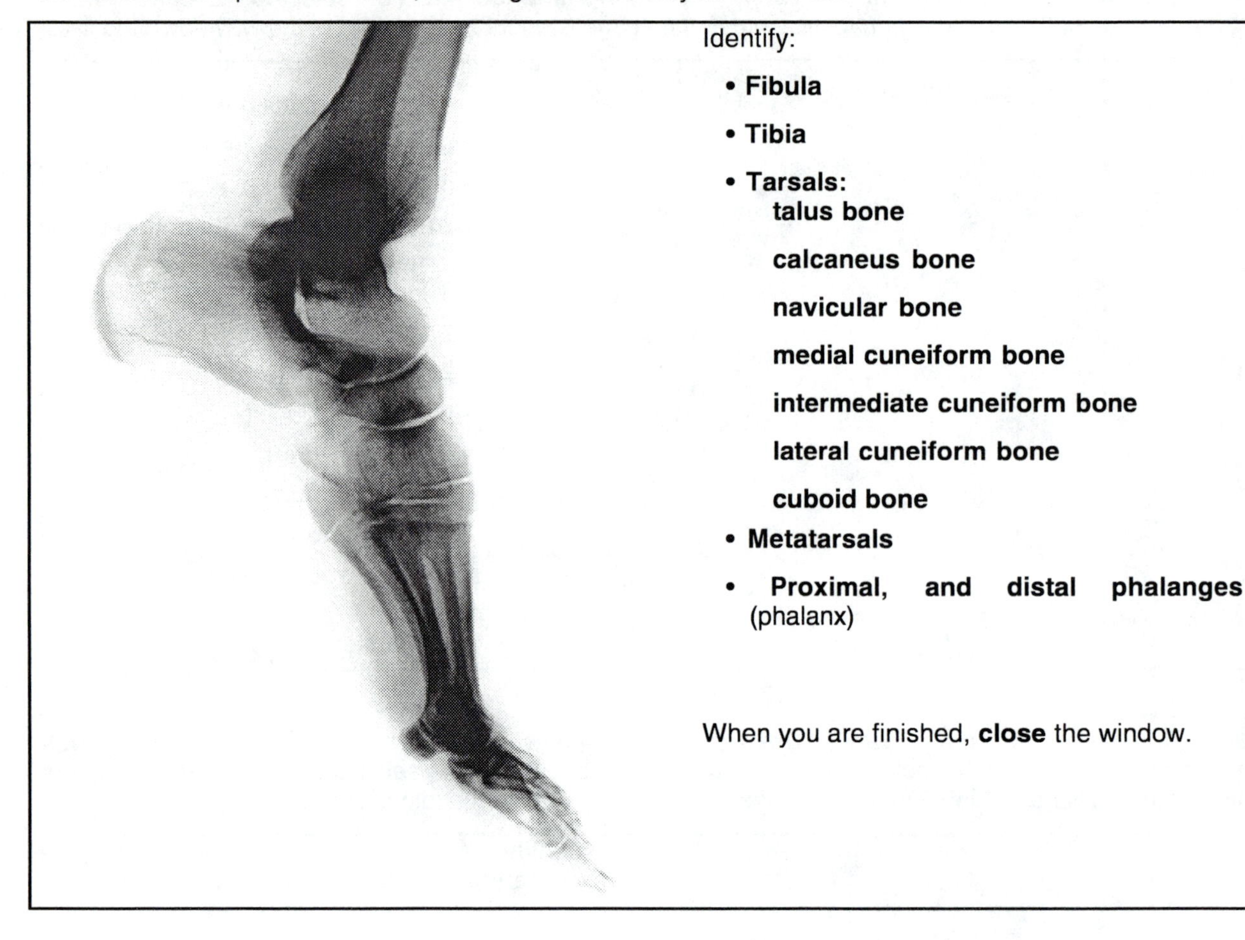

Identify:

- **Fibula**
- **Tibia**
- **Tarsals:**
  - **talus bone**
  - **calcaneus bone**
  - **navicular bone**
  - **medial cuneiform bone**
  - **intermediate cuneiform bone**
  - **lateral cuneiform bone**
  - **cuboid bone**
- **Metatarsals**
- **Proximal, and distal phalanges** (phalanx)

When you are finished, **close** the window.

## Objectives: Lower Appendicular Skeleton

Student will be able to identify the following the bones and bone markings of the lower appendicular skeleton:

**Bones**

- Coxal
    - Pubis
    - Ilium
    - Ischium
- Femur
- Patella
- Fibula
- Tibia
- Tarsals:
    - Talus bone
    - Calcaneus bone
    - Navicular bone
    - Medial cuneiform bone
    - Intermediate cuneiform bone
    - Lateral cuneiform bone
    - Cuboid bone
- Metatarsals
- Proximal, middle, and distal phalanges (phalanx)

**Other**

- Sacroiliac ligament
- Iliofemoral ligament
- Body of L5 vertebra
- Transverse process of L5 vertebra
- Intervertebral disc
- Sacrum
- Coccyx
- Synovial capsules (various)
- Articular cartilage
- Interosseous membrane of leg
- Ligaments (various)
- Bursa
- Tendon sheath
- Fibrous capsules (various)
- Medullary cavities (various)

**Bone Markings:**

On coxal bone:

- Iliac crest
- Pubic symphysis
- Anterior superior iliac spine
- Obturator foramen
- Acetabulum
- Body of pubis
- Posterior superior iliac spine
- Posterior inferior iliac spine
- Ischial tuberosity
- Ramus of ischium
- Superior ramus of pubis

On femur:

- Head
- Neck
- Greater trochanter
- Lesser trochanter
- Gluteal tuberosity
- Lateral condyle
- Medial condyle
- Lateral epicondyle
- Medial epicondyle

On fibula:

- Head
- Lateral malleolus

On Tibia:

- Tibial tuberosity
- Medial condyle
- Lateral condyle
- Medial malleolus

**Joints**

- Sacroiliac joint
- Hip (coxal) joint
- Knee (tibiofemoral) joint
- Femoropatellar joint
- Proximal tibiofibular joint
- Distal tibiofibular joint
- Ankle joint
- Intertarsal joints
- Tarsometatarsal joints
- Metatarsophalangeal joint
- Interphalangeal joints

# Knee Joint

## Opening A.D.A.M Interactive Anatomy

Open A.D.A.M. Interactive Anatomy according to the directions in the Tutorial.

## Anterior Dissection of Knee Joint

Choose **Dissectible Anatomy**.

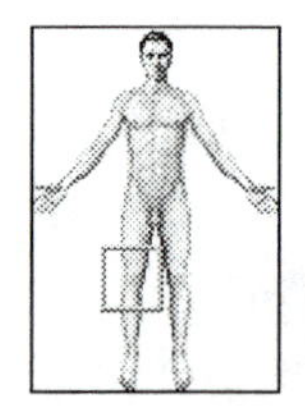

Choose either male or female. Select **Anterior.** Click **Open**.

Enlarge the window and position the navigation rectangle over the right knee.

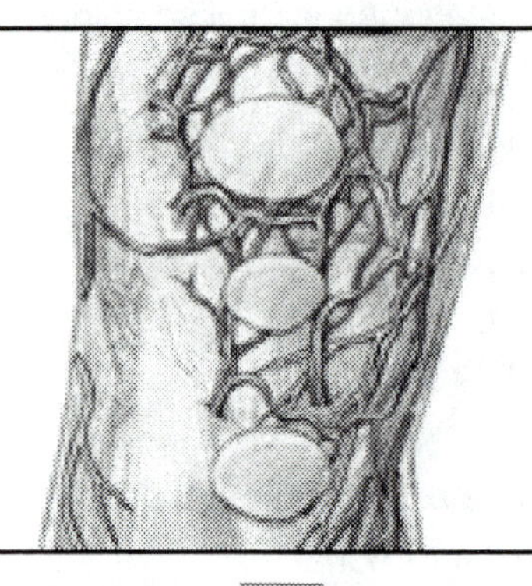

Click several times on the **Depth Bar** ▼ until you reach **layer 4**. Identify:

- **Bursa** (subcutaneous prepatellar bursa, subcutaneous infrapatellar bursa, subcutaneous bursa of tuberosity of tibia)

What is the function of a bursa?

______________________________

Click on the **Depth Bar** ▼ until you reach **layer 188**. No graphic has been provided for this layer. Identify :

- **Patellar ligament**

What is the name of the structure that connects the patella to the tibia?

______________________________

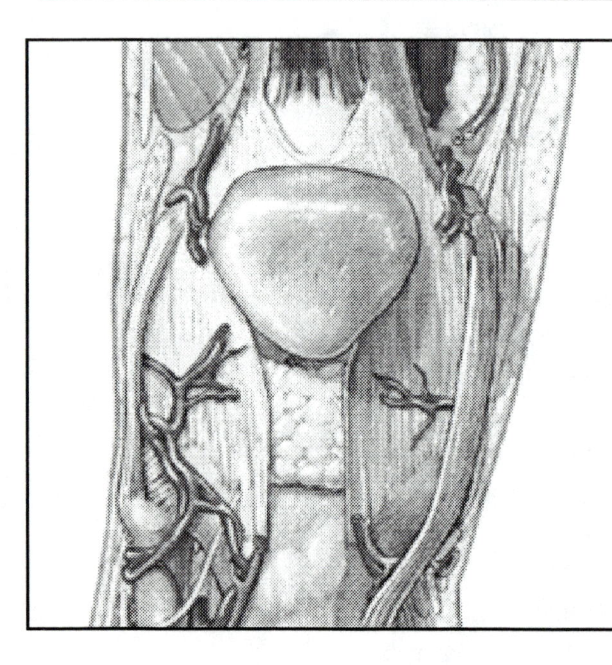

Click down to **layer 293**. Identify:

- **Patella**
- **Fat pad** (Infrapatellar)
- **Fibrous capsule of the knee joint**
- **Bursa** (suprapatellar bursa)
- **Ligaments** (various)
- **Femur**
- **Fibula**
- **Tibia**

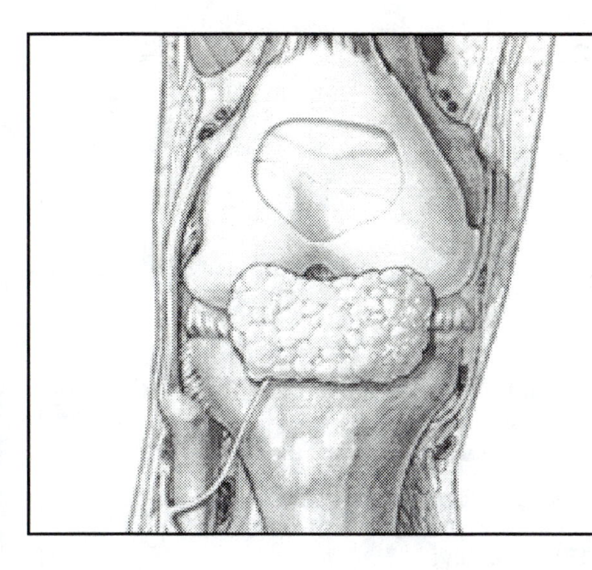

Click on the **Depth Bar** [▼] to remove the patella until you reach **layer 298**. Identify:

- **Synovial capsule of the knee joint**
- **Fat pad** (Infrapatellar)
- **Ligaments** (various)
- **Femur**
- **Fibula**
- **Tibia**
- **Interosseous membrane of the leg**
- **Articular cartilage** of the femoral condyles

What is the function of the interosseous membrane of the leg?

---

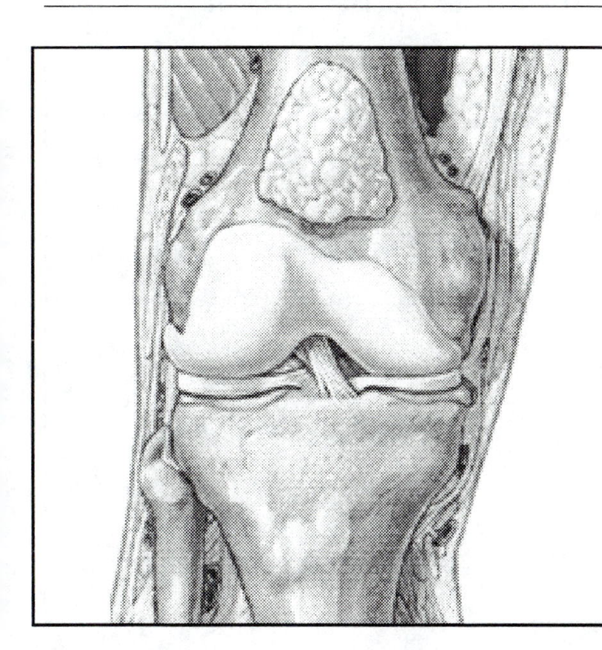

Click on the **Depth Bar** [▼] to **layer 321**. Identify:

- **Articular cartilage** of the femoral condyles, tibial condyles, and the tibiofibular joint.
- **Fat pad** (Suprapatellar)
- **Anterior cruciate ligament**
- **Medial meniscus**
- **Lateral meniscus**

What substance is found between the synovial membrane and the articular cartilage?

---

Click on the **Depth Bar** [▼] until you reach **layer 330**. No graphic has been provided for this layer. Identify:

- **Anterior cruciate ligament**
- **Lateral meniscus**
- **Medial meniscus**

## Posterior Dissection of the Knee Joint

Click on the **View** button and choose **Posterior**.

Windows — Mac

Click on the **Normal** button.

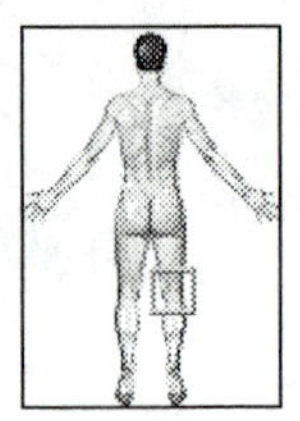

Drag the **Depth Bar** to **layer 0.**

Position the navigator rectangle over the posterior right knee.

Click on the **Depth Bar** or scroll until you reach **layer 160**. No graphic has been provided for this layer. Note the **fibrous capsule of the knee joint** and various **ligaments.** Click on the **Depth Bar** to **layer 169**. No graphic has been provided for this layer. Note the **synovial capsule of the knee joint.**

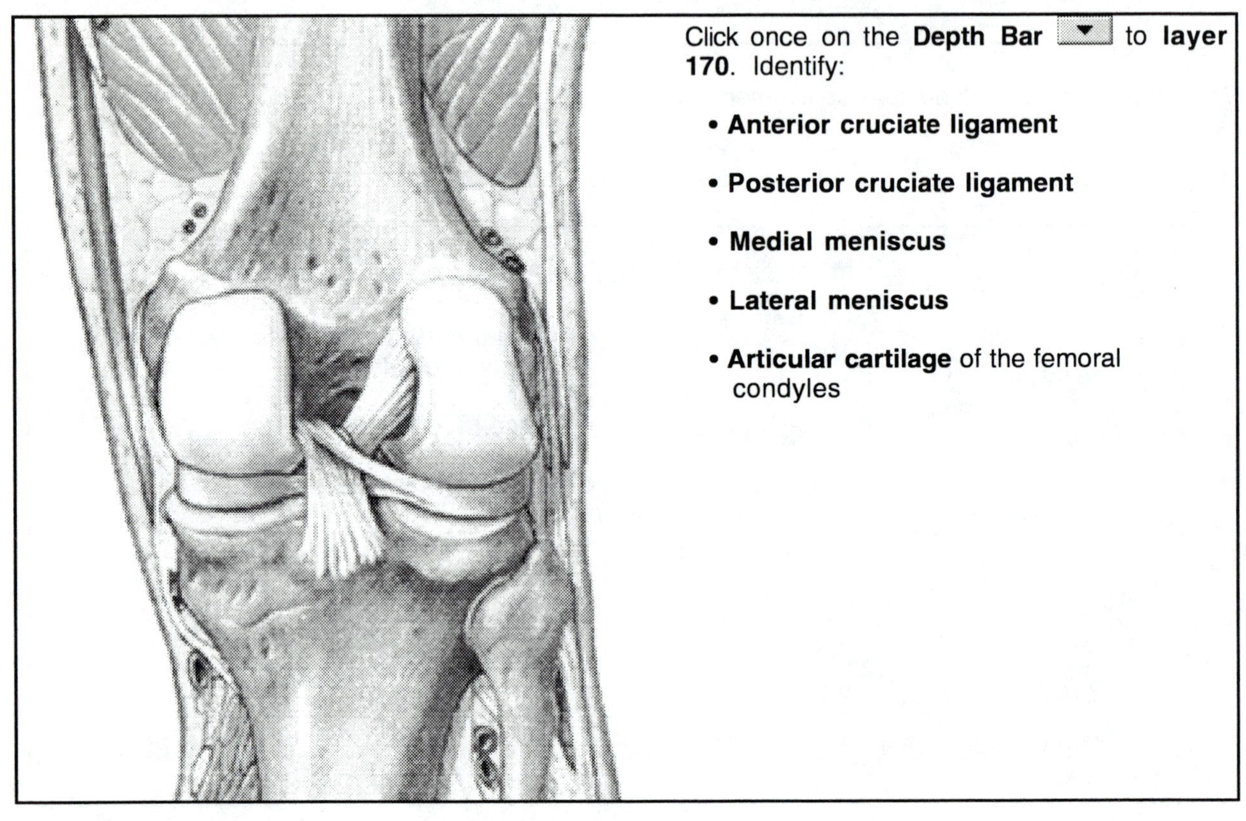

Click once on the **Depth Bar** to **layer 170**. Identify:

- **Anterior cruciate ligament**
- **Posterior cruciate ligament**
- **Medial meniscus**
- **Lateral meniscus**
- **Articular cartilage** of the femoral condyles

The lateral and medial meniscus are made of what type of tissue?

_______________________________________________

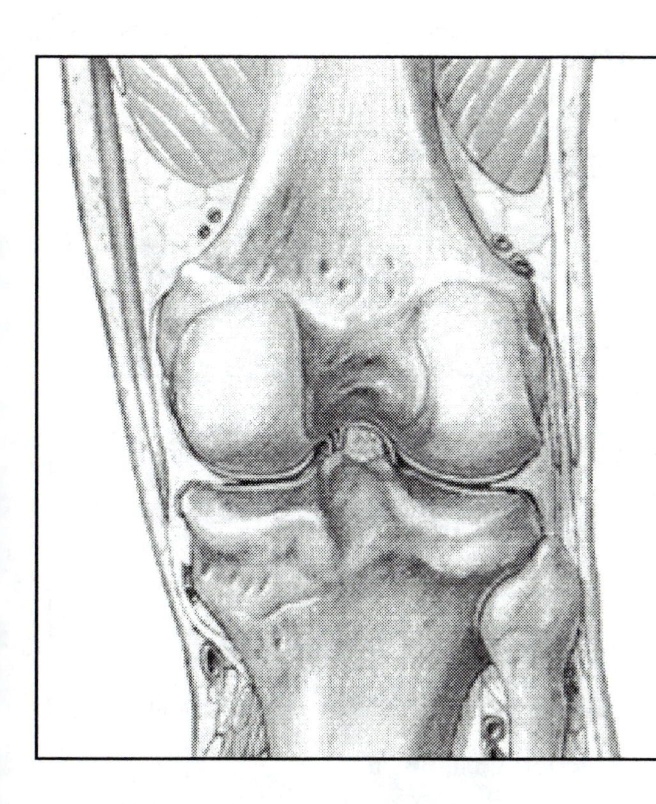

Click on the **Depth Bar** [▼] until you reach **layer 185**. Identify:

- **Anterior cruciate ligament**
- **Medial meniscus**
- **Lateral meniscus**
- **Articular cartilage** of the femoral condyles

What type of cartilage is articular cartilage?

___

## Lateral Dissection of the Knee Joint

Click on the **View** button and choose **Lateral**.

Click on the **Normal** button. Windows—Mac

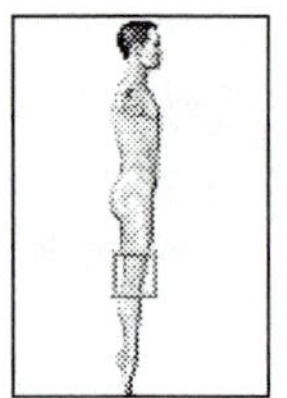

Drag the **Depth Bar** to **layer 0.**

Position the navigator rectangle over the right knee.

Click a few times on the **Depth Bar** until you reach **layer 2**. Identify:

- **Bursa** (subcutaneous prepatellar bursa, subcutaneous infrapatellar bursa, subcutaneous bursa of tuberosity of tibia)

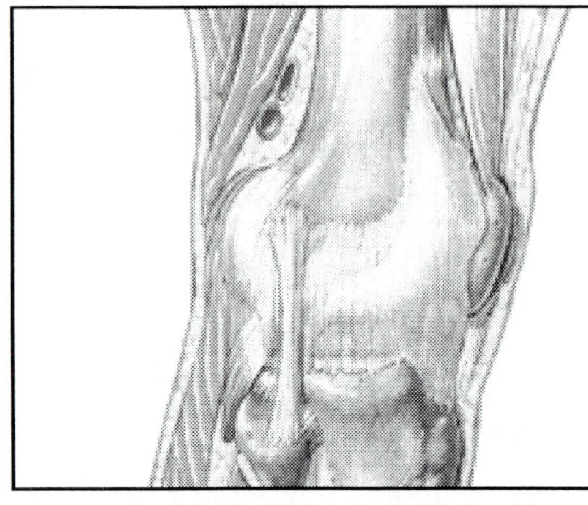

Scroll down on the **Depth Bar** to **layer 266**. Identify:

- **Ligaments** (various)
- **Fibrous capsule of the knee**
- **Patella**
- **Bursa** (subcutaneous prepatellar bursa)
- **Femur, tibia, fibula**
- **Interosseous membrane of the leg**

Click on the **Depth Bar** until you reach **layer 270**. No graphic has been provided for this layer. Identify the **synovial capsule of the knee joint** and the **infrapatellar fat pad**.

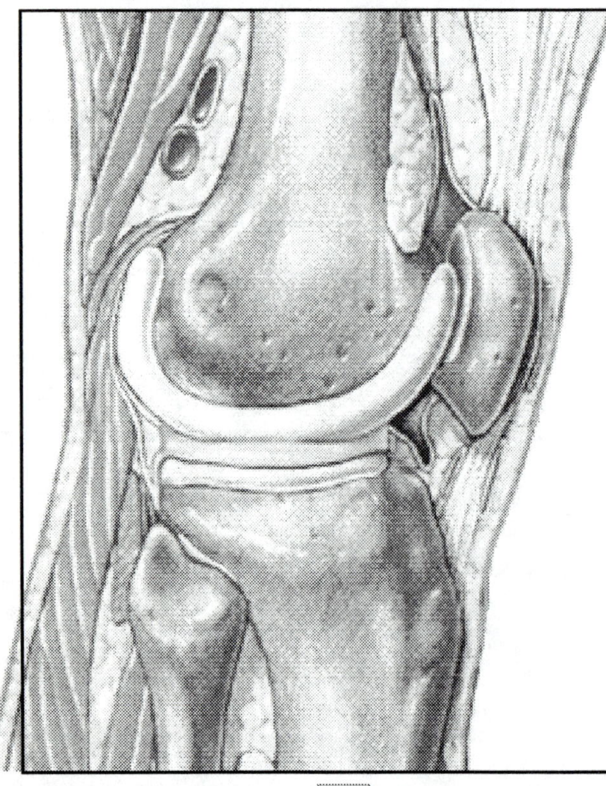

Then click on the **Depth Bar** to **layer 284**. Identify:

- **Patellar ligament**
- **Patella**
- **Bursa** (various)
- **Femur, tibia, fibula**
- **Articular cartilage of the femoral condyle**
- **Articular cartilage of the tibial condyle**
- **Articular cartilage of the tibiofibular joint**
- **Articular cartilage of the patella**
- **Lateral meniscus**
- **Synovial capsule of knee joint**
- **Fat pad** (infrapatellar, suprapatellar)

Click on the **Depth Bar** until you reach **layer 285** and identify the **medial meniscus.** No graphic has been provided for this layer.

When you are finished, **close** dissectible anatomy.

## Inferior and Superior Images of the Knee Joint

Under the **File** menu, drag down to **Open** (then to **Content** if you are working in Windows). Click on **Atlas Anatomy**. Select **Region** and **Lower Limb.** Choose **Articular Surface of Femur**. If the list is not in alphabetical order, the image is about half of the way down the list.

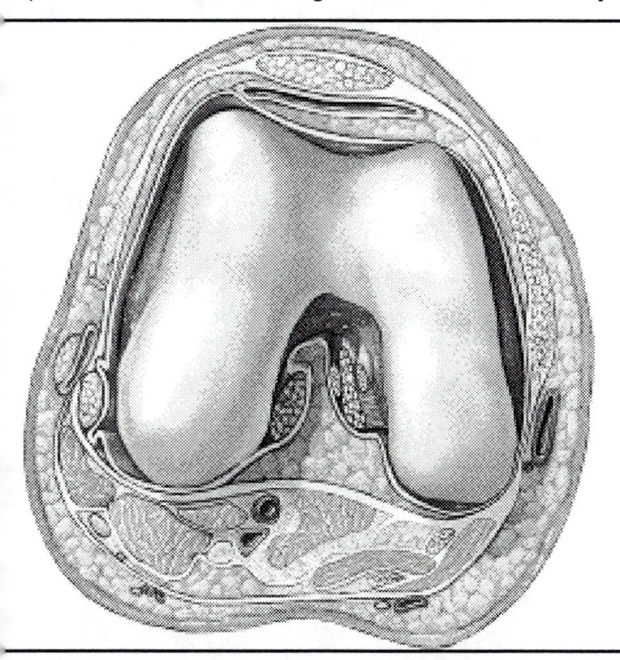

You are looking superiorly at the femur from the knee. Identify:

- **Patellar ligament**
- **Bursa** (including suprapatellar bursa)
- **Lateral condyle of femur**
- **Medial condyle of femur**
- **Synovial capsule of knee joint**
- **Anterior & posterior cruciate ligaments**
- **Ligaments** (various)
- **Muscles** (various)

When you are finished, **close** the window.

Under the **File** menu, drag down to **Open** (then to **Content** if you are working in Windows). **Atlas Anatomy** should still be selected. **Region** and **Lower Limb** should still be selected. Choose **Articular Surface of Tibia**. If the list is not in alphabetical order, the image is about half of the way down the list.

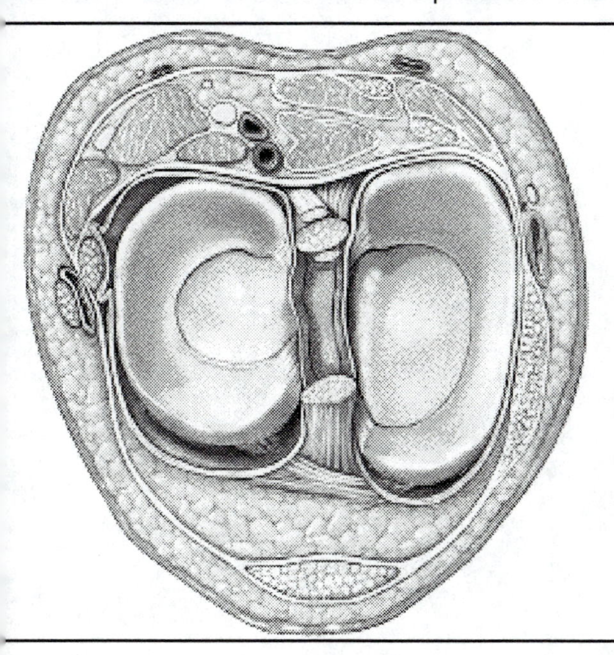

You are looking inferiorly at the tibia from the knee. Identify:

- **Patellar ligament**
- **Lateral meniscus**
- **Medial meniscus**
- **Lateral condyle of tibia**
- **Medial condyle of tibia**
- **Synovial capsule of knee joint**
- **Anterior cruciate ligaments**
- **Posterior cruciate ligaments**
- **Ligaments** (various)
- **Muscles** (various)

When you are finished, **close** the window.

## Objectives: Knee Joint

Student will be able to identify the following structures of the knee joint.

**Bones**

- Patella
- Femur
- Tibia
- Fibula

**Bone Markings**

- Lateral condyle of femur
- Medial condyle of femur

**Other**

- Bursa (various)
- Fat pad
- Fibrous capsule of the knee joint
- Synovial capsule of the knee joint
- Ligaments (various)
- Interosseous membrane of the leg
- Articular cartilage
- Anterior cruciate ligament
- Posterior cruciate ligament
- Patellar ligament
- Medial meniscus
- Lateral meniscus

# Face & Neck Muscles

## Opening A.D.A.M Interactive Anatomy

Open A.D.A.M. Interactive Anatomy according to the directions in the Tutorial.

## Anterior Dissection of Face & Neck Muscles

Choose **Dissectible Anatomy**.

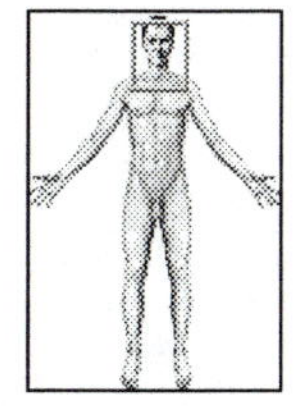

Choose either male or female. Select **Anterior.** Click **Open**.

Enlarge the window and position the image over the face & neck.

The muscles of the head and neck can be organized into several groups:

1. **Muscles of facial expression**
2. **Muscles of mastication**
3. **Muscles of the tongue**
4. **Muscles that move the head**
5. **Muscles of speech**

Click once on the **Depth Bar** ▼ to remove the skin. You are now at **layer 1**. Identify:

- **Fat** (superficial fascia).

Now click on the **Depth Bar** ▼ to remove the fat (you should be on **layer 5**). No graphic has been provided for this layer. Identify:

- **Veins** (blue)
- **Arteries** (red)
- **Nerves** (yellow structures)
- **Fascia** (translucent white structures over the muscle)

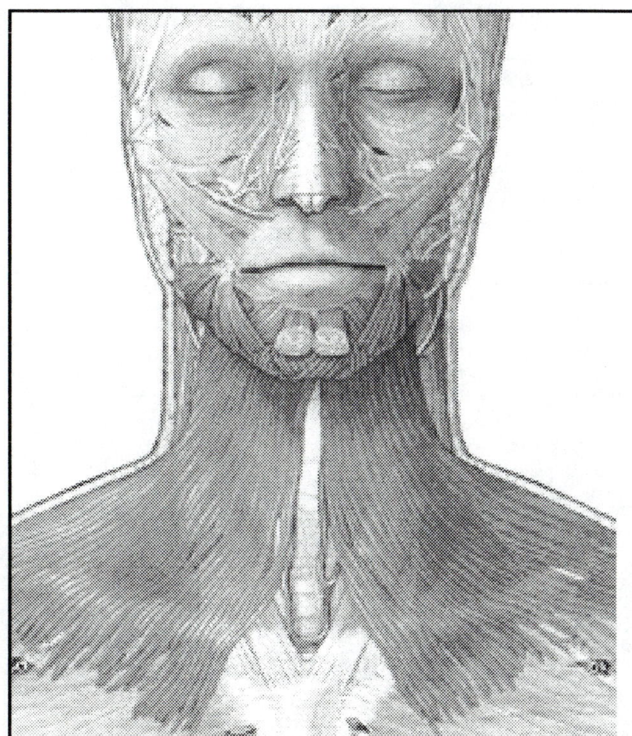

Click on the **Depth Bar** ▼ until you reach **layer 8**. Identify the following muscles of facial expression:

- **Platysma muscle**
- **Depressor anguli oris muscle**
- **Mentalis muscle**
- **Depressor labii inferioris muscle**

What is the origin and insertion of the platysma muscle?

________________________________________

________________________________________

Based on the origin and insertion of the platysma muscle, what would its action be?

________________________________________

Which of the muscles listed on the previous page is the "frowning" muscle?

---

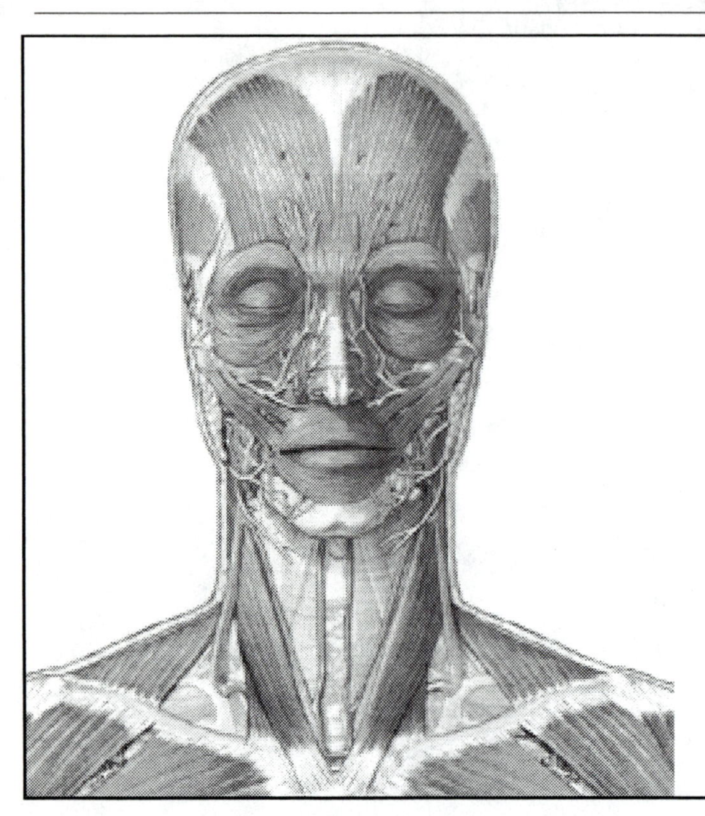

Click on the **Depth Bar** ▼ until you reach **layer 11**. Identify the following muscles of facial expression:

- **Frontalis muscle**
- **Procerus muscle**
- **Epicranial aponeurosis**
- **Orbicularis oculi muscle**
- **Nasalis muscle**
- **Zygomaticus minor muscle**
- **Zygomaticus major muscle**
- **Orbicularis oris muscle**
- **Levator labii superioris muscle**

Identify these muscles that move the head:

- **Sternocleidomastoid muscle** (sternal and clavicular head)
- **Trapezius muscle**

What is the origin and insertion of the zygomaticus major and the zygomaticus minor muscles?

---

Which of the muscles listed above are the "smiling" muscles?

---

What is the "winking" muscle?

---

What is the "kissing" muscle?

---

What muscle wrinkles the forehead?

---

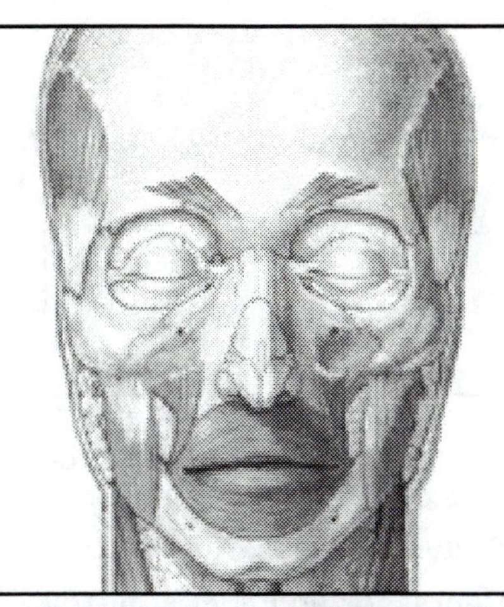

Click on the **Depth Bar** [▼] until you reach **layer 34**. Identify these muscles of mastication:

- **Temporalis muscle**
- **Masseter muscle**

Identify this muscle that moves the head:

- **Sternocleidomastoid muscle** (sternal and clavicular head)

Identify this muscle of facial expression:

- **Orbicularis oris muscle**

Click on the **Depth Bar** [▼] until you reach **layer 53**. Identify the following muscles of speech:

- **Sternohyoid muscle**
- **Omohyoid muscle** (inferior and superior bellies)

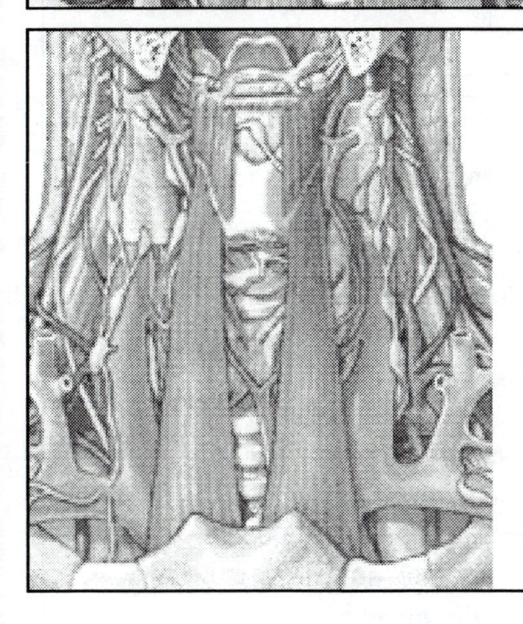

Click on the **Depth Bar** [▼] to **layer 62**. Identify the following muscles of speech:

- **Thyrohyoid muscle**
- **Sternothyroid muscle**

What is the origin and insertion for the thyrohyoid muscle?

________________________________________

________________________________________

## Lateral Dissection of Face and Neck Muscles

Click on the **View** button and choose **Lateral**.

Click on the **Normal** button. Windows— —Mac

Drag the **Depth Bar** to **layer 0.**

Position the navigator rectangle over the face & neck.

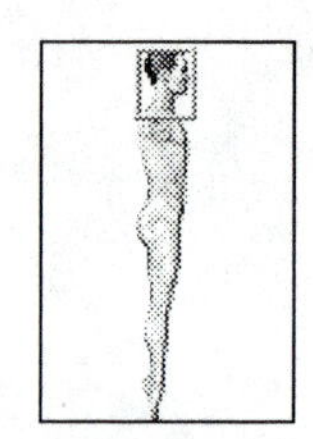

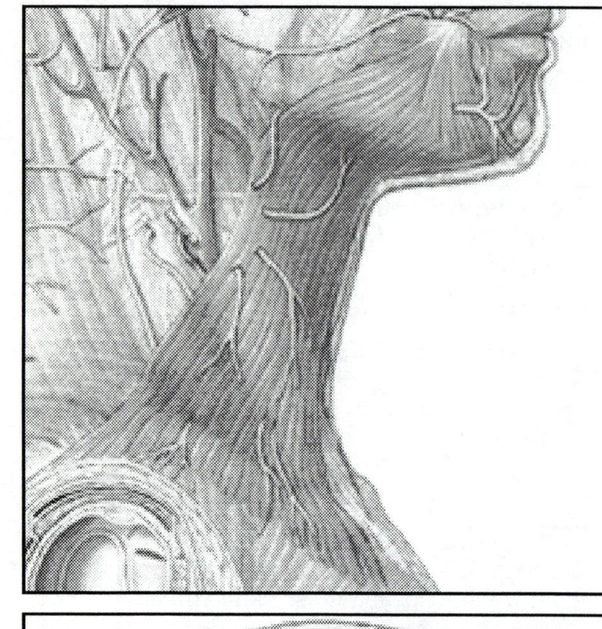

Click a few times on the **Depth Bar** until you reach **layer 5**. Identify the following muscles of facial expression:

- **Platysma muscle**
- **Depressor anguli oris muscle**
- **Depressor labii inferioris muscle**

These muscles of facial expression are innervated by what nerves?

______________________________

______________________________

These are branches of the Facial Nerve (Cranial Nerve VII).

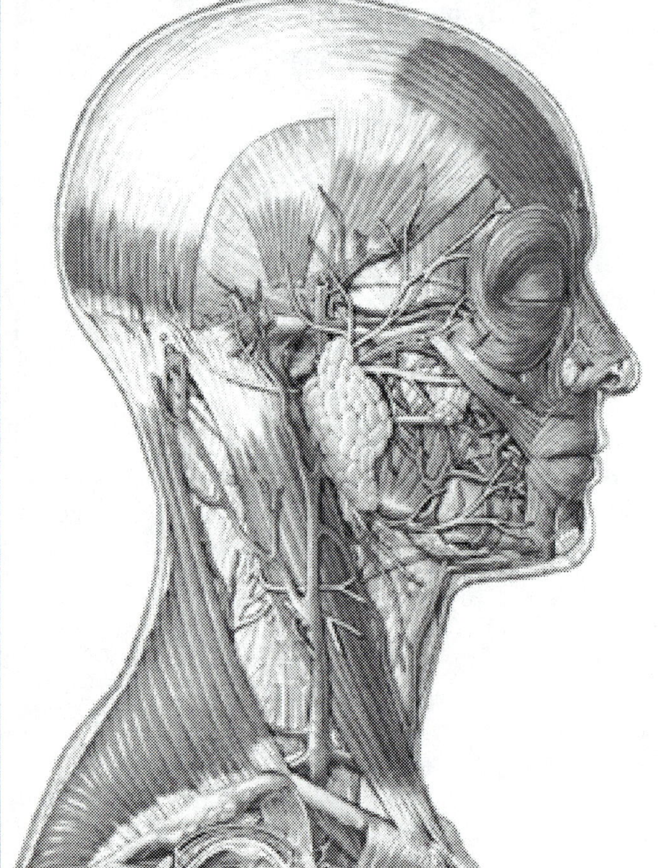

Click on the **Depth Bar** until you reach **layer 10**. Identify the following muscles of facial expression:

- **Frontalis muscle**
- **Occipitalis muscle**
- **Epicranial aponeurosis**
- **Orbicularis oculi muscle**
- **Zygomaticus major muscle**
- **Orbicularis oris muscle**
- **Procerus muscle**
- **Nasalis muscle**
- **Zygomaticus minor muscle**
- **Levator labii superioris muscle**
- **Mentalis muscle**
- **Depressor labii inferioris muscle**

Identify these muscles that move the head:

- **Sternocleidomastoid muscle** (sternal and clavicular head)
- **Trapezius muscle**

Identify this muscle of mastication:

- **Masseter muscle**

Name the two muscles which make up the epicranius of the scalp?

---

What is the origin of the sternocleidomastoid muscle?

---

What is the insertion of the sternocleidomastoid muscle?

---

Based on the origin and insertion of the sternocleidomastoid muscle, what would its action be?

---

What is the origin of the trapezius muscle?

---

What is the insertion of the trapezius muscle?

---

Based on the origin and insertion of the trapezius muscle, what would its action be?

---

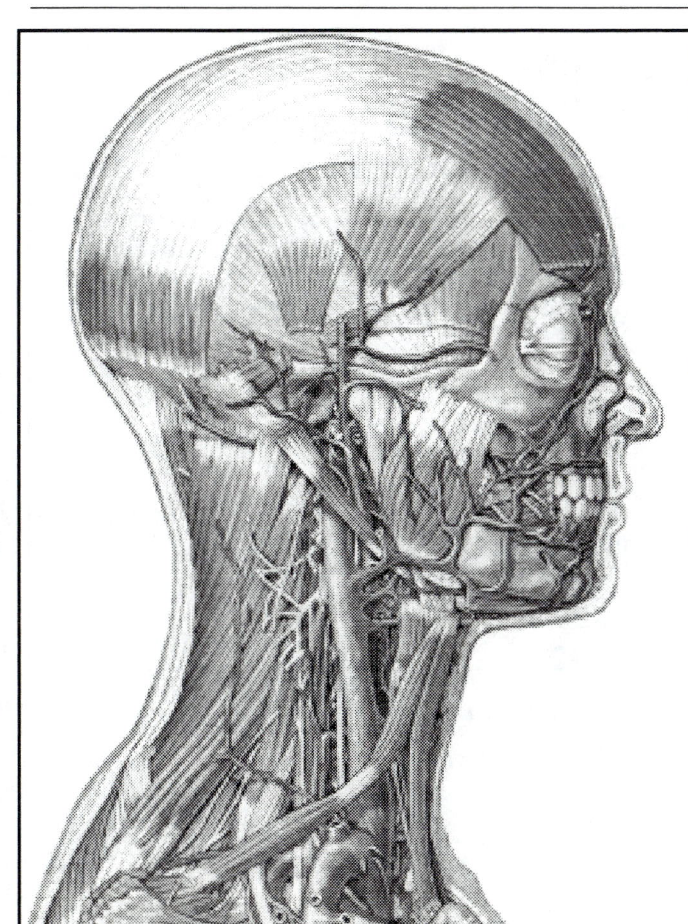

Click several times on the **Depth Bar** ▼ until you reach **layer 42**. Identify the following muscles of facial expression:

- **Frontalis muscle**
- **Epicranial aponeurosis**
- **Occipitalis muscle**
- **Temporoparietalis muscle**
- **Buccinator muscle**

Identify these muscles of mastication:

- **Masseter muscle**
- **Anterior belly of digastric muscle**
- **Posterior belly of digastric muscle**

Identify the following muscles of speech:

- **Sternohyoid muscle**
- **Omohyoid muscle**

What are the bones of origin and insertion of the sternohyoid muscle?

---

Based on the origin and insertion of the sternohyoid muscle, what would its action be?

---

What is one origin of the digastric muscle?

---

If you want to "wiggle" your ears, you must move what muscle?

---

Click on the **Depth Bar** until you reach **layer 43**. Identify the following muscles of speech:

- **Thyrohyoid muscle**
- **Sternothyroid muscle**

Scroll using the **Depth Bar** until you reach **layer 50**. Identify this muscle of mastication:

- **Masseter muscle**

What is the origin and insertion of the masseter muscle?

---

---

Based on the origin and insertion of the masseter muscle, what would its action be?

---

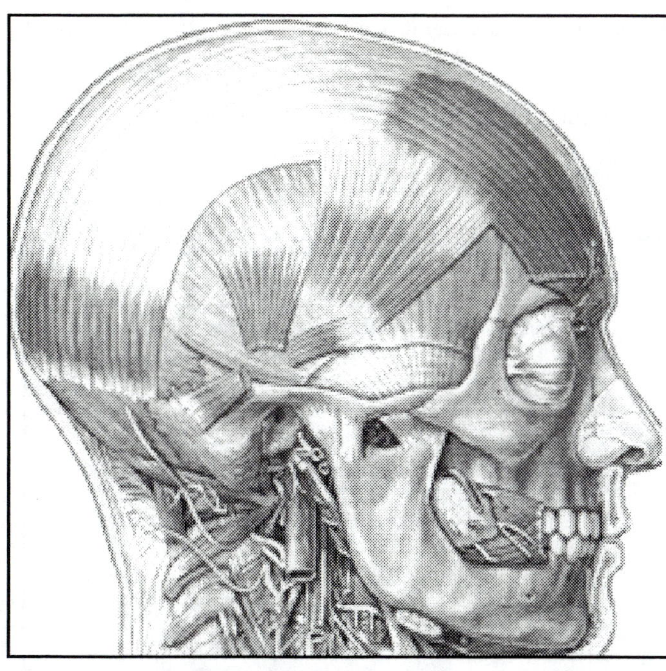

Click on the **Depth Bar** ▼ until you reach **layer 110**. Identify these muscles of facial expression:

- **Buccinator muscle**
- **Frontalis muscle**
- **Temporoparietalis muscle**
- **Occipitalis muscle**

Which muscle above would be well-developed in the nursing infant?

____________________________________

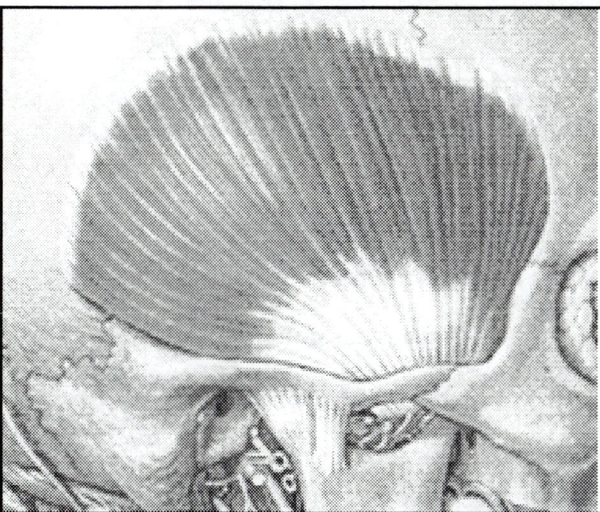

Click a few times on the **Depth Bar** ▼ until you reach **layer 115**. Identify this muscle of mastication:

- **Temporalis muscle**

What is the origin and insertion of the temporalis muscle?

____________________________________

____________________________________

Based on the origin and insertion of the temporalis muscle, what would its action be?

____________________________________

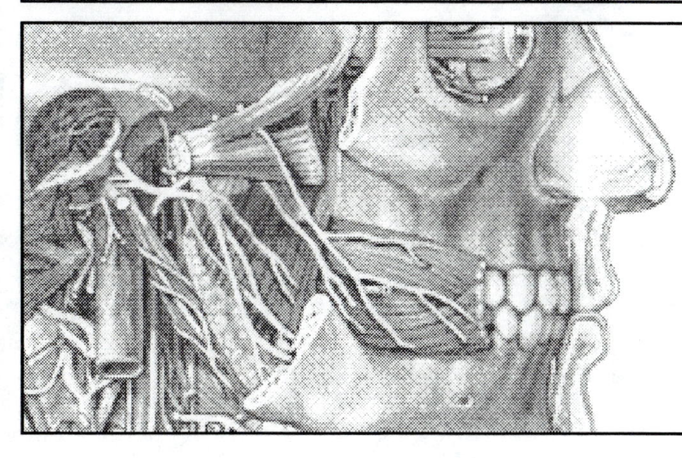

Click on the **Depth Bar** ▼ until you reach **layer 122**. Identify these muscles of mastication:

- **Medial pterygoid muscle**
- **Lateral pterygoid muscle**

The muscles of the tongue all end in the suffix *-glossus*. They are easy to remember if you remember their prefixes:

- The **palatoglossus muscle** originates at the palate.
- The **styloglossus muscle** originates at the styloid process.
- The **genioglossus muscle** originates at the chin.
- The **hyoglossus muscle** originates at the hyoid bone.

The muscles that move the tongue in the delicate and complex patterns necessary for speech also manipulate food within the mouth in preparation for swallowing.

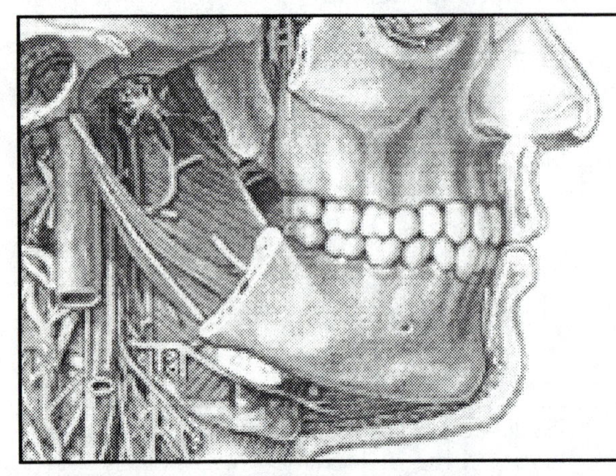

Click a few times on the **Depth Bar** ▼ until you reach **layer 128**. Identify this muscle that moves the tongue:

- **Styloglossus muscle**

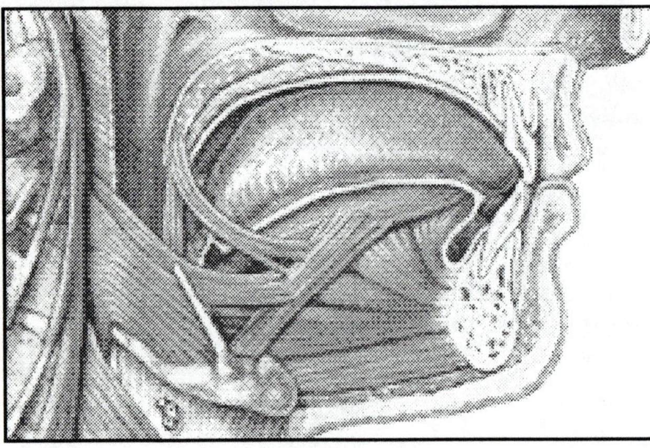

Click several times on the **Depth Bar** ▼ until you reach **layer 237**. Identify this muscle that moves the tongue:

- **Genioglossus muscle**
- **Hyoglossus muscle** (check layer 234)
- **Palatoglossus muscle**

Give two functions of the glossus muscle group.

________________________________________

________________________________________

## Posterior Dissection of Neck Muscles

Click on the **View** button and choose **Posterior**.

Click on the **Normal** button. Windows— —Mac

Drag the **Depth Bar** to **layer 0.**

Position the navigator rectangle over the neck.

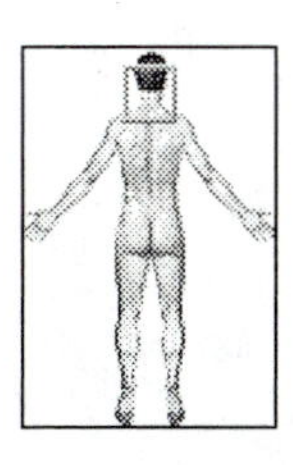

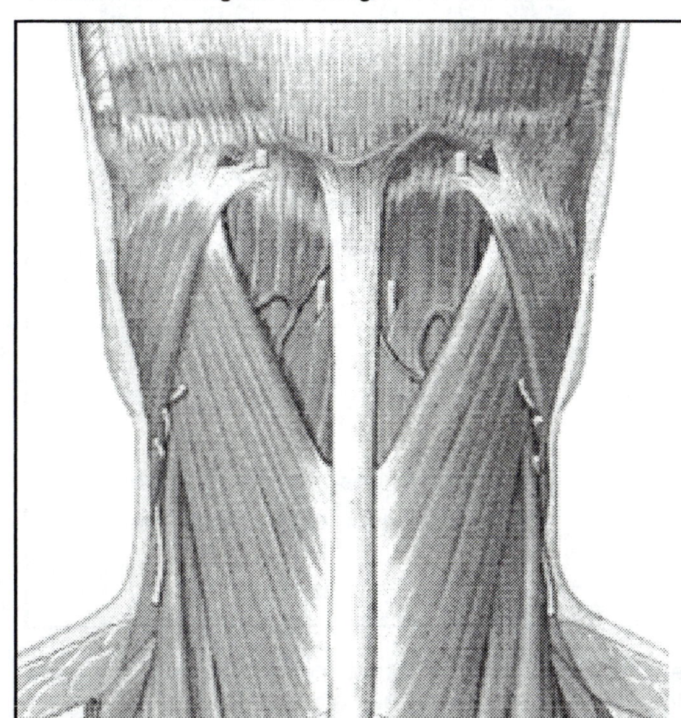

Click a few times on the **Depth Bar** until you reach **layer 12**. Identify the following muscles that move the head:

- **Occipitalis muscle**
- **Sternocleidomastoid muscle**
- **Semispinalis capitis muscle**
- **Splenius capitis muscle**
- **Epicranial aponeurosis**

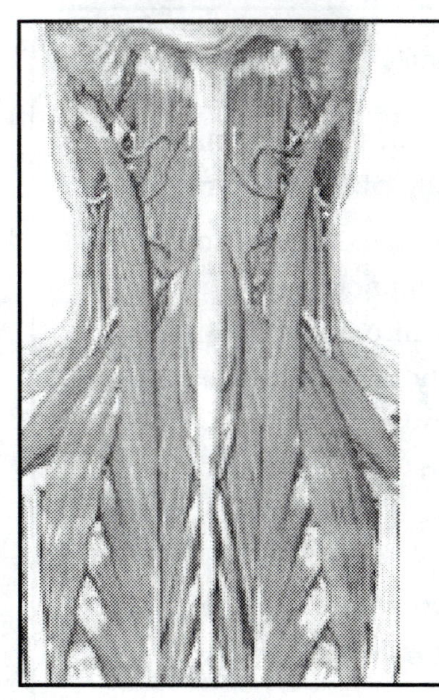

Dissect down to **layer 76**. Identify this muscle that moves the head:

- **Longissimus capitis muscle**
- **Semispinalis capitis muscle**

Based on the origin and insertion of these two muscles, what would their action be?

_______________________________________________

When you are done, close the window.

## Atlas Anatomy of Muscles of the Face and Neck

Under the **File** menu, drag down to **Open** (then to **Content** if you are working in Windows). Click on **Atlas Anatomy**. Select **Region** and **Head & Neck.** Choose **Hyoid Muscles (Lat)**. If the list is not in alphabetical order, the image is close to the bottom of the list.

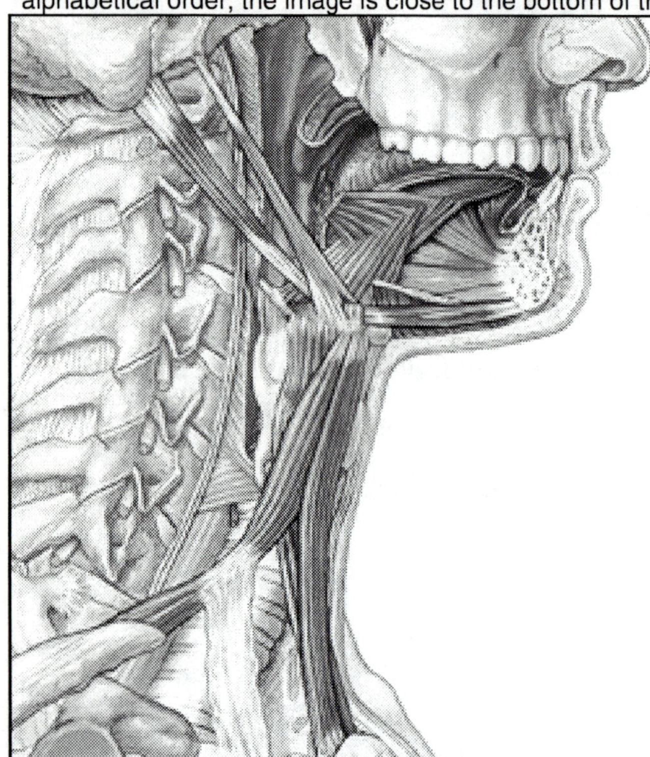

You may need to scroll to find the structures listed below. Identify the following muscles of mastication:

- **Anterior belly of digastric muscle**
- **Posterior belly of digastric muscle**

Identify the following muscles of speech:

- **Sternohyoid muscle**
- **Inferior belly of omohyoid muscle**
- **Superior belly of omohyoid muscle**
- **Thyrohyoid muscle**

Identify the following muscles of the tongue:

- **Genioglossus muscle**
- **Styloglossus**
- **Hyoglossus**

Identify these bones & bone markings:

- **Hyoid bone**
- **Mastoid process of temporal bone**
- **Styloid process of temporal bone**

When you are finished, **close** the window.

Under the **File** menu, drag down to **Open** (then to **Content** if you are working in Windows). **Atlas Anatomy** should still be selected. **Region** and **Head & Neck** should still be selected. Choose **Triangles of Neck (Ant)**. If the list is not in alphabetical order, the image is close to the bottom.

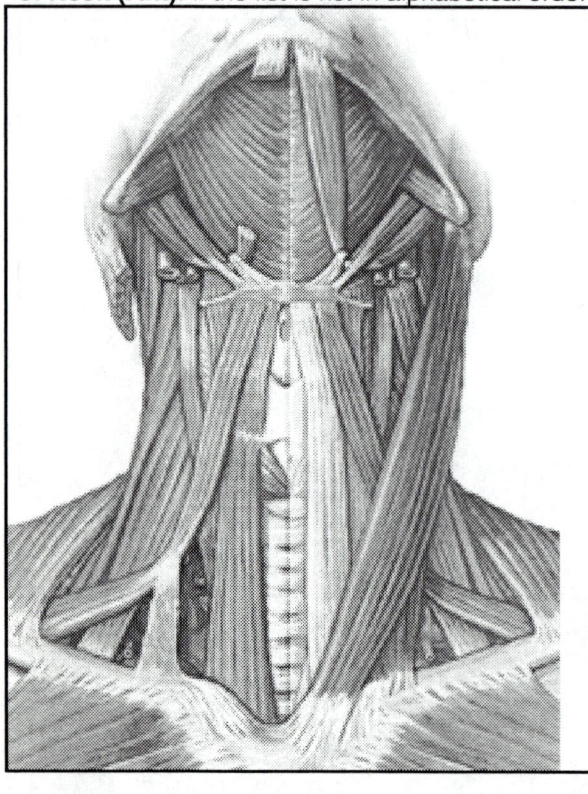

You may need to scroll to find the features listed below. Identify the following muscles of mastication:

- **Anterior belly of digastric muscle**
- **Posterior belly of digastric muscle**

Identify the following muscles of speech:

- **Sternohyoid muscle**
- **Inferior belly of omohyoid muscle**
- **Superior belly of omohyoid muscle**
- **Thyrohyoid muscle**
- **Sternothyroid**

Identify the following muscles that move the head and neck:

- **Sternocleidomastoid muscle**
- **Trapezius muscle**

Identify these bones:

- **Hyoid bone**
- **Manubrium**

When you are finished, **close** the window.

## Cadaver Images

Under the **File** menu, drag down to **Open** (then to **Content** if you are working in Windows). **Atlas Anatomy** should still be selected. **Region** and **Head & Neck** should still be selected. Choose **Muscles of Facial Expression**. If the list is not in alphabetical order, the image is half way down the list.

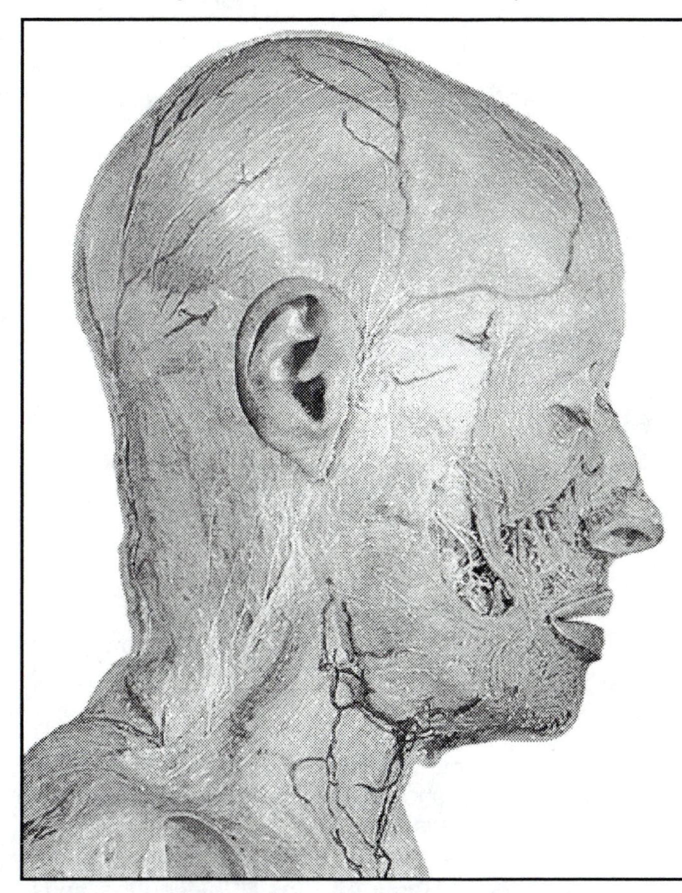

Click on the **Show All Pins** button and scroll to **Muscular**. Identify:

- **Frontalis muscle** (frontal belly of epicranius muscle)
- **Occipitalis muscle** (occipital belly of epicranius muscle)
- **Epicranial aponeurosis**
- **Orbicularis oculi muscle**
- **Major zygomatic muscle**
- **Orbicularis oris muscle**
- **Trapezius muscle**
- **Platysma muscle**
- **Nasalis muscle**

When you are finished, **close** the window.

Under the **File** menu, drag down to **Open** (then to **Content** if you are working in Windows). **Atlas Anatomy** should still be selected. **Region** and **Head & Neck** should still be selected. Choose **Dissection of Facial Nerve**. If the list is not in alphabetical order, the image is half way down the list.

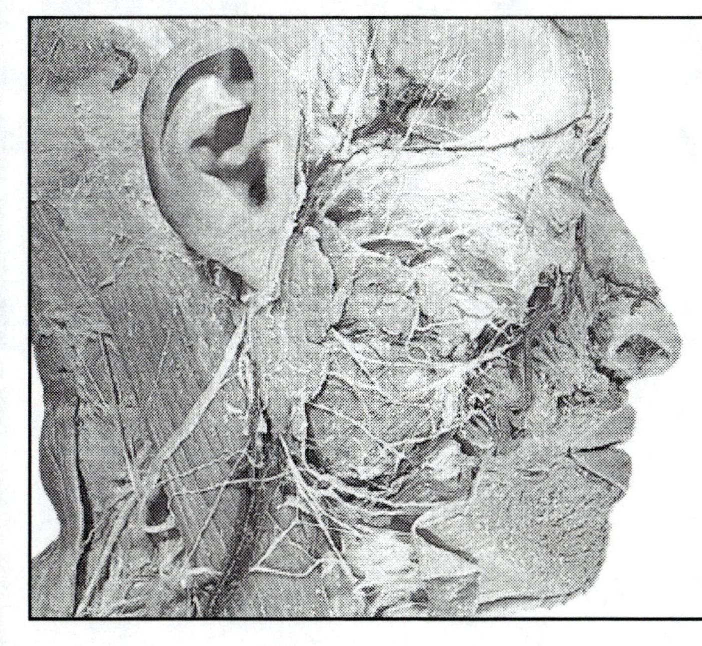

Click on the **Show All Pins** button and scroll to **Muscular**. Identify:

- **Masseter muscle**
- **Occipitalis muscle** (occipital belly of epicranius muscle)
- **Orbicularis oculi muscle**
- **Orbicularis oris muscle**
- **Trapezius muscle**
- **Sternocleidomastoid muscle**
- **Platysma muscle**

When you are finished, **close** the window.

Under the **File** menu, drag down to **Open** (then to **Content** if you are working in Windows). **Atlas Anatomy** should still be selected. **Region** and **Head & Neck** should still be selected. Choose **Facial Nerve (Lat) 2**. If the list is not in alphabetical order, the image is half way down the list.

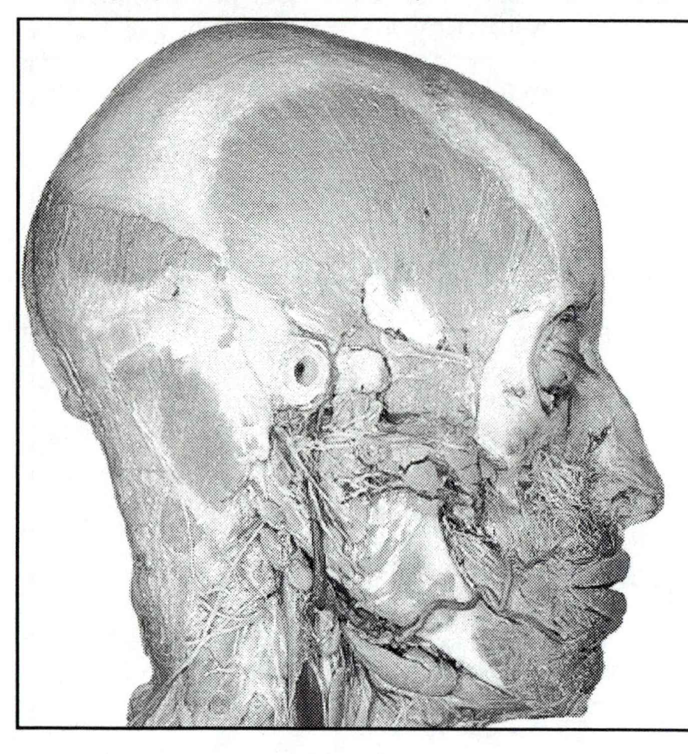

Click on the **Show All Pins** button and scroll to **Muscular**. Identify:

- **Anterior belly of digastric muscle**
- **Masseter muscle**
- **Occipitalis muscle** (occipital belly of epicranius muscle)
- **Splenius capitis muscle**
- **Temporalis muscle**

When you are finished, **close** the window.

Under the **File** menu, drag down to **Open** (then to **Content** if you are working in Windows). **Atlas Anatomy** should still be selected. **Region** and **Head & Neck** should still be selected. Choose **Infratemporal Fossa 2**. If the list is not in alphabetical order, the image is a little over halfway down.

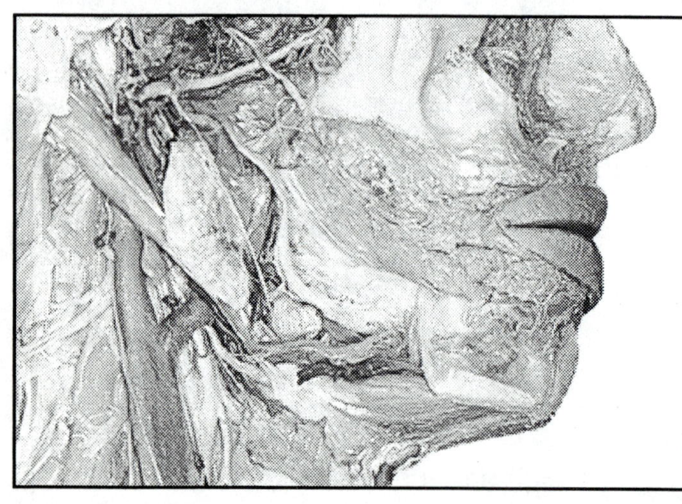

Click on the **Show All Pins** button and scroll to **Muscular**. Identify:

- **Anterior belly of digastric muscle**
- **Posterior belly of digastric muscle**

When you are finished, **close** the window.

Under the **File** menu, drag down to **Open** (then to **Content** if you are working in Windows). **Atlas Anatomy** should still be selected. **Region** and **Head & Neck** should still be selected. Choose **Orbicularis Oculi Muscle (Ant)**. If the list is not in alphabetical order, the image is a little over half way down the list.

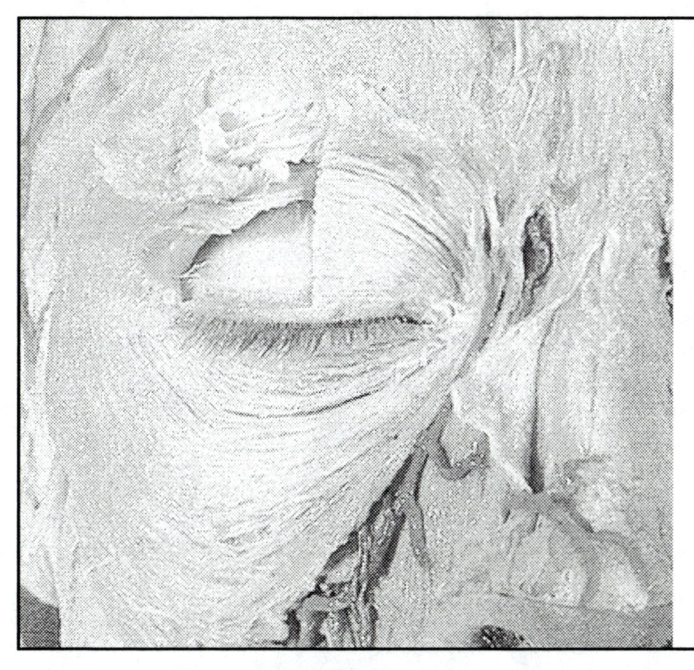

Click on the **Show All Pins** button and scroll to **Muscular**. Identify:

- **Orbicularis oculi muscle**
- **Nasalis muscle**

When you are finished, **close** the window.

Under the **File** menu, drag down to **Open** (then to **Content** if you are working in Windows). **Atlas Anatomy** should still be selected. **Region** and **Head & Neck** should still be selected. Choose **Superf. Muscles of Neck (Ant)**. If the list is not in alphabetical order, the image is near the bottom.

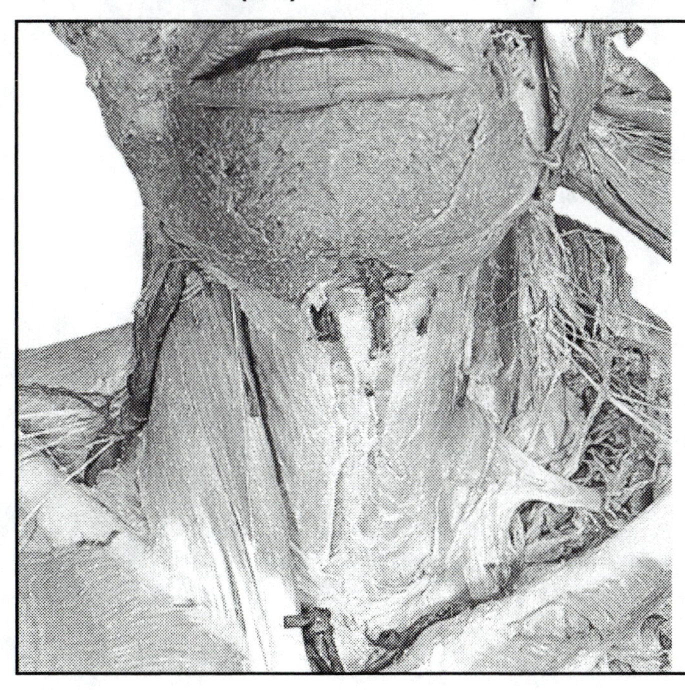

Click on the **Show All Pins** button and scroll to **Muscular**. Identify:

- **Sternocleidomastoid muscle** (Clavicular & Sternal head)
- **Masseter muscle**
- **Platysma muscle**
- **Sternohyoid muscle**
- **Superior belly of omohyoid muscle**
- **Trapezius muscle**

When you are finished, **close** the window.

# Objectives: Face & Neck Muscles

Student will identify the important muscles of the head and neck.

**Muscles of Facial Expression:**

- Platysma muscle
- Depressor anguli oris muscle
- Mentalis muscle
- Depressor labii inferioris muscle
- Frontalis muscle
- Procerus muscle
- Orbicularis oculi muscle
- Nasalis muscle
- Zygomaticus minor muscle
- Zygomaticus major muscle
- Orbicularis oris muscle
- Levator labii superioris muscle
- Occipitalis muscle
- Temporoparietalis muscle
- Buccinator muscle

**Muscles of Mastication:**

- Temporalis muscle
- Masseter muscle
- Anterior belly of digastric muscle
- Posterior belly of digastric muscle
- Medial pterygoid muscle
- Lateral pterygoid muscle

**Muscles of Tongue:**

- Styloglossus muscle
- Genioglossus muscle
- Hyoglossus muscle
- Palatoglossus muscle

**Muscles that Move the Head:**

- Sternocleidomastoid muscle
- Trapezius muscle
- Semispinalis capitis muscle
- Splenius capitis muscle
- Longissimus capitis muscle

**Muscles of Speech:**

- Sternohyoid muscle
- Omohyoid muscle
- Thyrohyoid muscle
- Sternothyroid muscle

**Other**

- Epicranial aponeurosis
- Hyoid bone
- Mastoid process of temporal bone
- Styloid process of temporal bone
- Manubrium

# Chest and Back Muscles

## Opening A.D.A.M Interactive Anatomy

Open A.D.A.M. Interactive Anatomy according to the directions in the Tutorial.

## Anterior Dissection of Chest Muscles

Choose **Dissectible Anatomy**.

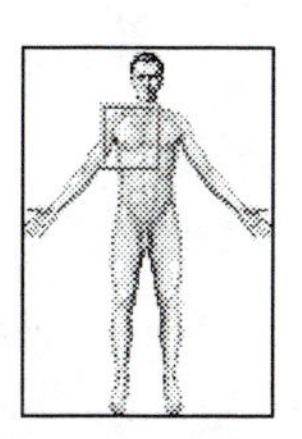

Choose **Male**. Select **Anterior.** Click **Open**.

Enlarge the window and position the navigation rectangle over the right shoulder.

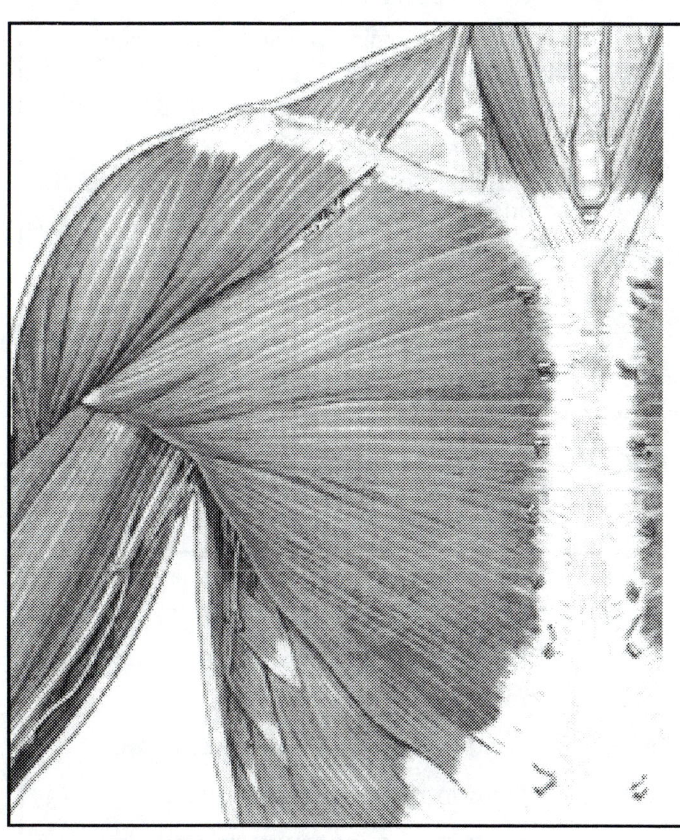

Click on the **Depth Bar** [▼] to remove skin, fat, blood vessels, and muscles until you reach **layer 19**. Identify:

- **Pectoralis major muscle** (all heads)
- **Deltoid muscle**
- **Trapezius muscle**
- **Serratus anterior muscle**

The deltoid muscle originates anteriorly on what bones?

______________________________________________

The pectoralis major originates on what bone?

______________________________________________

Click down to **level 20**. The pectoralis major inserts into what bone?

______________________________________________

Based on the origin and insertion of the pectoralis major muscle, what is its action?

______________________________________________

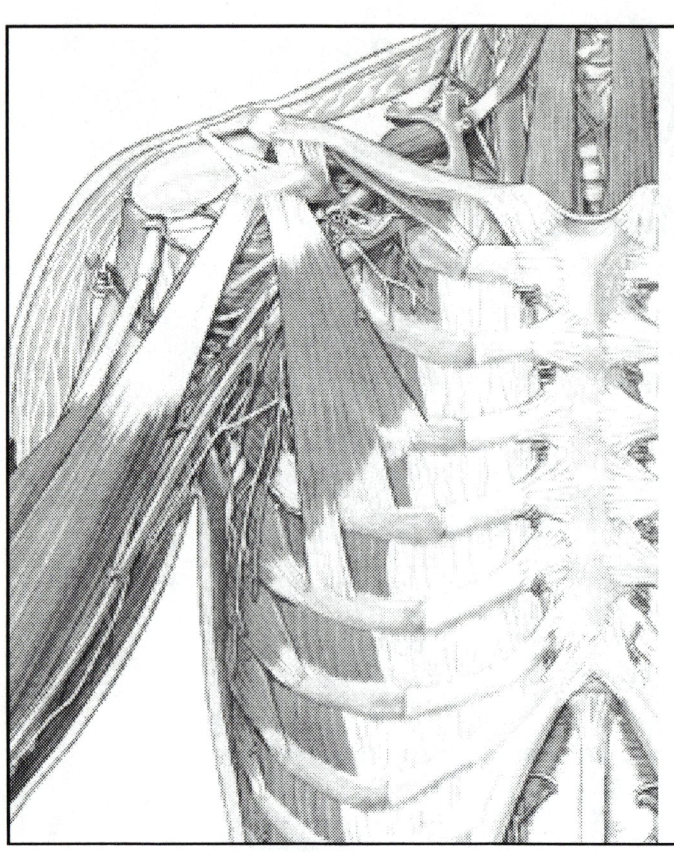

Click on the **Depth Bar** [▼] until you reach **layer 53**. Identify:

- **Pectoralis minor muscle** (all heads)
- **Serratus anterior muscle**
- **External intercostal muscle**
- **Subclavius muscle**

The origins of the pectoralis minor muscle are located on which bones?

______________________________

The pectoralis minor muscle inserts on what bone?

______________________________

Based on its origin and insertion, what is the action of the pectoralis minor muscle?

______________________________

Click on the **Depth Bar** [▼] until you reach **layer 87**. No graphic has been provided for this layer. Identify :

- **Coracobrachialis muscle**

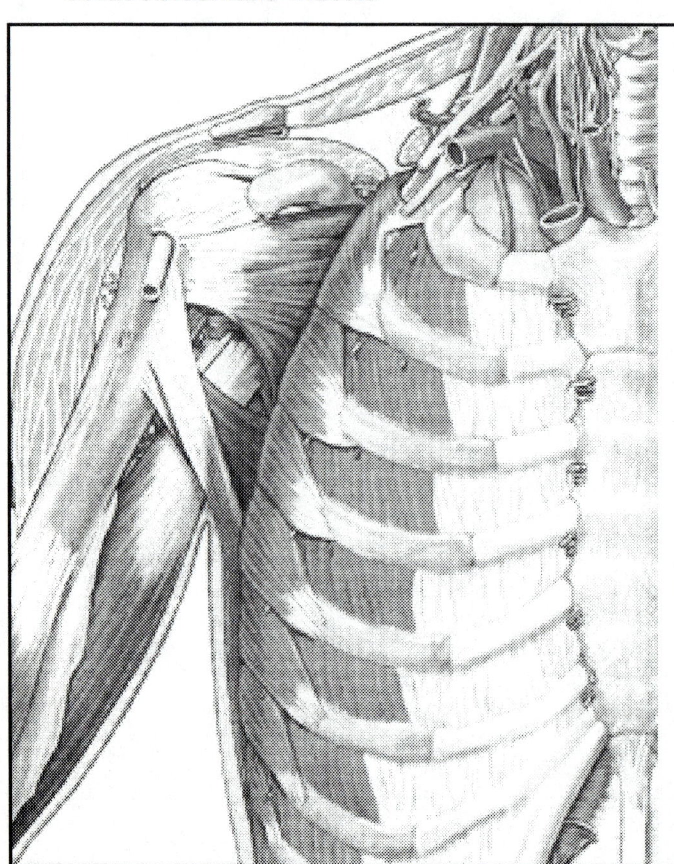

Click on the **Depth Bar** [▼] until you reach **layer 91**. Identify:

- **Serratus anterior muscle**
- **External intercostal muscle**
- **Tendon of latissimus dorsi muscle**
- **Tendon of teres major muscle**
- **Tendon of subscapularis muscle**

Into what bone do the latissimus dorsi, teres major, and subscapularis muscles insert ?

______________________________

Based on the insertion of the latissimus dorsi, teres major, and subscapularis muscles, what part of the body would move?

______________________________

# Posterior Dissection of the Back Muscles

Click on the **View** button  and choose **Posterior**.

Click on the **Normal** button. Windows— —Mac

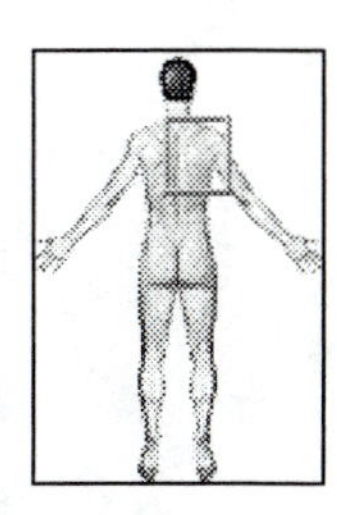

Drag the **Depth Bar** to **layer 0.**

Position the navigator rectangle over the right back area.

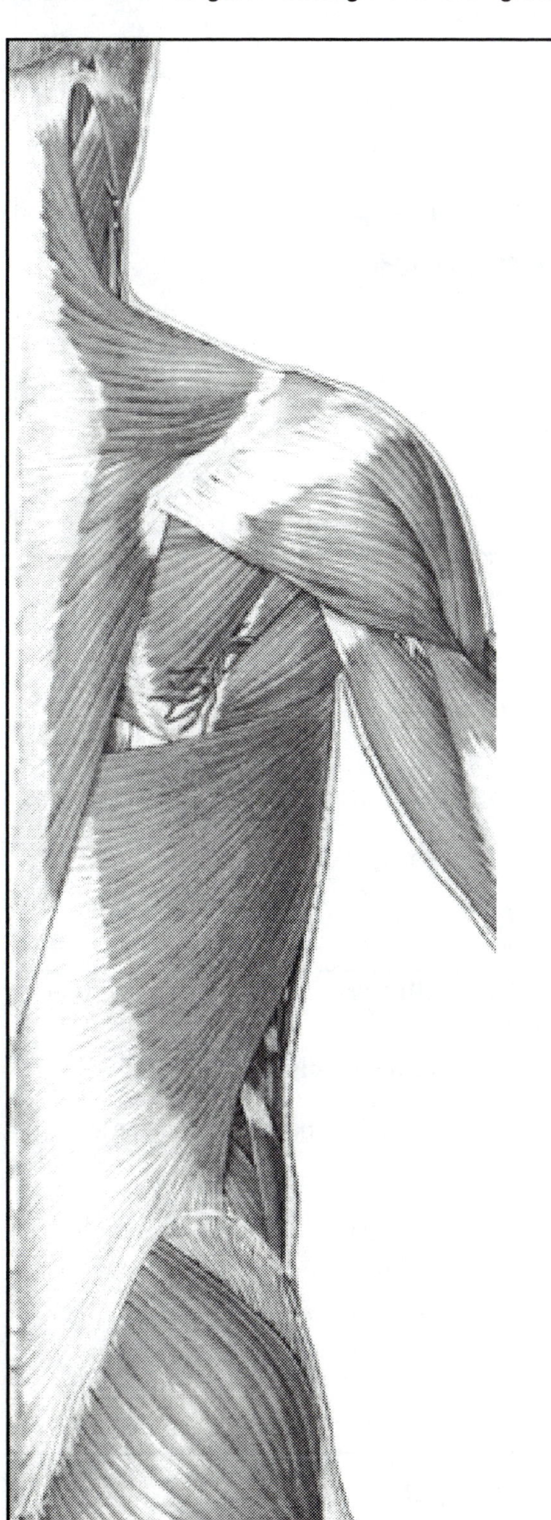

Click on the **Depth Bar** until you reach **layer 9**. You will need to scroll to find all of the structures. Identify:

- **Deltoid muscle**
- **Trapezius muscle**
- **Latissimus dorsi muscle**
- **Teres major muscle**
- **Infraspinatus muscle**

The trapezius muscle originates on what bones?

______________________________

Into what bone does the trapezius muscle insert?

______________________________

Based on the origin and insertion of the trapezius muscle, what is its action?

______________________________

The origin of the latissimus dorsi muscle is located on what bone?

______________________________

Based on the origin and insertion (from the previous graphic) of the latissimus dorsi muscle, what is its action?

______________________________

The deltoid muscle originates posteriorly on what bone?

______________________________

Based on the origin and insertion of the deltoid muscle, what would its action be?

______________________________

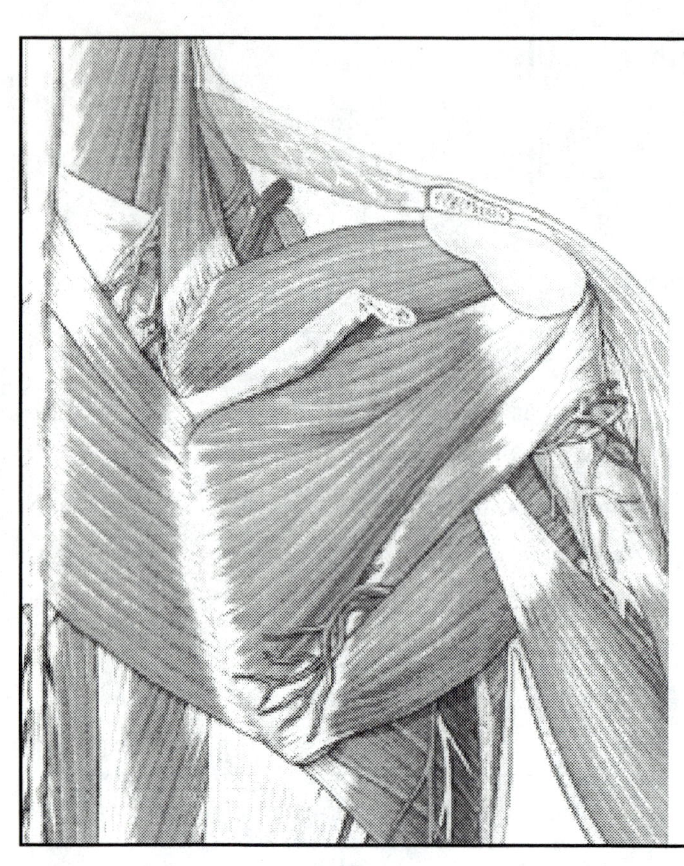

Click on the **Depth Bar** [▼] until you reach **layer 16**. Identify:

- **Supraspinatus muscle**
- **Teres major muscle**
- **Teres minor muscle**
- **Infraspinatus muscle**
- **Tendon of infraspinatus muscle** (proximal and distal)
- **Serratus anterior muscle**
- **Levator scapulae**
- **Rhomboideus major**

Into what bones does the infraspinatus muscle originate and insert?

___

Based on the origin and insertion of the infraspinatus muscle, what would its action be?

___

Into what bones does the rhomboideus major muscle originate and insert?

___

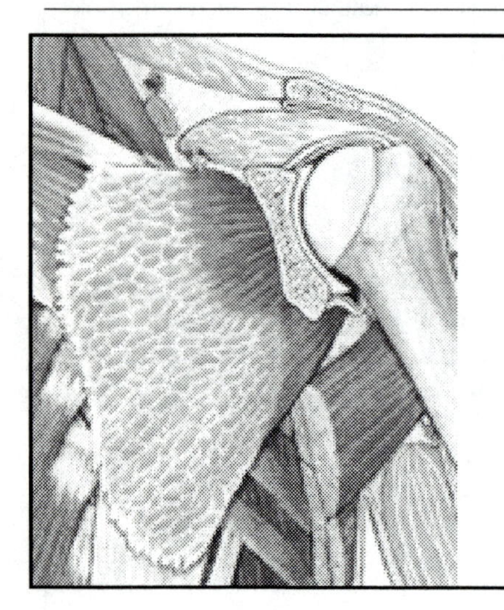

Click on the **Depth Bar** [▼] until you reach **layer 73**. Identify:

- **Subscapularis muscle**

Why is this muscle called the "subscapularis"?

___

# Lateral Dissection of the Chest and Back Muscles

Click on the **View** button and choose **Lateral**.

Click on the **Normal** button. Windows— —Mac

Drag the **Depth Bar** to **layer 0.**

Position the navigator rectangle over the upper chest.

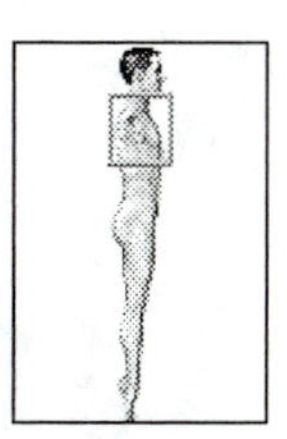

Click on the **Depth Bar** until you reach **layer 8**. You will need to scroll to see all of the muscles below. Identify:

- **Pectoralis major muscle** (all heads)
- **Deltoid muscle** (cut)
- **Trapezius muscle**
- **Serratus anterior muscle**
- **Latissimus dorsi muscle**
- **Teres major muscle**
- **Infraspinatus muscle**

Based on this diagram, which of the muscles above would move the arm?

______________________________

______________________________

______________________________

______________________________

______________________________

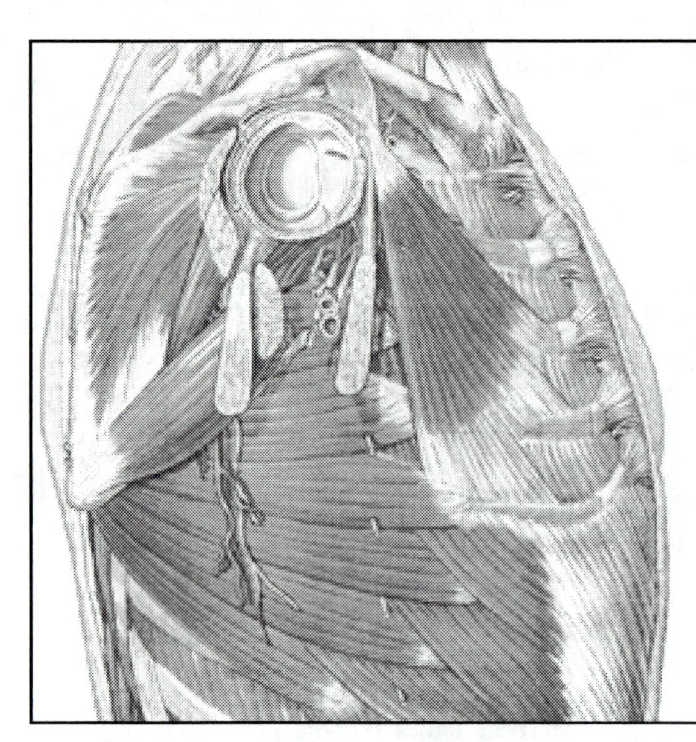

Click several times on the **Depth Bar** ▼ until you reach **layer 13**. Identify:

- **Serratus anterior muscle**
- **Teres major muscle**
- **Teres minor muscle**
- **Infraspinatus muscle**
- **Pectoralis minor muscle**
- **Subclavius muscle**

Click several times on the **Depth Bar** ▼ until you reach **layer 41**. No graphic has been provided for this layer. Note the **tendon of the subscapularis muscle** emerging from beneath the scapula. Click once to remove the scapula. You should be at **layer 42**. Identify the **subscapularis muscle**.

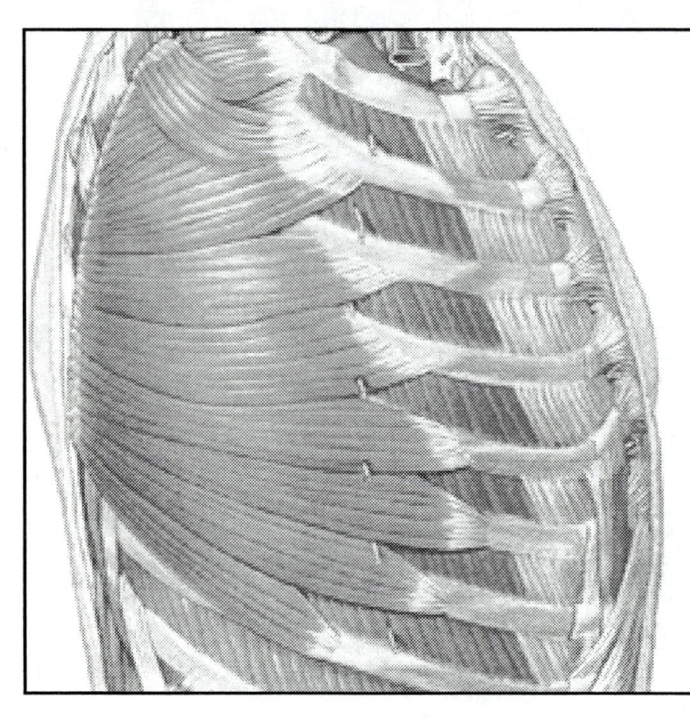

Click on the **Depth Bar** ▼ until you reach **layer 61**. No graphic has been provided for this layer. Identify:

- **Serratus anterior muscle**
- **External intercostal muscle**

When you have finished, **close** the window.

# Cross Sections of the Muscles of the Chest & Back

Under the **File** menu, drag down to **Open** (then to **Content** if you are working in Windows). Click on **Atlas Anatomy**. Select **System** and **Muscular.** Choose **Nerves of Thoracic Wall**. If the list is not in alphabetical order, the image is about half of the way down the list.

This is a superior view into an empty thoracic cavity, showing the musculature. You may need to scroll to find all of the structures. If you are working in Mac, click on the **Show All Pins** button and scroll to "All Systems".

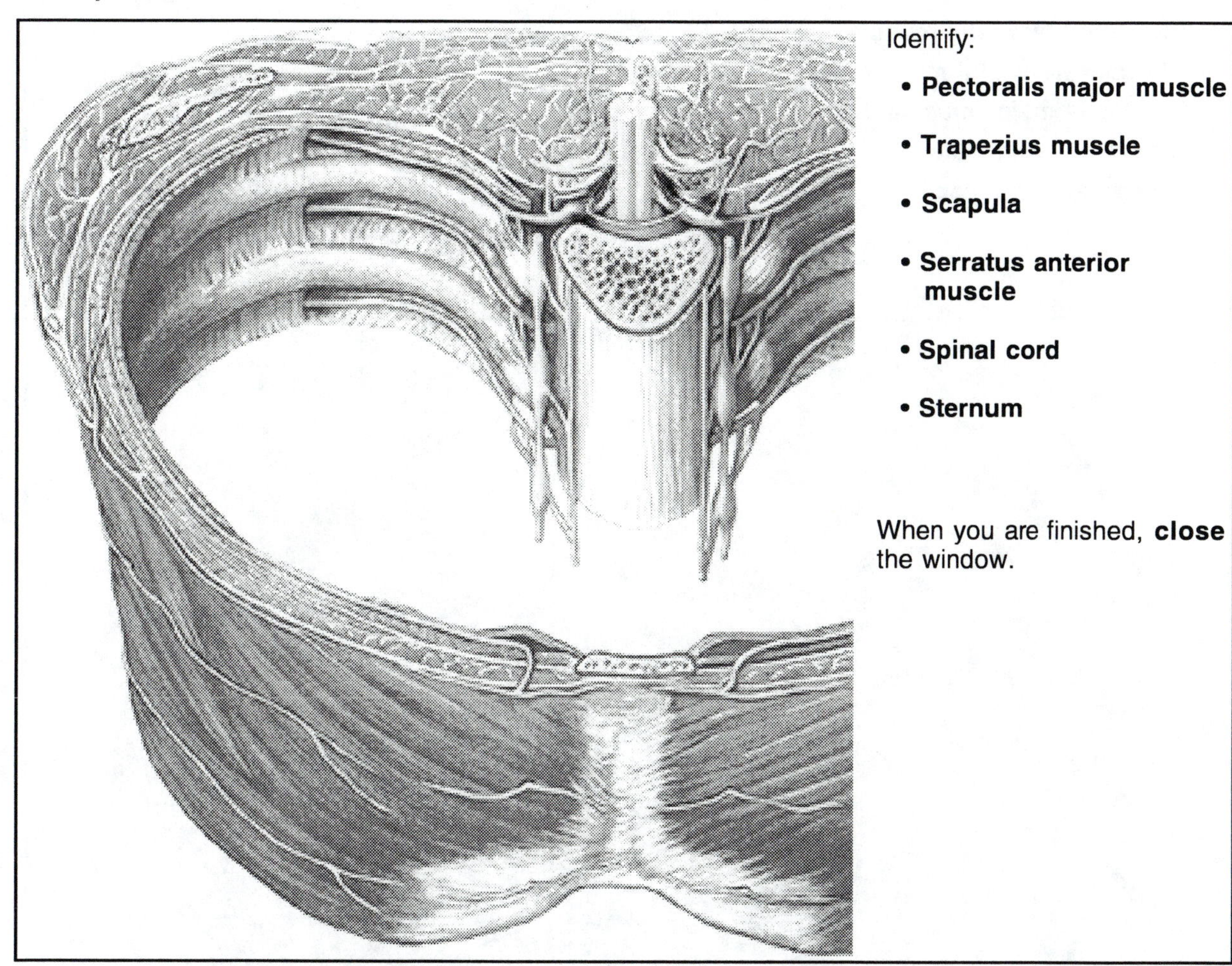

Identify:

- **Pectoralis major muscle**
- **Trapezius muscle**
- **Scapula**
- **Serratus anterior muscle**
- **Spinal cord**
- **Sternum**

When you are finished, **close** the window.

Under the **File** menu, drag down to **Open** (then to **Content** if you are working in Windows). **Atlas Anatomy** should still be selected. **System** and **Muscular** should still be selected. Choose **T8 Vertebra (Inf).** If the list is not in alphabetical order, the image is about half way down the list.

This is a cross section of the thorax. You may need to scroll to find all the structures. If you are working in Mac, click on the **Show All Pins** button and scroll to "All Systems".

Identify:

- **Latissimus dorsi muscle**
- **External intercostal muscle**
- **Internal intercostal muscle**
- **Pectoralis major muscle**
- **Infraspinatus muscle**
- **Subscapularis muscle**
- **Pectoralis minor muscle**
- **Scapula**
- **Serratus anterior muscle**
- **Sternum**
- **Teres major muscle**
- **Trapezius muscle**

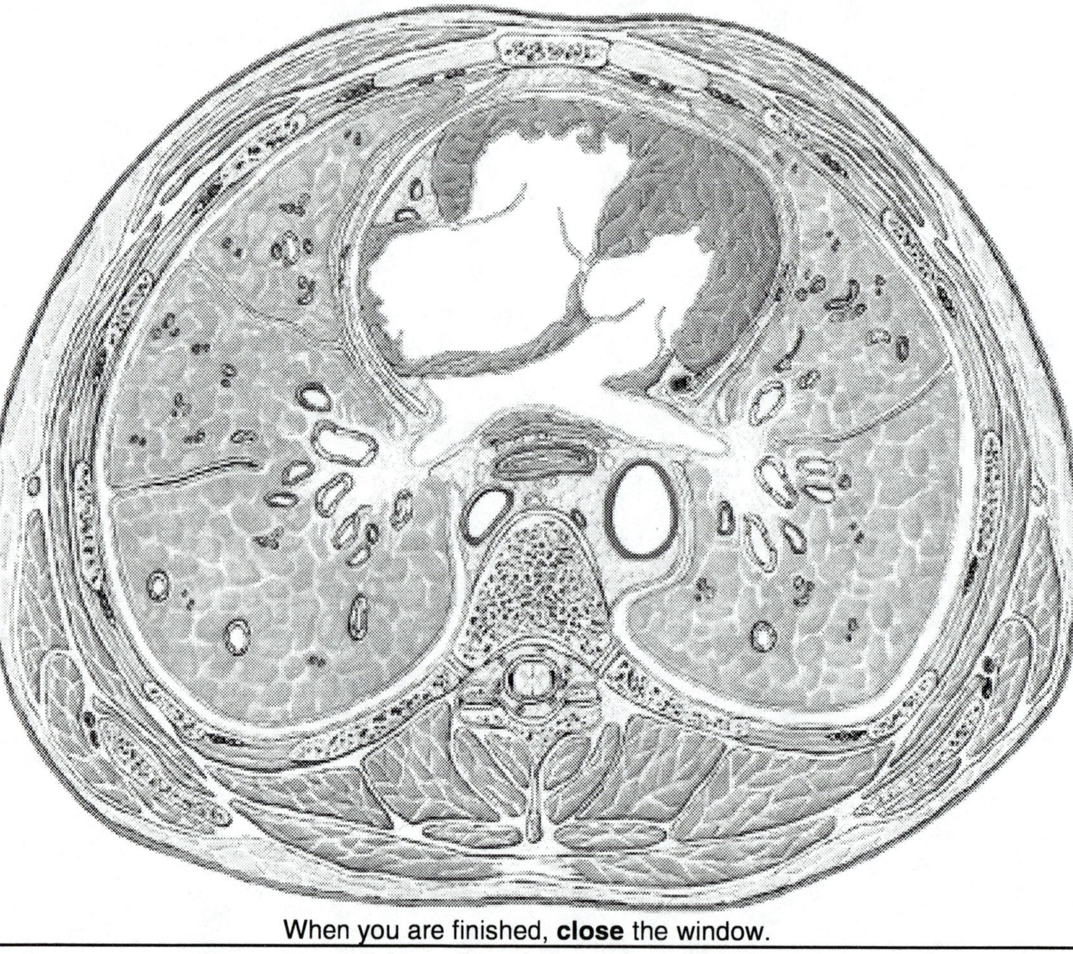

When you are finished, **close** the window.

List the following muscles from most superficial to deepest:

_____ **External intercostal muscle**

_____ **Pectoralis major muscle**

_____ **Internal intercostal muscle**

_____ **Pectoralis minor muscle**

## Objectives: Chest and Back Muscles:

The student will be able to identify the anatomy of the important chest and back muscles, particularly those that move the shoulder girdle and the arm at the shoulder.

**Muscles**

- Pectoralis minor muscle
- Pectoralis major muscle
- Deltoid muscle
- Trapezius muscle
- Serratus anterior muscle
- External intercostal muscle
- Subclavius muscle
- Latissimus dorsi muscle
- Teres major muscle
- Teres minor muscle
- Infraspinatus muscle
- Supraspinatus muscle
- Levator scapulae muscle
- Rhomboideus major muscle

**Tendons**

- Tendon of latissimus dorsi muscle
- Tendon of teres major muscle
- Tendon of subscapularis muscle
- Tendon of infrascapularis muscle

**Other**

- Spinal cord
- Sternum
- Scapula

# Abdominal Muscles & Muscles of Respiration

## Opening A.D.A.M Interactive Anatomy

Open A.D.A.M. Interactive Anatomy according to the directions in the Tutorial.

## Anterior Dissection of Lower Appendicular Skeleton

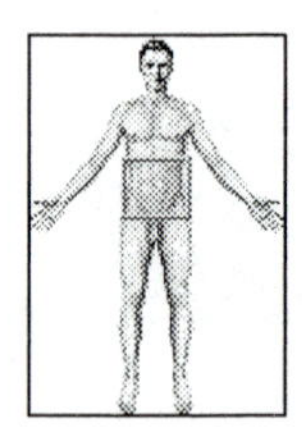

Choose **Dissectible Anatomy**.

Choose **Male**. Select **Anterior.** Click **Open**.

Enlarge the window and position the navigator rectangle over the abdomen.

## Anterior Dissection of Abdominal Muscles

The muscles of the anterior and lateral abdominal wall are arranged in three layers, with the fibers of each layer running in a different direction.

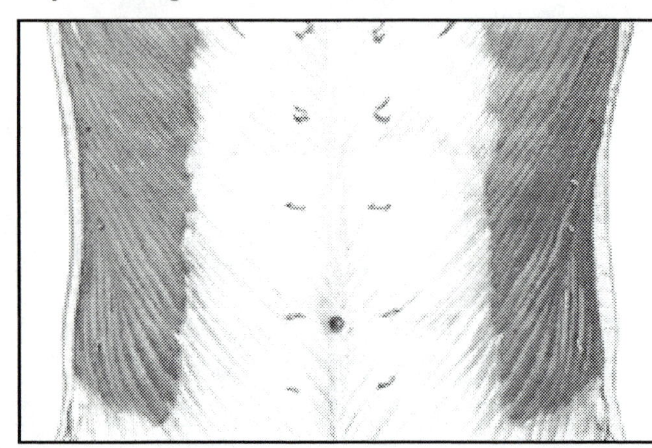

Click on the **Depth Bar** ▼ to remove skin, fat, blood vessels, and muscles until you reach **layer 20**. Identify:

- **External abdominal oblique muscle**
- **Linea alba**
- **External abdominal oblique aponeurosis**

What is the insertion of the external abdominal oblique muscle?

______________________________________________

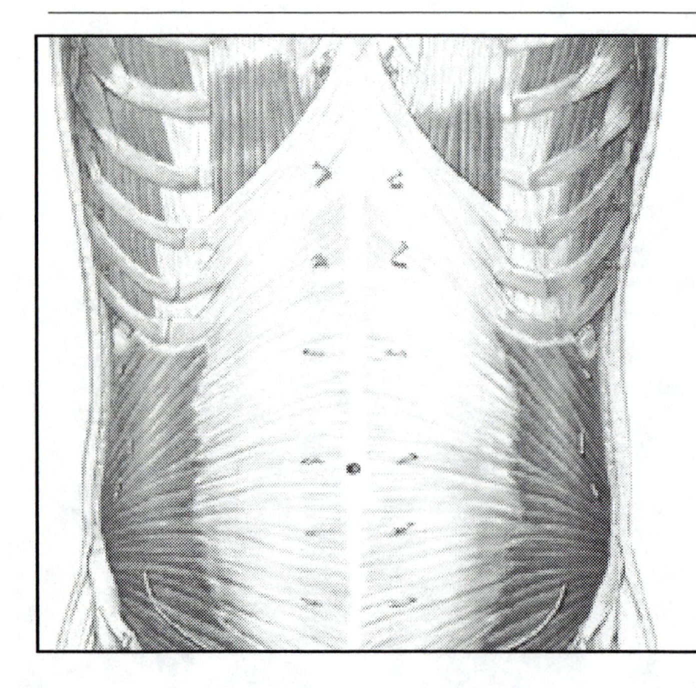

Click on the **Depth Bar** ▼ until you reach **layer 26**. Identify:

- **Linea alba**
- **Umbilicus**
- **Internal abdominal oblique aponeurosis**
- **Internal abdominal oblique muscle**
- **Costal cartilage**
- **Iliac crest**
- **External abdominal oblique muscle** (cut)
- **Rectus abdominis muscle**
- **External intercostal muscle**

What major difference do you see between external and internal abdominal oblique muscles?

---

What are the insertions for the internal abdominal oblique muscle?

---

What's the difference between a tendon and an aponeurosis?

---

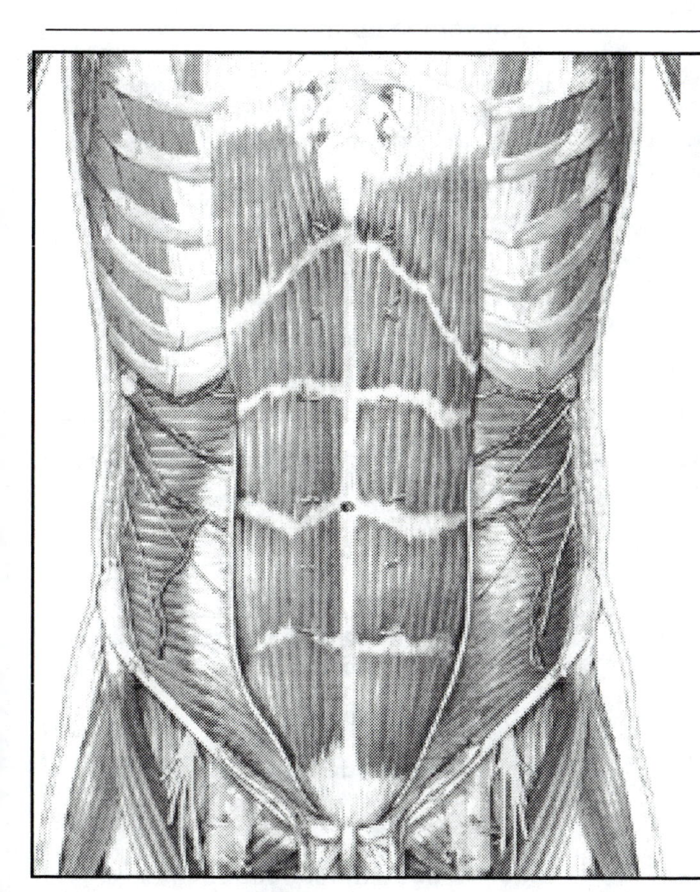

Click on the **Depth Bar** [▼] until you reach **layer 29**. Identify:

- **Rectus abdominis muscle**
- **Tendinous intersection of rectus abdominis muscle**
- **Transversus abdominis aponeurosis**
- **Transversus abdominis muscle**
- **Internal abdominal oblique muscle** (cut)
- **External abdominal oblique muscle** (cut)
- **External intercostal muscle**
- **Linea alba**

What is the origin (inferior) of the rectus abdominis muscle?

---

What is the insertion (superior) of the rectus abdominis muscle?

---

Based on its origin and insertion, what is the action of the rectus abdominis muscle?

---

 

Click on the **Depth Bar** [▼] or scroll until you reach **layer 146**. No graphic has been provided for this layer. Identify the **internal intercostal muscle**.

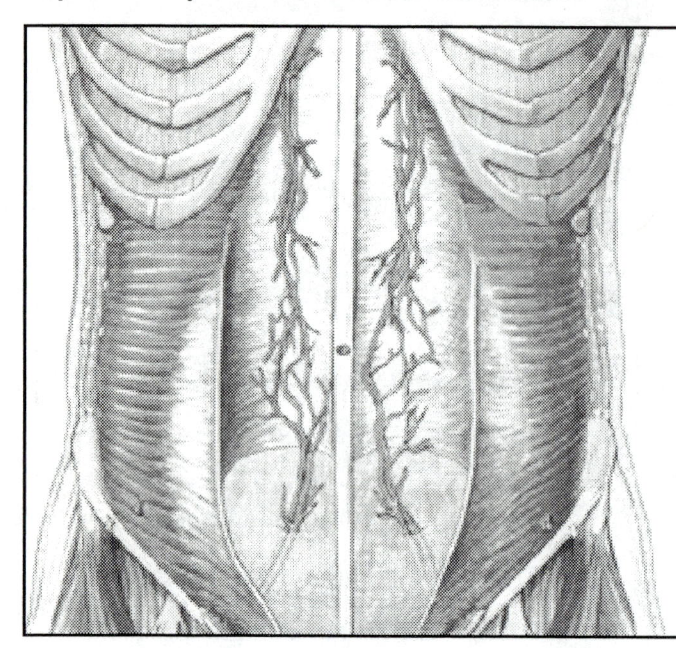

Click on the **Depth Bar** [▼] until you reach **layer 150**. Identify:

- **Posterior layer of rectus sheath**
- **Transversus abdominis aponeurosis**
- **Transversus abdominis muscle**
- **Internal abdominal oblique muscle** (cut)
- **External abdominal oblique muscle** (cut)
- **Linea alba**

Compare the appearance of the external and internal oblique muscles to the transversus abdominis muscle.

---

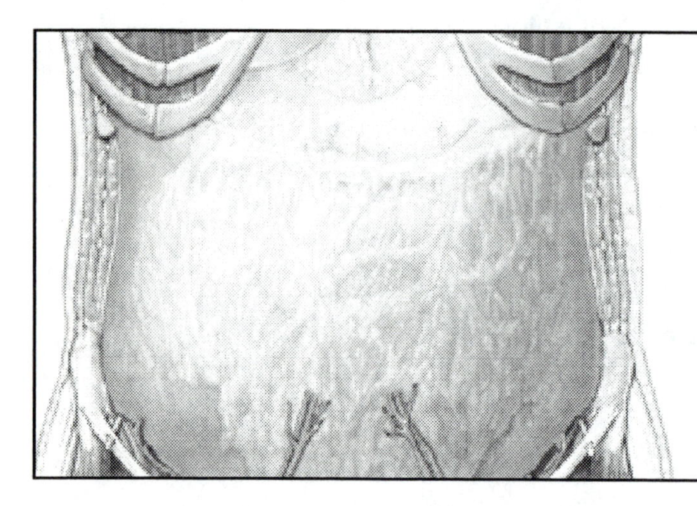

Click on the **Depth Bar** [▼] or scroll until you reach **layer 158**. Identify:

- **Parietal peritoneum**
- **Transversus abdominis muscle** (cut)
- **Internal abdominal oblique muscle** (cut)
- **External abdominal oblique muscle** (cut)

Click on the **Depth Bar** [▼] or scroll until you reach **layer 166**. No graphic has been provided for this layer. Identify the **diaphragm muscle**.

# Lateral Dissection of the Abdominal Muscles

Click on the **View** button and choose **Lateral**.

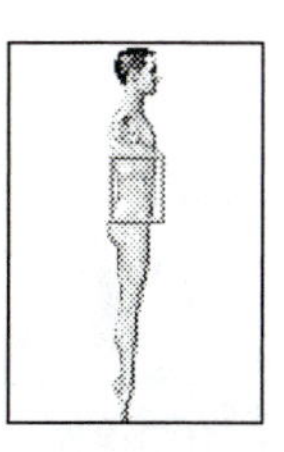

Windows—Mac

Click on the **Normal** button.

Drag the **Depth Bar** to **layer 0.**

Position the navigator rectangle over the abdomen.

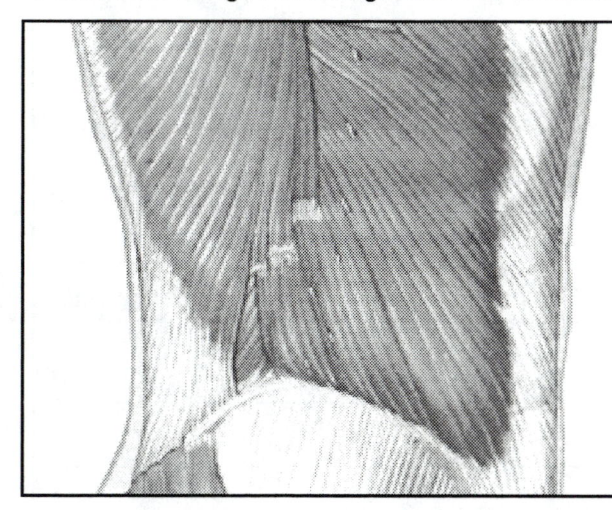

Click on the **Depth Bar** to remove the blood vessels, nerves and fascia until you reach **layer 11**. Identify:

- **External abdominal oblique aponeurosis**
- **External abdominal oblique muscle**

Click on the **Depth Bar** once to **layer 12**. What is the origin of the external abdominal oblique muscle?

---

Based on its origin and insertion, what is the action of the external abdominal oblique muscle?

---

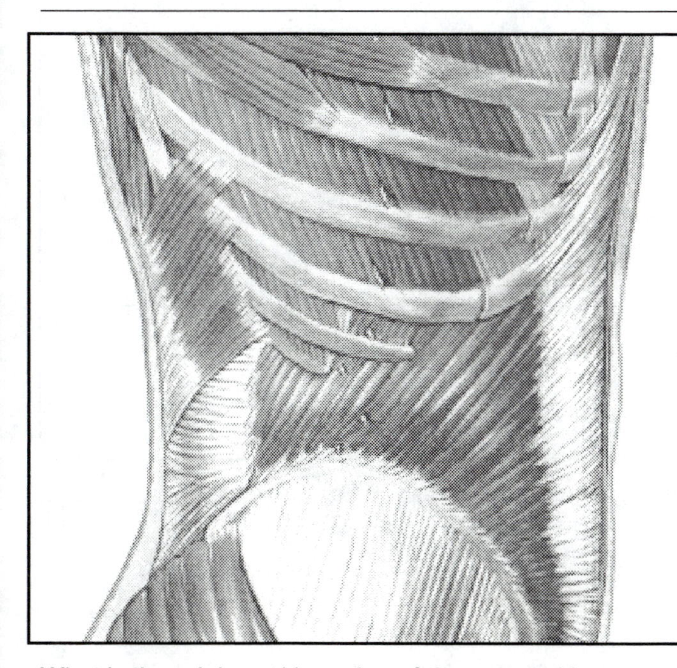

Click on the **Depth Bar** or scroll until you reach **layer 62**. Identify:

- **Internal abdominal oblique aponeurosis**
- **Internal abdominal oblique muscle**
- **Tendon of internal abdominal oblique muscle**
- **Linea alba**
- **External intercostal muscle**
- **Ribs**
- **Iliac crest**

What is the origin and insertion of the external intercostal muscle?

 

What is the origin of the internal abdominal oblique muscle?

---

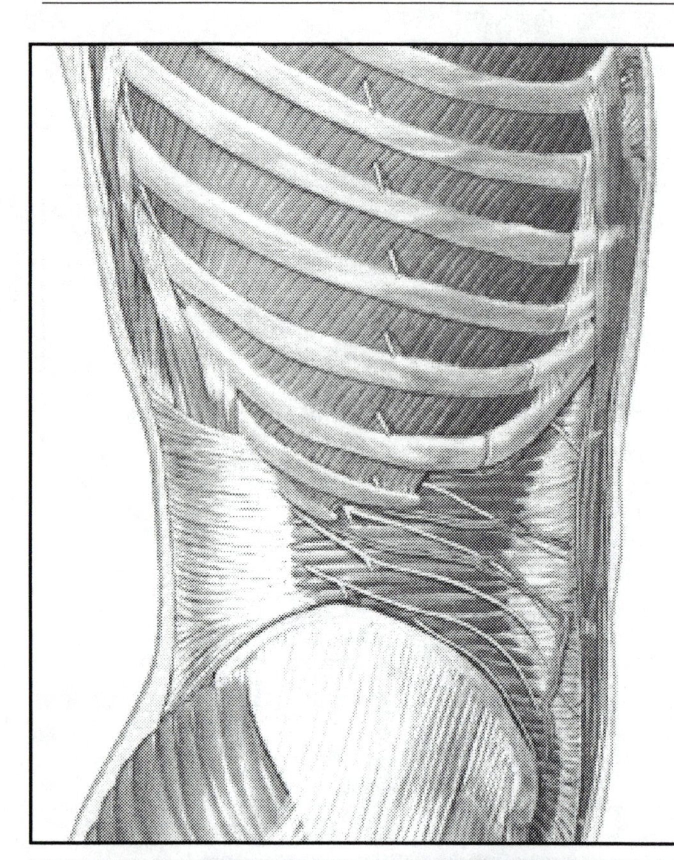

Click on the **Depth Bar** ▼ until you reach **layer 67**. Identify:

- **Rectus abdominis muscle**
- **Tendinous intersection of rectus abdominis muscle (several)**
- **Transversus abdominis aponeurosis**
- **Transversus abdominis muscle**
- **Internal intercostal muscle**

Compare the appearance of the external and internal intercostal muscles.

---

---

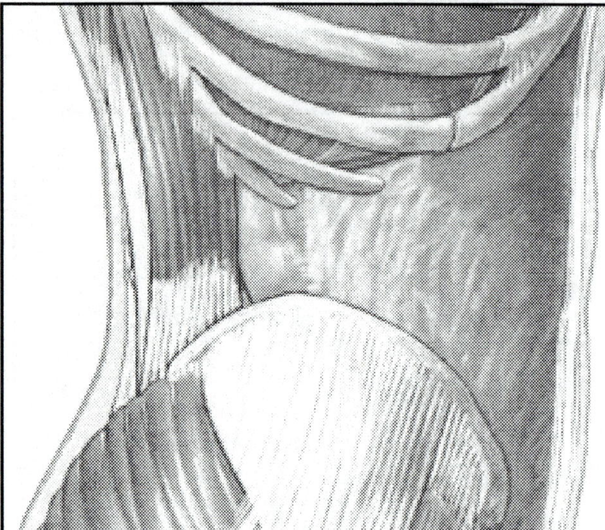

Click on the **Depth Bar** ▼ to **layer 75**. Identify:

- **Linea alba**
- **Parietal peritoneum**
- **Diaphragm**

Why does the linea alba look different in this view than it did previously?

---

Scroll down to **layer 149** and identify the **diaphragm muscle**. Also note the phrenic nerve that innervates the diaphragm.

When you are finished, **close** the window.

## Cadaver Image of the Muscles of the Abdomen

Under the **File** menu, drag down to **Open** (then to **Content** if you are working in Windows). Click on **Atlas Anatomy**. Select **System** and **Muscular.** Choose **Abdominal Viscera**. If the list is not in alphabetical order, the image is about two-thirds of the way down the list.

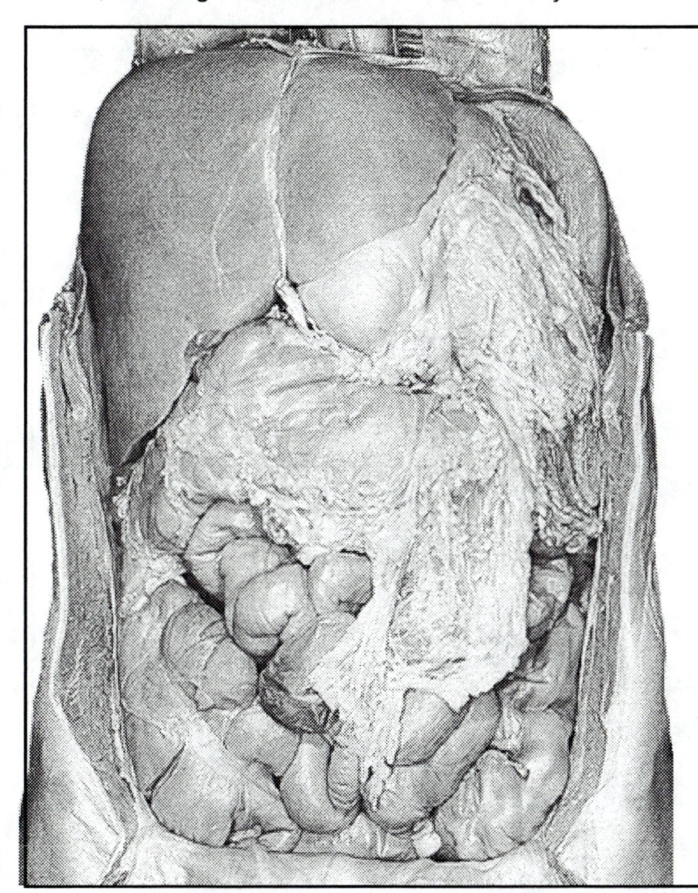

You may need to scroll to identify all of the structures below. Identify :

- **External abdominal oblique muscle**
- **Internal abdominal oblique muscle**
- **Transversus abdominis muscle**
- **Diaphragm muscle**

When you are finished, **close** the window.

## Cross Section of the Muscles of the Abdomen

Under the **File** menu, drag down to **Open** (then to **Content** if you are working in Windows). **Atlas Anatomy** should still be selected. **System** and **Muscular** should still be selected. Choose **Viscera of Female Pelvis (Sup)**. If the list is not in alphabetical order, the image is about fourth-fifths of the way down the list.

This is a view into the pelvic cavity of the female. You may need to scroll to see all of the structures. Identify:

- **External abdominal oblique muscle**
- **Internal abdominal oblique muscle**
- **Transversus abdominis muscle**
- **Linea alba**
- **Rectus abdominis muscle**
- **External abdominal oblique aponeurosis**
- **Internal abdominal oblique aponeurosis**
- **Transversus abdominis aponeurosis**

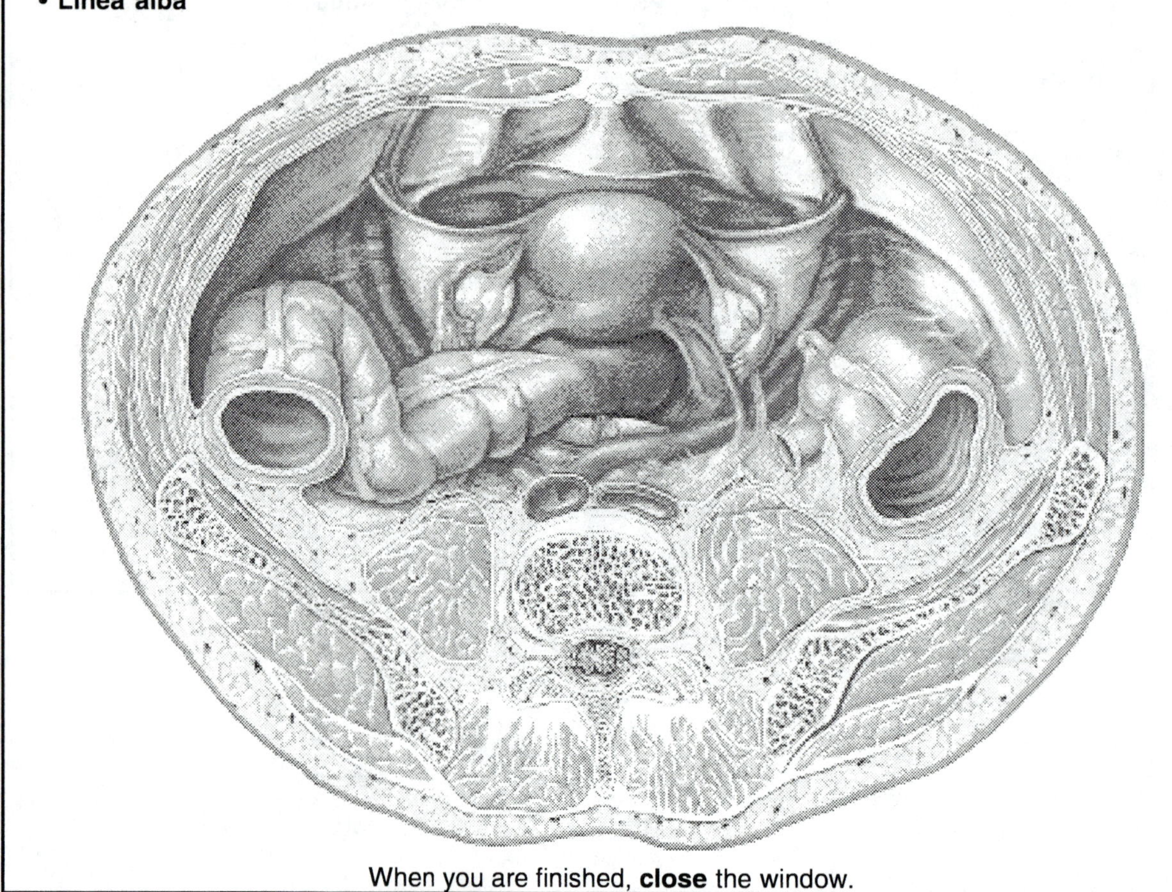

When you are finished, **close** the window.

List the following muscles from most superficial to deepest:

_____ Internal abdominal oblique muscle

_____ External abdominal oblique muscle

_____ Transversus abdominis muscle

## Objectives: Abdominal Muscles & Muscles of Respiration

To learn the anatomy of the abdominal muscles, the diaphragm, and intercostal muscles.

**Muscles**

- External abdominal oblique muscle
- Internal abdominal oblique muscle
- Rectus abdominis muscle
- External intercostal muscle
- Transversus abdominis muscle
- Diaphragm muscle
- Internal intercostal muscle

**Other**

- Linea alba
- External abdominal oblique aponeurosis
- Internal abdominal oblique aponeurosis
- Umbilicus
- Ribs
- Costal cartilage
- Iliac crest
- Tendinous intersection of rectus abdominis
- Transversus abdominis aponeurosis
- Rectus sheath
- Parietal peritoneum
- Tendon of internal abdominal oblique

# Arm & Hand Muscles

## Opening A.D.A.M Interactive Anatomy

Open A.D.A.M. Interactive Anatomy according to the directions in the Tutorial.

## Anterior Dissection of Upper Arm Muscles

Choose **Dissectible Anatomy**.

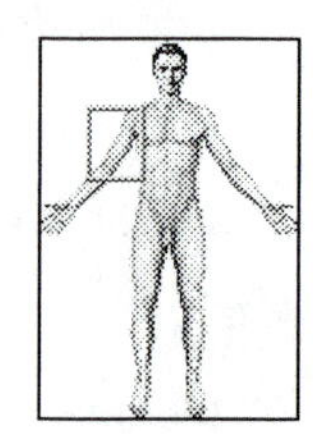

Choose either male or female. Select **Anterior.** Click **Open**.

Enlarge the window and position the image over the upper right arm.

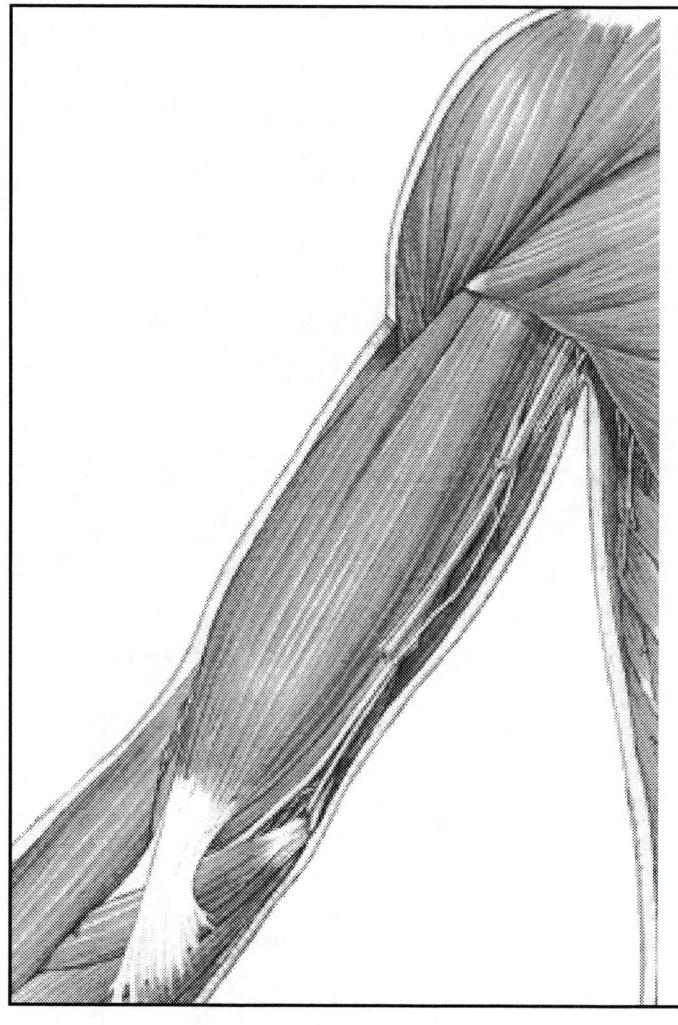

Click on the **Depth Bar** until you reach **layer 19**. Identify:

- **Biceps brachii muscle** (short and long head)
- **Brachioradialis muscle**
- **Pronator teres muscle**
- **Deltoid muscle**

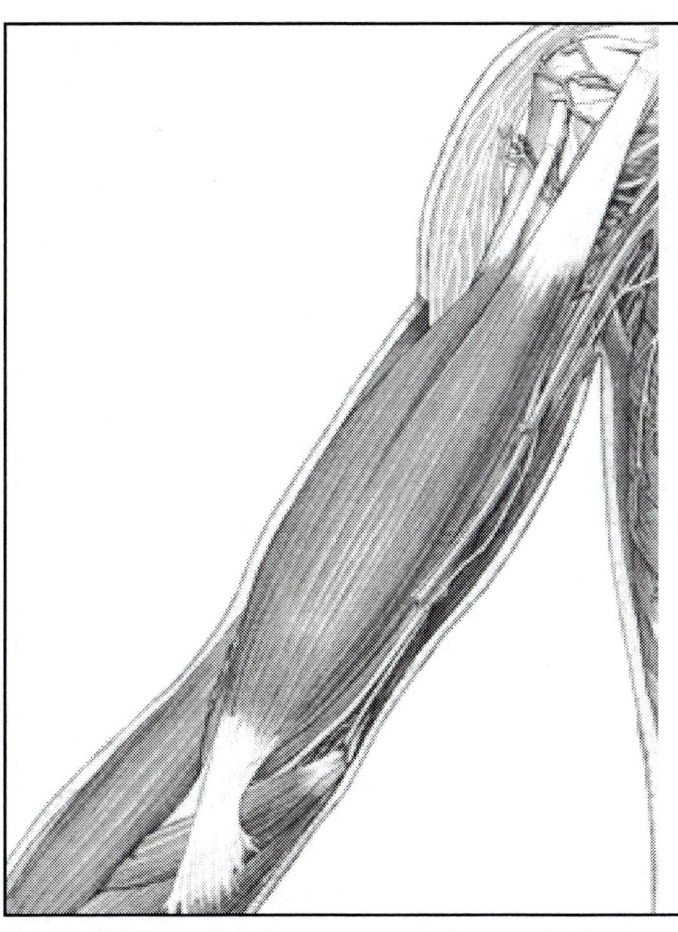

Click on the **Depth Bar** [▼] until you reach **layer 81.** Identify:

- **Biceps brachii muscle** (short and long head)
- **Brachioradialis muscle**
- **Pronator teres muscle**

What is the origin of the short head of the biceps brachii muscle?

_______________________________________________

What is the origin of the long head of the biceps brachii muscle?

_______________________________________________

The insertion of the biceps brachii muscle is the radius. Based on the origin and insertion, what would be the action of this muscle?

_______________________________________________

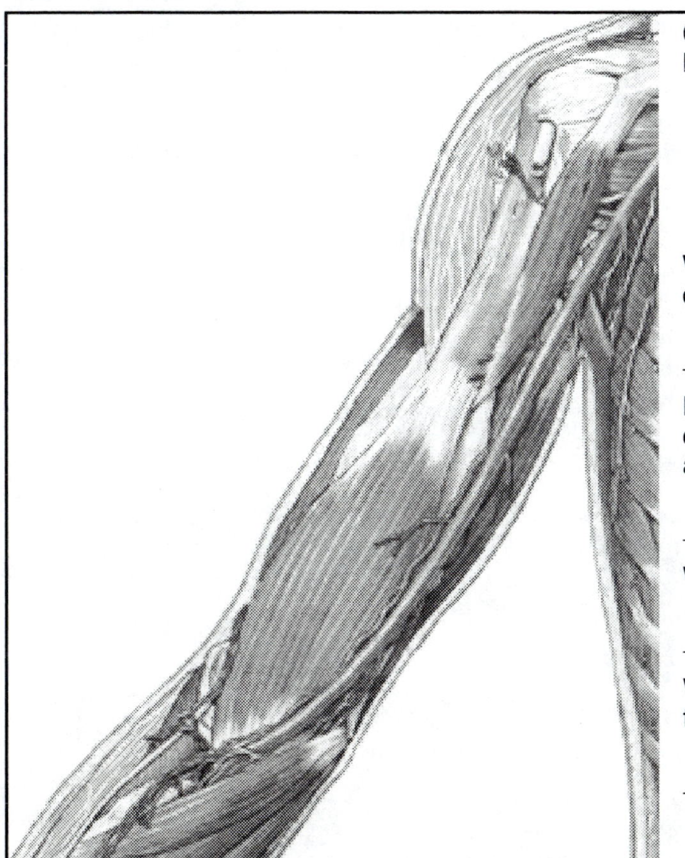

Click on the **Depth Bar** [▼] until you reach **layer 87.** Identify:

- **Brachialis muscle**
- **Coracobrachialis muscle**
- **Pronator teres muscle**

What is the origin and insertion of the coracobrachialis muscle?

_______________________________________________

Based on the origin and insertion of the coracobrachialis muscle, what would be its action?

_______________________________________________

What is the origin of the brachialis muscle?

_______________________________________________

What do you think the action of the pronator teres muscle is?

_______________________________________________

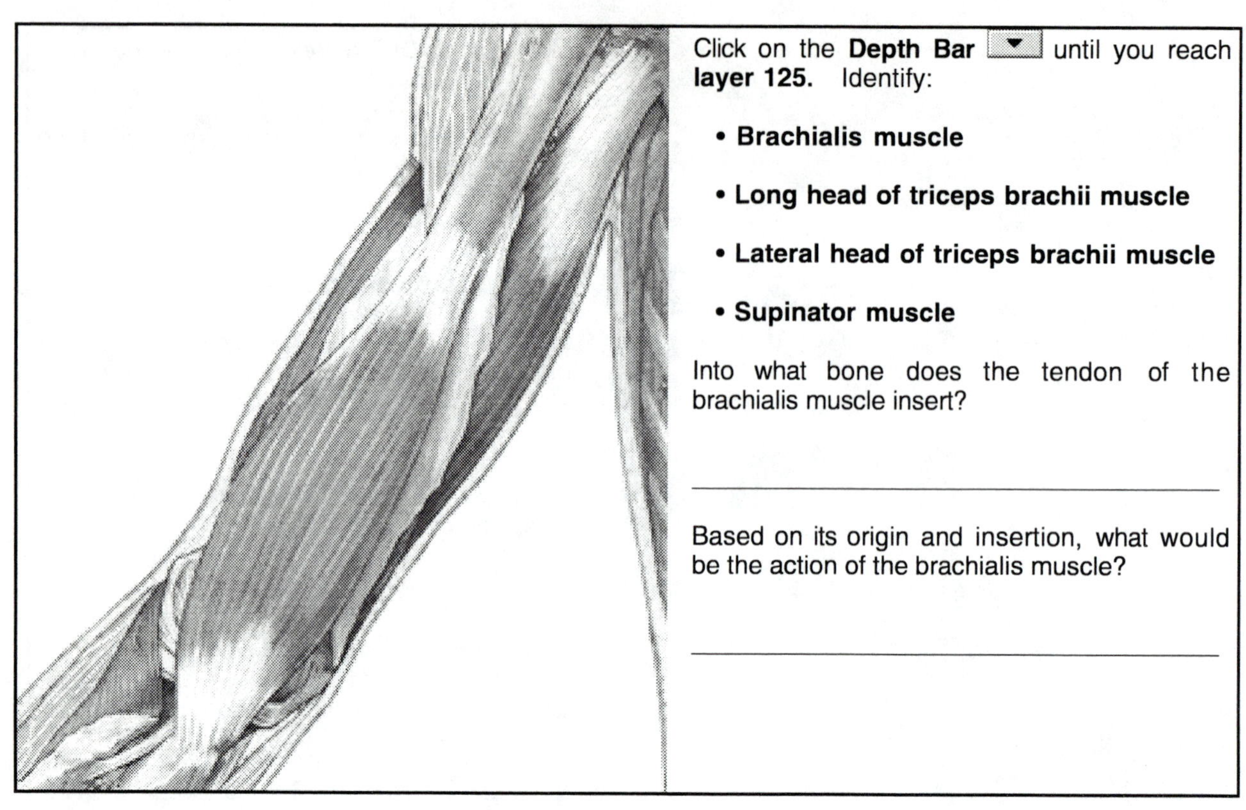

Click on the **Depth Bar** [▼] until you reach **layer 125.** Identify:

- **Brachialis muscle**
- **Long head of triceps brachii muscle**
- **Lateral head of triceps brachii muscle**
- **Supinator muscle**

Into what bone does the tendon of the brachialis muscle insert?

___________________________________

Based on its origin and insertion, what would be the action of the brachialis muscle?

___________________________________

What do you think the action of the supinator muscle is?

___________________________________

Click on the **Depth Bar** [▼] until you reach **layer 127.** No graphic has been provided for this layer. Identify:

- **Long head of triceps brachii muscle**
- **Lateral head of triceps brachii muscle**
- **Medial head of triceps brachii muscle**

 

# Posterior Dissection of the Upper Arm Muscles

Click on the **View** button and choose **Posterior**.

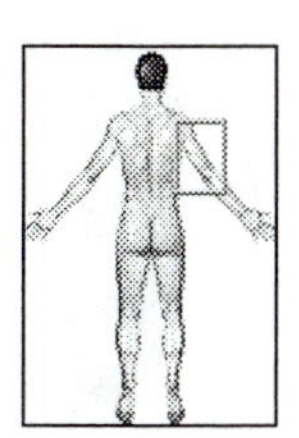

Click on the **Normal** button. Windows— —Mac

Drag the **Depth Bar** to **layer 0.**

Position the navigator rectangle over the upper right arm.

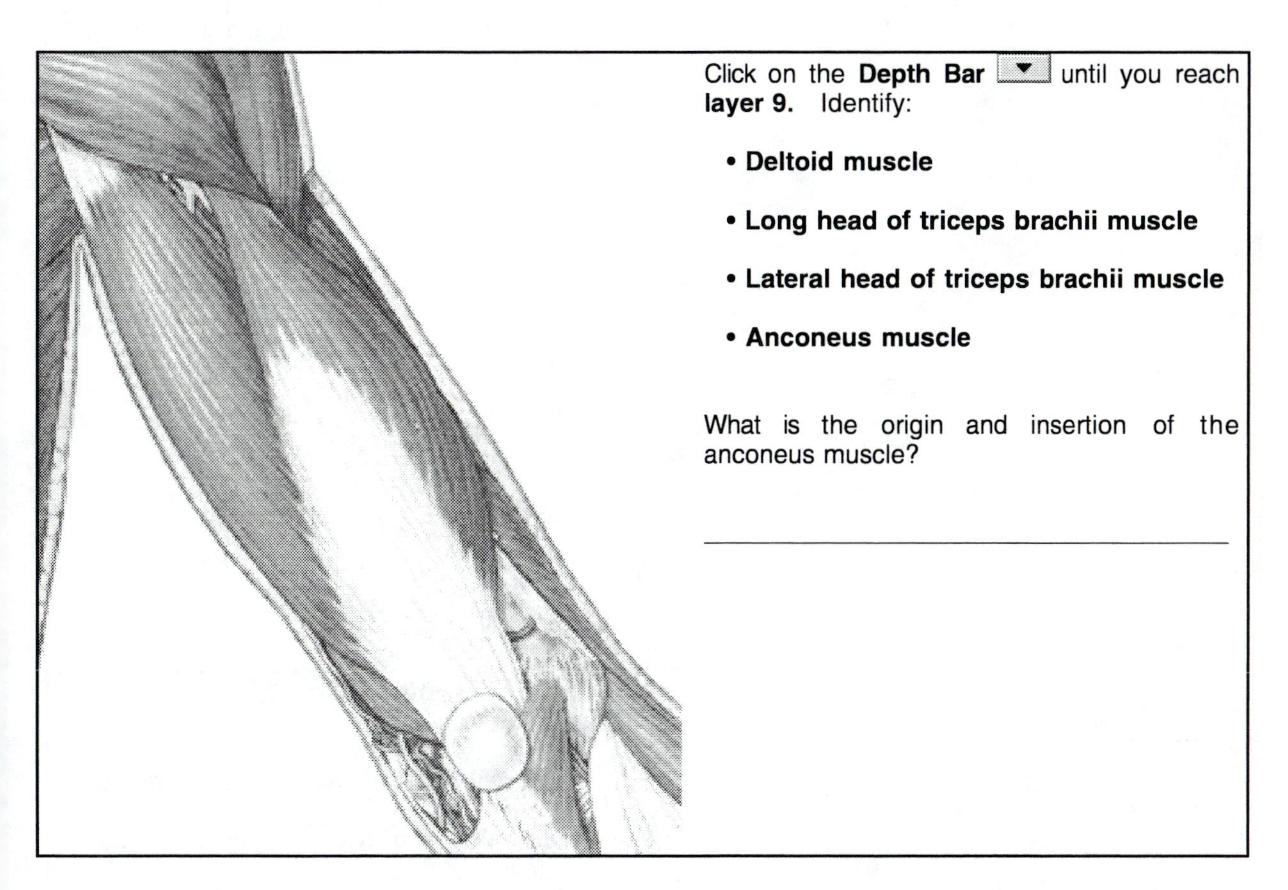

Click on the **Depth Bar** until you reach **layer 9.** Identify:

- **Deltoid muscle**
- **Long head of triceps brachii muscle**
- **Lateral head of triceps brachii muscle**
- **Anconeus muscle**

What is the origin and insertion of the anconeus muscle?

______________________________________

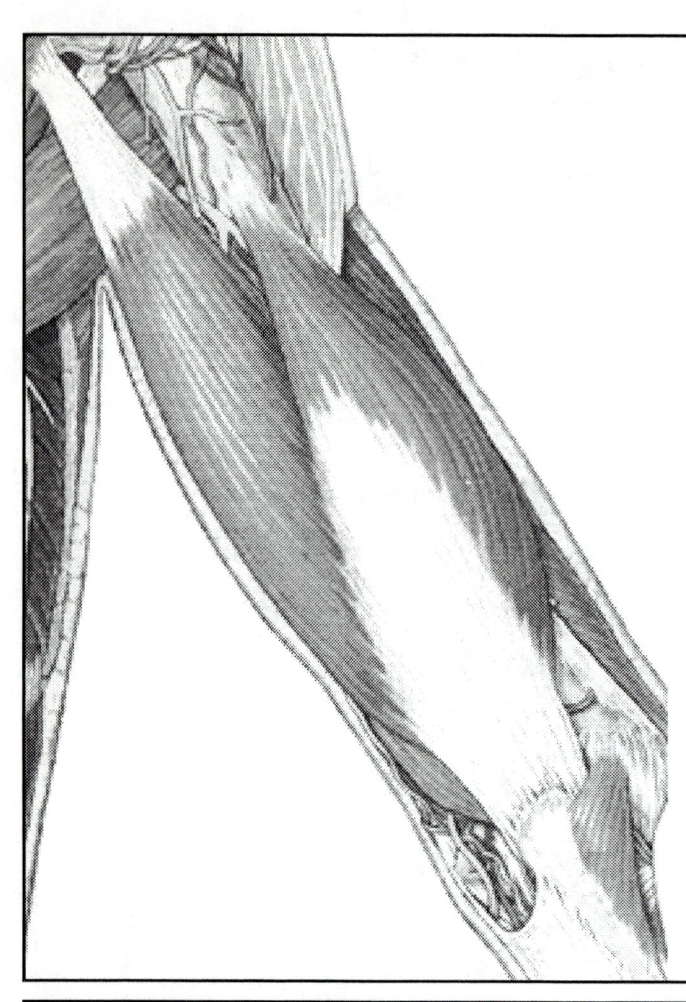

Click on the **Depth Bar** [▼] until you reach **layer 28.** Identify:

- **Long head of triceps brachii muscle**
- **Lateral head of triceps brachii muscle**
- **Anconeus muscle**

What is the origin and insertion of the long head and the lateral head of the triceps brachii muscle?

______________________________________

______________________________________

What is the insertion of the tendon of the triceps brachii muscle?

______________________________________

Based on its origin and insertion, what is the action of the triceps brachii muscle?

______________________________________

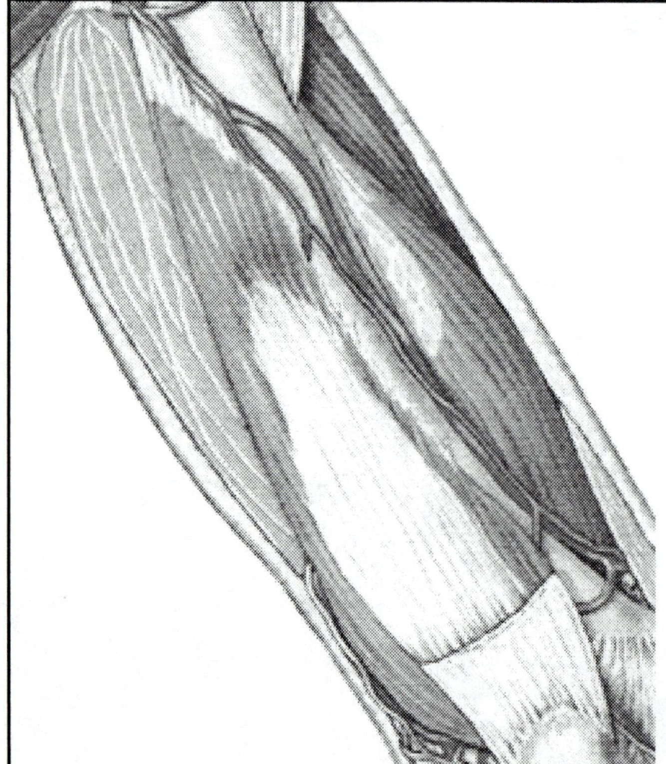

Click on the **Depth Bar** [▼] until you reach **layer 55.** Identify:

- **Medial head of triceps brachii muscle**
- **Long head of triceps brachii muscle** (cut)
- **Brachialis muscle**
- **Biceps brachii muscle**

What is the origin of the medial head of the triceps brachii muscle?

______________________________________

# Anterior Dissection of Lower Arm & Hand Muscles

Click on the **View** button and choose **Anterior**.

Click on the **Normal** button.

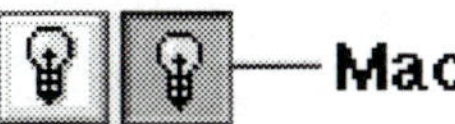


Drag the **Depth Bar** to **layer 0.**

Position the navigator rectangle over the lower right arm.

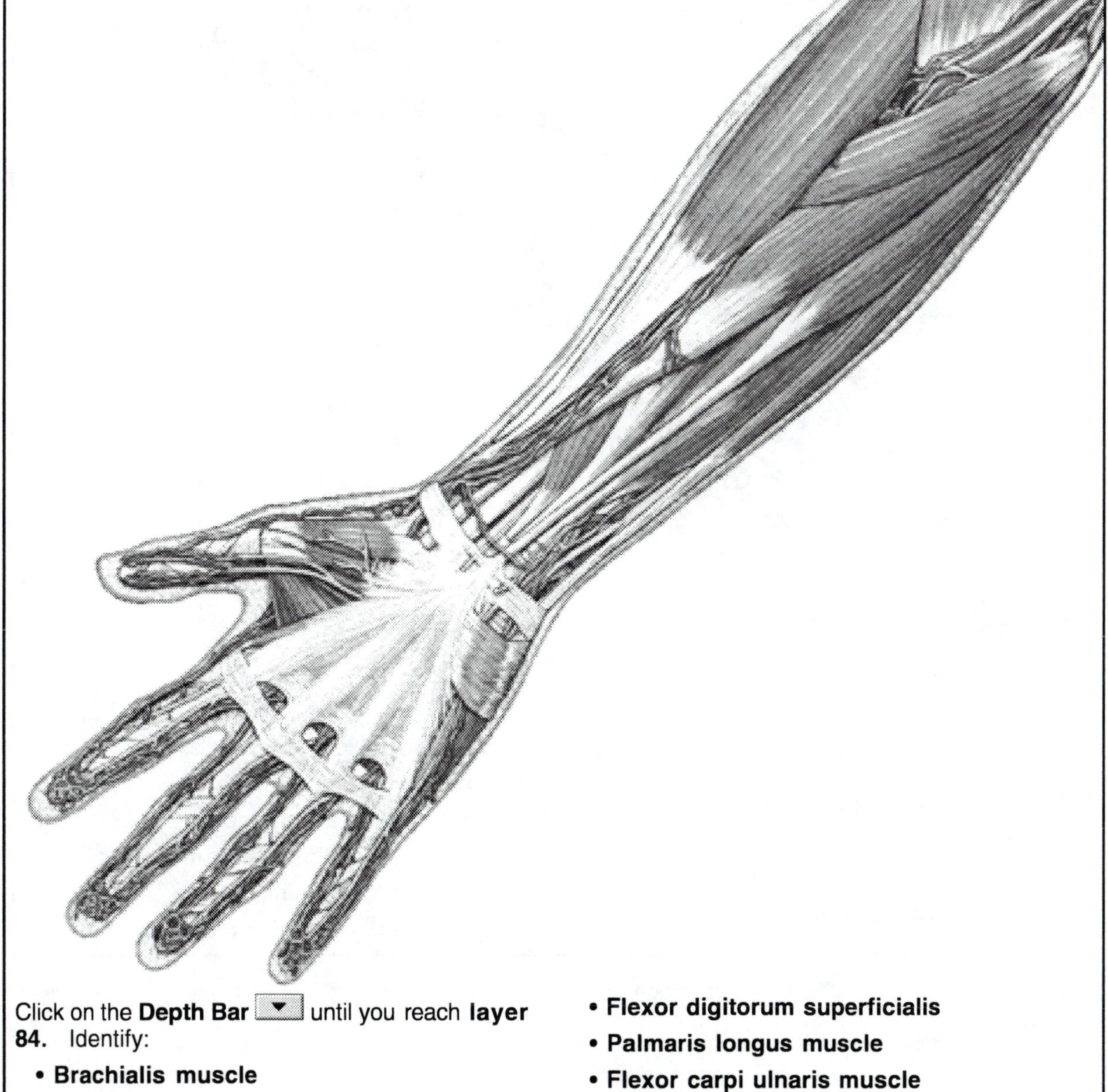

Click on the **Depth Bar** until you reach **layer 84.** Identify:

- **Brachialis muscle**
- **Brachioradialis muscle**
- **Pronator teres muscle**
- **Flexor carpi radialis muscle**
- **Flexor digitorum superficialis**
- **Palmaris longus muscle**
- **Flexor carpi ulnaris muscle**
- **Palmar aponeurosis**
- **Palmaris brevis muscle**
- **Abductor pollicis brevis**

The tendon of what muscle forms the palmar aponeurosis?

---

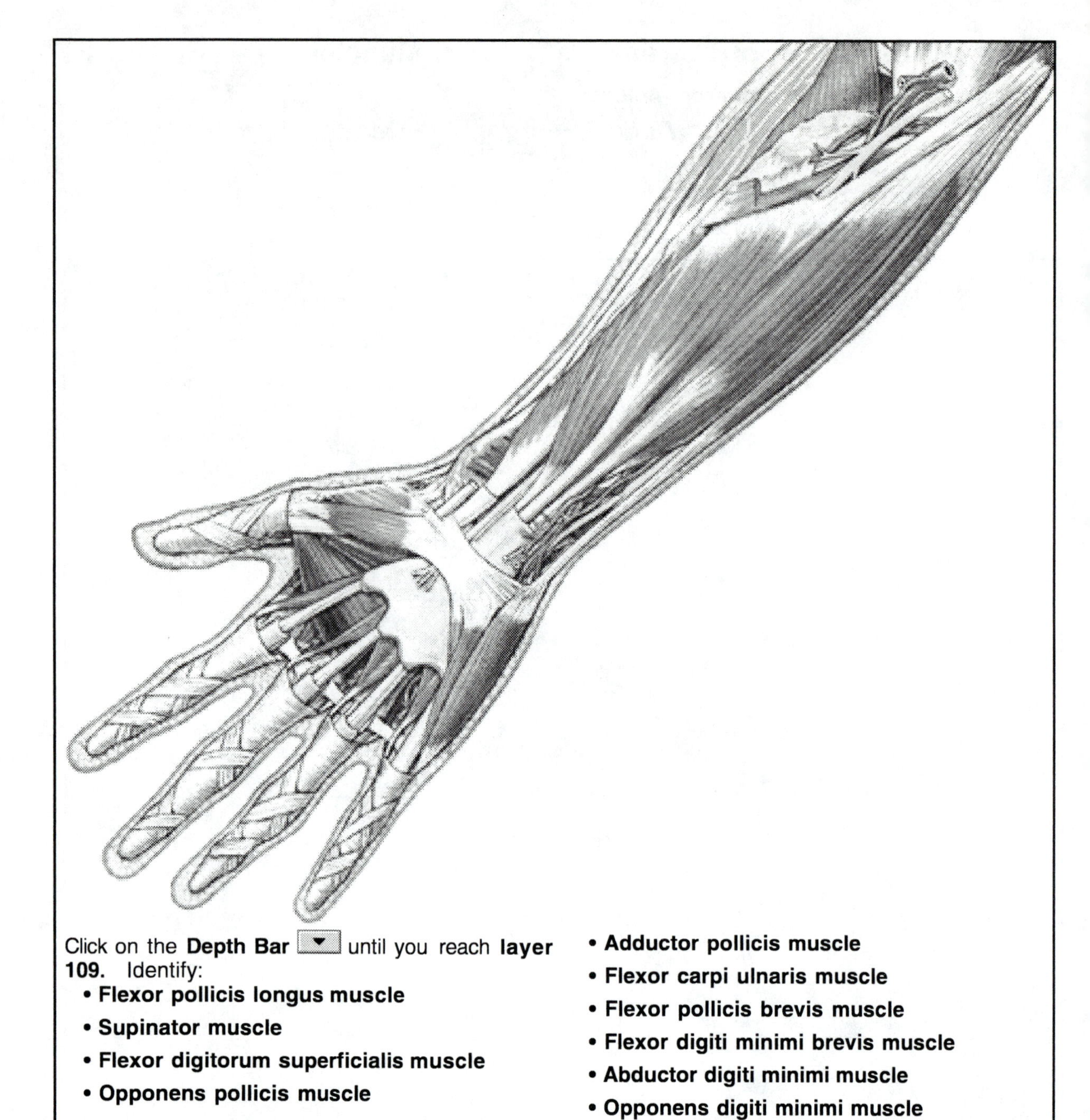

Click on the **Depth Bar** [▼] until you reach **layer 109.** Identify:

- **Flexor pollicis longus muscle**
- **Supinator muscle**
- **Flexor digitorum superficialis muscle**
- **Opponens pollicis muscle**
- **Adductor pollicis muscle**
- **Flexor carpi ulnaris muscle**
- **Flexor pollicis brevis muscle**
- **Flexor digiti minimi brevis muscle**
- **Abductor digiti minimi muscle**
- **Opponens digiti minimi muscle**

Click on the **Depth Bar** [▼] until you reach **layer 111.** No graphic has been provided for this layer. Identify:

- **Opponens digiti minimi** muscle.

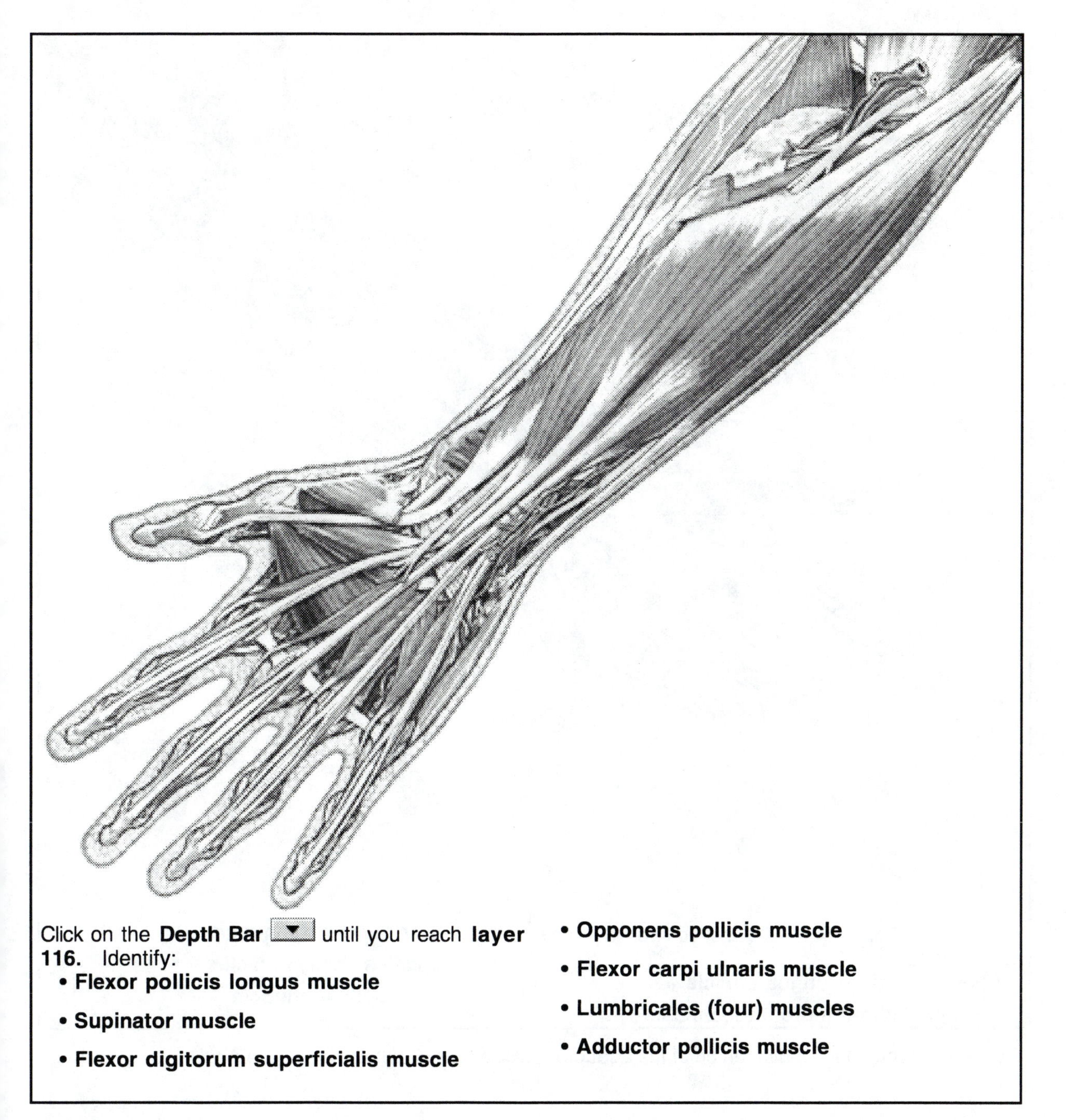

Click on the **Depth Bar** [▼] until you reach **layer 116.** Identify:

- **Flexor pollicis longus muscle**
- **Supinator muscle**
- **Flexor digitorum superficialis muscle**
- **Opponens pollicis muscle**
- **Flexor carpi ulnaris muscle**
- **Lumbricales (four) muscles**
- **Adductor pollicis muscle**

Based on the location of the distal tendon, which muscle located in the anterior forearm will flex the thumb?

---

Based on the location of the distal tendon, which muscle located in the anterior forearm will flex the fingers?

---

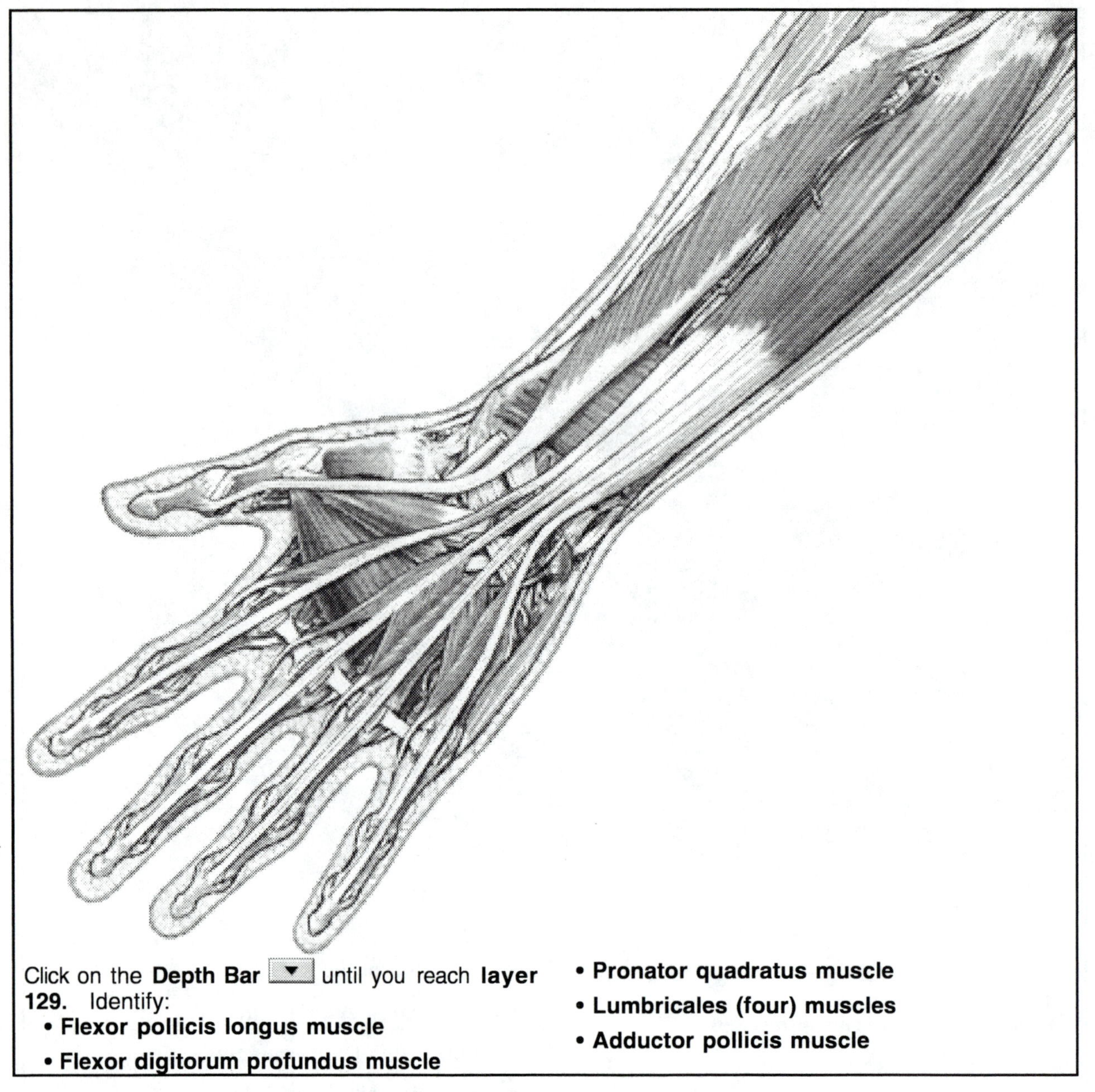

Click on the **Depth Bar** [▼] until you reach **layer 129.** Identify:

- **Flexor pollicis longus muscle**
- **Flexor digitorum profundus muscle**
- **Pronator quadratus muscle**
- **Lumbricales (four) muscles**
- **Adductor pollicis muscle**

Considering the origin and insertion of the adductor pollicis muscle, what would its action be?

---

Click on the **Depth Bar** [▼] until you reach **layer 132.** No graphic has been provided for this layer. Identify the **pronator quadratus muscle** and the **adductor pollicis muscle**.

Drag the **Depth Bar** [▼] down to **layer 325.** No graphic has been provided for this layer. Identify:

- **Palmar interosseus muscles** (identify all four)
- **Dorsal interosseus muscles** (identify all four)

# Posterior Dissection of Lower Arm & Hand Muscles

Click on the **View** button and choose **Posterior**.

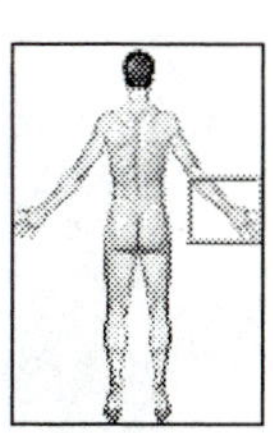

Click on the **Normal** button. Windows— —Mac

Drag the **Depth Bar** to **layer 0.**

Position the navigator rectangle over the lower right arm.

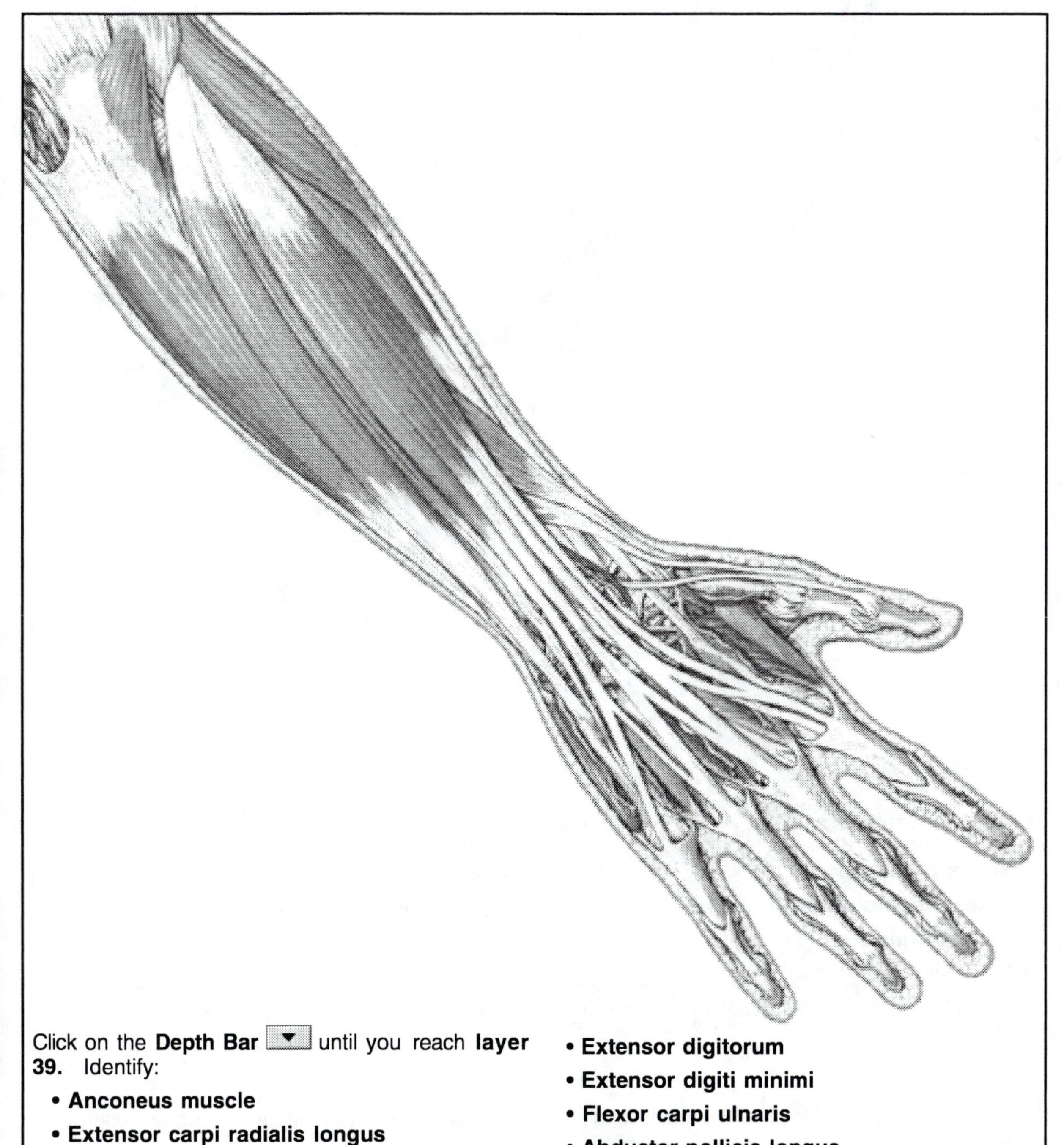

Click on the **Depth Bar** until you reach **layer 39.** Identify:

- **Anconeus muscle**
- **Extensor carpi radialis longus**
- **Extensor carpi radialis brevis**
- **Extensor digitorum**
- **Extensor digiti minimi**
- **Flexor carpi ulnaris**
- **Abductor pollicis longus**
- **Extensor carpi ulnaris**

Based on the location of the distal tendon, which muscle located over the posterior forearm will extend the fingers?

---

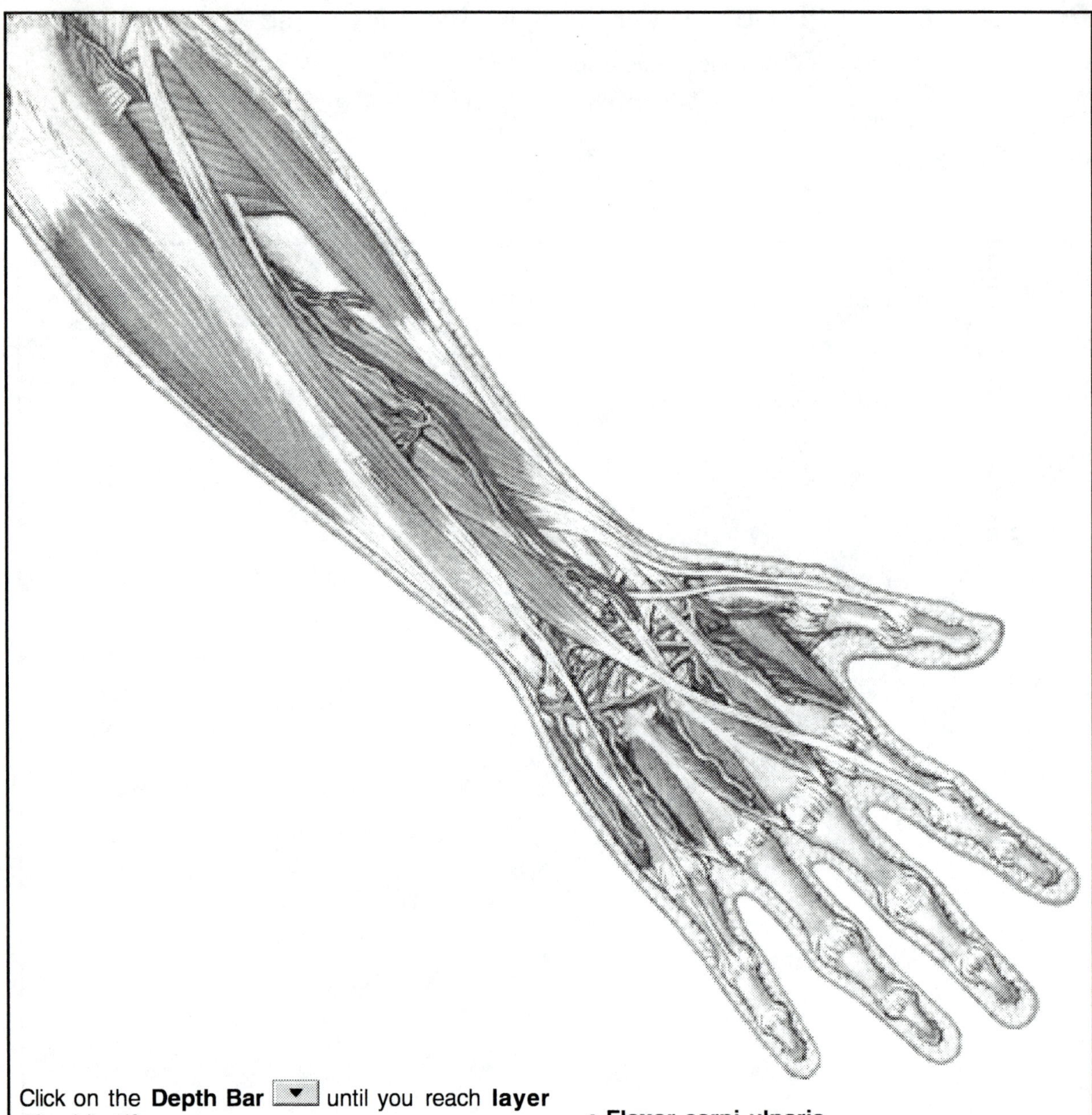

Click on the **Depth Bar** [▼] until you reach **layer 43.** Identify:

- **Supinator muscle**
- **Extensor carpi radialis brevis**
- **Extensor indicis**
- **Extensor digiti minimi**
- **Flexor carpi ulnaris**
- **Abductor pollicis longus**
- **Extensor pollicis brevis**
- **Extensor pollicis longus**

Based on the location of the distal tendon, which muscle located over the posterior forearm will extend the little finger?

---

Based on the location of the distal tendon, which muscle located over the posterior forearm will extend the index finger?

---

Based on the location of the distal tendon, which muscles located over the posterior forearm will extend the thumb?

---

## Lateral Dissection of Arm Muscles

Click on the **View** button and choose **Lateral Arm**.

Click on the **Normal** button. Windows— —Mac

Drag the **Depth Bar** to **layer 0.**

Position the navigator rectangle over the right arm.

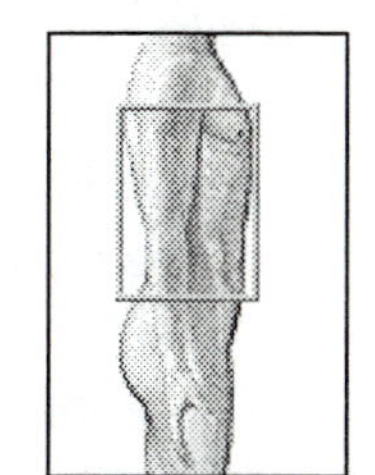

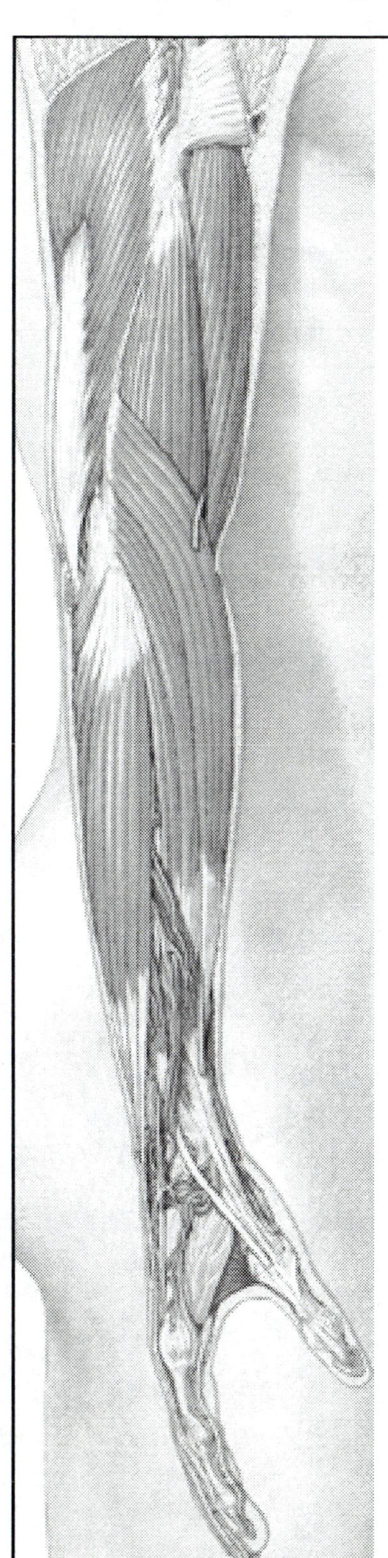

Click on the **Depth Bar** until you reach **layer 14.** You will need to scroll to see all the muscles. Identify the following on the diagram to the left:

- **Triceps brachii muscle** (long and lateral head)
- **Brachialis muscle**
- **Brachioradialis muscle**
- **Biceps brachii muscle** (long head)
- **Extensor carpi radialis longus muscle**
- **Extensor carpi radialis brevis muscle**
- **Extensor digitorum muscle**
- **Anconeus muscle**

Click on the **Depth Bar** until you reach **layer 28.** You will need to scroll to see all the muscles. Identify the following on the diagram to the right:

- **Triceps brachii muscle** (medial head)
- **Proximal and distal tendon of triceps brachii muscle** (medial head)
- **Brachialis muscle**
- **Biceps brachii muscle** (long head)
- **Supinator muscle**
- **Pronator teres muscle**
- **Adductor pollicis longus muscle**
- **Extensor pollicis brevis muscle**
- **Extensor pollicis longus muscle**
- **Extensor indicus muscle**

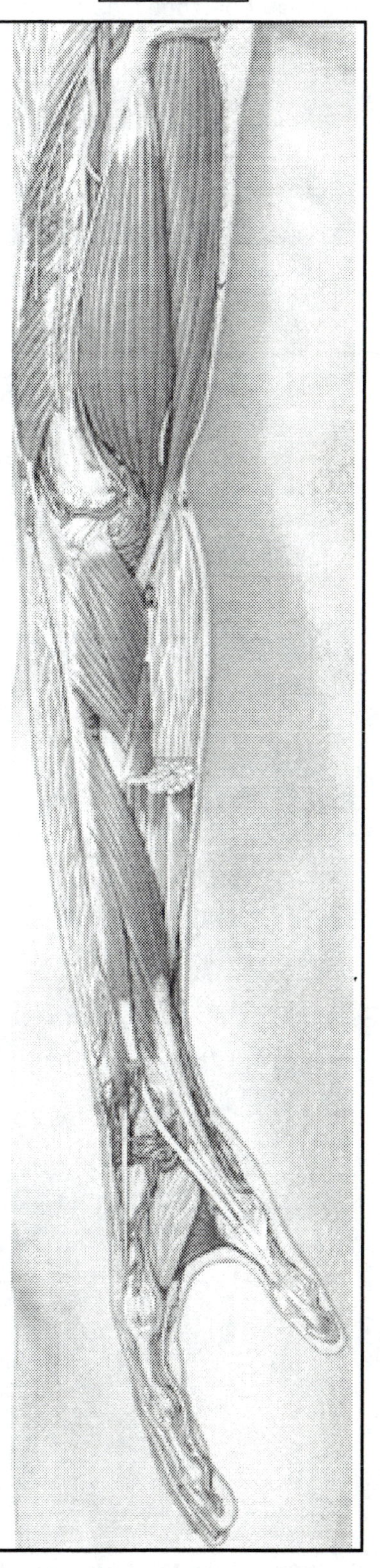

Click on the **Depth Bar** ▼ until you reach **layer 40.** Into what bone does the biceps brachii insert?

_______________________________________________

Into what bone does the proximal tendon of the lateral branch of the triceps brachii muscle insert?

_______________________________________________

Into what bone does the proximal tendon of the brachioradialis muscle insert?

_______________________________________________

The tendon of the medial head of the triceps brachii muscle originates on what bone?

_______________________________________________

Into what bone does the tendon of the medial head of the triceps brachii muscle insert?

_______________________________________________

## Medial Dissection of Arm Muscles

Click on the **View** button and choose **Medial Arm**.

Click on the **Normal** button. Windows — — Mac

Drag the **Depth Bar** to **layer 0.**

Position the navigator rectangle over the upper arm.

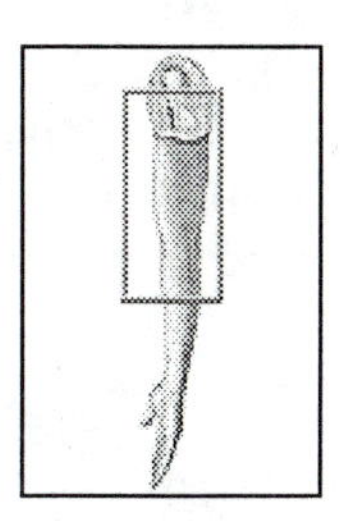

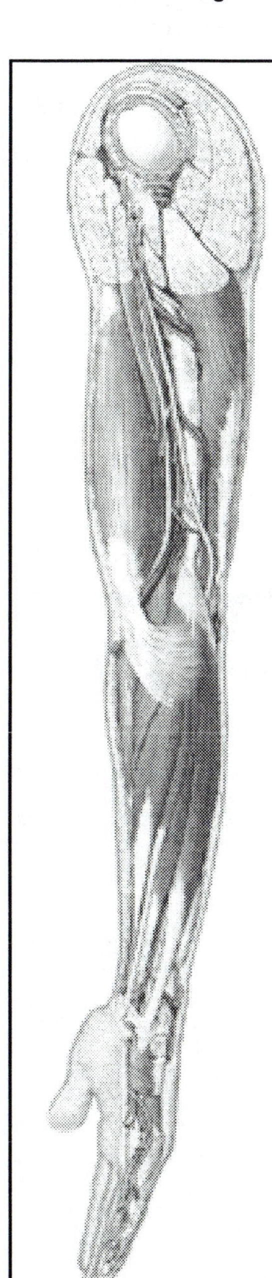

Click on the **Depth Bar** until you reach **layer 7.** You will need to scroll to see all of the muscles. Identify the following on the diagram to the left:

- **Triceps brachii muscle** (long head)
- **Biceps brachii muscle** (short head)
- **Coracobrachialis muscle**
- **Pronator teres muscle**
- **Brachialis muscle**
- **Brachioradialis muscle**
- **Flexor carpi radialis muscle**
- **Palmaris longus muscle**
- **Flexor digitorum superficialis muscle**
- **Flexor carpi ulnaris muscle**
- **Palmaris brevis muscle**

Click on the **Depth Bar** until you reach **layer 40.** You will need to scroll to see all of the muscles. Identify the following on the diagram to the right:

- **Triceps brachii muscle (long & medial heads)**
- **Biceps brachii muscle (short head)**
- **Coracobrachialis muscle**
- **Brachialis muscle**
- **Brachioradialis muscle**
- **Flexor pollicis longus muscle**
- **Flexor digitorum profundus muscle**
- **Pronator quadratus muscle**
- **Supinator muscle**

When you are finished, **close** the window.

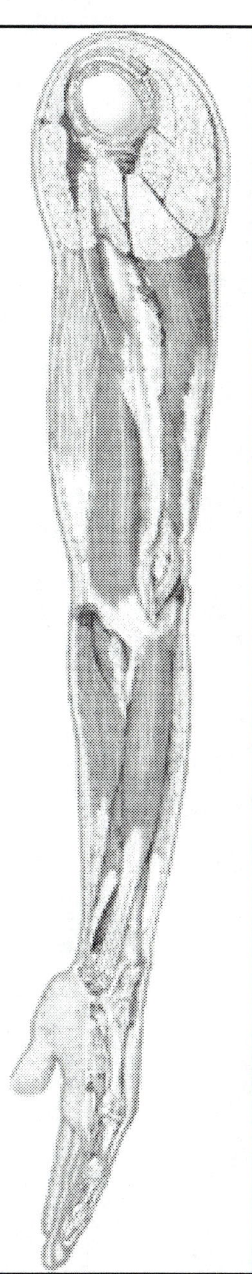

## Cadaver Image of the Muscles of the Upper arm

Under the **File** menu, drag down to **Open** (then to **Content** if you are working in Windows). Click on **Atlas Anatomy**. Select **Region** and **Upper Limb.** Choose **Anterior Arm**. If the list is not in alphabetical order, the image is about one-fourth of the way down the list.

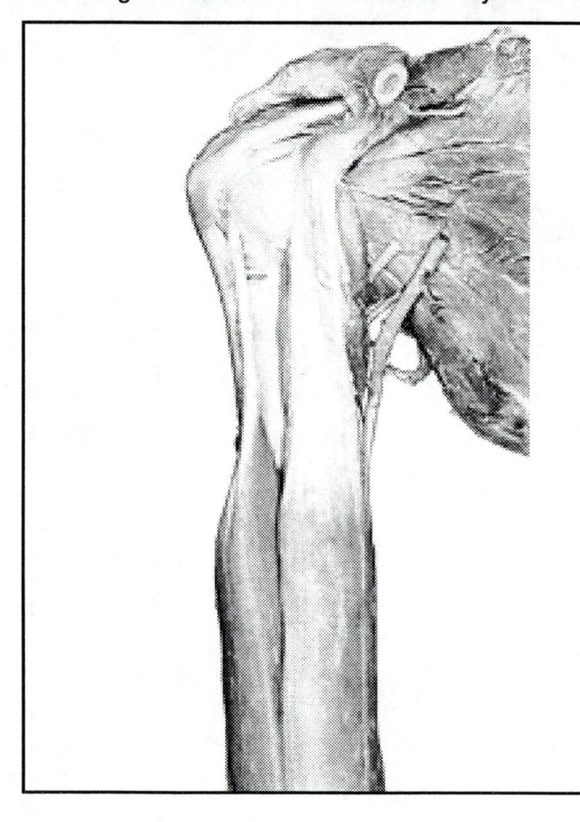

This is an image of the arm and shoulder of a cadaver. You may need to scroll to find the structures listed below. Identify:

- **Biceps brachii muscle** (short and long head)
- **Distal tendon of biceps brachii muscle**
- **Tendons of biceps brachii muscle** (long and short head)
- **Body of humerus**
- **Coracoid process of scapula**

When you are finished, **close** the window.

## Cross Sections of the Arm

Under the **File** menu, drag down to **Open** (then to **Content** if you are working in Windows). **Atlas Anatomy** should still be selected. **Region** and **Upper Limb** should still be selected. Choose **Midarm (Inf)**. If the list is not in alphabetical order, the image is about one-fourth of the way down the list.

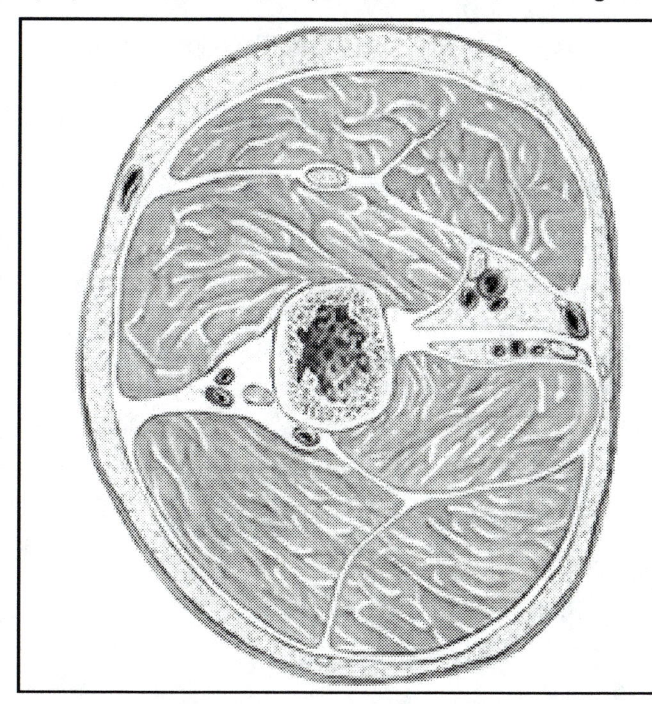

This is a cross section of the upper arm. Identify:

- **Short head of biceps brachii muscle**
- **Long head of biceps brachii muscle**
- **Brachialis muscle**
- **Long head of triceps brachii muscle**
- **Medial head of triceps brachii muscle**
- **Lateral head of triceps brachii muscle**
- **Humerus**

When you are finished, **close** the window.

Under the **File** menu, drag down to **Open** (then to **Content** if you are working in Windows). **Atlas Anatomy** should still be selected. **Region** and **Upper Limb** should still be selected. Choose **Forearm (Inf).** If the list is not in alphabetical order, the image is about half way down the list.

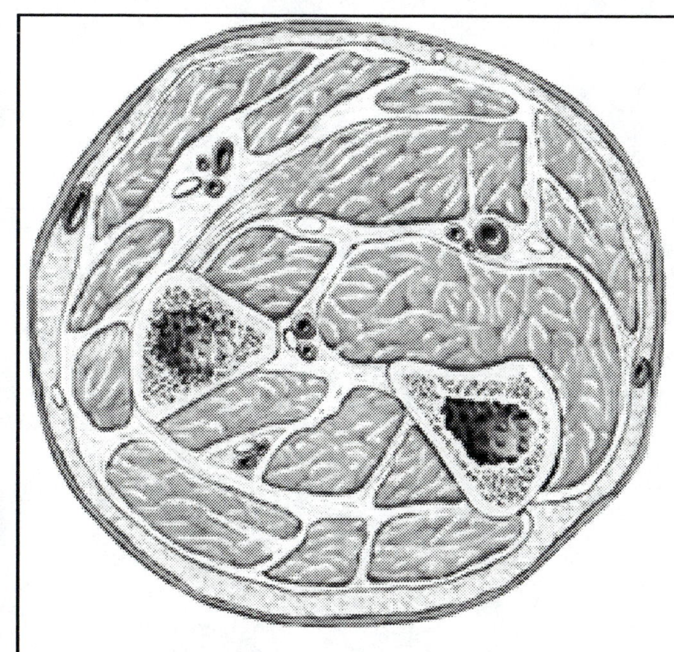

This is a cross section of the forearm. Identify:

- **Ulna**
- **Radius**
- **Abductor pollicis longus muscle**
- **Brachioradialis muscle**
- **Extensor carpi radialis brevis muscle**
- **Extensor carpi radialis longus muscle**
- **Extensor carpi ulnaris muscle**
- **Extensor digiti minimi muscle**
- **Extensor digitorum muscle**
- **Extensor pollicis brevis muscle**
- **Extensor pollicis longus muscle**
- **Flexor carpi radialis muscle**
- **Flexor carpi ulnaris muscle**
- **Flexor digitorum profundus muscle**
- **Flexor digitorum superficialis muscle**
- **Flexor pollicis longus muscle**
- **Palmaris longus muscle**

When you are finished, **close** the window.

## Objectives: Arm & Hand Muscles

To learn the anatomy of the important arm muscles, particularly those that move the arm at the elbow, hand, and fingers.

**Muscles of Lower Arm**

- Abductor digiti minimi muscle
- Abductor pollicis brevis muscle
- Abductor pollicis longus muscle
- Anconeus muscle
- Adductor pollicis longus muscle
- Dorsal interosseus muscle
- Extensor carpi radialis brevis muscle
- Extensor carpi radialis longus muscle
- Extensor carpi ulnaris muscle
- Extensor digiti minimi muscle
- Extensor digitorum muscle
- Extensor indicis muscle
- Extensor pollicis brevis muscle
- Extensor pollicis longus muscle
- Flexor carpi radialis muscle
- Flexor carpi ulnaris muscle
- Flexor digiti minimi muscle
- Flexor digitorum profundus muscle
- Flexor digitorum superficialis muscle
- Flexor pollicis brevis muscle
- Flexor pollicis longus muscle
- Lumbricalis muscle
- Opponens digiti minimi muscle
- Opponens pollicis muscle
- Palmar interosseus muscle
- Palmaris brevis muscle
- Palmaris longus muscle
- Pronator quadratus muscle
- Supinator muscle

**Muscles of Upper Arm**

- Biceps brachii muscle (short and long heads)
- Brachialis muscle
- Brachioradialis muscle
- Coracobrachialis muscle
- Deltoid muscle
- Triceps brachii muscle (long, medial, and lateral heads)

**Other**

- Palmar aponeurosis
- Body of humerus
- Coracoid process of scapula
- Ulna
- Radius

# Upper Leg Muscles

## Opening A.D.A.M Interactive Anatomy

Open A.D.A.M. Interactive Anatomy according to the directions in the Tutorial.

## Anterior Dissection of Upper Leg Muscles

Choose **Dissectible Anatomy**.

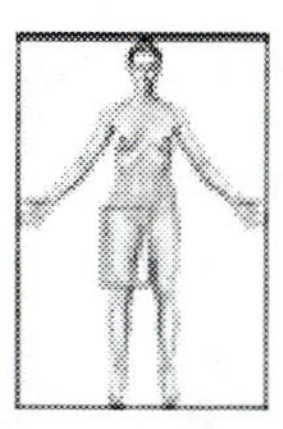

Choose female. Select **Anterior.** Click **Open**.

Enlarge the window and position the navigation rectangle over the upper leg/lower pelvis area.

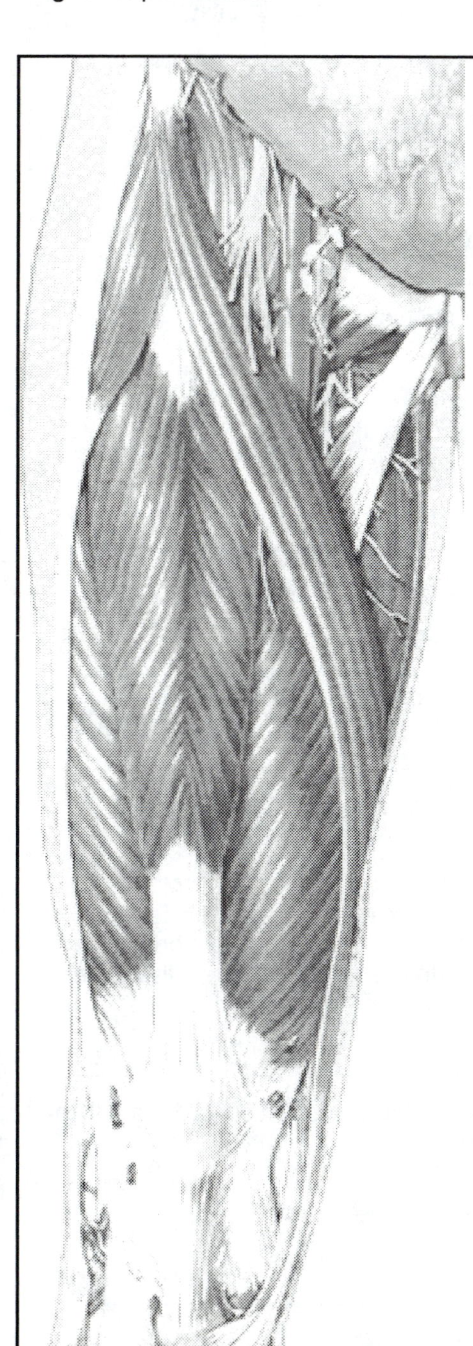

Click several times on the **Depth Bar** ▼ to remove the skin, fascia, and blood vessels until you reach **layer 23**. Then scroll down to **layer 179**. You will not be able to see the entire image to the left on screen and will need to scroll to see all the parts. Identify:

- **Iliotibial tract**
- **Tensor fasciae latae muscle**
- **Sartorius muscle**
- **Rectus femoris muscle**
- **Vastus lateralis muscle**
- **Vastus medialis muscle**
- **Gracilis muscle**
- **Adductor magnus muscle** (ischiocondylar part)
- **Adductor longus muscle**
- **Pectineus muscle**

Scroll down to the knee area and identify:

- **Patellar ligament**

The origin of the sartorius muscle is located on what bone?

________________________________________

Into what bone does the sartorius muscle insert?

________________________________________

Based on its origin and insertion, what is the action of the sartorius muscle?

________________________________________

Which four muscles form the quadriceps femoris muscle group?

_______________________________ _______________________________

_______________________________ _______________________________

Scroll to hip area. Click several times on the **Depth Bar** [▼] to remove the sartorius and tensor fasciae latae muscles until you reach **layer 186**. No graphic has been provided for this layer. Identify:

- **Gluteus medius muscle**

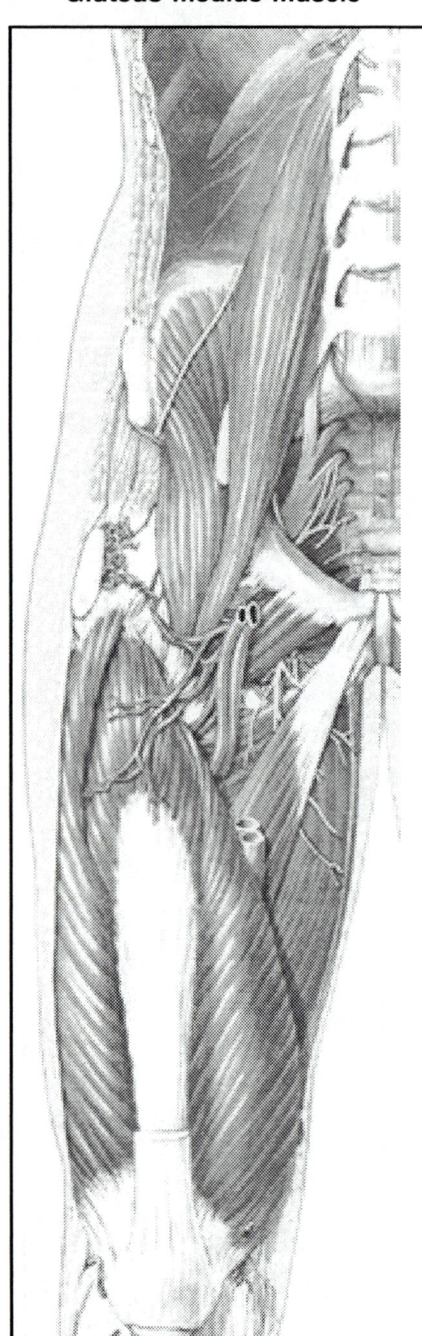

Click on the **Depth Bar** [▼] to remove the rectus femoris muscles and other structures until you reach **layer 266**. Identify:

- **Psoas major muscle**
- **Iliacus muscle**
- **Gluteus medius muscle** (cut)
- **Vastus lateralis muscle**
- **Vastus medialis muscle**
- **Vastus intermedius muscle**
- **Gracilis muscle**
- **Adductor magnus muscle** (ischiocondylar part)
- **Adductor longus muscle**
- **Adductor brevis muscle**
- **Pectineus muscle**

What is the origin of the vastus lateralis muscle?

_______________________________

What is the insertion of the vastus lateralis muscle?

_______________________________

What is the origin of the vastus intermedius muscle?

_______________________________

What is the insertion of the vastus intermedius muscle?

_______________________________

Scroll back up to **layer 185** for a moment and note the tendons of the vastus lateralis, vastus medialis, and vastus intermedius which insert into the patellar tendon. Into what bone does the patellar tendon insert?

_______________________________

Based on the information above, what is the action of these three muscles?

_______________________________

Scroll back down to **layer 266**, then click several times on the **Depth Bar** to remove the psoas major muscle and other structures until you reach **layer 280**. (No graphic has been provided for this level.) Identify:

- **Iliacus muscle**
- **Vastus lateralis muscle**
- **Pectineus muscle**
- **Adductor longus muscle**

Click several times on the **Depth Bar** to remove the vastus lateralis and vastus medialis muscles until you reach **layer 290**. No graphic has been provided for this layer. Identify:

- **Vastus intermedius muscle**

Click several times on the **Depth Bar** to remove the vastus intermedius muscles and other structures until you reach **layer 299**. No graphic has been provided for this layer. Identify:

- **Adductor longus muscle**

What is the origin and insertion of the adductor longus muscle?

______________________________________________

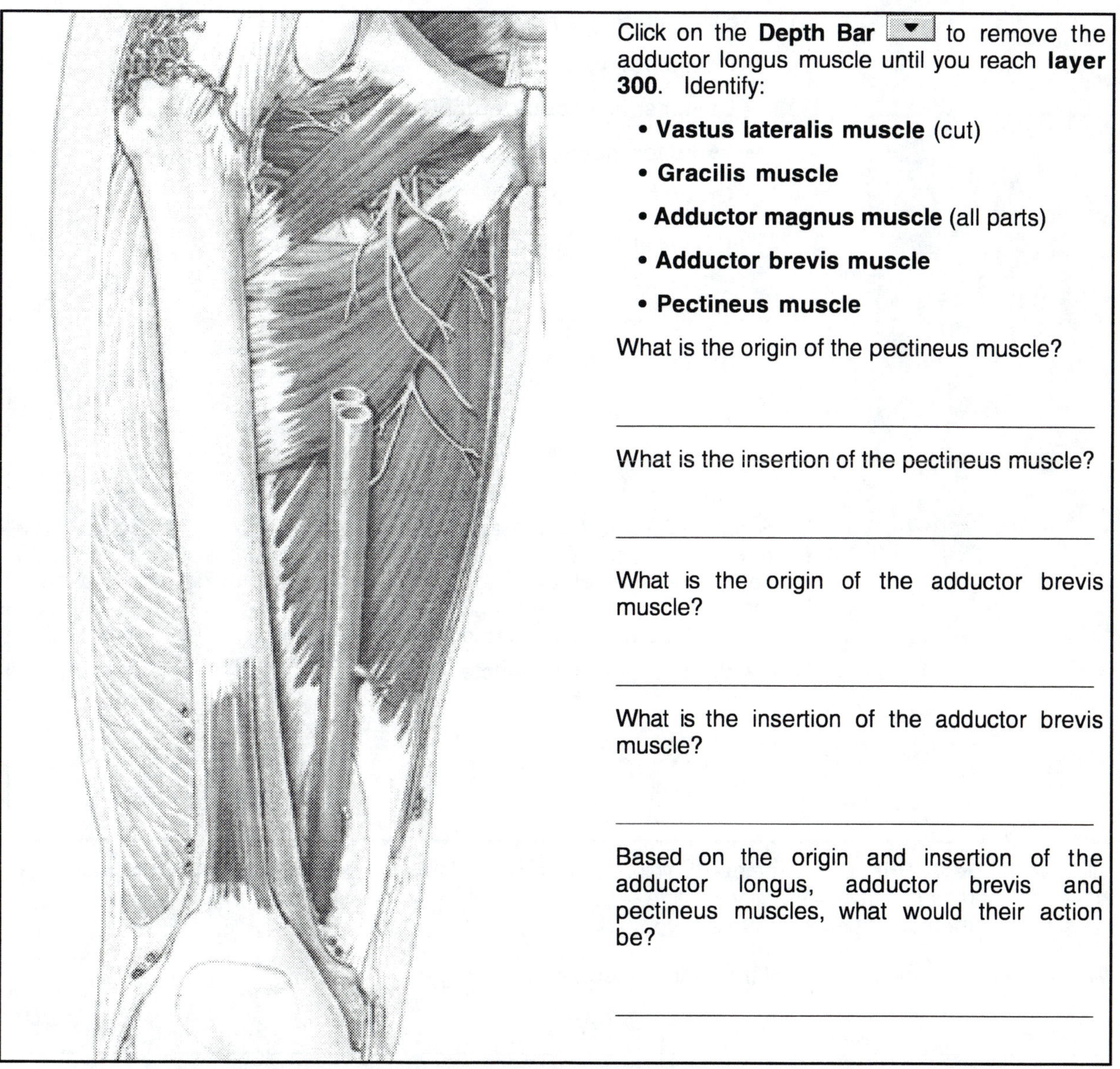

Click on the **Depth Bar** to remove the adductor longus muscle until you reach **layer 300**. Identify:

- **Vastus lateralis muscle** (cut)
- **Gracilis muscle**
- **Adductor magnus muscle** (all parts)
- **Adductor brevis muscle**
- **Pectineus muscle**

What is the origin of the pectineus muscle?

______________________________________________

What is the insertion of the pectineus muscle?

______________________________________________

What is the origin of the adductor brevis muscle?

______________________________________________

What is the insertion of the adductor brevis muscle?

______________________________________________

Based on the origin and insertion of the adductor longus, adductor brevis and pectineus muscles, what would their action be?

______________________________________________

Click on the **Depth Bar** until you reach **layer 304**. You will remove the adductor brevis and pectineus muscles. No graphic has been provided for this layer. Identify:

- **Adductor magnus muscle** (all parts)

## Posterior Dissection of Upper Leg Muscles

Click on the **View** button and choose **Posterior**.

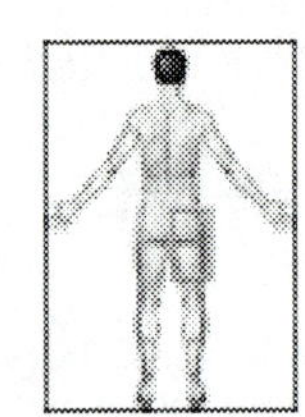

Click on the **Normal** button. Windows— —Mac

Drag the **Depth Bar** to **layer 0.**

Position the navigator rectangle over the right hip and upper thigh.

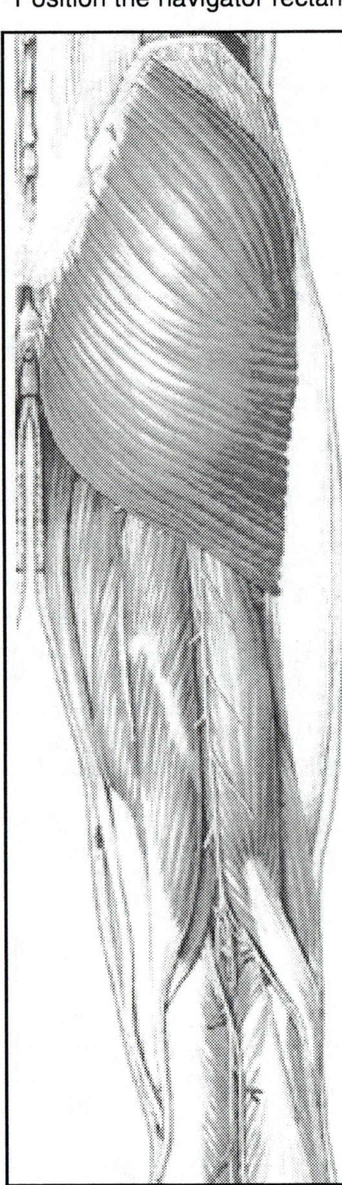

Click several times on the **Depth Bar** to remove the skin, fascia, and blood vessels until you reach **layer 81**. You will need to scroll from the hip to the knee area as you proceed. Identify:

- **Gluteus maximus muscle**
- **Iliotibial tract**
- **Gluteal fascia over gluteus medius muscle**
- **Biceps femoris muscle** (long and short head)
- **Semitendinosus muscle**
- **Semimembranosus muscle**
- **Adductor magnus muscle** (ischiocondylar part)
- **Gracilis muscle**

What is the origin of the gluteus maximus muscle?

_______________________________________________

Scroll back up to the hip area. Click once on the **Depth Bar** to remove the fascia covering the gluteus medius and iliotibial tract. You should be at **level 82**. No graphic has been provided for this layer. Identify:

- **Gluteus medius muscle**
- **Vastus lateralis muscle**

Click on the **Depth Bar** to remove the gluteus maximus. You should be at **level 83**. No graphic has been provided for this layer. Identify:

- **Gluteus medius muscle**

What is the origin and insertion of the gluteus medius muscle?

_______________________________________________

Based on the origin and insertion of the gluteus medius muscle, what bone is moved when these muscles contract and how is that bone moved?

_______________________________________________

Scroll down to the thigh and identify:

- **Biceps femoris muscle** (long head, proximal)
- **Semitendinosus muscle** (proximal)

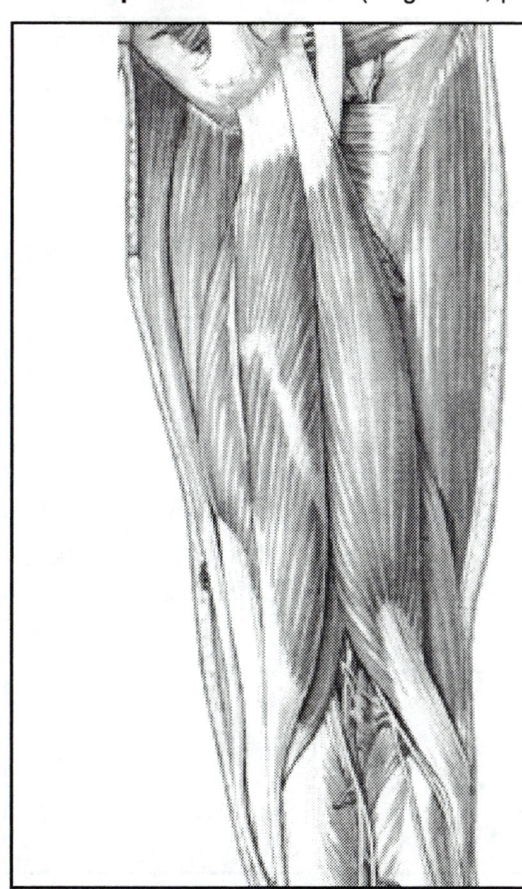

Click on the **Depth Bar** ▼ until you reach **layer 113**. Identify:

- **Vastus lateralis muscle**
- **Biceps femoris muscle** (long and short head)
- **Semitendinosus muscle**
- **Semimembranosus muscle**
- **Adductor magnus muscle** (ischiocondylar part)
- **Gracilis muscle**

What is the bone of origin for the semitendinosus muscle?

___________________________________

What three muscles compose the hamstrings muscle group?

___________________________________

___________________________________

___________________________________

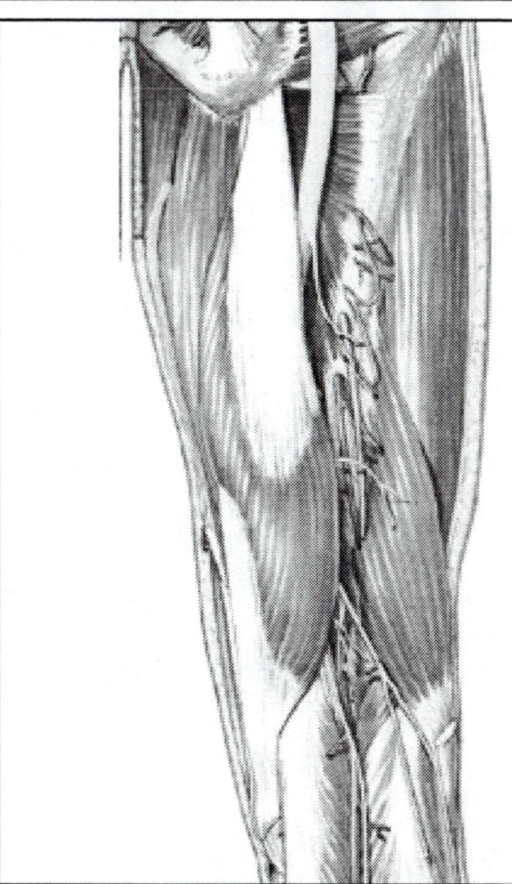

Click on the **Depth Bar** ▼ to remove the gracilis muscle, long head of the biceps femoris muscle, and semitendinosus muscles until you reach **layer 116**. Identify:

- **Vastus lateralis muscle**
- **Biceps femoris muscle** (short head)
- **Semimembranosus muscle**
- **Adductor magnus muscle** (ischiocondylar part)

What bone serves as the origin of the short head of the biceps femoris muscle?

___________________________________

What bone serves as the origin of the semimembranosus muscle?

___________________________________

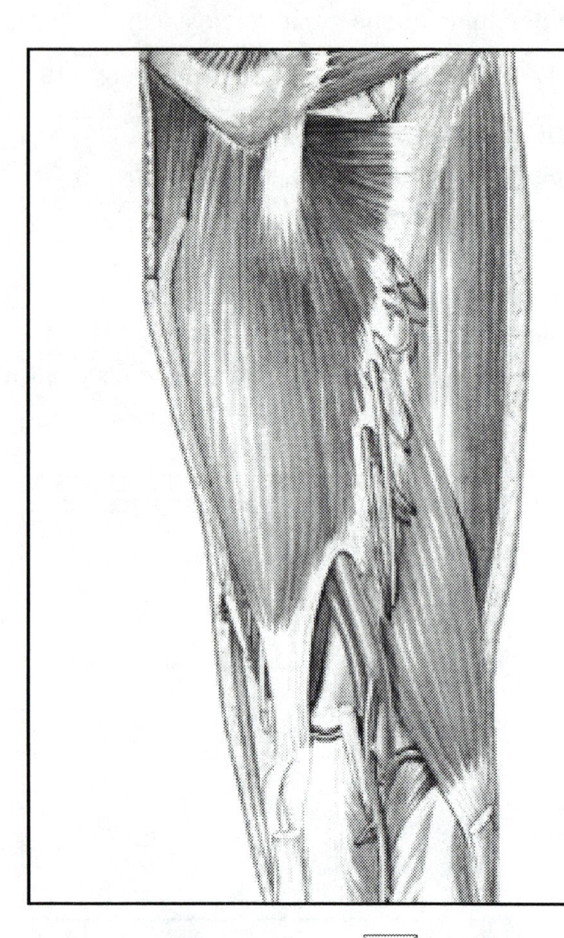

Click on the **Depth Bar** [▼] to remove the semimembranosus muscle until you reach **layer 121**. Identify:

- **Vastus lateralis muscle**
- **Biceps femoris muscle** (short head)
- **Adductor magnus muscle** (ischiocondylar part)

What is the origin of the adductor magnus muscle?

______________________________________________

What is the insertion of the adductor magnus muscle?

______________________________________________

Based on the origin and insertion of the adductor magnus muscle, what would its action be?

______________________________________________

Click once on the **Depth Bar** [▼] to remove the adductor magnus muscle to **level 122**. No graphic has been provided for this layer. Identify:

- **Adductor longus**
- **Adductor brevis**

Click on the **Depth Bar** [▼] to remove the adductor longus and adductor brevis muscles until you reach **level 124**. No graphic has been provided for this layer. Identify:

- **Vastus medialis muscle**

## Lateral Dissection of Upper Leg Muscles

Click on the **View** button and choose **Lateral**.

Click on the **Normal** button. Windows— —Mac

Drag the **Depth Bar** to **layer 0.**

Position the navigator rectangle over the hip and upper thigh.

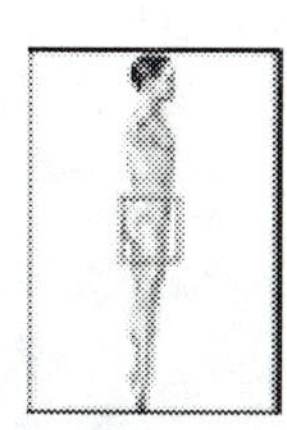

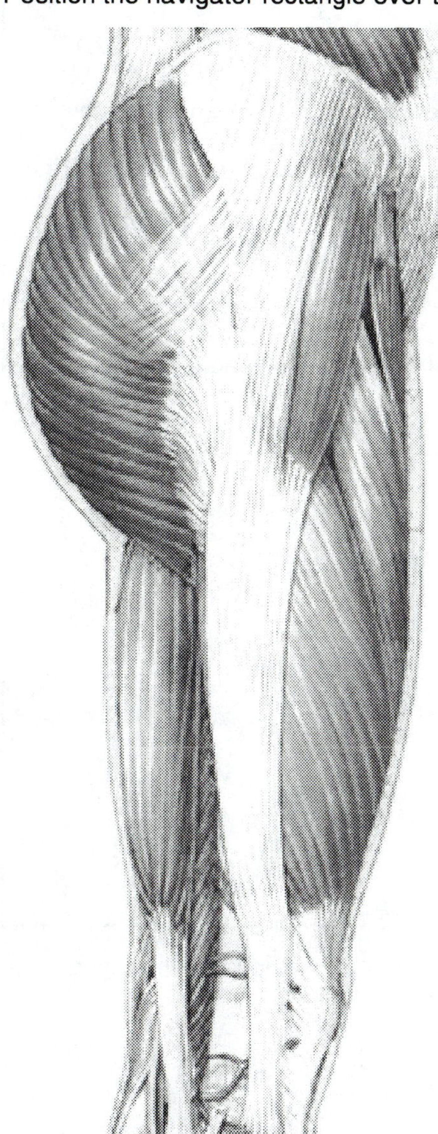

Click several times on the **Depth Bar** to remove the skin, fascia, and blood vessels until you reach **layer 12**. You may need to scroll downward to see all the muscles below. Identify:

- **Gluteus maximus muscle**
- **Iliotibial tract**
- **Gluteal fascia over gluteus medius muscle**
- **Tensor fasciae latae muscle**
- **Vastus lateralis muscle**
- **Rectus femoris muscle**
- **Sartorius muscle**
- **Biceps femoris muscle** (long and short head)

What is the origin of the tensor fasciae latae muscle?

______________________________

What two muscles connect into the iliotibial tract?

______________________________

______________________________

Scroll down to the knee area. Into what bone does the iliotibial tract insert?

______________________________

When these two muscles contract at the same time, what motion occurs?

______________________________

Click several times on the **Depth Bar** to **layer 87**. Note the removal of the tensor fasciae latae muscle and the iliotibial tract. No graphic has been provided for this level. Identify:

- **Vastus lateralis muscle**

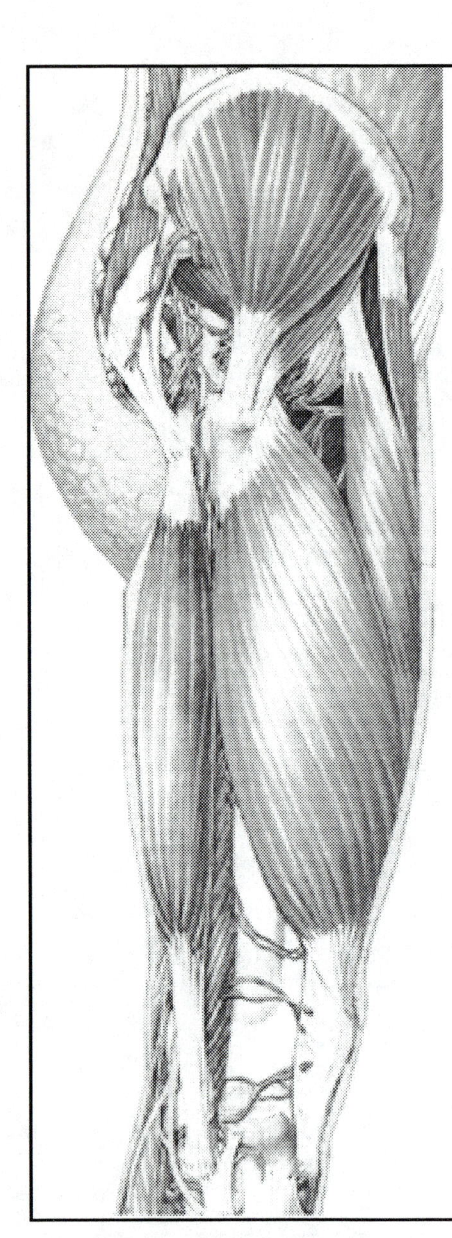

Click several times on the **Depth Bar** [▼] to **layer 94**. Note the removal of the gluteus maximus muscle. Identify:

- **Gluteus medius muscle**
- **Vastus lateralis muscle**
- **Rectus femoris muscle**
- **Sartorius muscle**
- **Biceps femoris muscle** (long and short head)

What is the origin of the long head of the biceps femoris muscle?

_______________________________________________

What is the insertion of the long head of the biceps femoris muscle?

_______________________________________________

Based on the origin and insertion of the long head of the biceps femoris muscle, what would its action be?

_______________________________________________

Click several times on the **Depth Bar** [▼] to **layer 103**. No graphic has been provided for this layer. Note the removal of the gluteus medius muscle and the vastus lateralis muscle. Identify:

- **Vastus intermedius muscle**

Click several times on the **Depth Bar** [▼] to **layer 106**. No graphic has been provided for this layer. Note the removal of the long head of the biceps femoris muscle. Identify:

- **Short head of biceps femoris muscle**

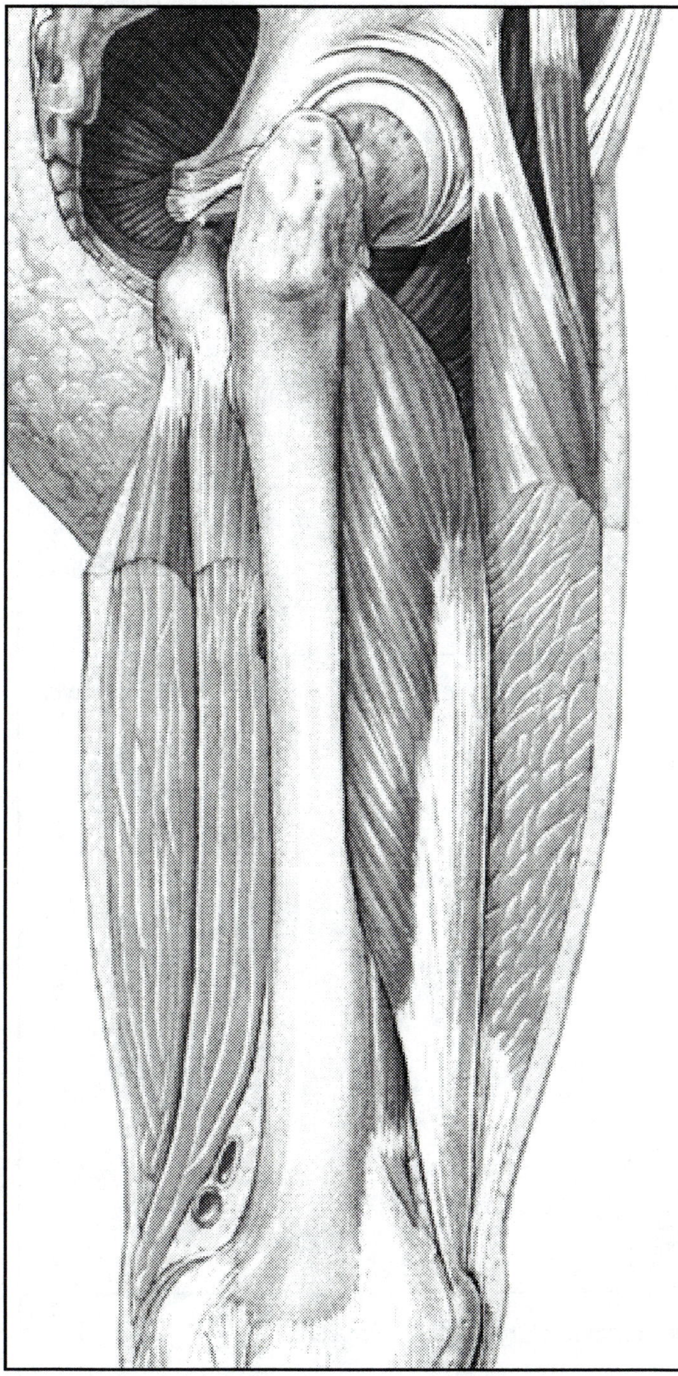

Click several times on the **Depth Bar** [▼] to **layer 267**. Note the removal of the short head of the biceps femoris muscle. Identify the following muscles from anterior to posterior:

- **Rectus femoris muscle**
- **Vastus intermedius muscle**
- **Semimembranosus muscle**
- **Semitendinosus muscle**

What is the origin and insertion of the rectus femoris muscle?

______________________________

______________________________

Base on its origin and insertion, what would be the action of the rectus femoris muscle?

______________________________

______________________________

Based on the location of the quadriceps femoris and hamstrings muscle groups, contrast their actions.

______________________________

______________________________

______________________________

______________________________

## Medial Dissection of Upper Leg Muscles

Click on the **View** button and choose **Medial**.

Click on the **Normal** button. 

Drag the **Depth Bar** to **layer 0.**

Position the navigator rectangle over the upper thigh.

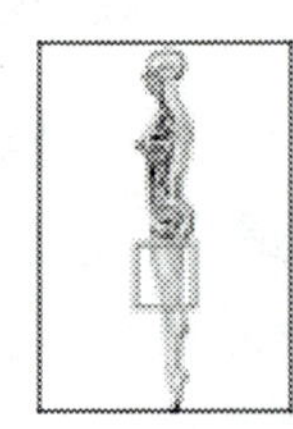

Click several times on the **Depth Bar** to remove the skin, fascia, and blood vessels until you reach **layer 9**. Identify:

- **Vastus medialis muscle**
- **Sartorius muscle**
- **Gracilis muscle**
- **Adductor magnus muscle** (ischiocondylar part)
- **Semimembranosus muscle**
- **Semitendinosus muscle**

Into what bone does the gracilis muscle insert?

_______________________________________

When the gracilis muscle contracts what motion occurs?

_______________________________________

Click on the **Depth Bar** to **level 13**. Note the removal of the sartorius muscle.

Click on the **Depth Bar** to **layer 15**. Note the removal of the gracilis muscle.

Click once on the **Depth Bar** to **layer 16**. Note the removal of the semimembranosus muscle.

Into what bone does the semitendinosus muscle insert?

_______________________________________

Click once on the **Depth Bar** [▼] to **level 17**. Note the removal of the semitendinosus muscle. No graphic has been provided for these layers. Identify:

- **Adductor longus muscle**

Click on the **Depth Bar** [▼] to **layer 21**. Note the removal of the adductor longus muscle.

Click on the **Depth Bar** [▼] to **layer 25**. Note the removal of the adductor magnus muscle and vastus medialis muscles. No graphic has been provided for these layers.

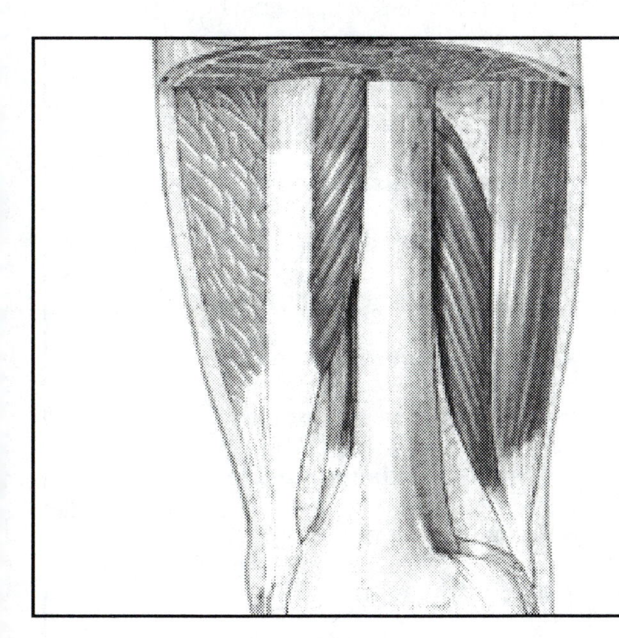

Click on the **Depth Bar** [▼] to **layer 71**. Identify the following muscles from anterior to posterior:

- **Rectus femoris muscle** (cut)
- **Vastus intermedius muscle**
- **Short head of biceps femoris muscle**
- **Long head of biceps femoris muscle**

When you are finished, **close** the window.

## Views of Upper Leg Muscles

Under the **File** menu, drag down to **Open** (then to **Content** if you are working in windows). Click on **Atlas Anatomy**. Select **System** and **Muscular.** Choose **Medial Thigh**. If the list is not in alphabetical order, the image is close to the bottom of the list.

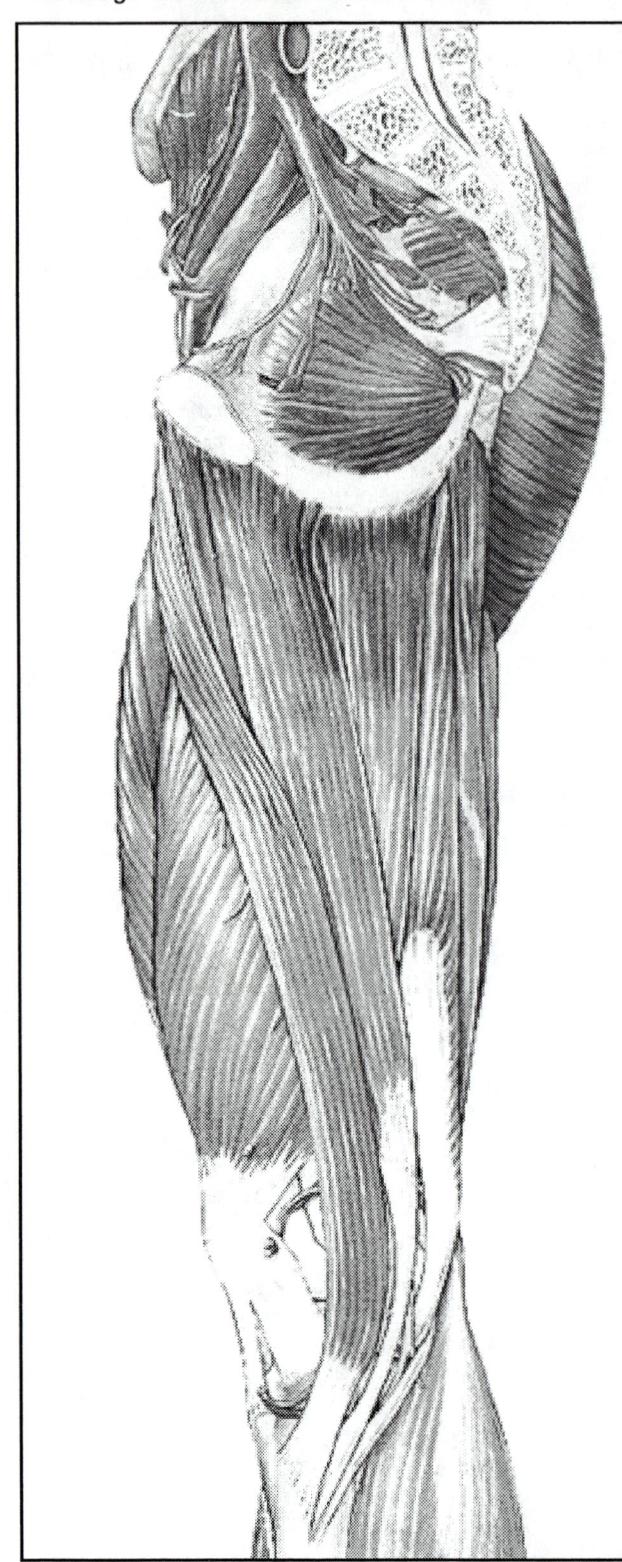

If you are working in Mac, click on the **Show All Pins** button and scroll to **All Systems**.

You may need to scroll to find all the muscles and tendons below. Identify the following by clicking on the heads of the pins.

- **Rectus femoris muscle**
- **Vastus medialis muscle**
- **Sartorius muscle**
- **Gracilis muscle**
- **Adductor magnus muscle**
- **Semimembranosus muscle**
- **Semitendinosus muscle**
- **Gluteus maximus muscle**
- **Pubis** (Inferior ramus of)
- **Iliacus muscle**
- **Patellar ligament**

Which two muscles above are a part of the adductor group?

______________________________________

______________________________________

When you are finished, **close** the window.

Under the **File** menu, drag down to **Open** (then to **Content** if you are working in windows). **Atlas Anatomy** should still be selected. **System** and **Muscular** should still be selected. Choose **Lumbar Plexus In Situ (Ant)**. If the list is not in alphabetical order, the image is close to the bottom of the list.

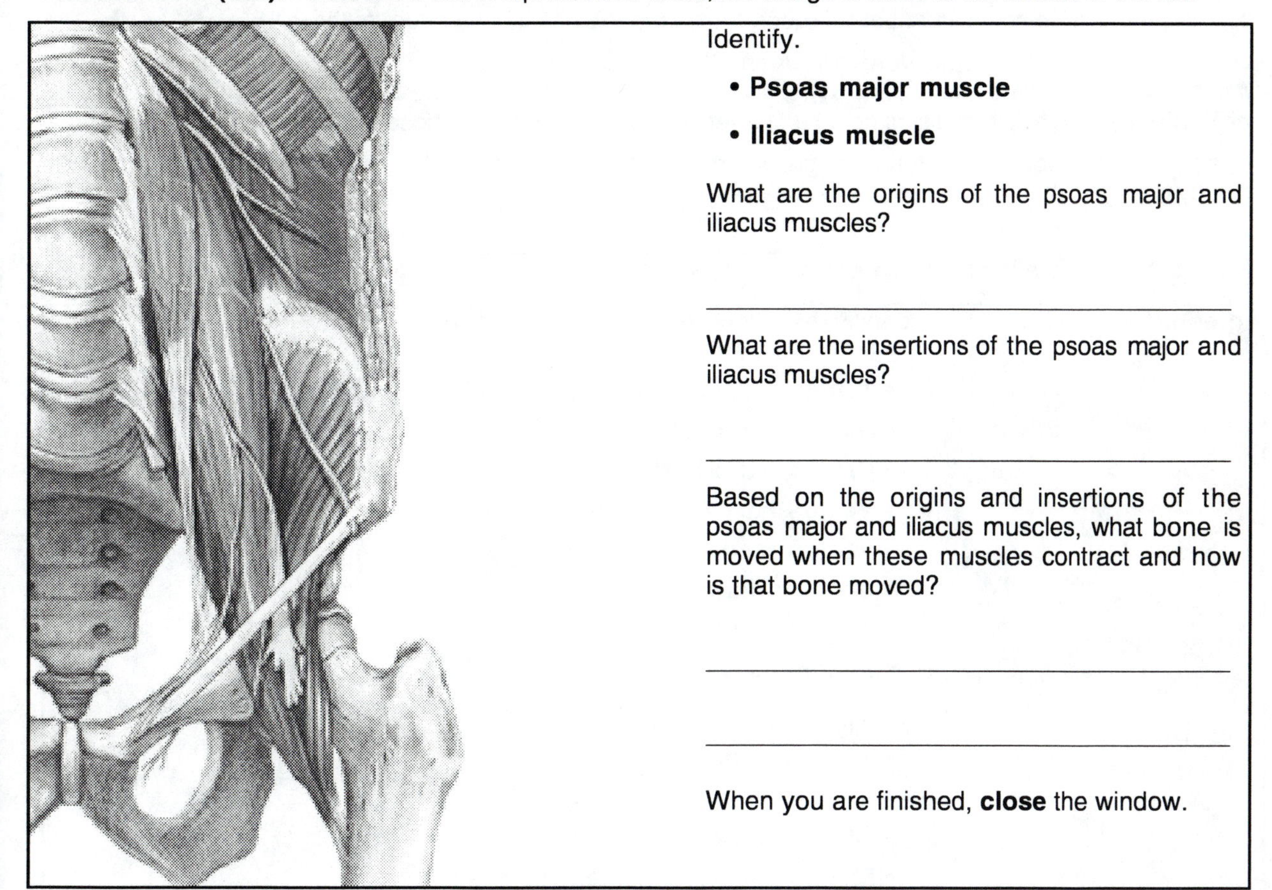

Identify.

- **Psoas major muscle**
- **Iliacus muscle**

What are the origins of the psoas major and iliacus muscles?

_______________________________________

What are the insertions of the psoas major and iliacus muscles?

_______________________________________

Based on the origins and insertions of the psoas major and iliacus muscles, what bone is moved when these muscles contract and how is that bone moved?

_______________________________________

_______________________________________

When you are finished, **close** the window.

## Cadaver Images of the Upper Leg Muscles

Under the **File** menu, drag down to **Open** (then to **Content** if you are working in windows). **Atlas Anatomy** should still be selected. **System** and **Muscular** should still be selected. Choose **Anterior Thigh**. If the list is not in alphabetical order, the image is close to the bottom of the list.

Under the **File** menu, drag down to **Open** (then to **Content** if you are working in windows). **Atlas Anatomy** should still be selected. **System** and **Muscular** should still be selected. Choose **Dissection of Posterior Thigh**. If the list is not in alphabetical order, the image is close to the bottom of the list.

Position the anterior and posterior images so they can both be seen at the same time on the screen. Do this by dragging the gray bar above the images:

If you are working in Windows, click Window, click Tile, and center the images in their respective windows.

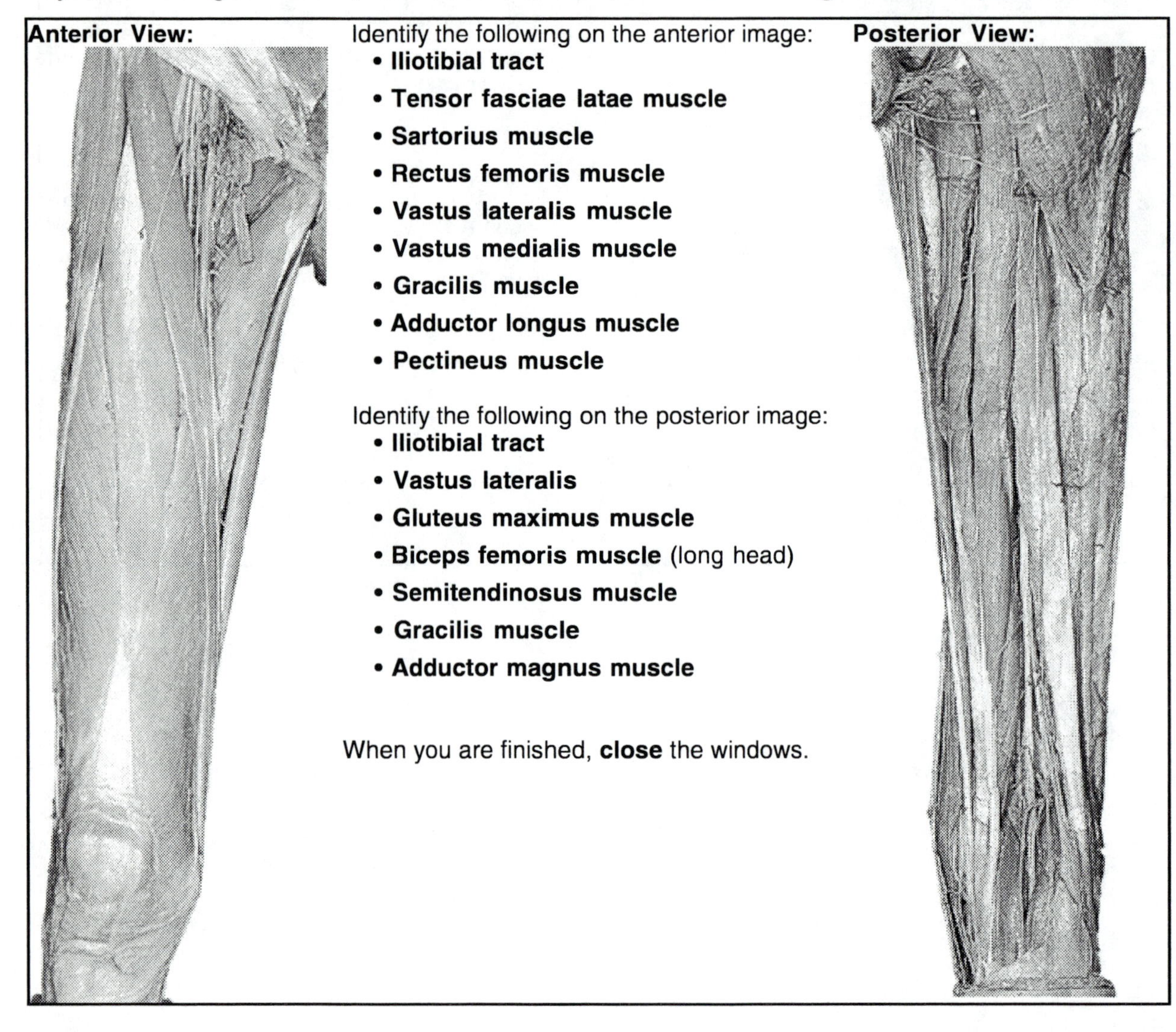

**Anterior View:**
Identify the following on the anterior image:
• **Iliotibial tract**
• **Tensor fasciae latae muscle**
• **Sartorius muscle**
• **Rectus femoris muscle**
• **Vastus lateralis muscle**
• **Vastus medialis muscle**
• **Gracilis muscle**
• **Adductor longus muscle**
• **Pectineus muscle**
Identify the following on the posterior image:
• **Iliotibial tract**
• **Vastus lateralis**
• **Gluteus maximus muscle**
• **Biceps femoris muscle** (long head)
• **Semitendinosus muscle**
• **Gracilis muscle**
• **Adductor magnus muscle**
When you are finished, **close** the windows.
**Posterior View:**

# Cross Sections of the Upper Leg Muscles

Under the **File** menu, drag down to **Open** (then to **Content** if you are working in windows). **Atlas Anatomy** should still be selected. **System** and **Muscular** should still be selected. Choose **Viscera of Female Pelvis (Sup)**. If the list is not in alphabetical order, the image is about three-fourths of the way down.

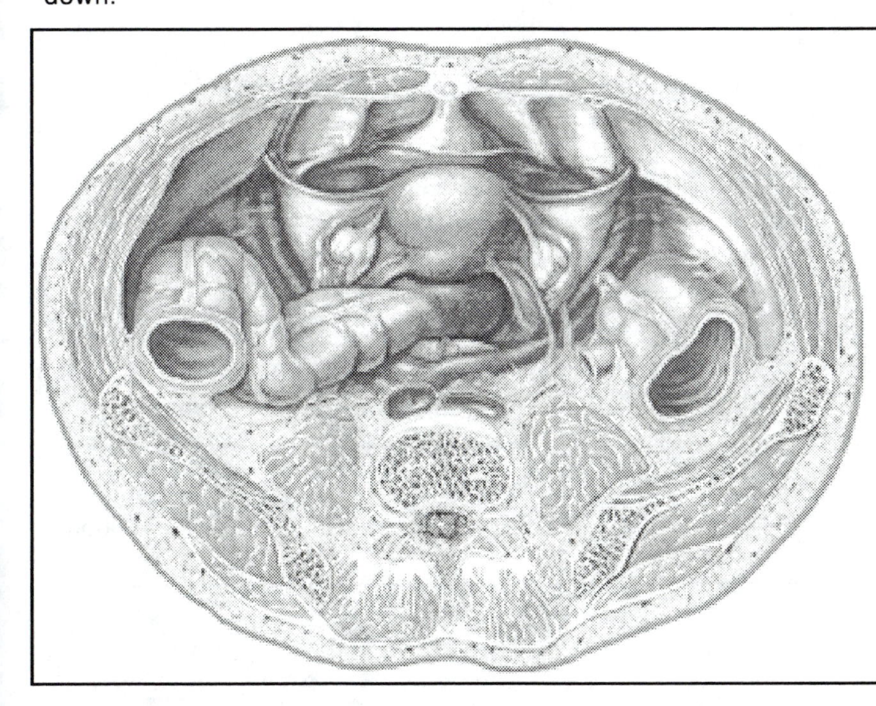

This is a view into the pelvic cavity of the female. Identify:

- **Psoas major muscle**
- **Iliacus muscle**
- **Gluteus medius muscle**
- **Gluteus maximus muscle**

When you are finished, **close** the window.

Under the **File** menu, drag down to **Open** (then to **Content** if you are working in windows). **Atlas Anatomy** should still be selected. **System** and **Muscular** should still be selected. Choose **Proximal Thigh (Inf)**. If the list is not in alphabetical order, the image is close to the bottom of the list.

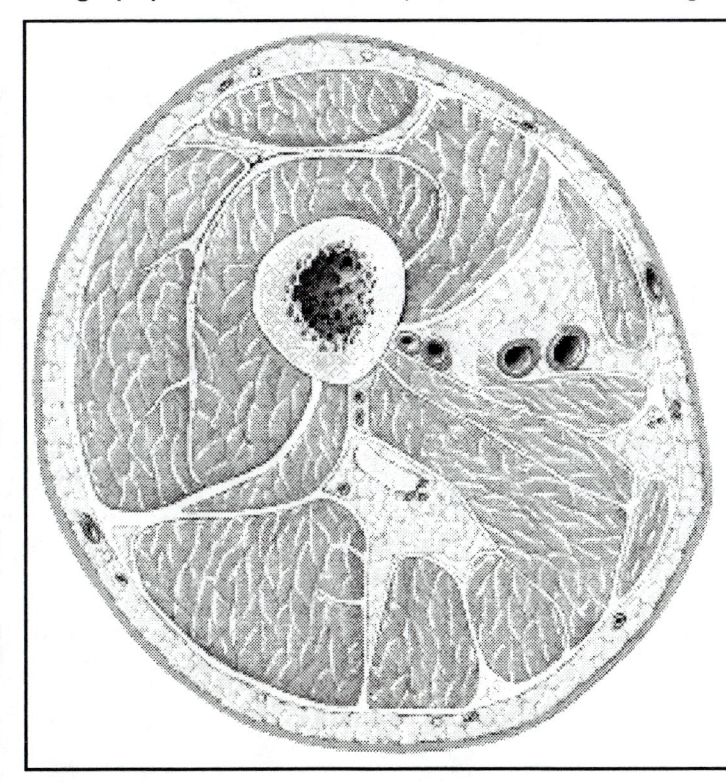

This is a cross section of the upper thigh. Identify:

- **Biceps femoris muscle** (long and short head)
- **Semitendinosus muscle**
- **Semimembranosus muscle**
- **Adductor magnus muscle**
- **Gracilis muscle**
- **Adductor brevis**
- **Adductor longus**
- **Sartorius muscle**
- **Vastus medialis muscle**
- **Rectus femoris**
- **Vastus intermedius**
- **Vastus lateralis**

Circle the following groups on this diagram:

**Hamstrings**
**Adductors**
**Quadriceps femoris**

When you are finished, **close** the window.

Under the **File** menu, drag down to **Open** (then to **Content** if you are working in windows). **Atlas Anatomy** should still be selected. **System** and **Muscular** should still be selected. Choose **Distal Thigh (Inf)**. If the list is not in alphabetical order, the image is close to the bottom of the list.

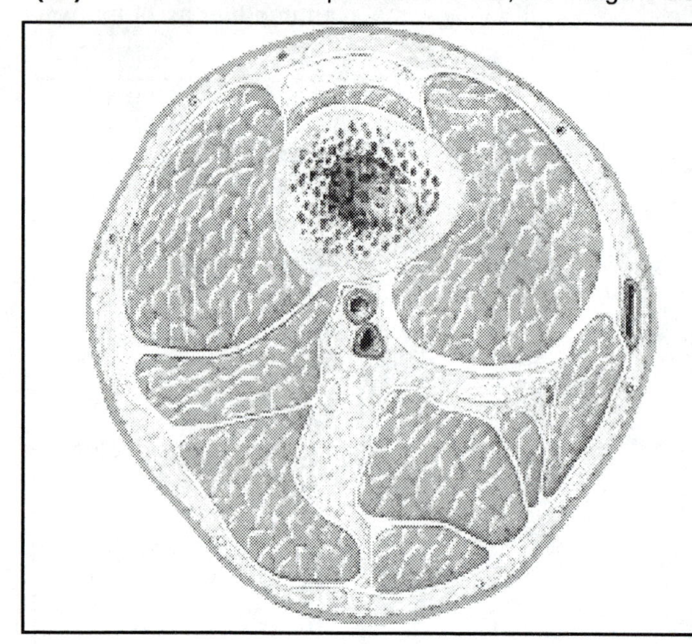

This is a cross section of the lower thigh. Identify:

**Hamstrings:**

- **Biceps femoris muscle** (long and short head)
- **Semitendinosus muscle**
- **Semimembranosus muscle**

**Adductors:**

- **Gracilis muscle**
- **Sartorius muscle**

**Quadriceps femoris:**

- **Vastus medialis muscle**
- **Tendon of quadriceps femoris**
- **Vastus lateralis**

**Iliotibial tract**

When you are finished, **close** the window.

# Objectives: Upper Leg Muscles

Student will learn the anatomy of the important leg muscles, particularly those that move the leg at the hip and knee.

**Muscles**

- Tensor fasciae latae muscle
- Gluteus medius muscle
- Psoas major muscle
- Gluteus maximus muscle
- Iliacus muscle
- Sartorius muscle

Hamstrings group:

- Biceps femoris muscle (long and short head)
- Semitendinosus muscle
- Semimembranosus muscle

Adductors group:

- Adductor magnus muscle
- Gracilis muscle
- Adductor brevis
- Adductor longus
- Pectineus muscle

Quadriceps femoris group:

- Vastus medialis muscle
- Rectus femoris
- Vastus intermedius
- Vastus lateralis

**Other Structures**

- Iliotibial tract
- Patellar ligament
- Gluteal fascia
- Pubis

# Lower Leg Muscles

## Opening A.D.A.M Interactive Anatomy

Open A.D.A.M. Interactive Anatomy according to the directions in the Tutorial.

## Anterior Dissection of Lower Leg Muscles

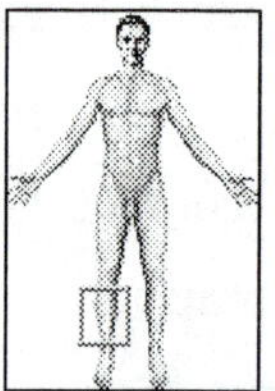

Choose **Dissectible Anatomy**.

Choose male or female. Select **Anterior.** Click **Open**.

Enlarge the window and position the navigation rectangle over the right lower leg.

Click several times on the **Depth Bar** ▼ to remove the skin, fascia, and blood vessels until you reach **layer 186**. Identify:

- **Gastrocnemius muscle**
- **Soleus muscle**
- **Tibia**
- **Tibialis anterior muscle**
- **Extensor digitorum longus muscle**
- **Peroneus longus** (fibularis) **muscle**

Observe the distal tendons of the tibialis anterior and the extensor digitorum longus muscles. Based on the location of these tendons, what is the action of these muscles?

________________________________________________

________________________________________________

________________________________________________

Click several times on the **Depth Bar** ▼ to remove the tibialis anterior muscle. You should now be at **layer 187**. No graphic has been provided for this layer. Identify:

- **Extensor digitorum longus muscle**

# Lateral Dissection of Lower Leg Muscles

Click on the **View** button and choose **Lateral**.

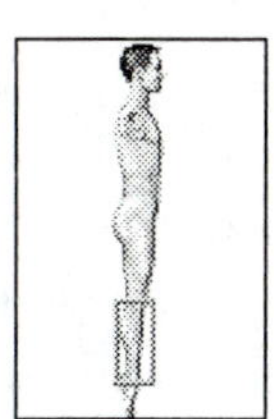

Click on the **Normal** button.

Drag the **Depth Bar** to **layer 0.**

Position the navigator rectangle over the right lower leg.

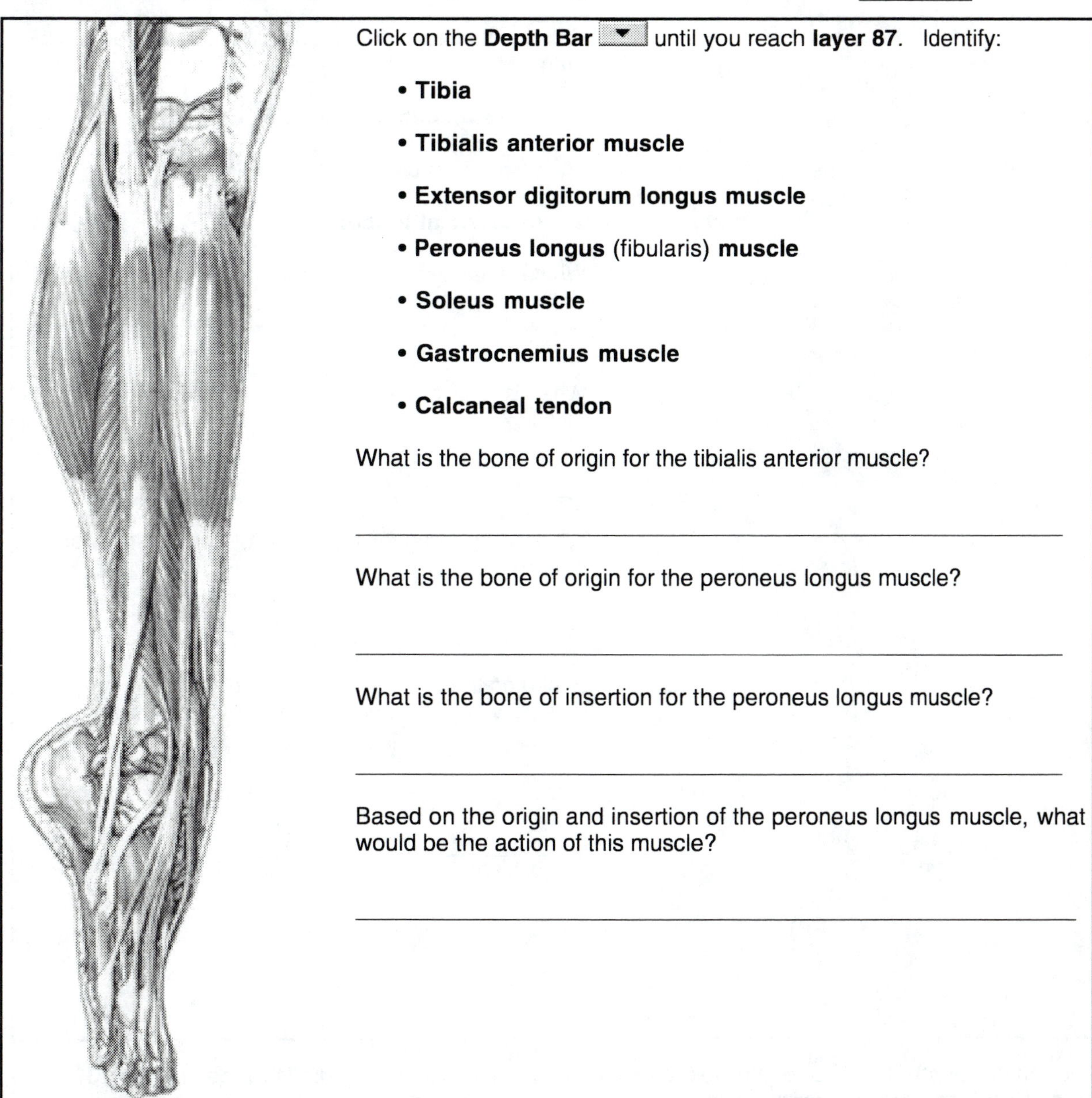

Click on the **Depth Bar** until you reach **layer 87**. Identify:

- **Tibia**
- **Tibialis anterior muscle**
- **Extensor digitorum longus muscle**
- **Peroneus longus** (fibularis) **muscle**
- **Soleus muscle**
- **Gastrocnemius muscle**
- **Calcaneal tendon**

What is the bone of origin for the tibialis anterior muscle?

______________________________

What is the bone of origin for the peroneus longus muscle?

______________________________

What is the bone of insertion for the peroneus longus muscle?

______________________________

Based on the origin and insertion of the peroneus longus muscle, what would be the action of this muscle?

______________________________

Click on the **Depth Bar** to remove the peroneus longus muscle. You should now be at **layer 88**. No graphic has been provided for this layer. Identify:

- **Lateral surface of fibula**

What is the bone of insertion for the extensor digitorum longus muscle?

______________________________

## Posterior Dissection of Lower Leg Muscles

Click on the **View** button and choose **Posterior**.

Click on the **Normal** button. Windows— —Mac

Drag the **Depth Bar** to **layer 0.**

Position the navigator rectangle over the right calf.

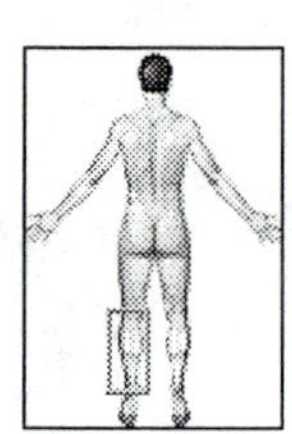

Click on the **Depth Bar** until you reach **layer 136**. Identify:

- **Gastrocnemius muscle**
- **Soleus muscle**
- **Calcaneal tendon**

What is the common name for the calcaneal tendon?

___

What is the bone of origin for the gastrocnemius muscle?

___

What is the bone of insertion for the gastrocnemius muscle?

___

Based on the origin and insertion of the gastrocnemius muscle, what would be the action of this muscle?

___

___

Click on the **Depth Bar** to remove the gastrocnemius muscle. You should now be at **layer 137**. (No graphic has been provided for this layer.) Identify:

- **Soleus muscle**
- **Calcaneal tendon**

Click several times on the **Depth Bar** to remove the soleus muscle. You should now be at **layer 141**. No graphic has been provided for this layer. Identify:

- **Flexor digitorum longus muscle**

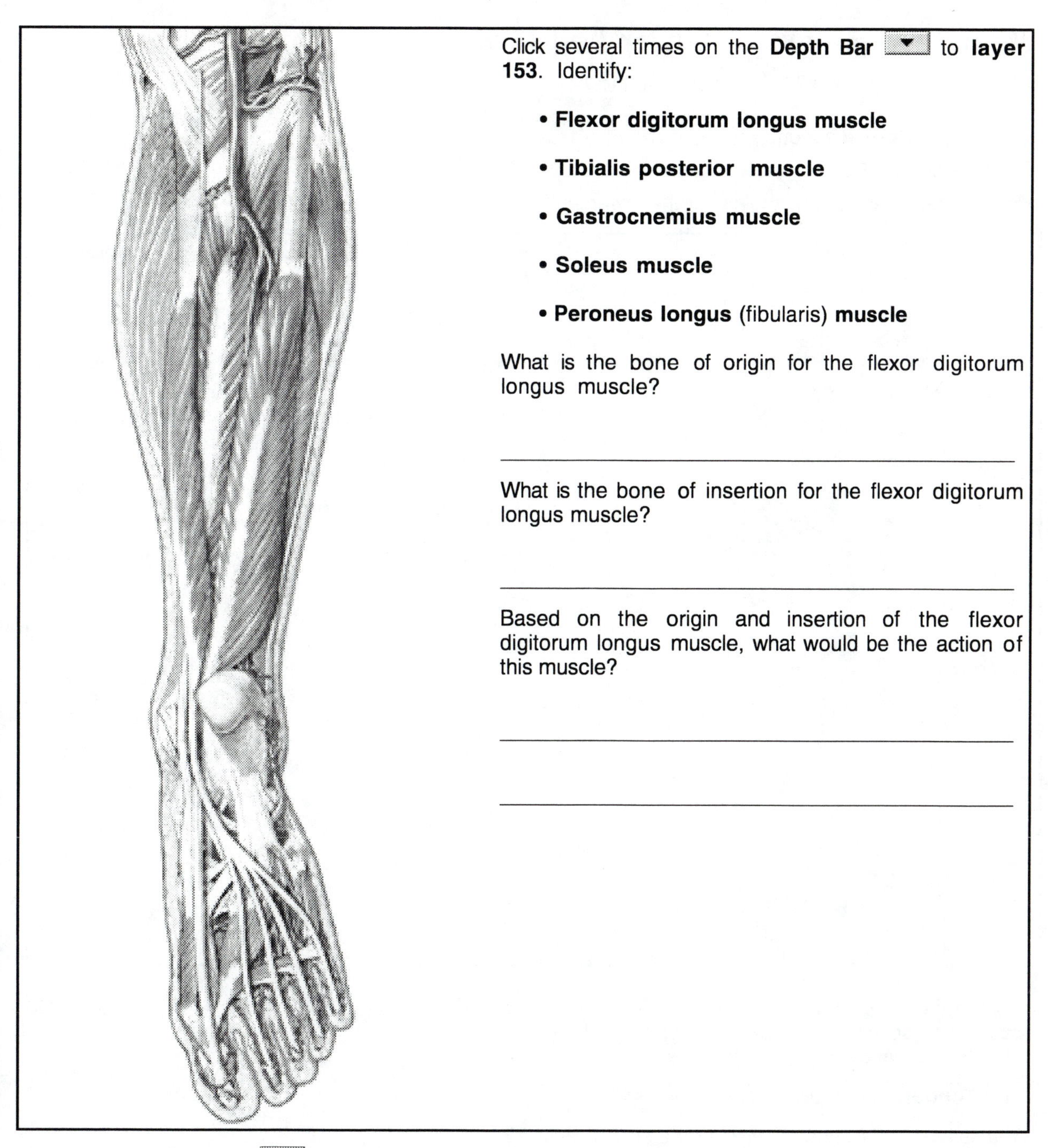

Click several times on the **Depth Bar** to **layer 153**. Identify:

- **Flexor digitorum longus muscle**
- **Tibialis posterior muscle**
- **Gastrocnemius muscle**
- **Soleus muscle**
- **Peroneus longus** (fibularis) **muscle**

What is the bone of origin for the flexor digitorum longus muscle?

______________________________________________

What is the bone of insertion for the flexor digitorum longus muscle?

______________________________________________

Based on the origin and insertion of the flexor digitorum longus muscle, what would be the action of this muscle?

______________________________________________

______________________________________________

Click on the **Depth Bar** until you reach **layer 163**. No graphic has been provided for this layer. Identify:

- **Tibialis posterior muscle**

## Medial Dissection of Lower Leg Muscles

Click on the **View** button and choose **Medial**.

Click on the **Normal** button. Windows— —Mac

Drag the **Depth Bar** to **layer 0.**

Position the navigator rectangle over the lower leg.

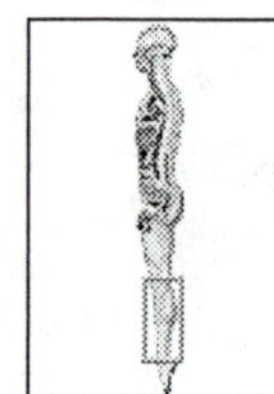

Click on the **Depth Bar** until you reach **layer 26**. Identify:

- **Tibialis anterior muscle**
- **Tibia**
- **Flexor digitorum longus muscle**
- **Soleus muscle**
- **Gastrocnemius muscle**
- **Calcaneal tendon**

Click on the **Depth Bar** to reach **layer 28**. Note the removal of the gastrocnemius muscle. No graphic has been provided for this layer. Identify:

- **Soleus muscle**
- **Tendon of soleus muscle**

What is the medial bone of origin for the soleus muscle?

_______________________________________________

What is the bone of insertion for the soleus muscle?

_______________________________________________

Based on the origin and insertion of the soleus muscle, what would be the action of this muscle?

_______________________________________________

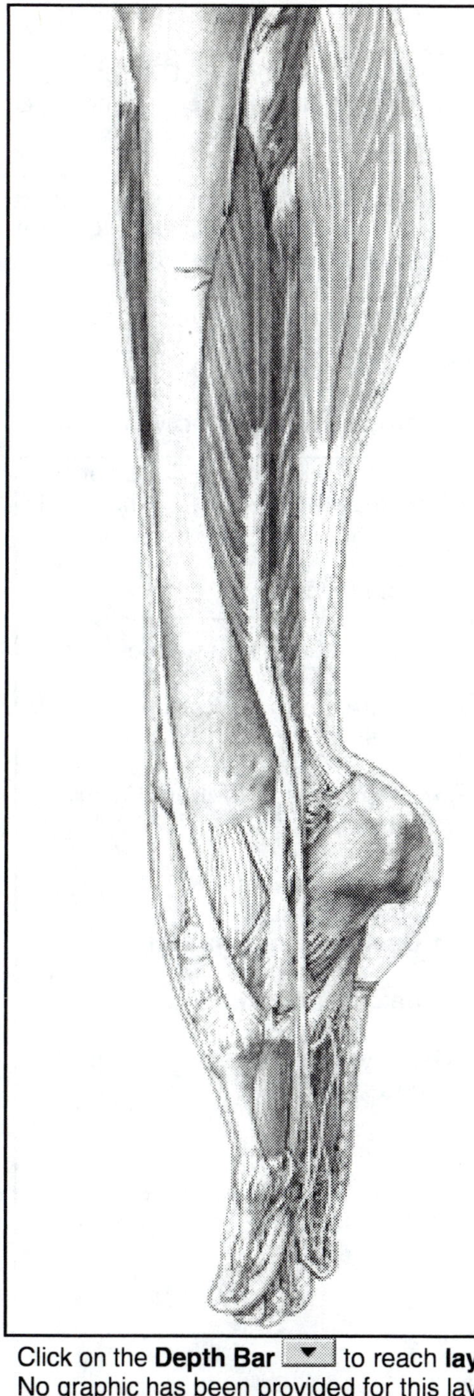

Click on the **Depth Bar** [▼] until you reach **layer 62**. Identify:

- **Tibialis anterior muscle**
- **Tibia**
- **Flexor digitorum longus muscle**
- **Soleus muscle**
- **Gastrocnemius muscle**
- **Calcaneal tendon**

Click on the **Depth Bar** [▼] to reach **layer 63**. Note the removal of the flexor digitorum longus muscle. No graphic has been provided for this layer. Identify:

- **Tibialis posterior muscle**

What is the medial bone of origin for the tibialis posterior muscle?

---

What is the bone of insertion for the tibialis posterior muscle?

---

Based on the origin and insertion of the tibialis posterior muscle, what would be the action of this muscle?

---

When you are done, **close** the window.

## Cadaver Images of Lower Leg Muscles

Under the **File** menu, drag down to **Open** (then to **Content** if you are working in windows). Click on **Atlas Anatomy**. Choose **Region** and **Lower Limb** should still be selected. Choose **Anterior Leg**. If the list is not in alphabetical order, the image is half way down the list.

Under the **File** menu, drag down to **Open** (then to **Content** if you are working in windows). **Atlas Anatomy** should still be selected. **Region** and **Lower Limb** should still be selected. Choose **Lateral Leg & Foot**. If the list is not in alphabetical order, the image is three-fourths of the way down the list.

Position the anterior and posterior images so they can both be seen at the same time on the screen. Do this by dragging the gray bar above the images:

If you are working in Windows, click Window, click Tile, and center the images in their respective windows.

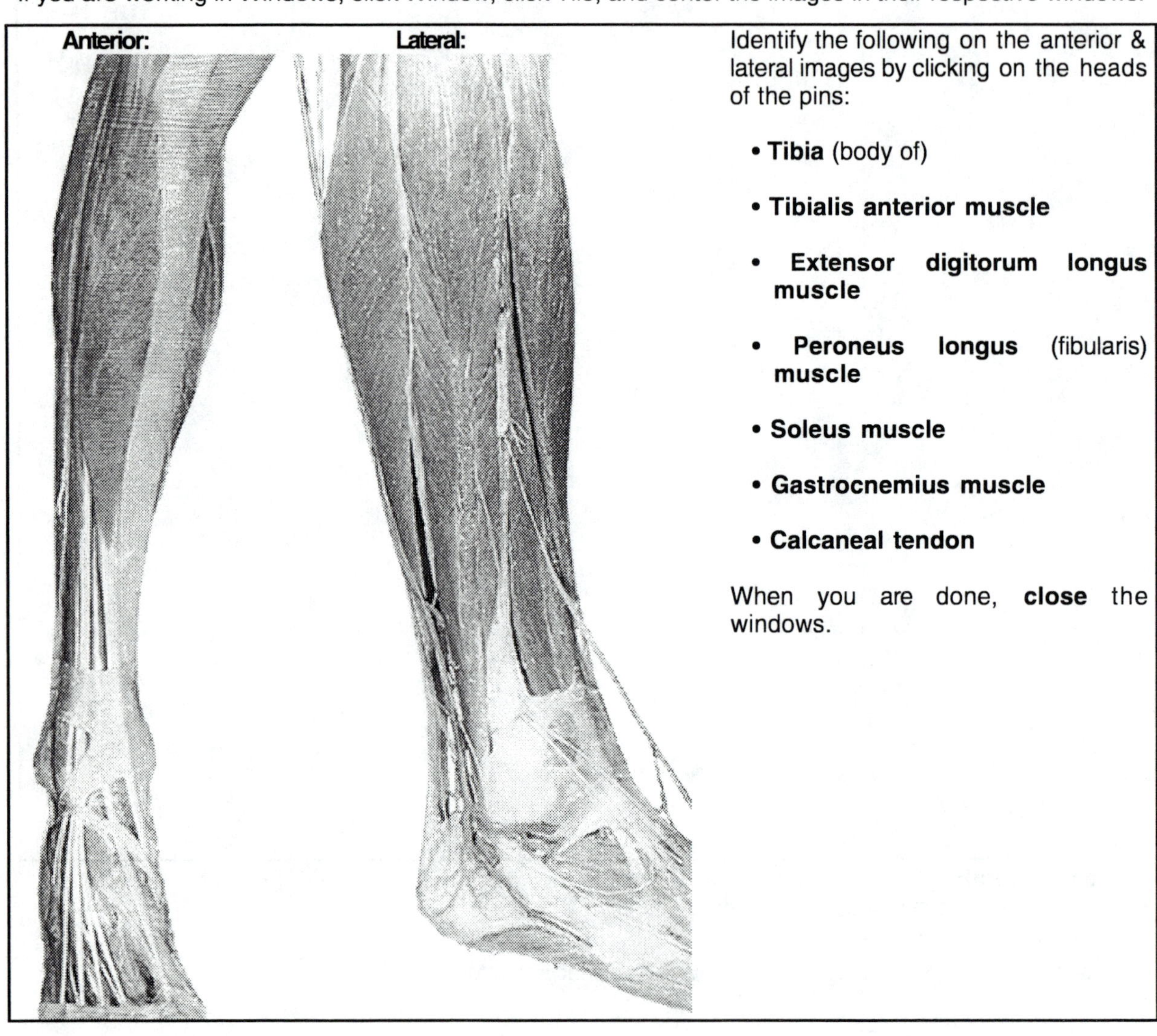


Identify the following on the anterior & lateral images by clicking on the heads of the pins:

- **Tibia** (body of)
- **Tibialis anterior muscle**
- **Extensor digitorum longus muscle**
- **Peroneus longus** (fibularis) **muscle**
- **Soleus muscle**
- **Gastrocnemius muscle**
- **Calcaneal tendon**

When you are done, **close** the windows.

Under the **File** menu, drag down to **Open** (then to **Content** if you are working in windows). **Atlas Anatomy** should still be selected. **Region** and **Lower Limb** should still be selected. Choose **Dissection of Medial Leg and Foot**. If the list is not in alphabetical order, the image is two-thirds of the way down the list.

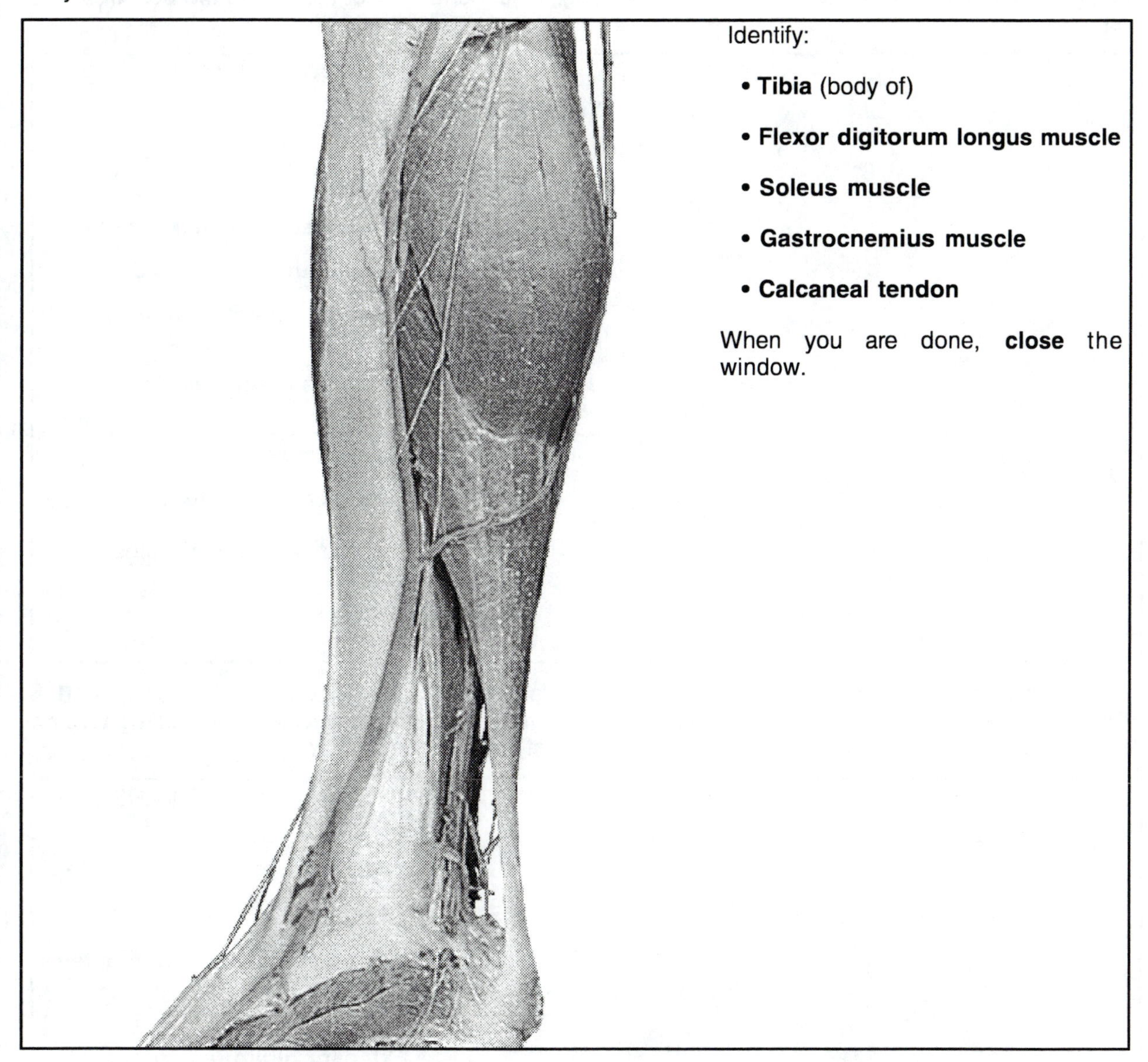

Identify:

- **Tibia** (body of)
- **Flexor digitorum longus muscle**
- **Soleus muscle**
- **Gastrocnemius muscle**
- **Calcaneal tendon**

When you are done, **close** the window.

## Cross Sections of the Lower Leg Muscles

Under the **File** menu, drag down to **Open** (then to **Content** if you are working in windows). **Atlas Anatomy** should still be selected. **Region** and **Lower Limb** should still be selected. Choose **Proximal Leg (Inf)**. If the list is not in alphabetical order, the image is three-fourths of the way to the bottom of the list.

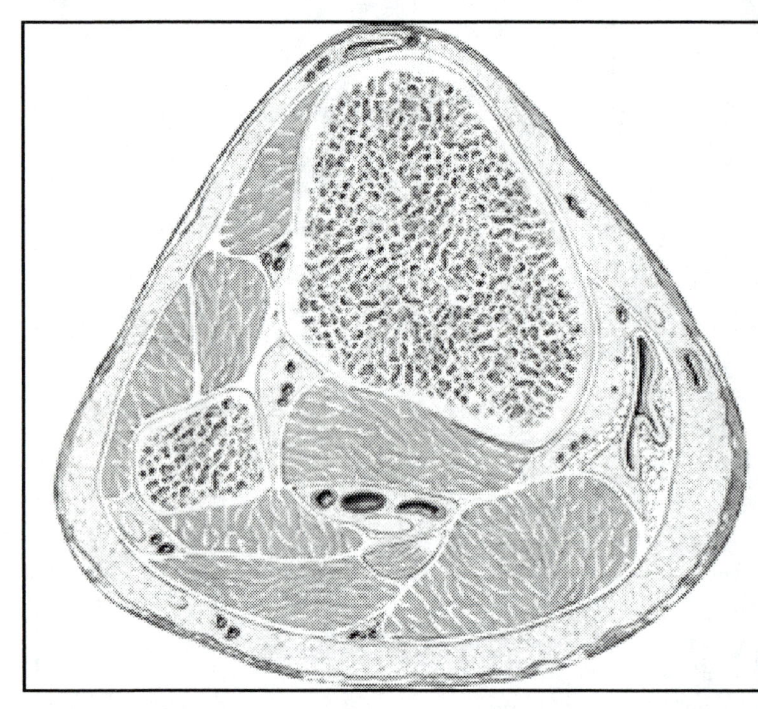

This is a cross section of the leg. Identify:

- **Tibia**
- **Fibula**
- **Interosseous membrane of leg**
- **Tibialis anterior muscle**
- **Extensor digitorum longus muscle**
- **Peroneus longus** (fibularis) **muscle**
- **Soleus muscle**
- **Gastrocnemius muscle**

When you are done, **close** the window.

Under the **File** menu, drag down to **Open** (then to **Content** if you are working in windows). **Atlas Anatomy** should still be selected. **Region** and **Lower Limb** should still be selected. Choose **Distal Leg (Inf)**. If list is not in alphabetical order, the image is almost three-fourths of the way down.

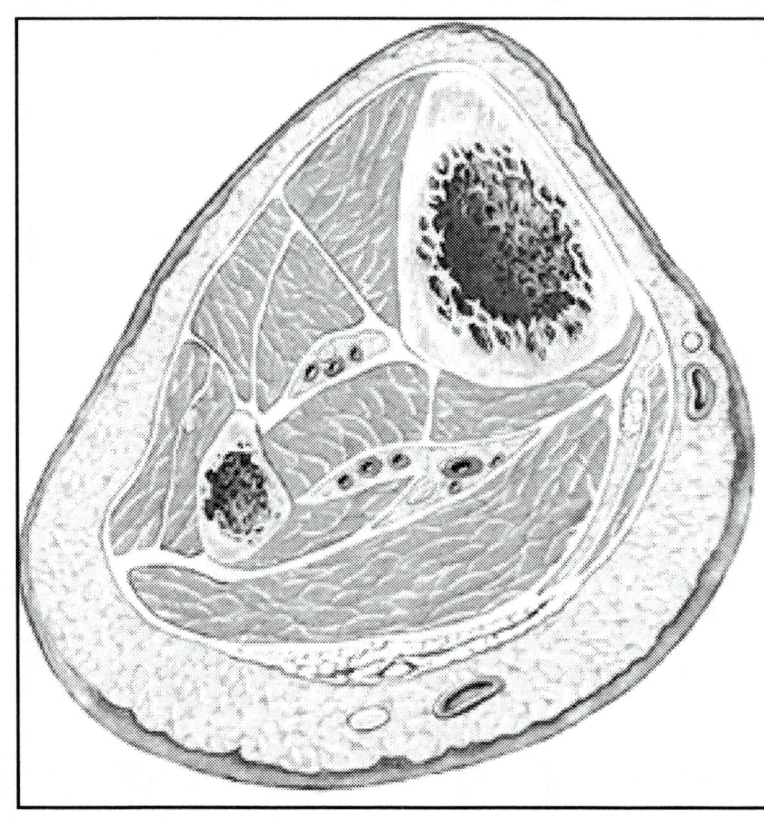

This is a cross section of the leg. Identify:

- **Tibia**
- **Fibula**
- **Interosseous membrane of leg**
- **Tibialis anterior muscle**
- **Extensor digitorum longus muscle**
- **Peroneus longus** (fibularis) **muscle**
- **Soleus muscle**
- **Tibialis posterior muscle**
- **Flexor digitorum longus muscle**
- **Calcaneal tendon**

When you are done, **close** the window.

## Objectives: Lower Leg Muscles:

Student will learn the anatomy of the important leg muscles, particularly those that move the foot.

**Muscles**

- Gastrocnemius muscle
- Soleus muscle
- Tibialis anterior muscle
- Extensor digitorum longus muscle
- Peroneus longus (fibularis) muscle
- Extensor digitorum longus muscle
- Flexor digitorum longus muscle
- Tibialis posterior muscle

**Other Structures**

- Tibia
- Calcaneal tendon
- Fibula
- Tendon of soleus muscle
- Interosseous membrane of leg

# Spinal Cord & Spinal Nerves

## Opening A.D.A.M Interactive Anatomy

Open A.D.A.M. Interactive Anatomy according to the directions in the Tutorial.

## Atlas Anatomy of Spinal Cord

Choose **Atlas Anatomy**. Select **System** and **Nervous.** Choose **T12 Vertebra (Sup)**. If the list is not in alphabetical order, the image is about half way down the list.

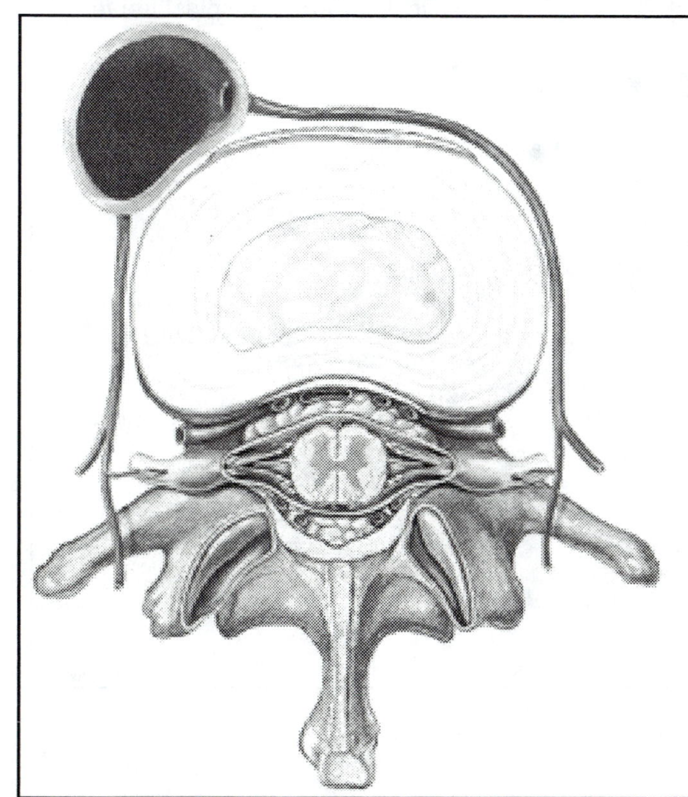

If you are working in Mac, click on the **Show All Pins** button and scroll to **All Systems**.

Identify:

- **Denticulate ligament**
- **Dorsal root** of spinal nerve
- **Dura mater**
- **Epidural fat**
- **Gray matter**
- **Spinal nerve**
- Spinal **Dorsal root ganglion** of spinal nerve
- **Subarachnoid space**
- **Transverse process of vertebrae**
- **Ventral root** of spinal nerve
- **White matter**

When you are finished, **close** the window.

Under the **File** menu, drag down to **Open** (then to **Content** if you are working in windows). **Atlas Anatomy** should still be selected. **System** and **Nervous** should still be selected. Choose **Spinal Cord Vessels & Meninges**. If the list is not in alphabetical order, the image is about half way down.

You may need to scroll to find the structures. Identify:

- **Arachnoid mater**
- **Dorsal root** of spinal nerve
- **Dura mater**
- **Epidural space**
- **Pia mater**
- **Gray matter**
- **Spinal nerve**
- **Dorsal root ganglion**
- **Subarachnoid space**
- **Subdural space**
- **Ventral root** of spinal nerve
- **White matter**

**Close** the window.

Under the **File** menu, drag down to **Open** (then to **Content** if you are working in windows). **Atlas Anatomy** should still be selected. **System** and **Nervous** should still be selected. Choose **Nerves of Thoracic Wall**. If list is not in alphabetical order the image is about half way down.

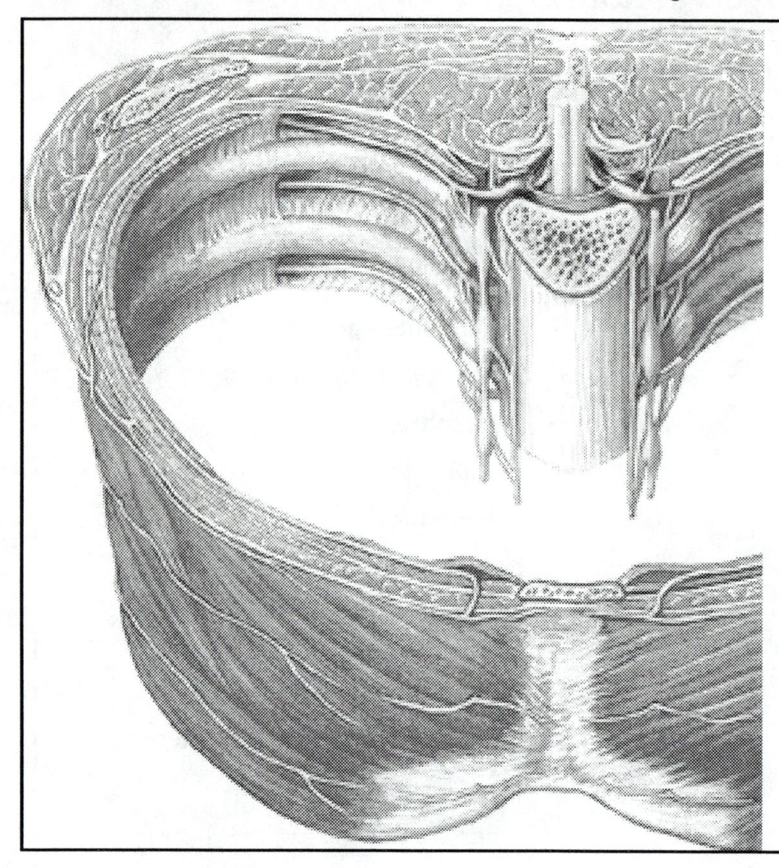

You may need to scroll to find the structures. Identify:

- **Dorsal** nerve **root** of spinal nerve
- **Dorsal ramus** of spinal nerve
- **Dura mater**
- Gray & white **rami communicans** of spinal nerve
- **Intercostal nerve**
- **Spinal cord**
- Spinal **Dorsal root ganglion** of spinal nerve
- **Ventral ramus** of spinal nerve
- **Ventral root** of spinal nerve

What is the difference between the dorsal root and the dorsal ramus?

___

Do the intercostal nerves emerge from the dorsal or ventral ramus?

___

When you are finished, **close** the window.

Under the **File** menu, drag down to **Open** (then to **Content** if you are working in windows). **Atlas Anatomy** should still be selected. **System** and **Nervous** should still be selected. Choose **Lumbosacral Spinal Cord (Post)**. If the list is not in alphabetical order, the image is about half way down.

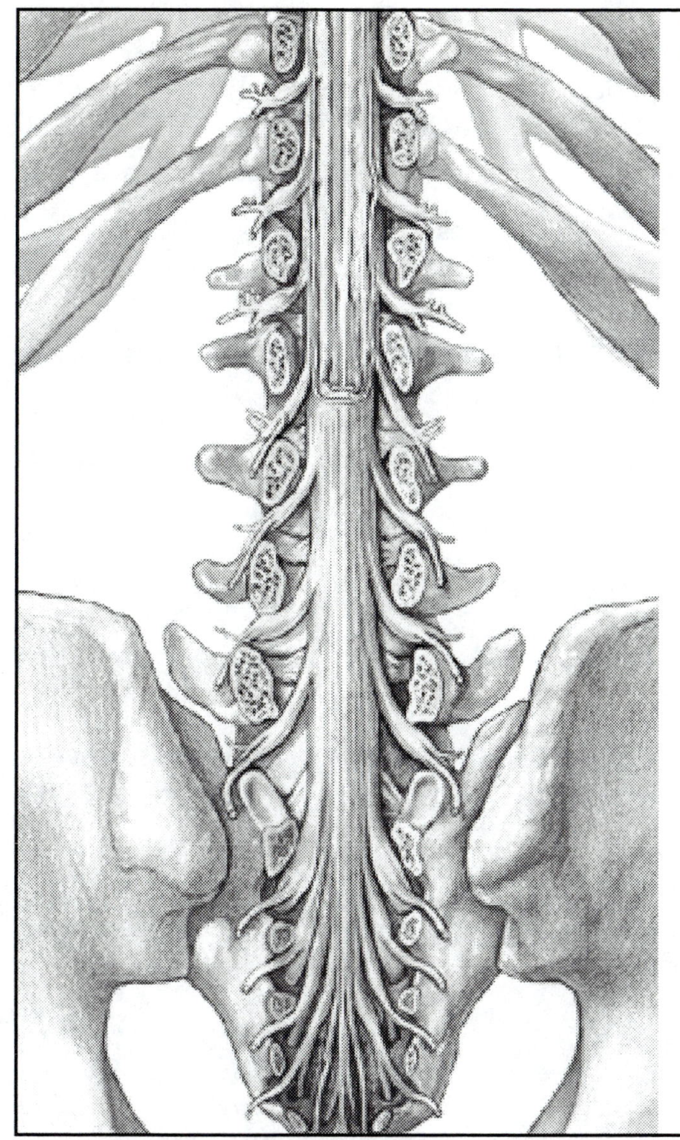

You may need to scroll to find the structures. Identify:

- **Arachnoid mater**
- **Cauda equina**
- **Dorsal rami** of T11 to S5 spinal nerves
- **Dura mater**
- External **Filum terminale**
- nternal **Filum terminale**
- Gray & white **rami communicans** of spinal nerve
- Medullary cone (**Conus medullaris**)
- **Pia mater**
- **Spinal cord**
- Spinal **Dorsal root ganglia** of T11 to S4 spinal nerves
- **Ventral rami** of L1 to S4 spinal nerves

The spinal cord ends at what vertebrae?

___

What is the difference between the cauda equina and the filum terminale?

___

___

When you are finished, **close** the window.

## Posterior Dissection of Spinal Cord & Spinal Nerves

Under the **File** menu, drag down to **Open** (then to **Content** if you are working in windows).

Choose **Dissectible Anatomy**.

Choose male or female. Select **Posterior.** Click **Open**.

Enlarge the window and position the navigation rectangle over the posterior neck.

Click on the **Depth Bar** ▼ until you reach **layer 6**. No graphic has been provided for this layer. Identify:

- **Cutaneous branch of dorsal ramus** of the various spinal nerves

Click on the **Depth Bar** ▼ until you reach **layer 176**. No graphic has been provided for this layer. Identify:

- **Ventral ramus** of the various spinal nerves
- **Dorsal ramus** of the various spinal nerves

Click on the **Depth Bar** ▼ until you reach **layer 178**. No graphic has been provided for this layer. Identify:

- **Dura mater**

Click on the **Depth Bar** ▼ until you reach **layer 179**. No graphic has been provided for this layer. Identify:

- **Arachnoid mater**

How many cervical spinal nerves are there on each side? ________

How many thoracic spinal nerves are there on each side? ________

How many lumbar spinal nerves are there on each side? ________

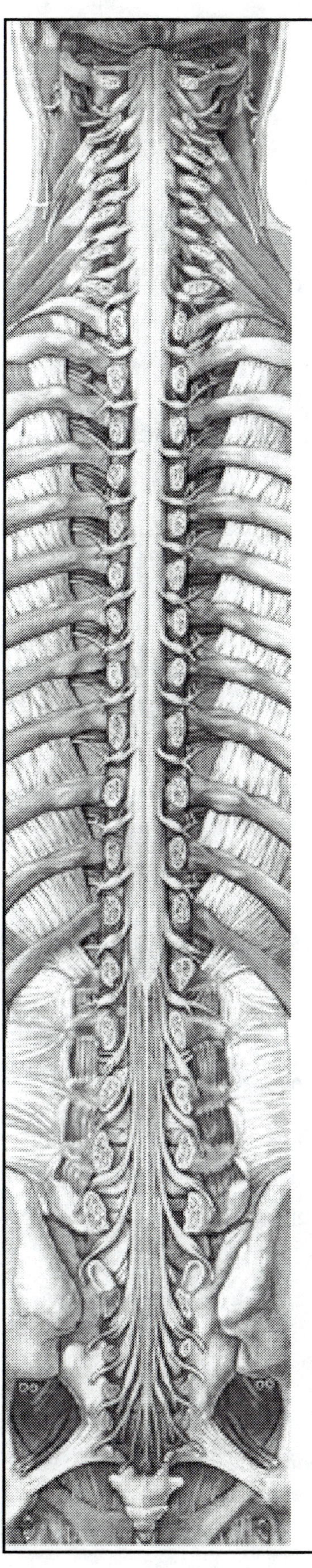

Click on the **Depth Bar** ▼ until you reach **layer 180**. You will need to scroll to view the entire spinal cord. Identify:

- **Cauda equina**
- **Spinal cord**
- **Dorsal ramus** of the various spinal nerves
- External **Filum terminale**
- Internal **Filum terminale**
- Medullary cone (**Conus medullaris**)
- Spinal **Dorsal root ganglion** of all spinal nerves
- **Ventral ramus** of the various spinal nerves
- Gray & white **rami communicans** of spinal nerve
- **Ventral nerve root** of the various spinal nerves
- **Dorsal nerve root** of the various spinal nerves

Note the angle that the spinal nerves emerge from the spinal cord. Which nerves emerge almost horizontally?

______________________________________________

Which nerves emerge from the bottom of the spinal cord?

______________________________________________

Which nerves emerge from the spinal cord at an angle?

______________________________________________

Do the C1 to C8 spinal nerves emerge above or below the C1 to C7 vertebrae?

______________________________________________

Do the T1 to T12 spinal nerves emerge above or below the T1 to T12 vertebrae?

______________________________________________

Do the L1 to L5 spinal nerves emerge above or below the L1 to L5 vertebrae?

______________________________________________

Click on the **Depth Bar** ▼ until you reach **layer 181**. No graphic has been provided for this layer. Note the location of the **intervertebral disks**. Consider the consequences if there is a slipped disk which impinges on a spinal nerve or the spinal cord.

## Lateral Dissection of Spinal Cord & Spinal Nerves

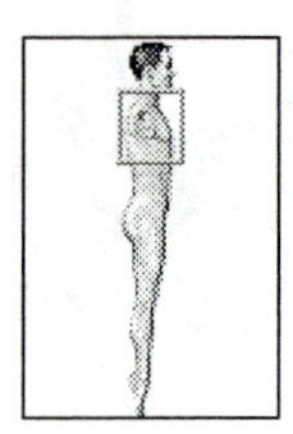

Click on the **View** button and choose **Lateral**.

Click on the **Normal** button. Windows— —Mac

Drag the **Depth Bar** to **layer 0.**

Position the navigator rectangle over the chest

Click on the **Depth Bar** until you reach **layer 4**. No graphic has been provided for this layer. Identify:

- **Cutaneous branches of the intercostal nerves**
- **Cutaneous branches of the dorsal ramus of the various spinal nerves**

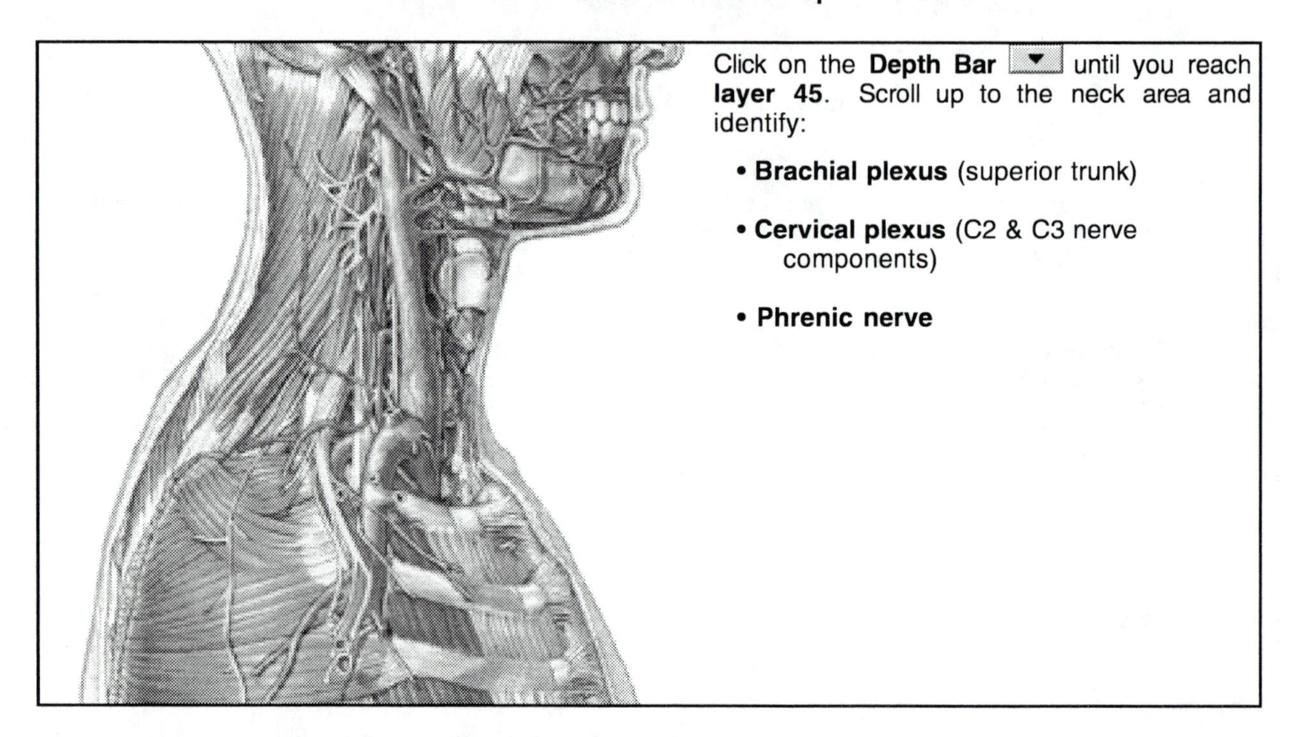

Click on the **Depth Bar** until you reach **layer 45**. Scroll up to the neck area and identify:

- **Brachial plexus** (superior trunk)
- **Cervical plexus** (C2 & C3 nerve components)
- **Phrenic nerve**

Click on the **Depth Bar** until you reach **layer 71**. No graphic has been provided for this layer. Identify:

- **The first through eleventh intercostal nerve**

Click on the **Depth Bar** until you reach **layer 149**. No graphic has been provided for this layer. Identify:

- **Right phrenic nerve**
- **Diaphragm muscle**

Scroll down to the leg area and identify:

- **Sciatic nerve**

Click on the **Depth Bar** until you reach **layer 288**. No graphic has been provided for this layer. Identify:

- **The first through tenth intercostal nerves**
- **The spinal nerves** emerging from the intervertebral foramen

## Medial Dissection of the Spinal Cord and Spinal Nerves

Click on the **View** button and choose **Medial**.

Click on the **Normal** button. Windows— —Mac

Drag the **Depth Bar** to **layer 0.**

Position the navigator rectangle over the chest

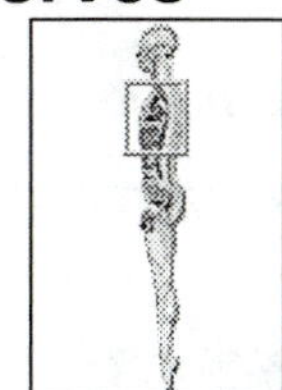

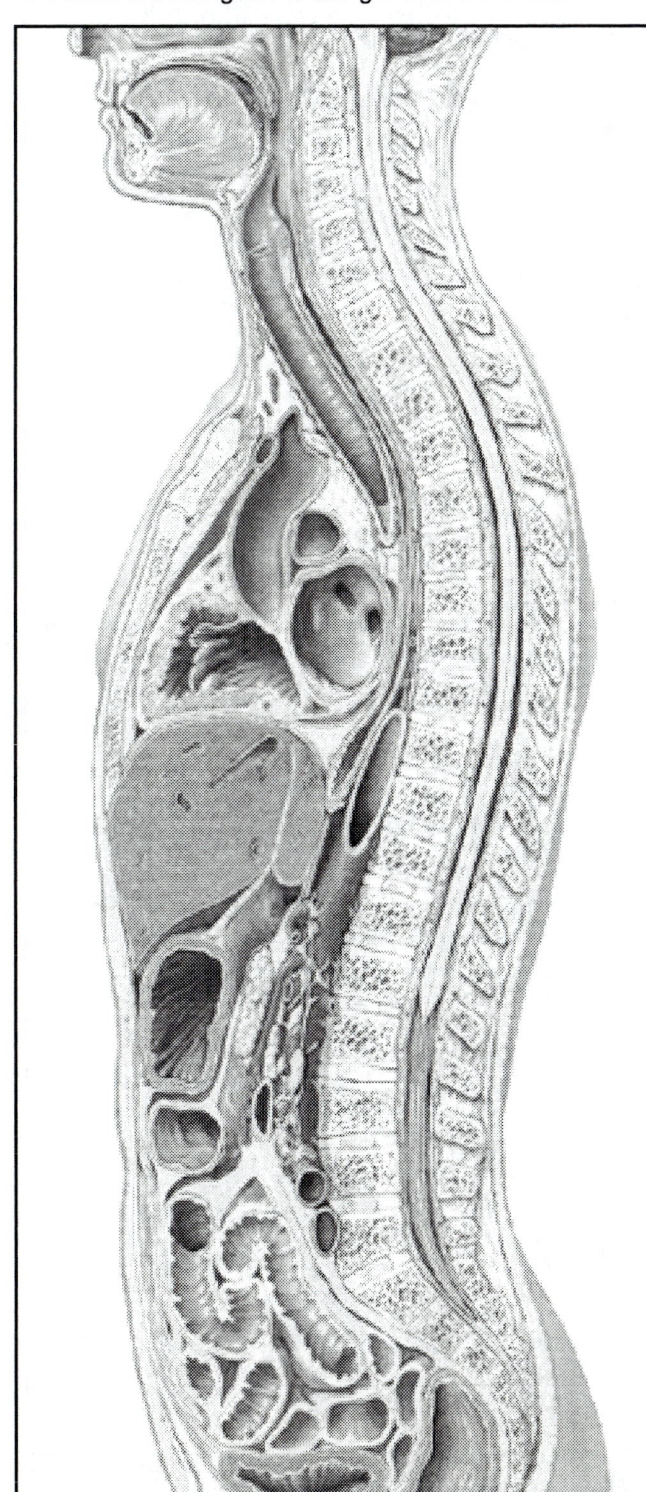

You will need to scroll up and down to view the entire spinal cord and cauda equina. Identify:

- **Dura mater**
- **Epidural fat**
- **Cauda equina**
- **Spinal cord**
- Internal **Filum terminale**
- Medullary cone (**Conus medullaris**)
- **Medulla oblongata**
- **Intervertebral disks** (nucleus pulposus & anulus fibrosus)

Scroll down to the leg area. Click on the **Depth Bar** [▼] until you reach **layer 57**. No graphic has been provided for this layer. Identify:

- **Sciatic nerve**

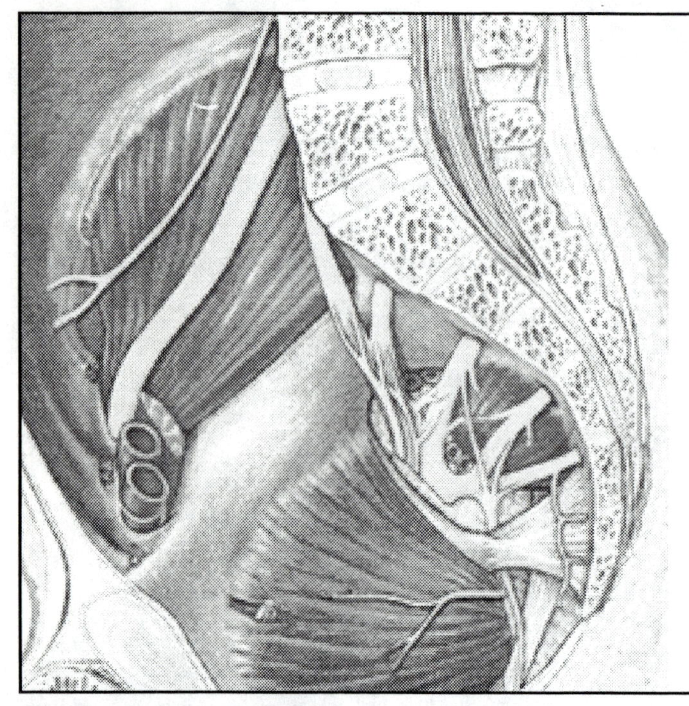

Scroll up to the pelvis area. Click on the **Depth Bar** [▼] until you reach **layer 99**. Identify:

- **Ventral rami of S1-S4**
- **Sciatic nerve**
- **Femoral nerve**

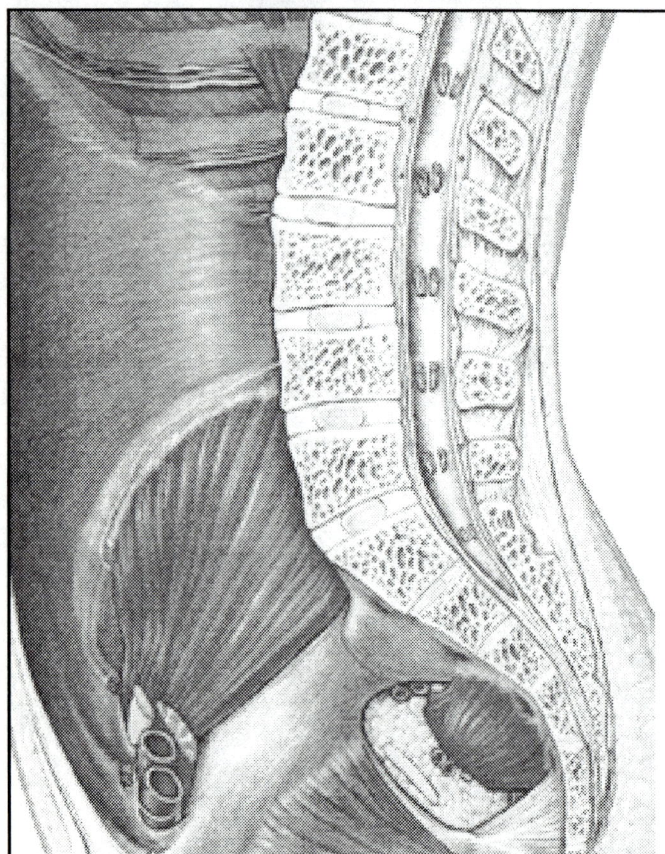

Click on the **Depth Bar** [▼] until you reach **layer 103**. Identify:

- **Dura mater**
- **Epidural fat**
- **Dorsal nerve root** of the various spinal nerves
- **Ventral nerve root** of the various spinal nerves
- Internal **Filum terminale**
- **Intercostal nerves**

## Anterior Dissection of Spinal Cord and Spinal Nerves

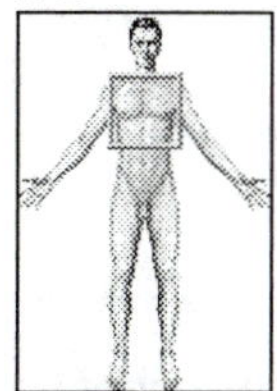

Click on the **View** button and choose **Anterior**.

Click on the **Normal** button. Windows— —Mac

Drag the **Depth Bar** to **layer 0.**

Position the navigator rectangle over the chest.

Click on the **Depth Bar** until you reach **layer 7.** No graphic has been provided for this layer. Identify:

- **Cutaneous branches of the intercostal nerves**

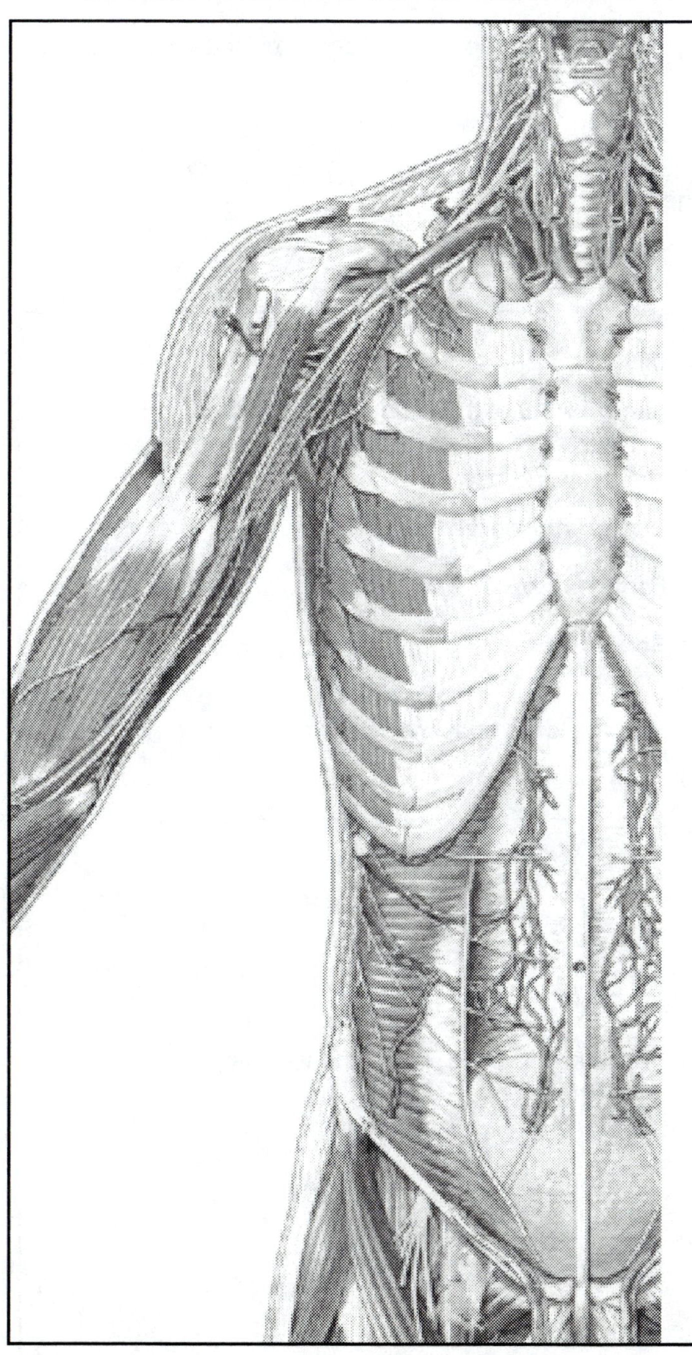

Click on the **Depth Bar** until you reach **layer 86.** Identify:

- **Brachial plexus** (superior trunk & medial cord)
- **Radial nerve** (superficial branch)
- **Ulnar nerve**

From what plexus does the ulnar nerve emerge?

______________________________________

Scroll down to the thorax. You are still at **layer 86**. No graphic has been provided for this layer. Identify:

- **Intercostal nerves**

Scroll down to the pelvis. You are still at **layer 86**. No graphic has been provided for this layer. Identify:

- **Femoral nerve**

From what plexus does the femoral nerve emerge?

---

Scroll up to the chest area. Click on the **Depth Bar** ▼ until you reach **layer 148.** No graphic has been provided for this layer. Identify:

- **Intercostal nerves**
- **Brachial plexus** (superior trunk & medial cord)

Click on the **Depth Bar** ▼ until you reach **layer 167.** No graphic has been provided for this level. Identify:

- **Phrenic nerve**
- **Diaphragm**

From what plexus does the phrenic nerve emerge?

---

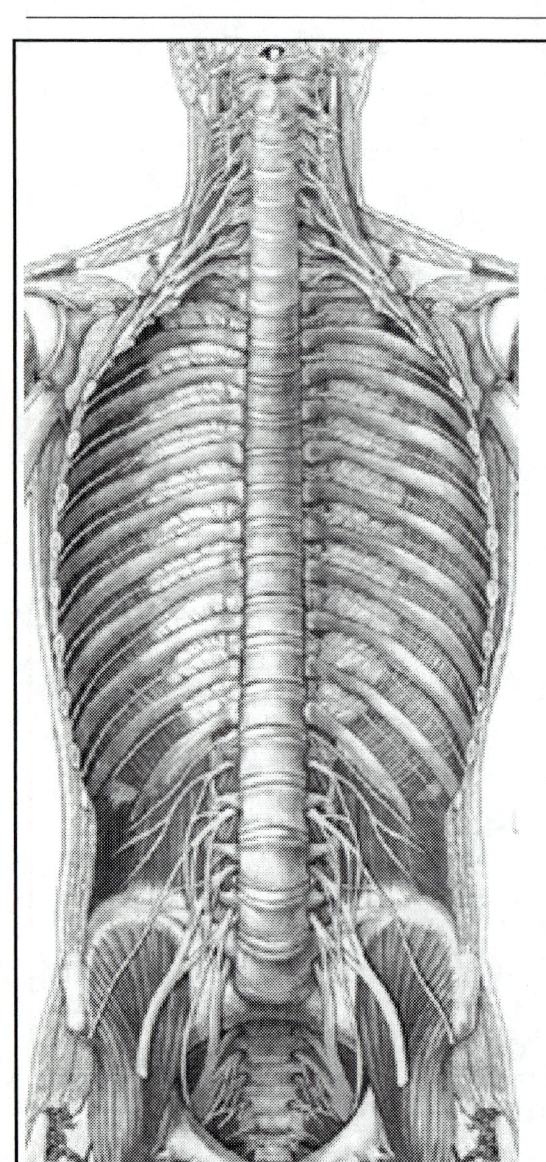

Click on the **Depth Bar** ▼ until you reach **layer 275.** Identify:

- **Ventral rami** of the various spinal nerves
- **Brachial plexus**
- **Intercostal nerves**
- **Femoral nerve**
- **Sciatic nerve**

From what plexus does the sciatic nerve emerge?

When you are finished, **close** the window.

## Atlas Anatomy of Spinal Nerves

Under the **File** menu, drag down to **Open** (then to **Content** if you are working in windows). **Atlas Anatomy** should still be selected. **System** and **Nervous** should still be selected. Choose **Nerves of Upper Limb (Ant).** If the list is not in alphabetical order, the image is about a third of the way from the top .

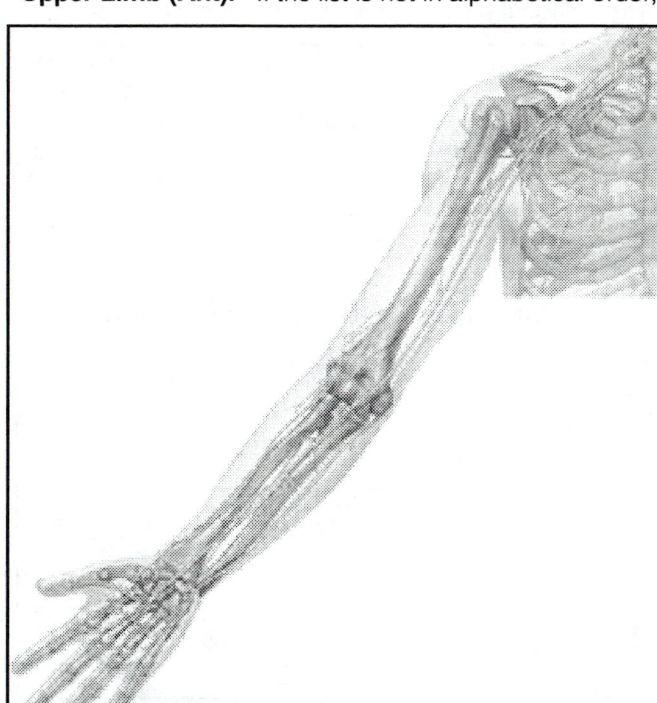

You may need to scroll to find the structures. Identify :

- **Brachial plexus** (inferior trunk, lateral cord, medial cord, middle trunk, posterior cord, superior trunk)
- **Median nerve**
- **Radial nerve**
- **Ulnar nerve**

When you are finished, **close** the window.

Under the **File** menu, drag down to **Open** (then to **Content** if you are working in windows). **Atlas Anatomy** should still be selected. **System** and **Nervous** should still be selected. Choose **Sacral Plexus (Ant).** If the list is not in alphabetical order, the image is toward the bottom of the list.

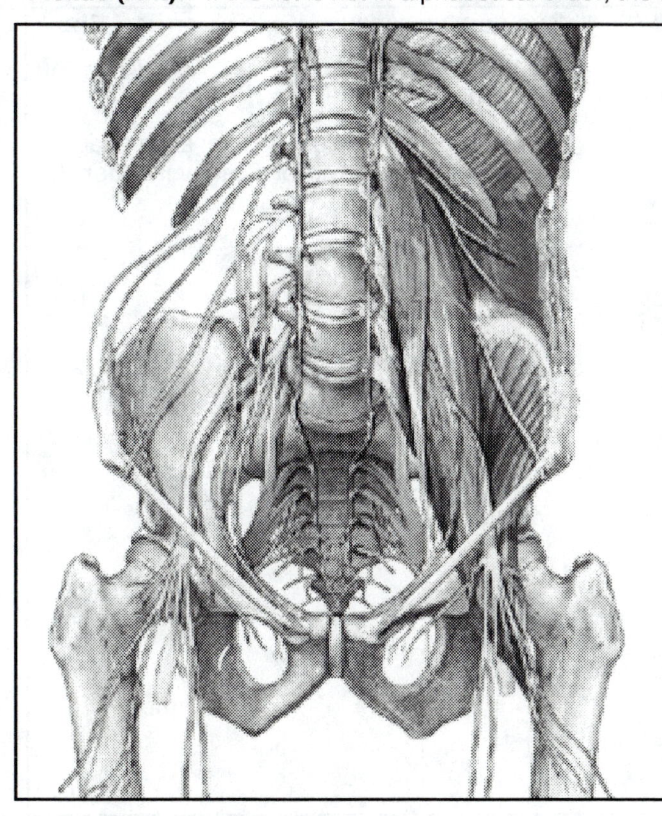

You may need to scroll to find the structures. Identify :

- **Intercostal nerves**
- **Femoral nerve**
- **Sacral plexus**
- **Sciatic nerve**
- **Ventral rami of T12-S4**

When you are finished, **close** the window.

Identify the **femoral plexus** on this diagram.

## Cadaver Images

Under the **File** menu, drag down to **Open** (then to **Content** if you are working in windows). **Atlas Anatomy** should still be selected. **System** and **Nervous** should still be selected. Choose **Brachial Plexus.** If the list is not in alphabetical order, the image is about a third of the way from the top of the list.

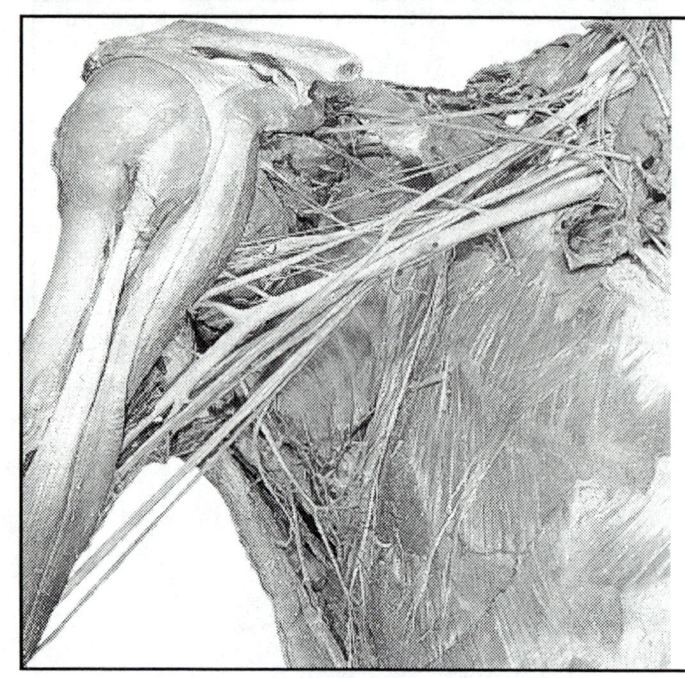

The image to the left shows the **brachial plexus** in a cadaver. You will not be able to see the entire image on screen. It is not necessary to label the various nerves in this image. Rather, identify the nerves as the **brachial plexus.**

When you are finished, **close** the window.

Under the **File** menu, drag down to **Open** (then to **Content** if you are working in windows). **Atlas Anatomy** should still be selected. **System** and **Nervous** should still be selected. Choose **Dissection of Left Mediastinum**. If the list is not in alphabetical order, the image is about half of the way from the top of the list.

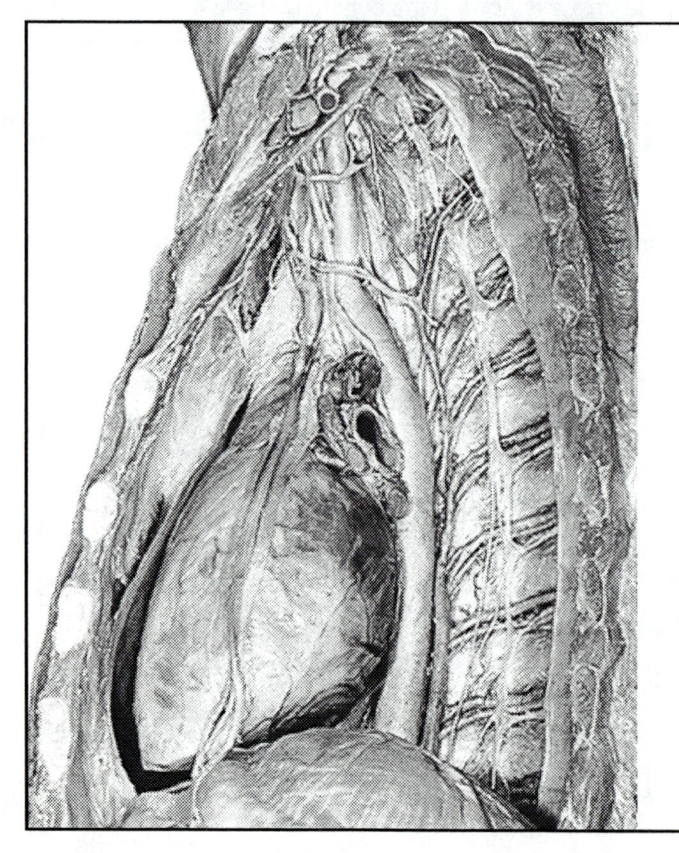

If you are working in Mac, click on the **Show All Pins** button and scroll to **All Systems**.

The image to the right shows a parasagittal view of the chest cavity of a cadaver. Identify:

- **Left phrenic nerve**
- **Diaphragm muscle**

When you are finished, **close** the window.

## Dermatomes

Under the **File** menu, drag down to **Open** (then to **Content** if you are working in windows). **Atlas Anatomy** should still be selected. **System** and **Nervous** should still be selected. Choose **Dermatomes of Head/Neck (Ant)**. If the list is not in alphabetical order, the image at the top of the list.

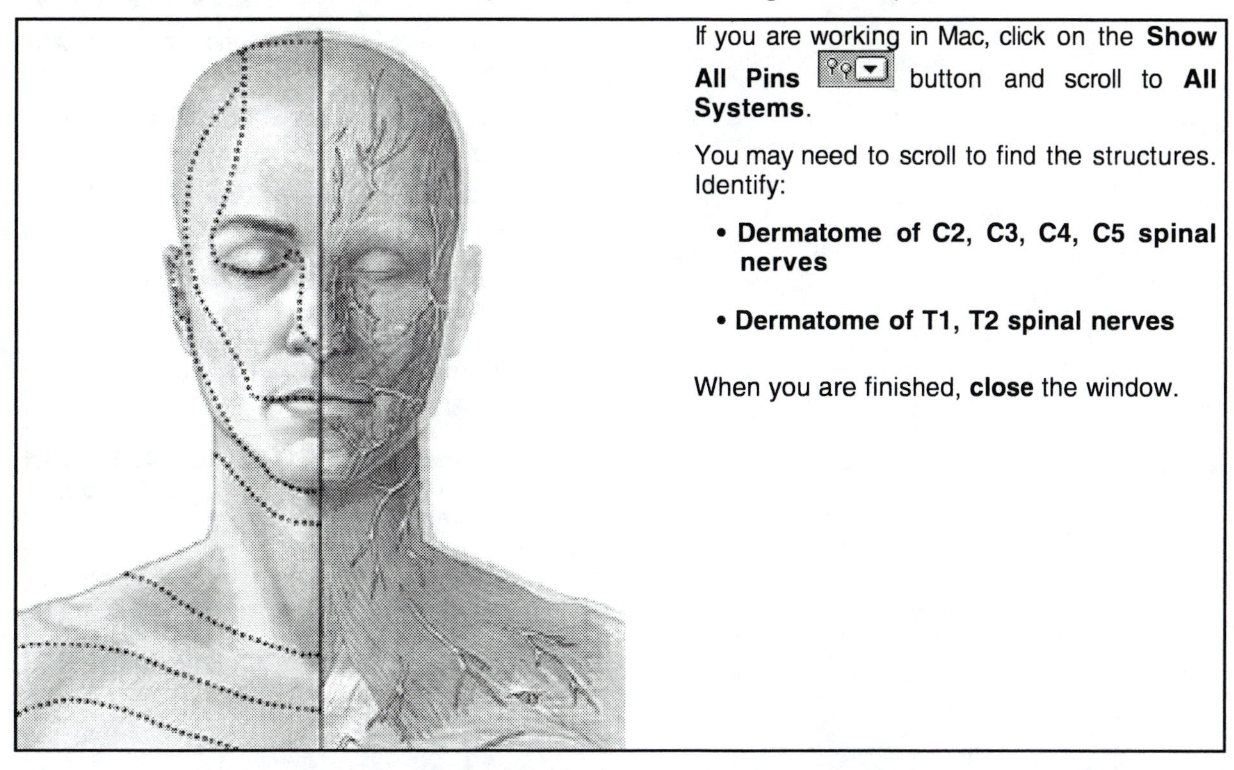

If you are working in Mac, click on the **Show All Pins** button and scroll to **All Systems**.

You may need to scroll to find the structures. Identify:

- **Dermatome of C2, C3, C4, C5 spinal nerves**
- **Dermatome of T1, T2 spinal nerves**

When you are finished, **close** the window.

Under the **File** menu, drag down to **Open** (then to **Content** if you are working in windows). **Atlas Anatomy** should still be selected. **System** and **Nervous** should still be selected. Choose **Dermatomes of Head/Neck (Post)**. If the list is not in alphabetical order, the image is close to the top of the list.

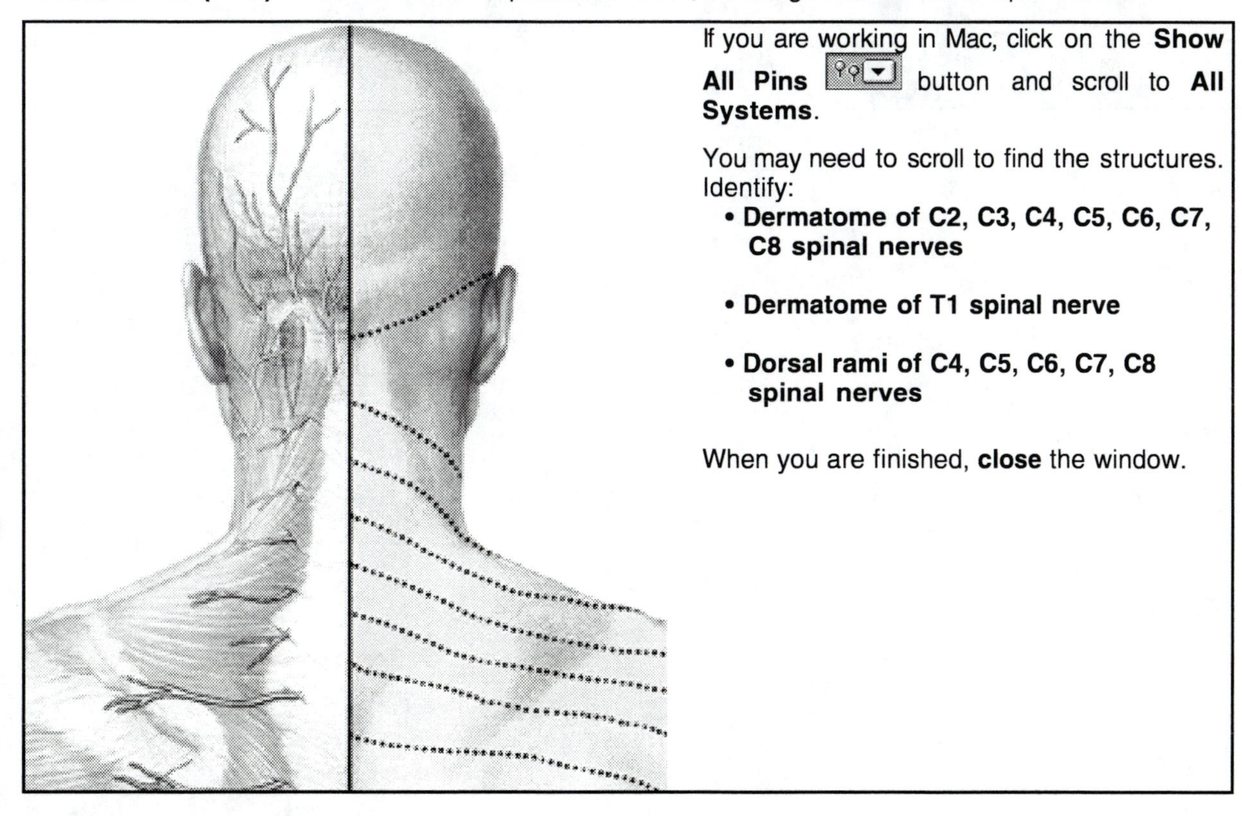

If you are working in Mac, click on the **Show All Pins** button and scroll to **All Systems**.

You may need to scroll to find the structures. Identify:

- **Dermatome of C2, C3, C4, C5, C6, C7, C8 spinal nerves**
- **Dermatome of T1 spinal nerve**
- **Dorsal rami of C4, C5, C6, C7, C8 spinal nerves**

When you are finished, **close** the window.

Under the **File** menu, drag down to **Open** (then to **Content** if you are working in windows). **Atlas Anatomy** should still be selected. **System** and **Nervous** should still be selected. Choose **Dermatomes of Trunk (Ant)**. If the list is not in alphabetical order, the image is about half way down the list.

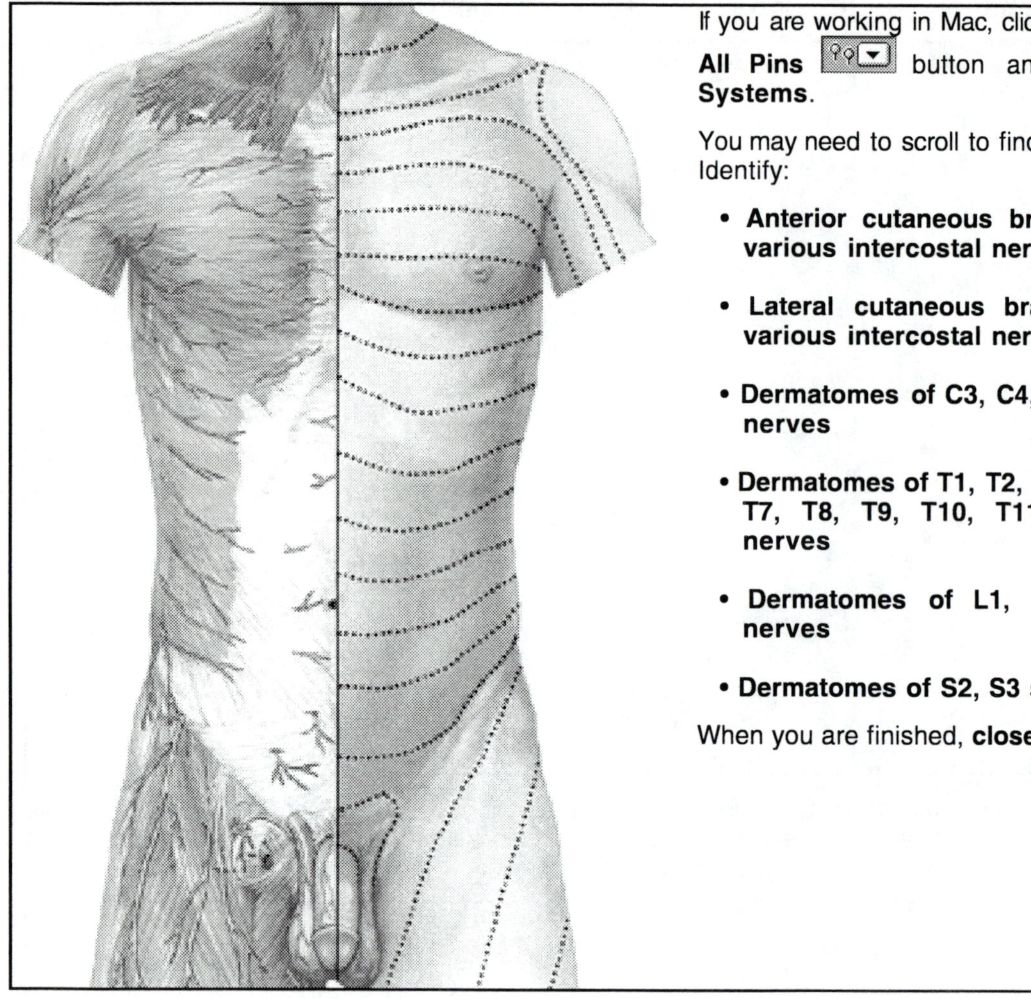

If you are working in Mac, click on the **Show All Pins** button and scroll to **All Systems**.

You may need to scroll to find the structures. Identify:

- **Anterior cutaneous branches of the various intercostal nerves**
- **Lateral cutaneous branches of the various intercostal nerves**
- **Dermatomes of C3, C4, C5, C6 spinal nerves**
- **Dermatomes of T1, T2, T3, T4, T5, T6, T7, T8, T9, T10, T11, T12 spinal nerves**
- **Dermatomes of L1, L2, L3 spinal nerves**
- **Dermatomes of S2, S3 spinal nerves**

When you are finished, **close** the window.

Under the **File** menu, drag down to **Open** (then to **Content** if you are working in windows). **Atlas Anatomy** should still be selected. **System** and **Nervous** should still be selected. Choose **Dermatomes of Trunk (Post)**. If the list is not in alphabetical order, the image is about half way down the list.

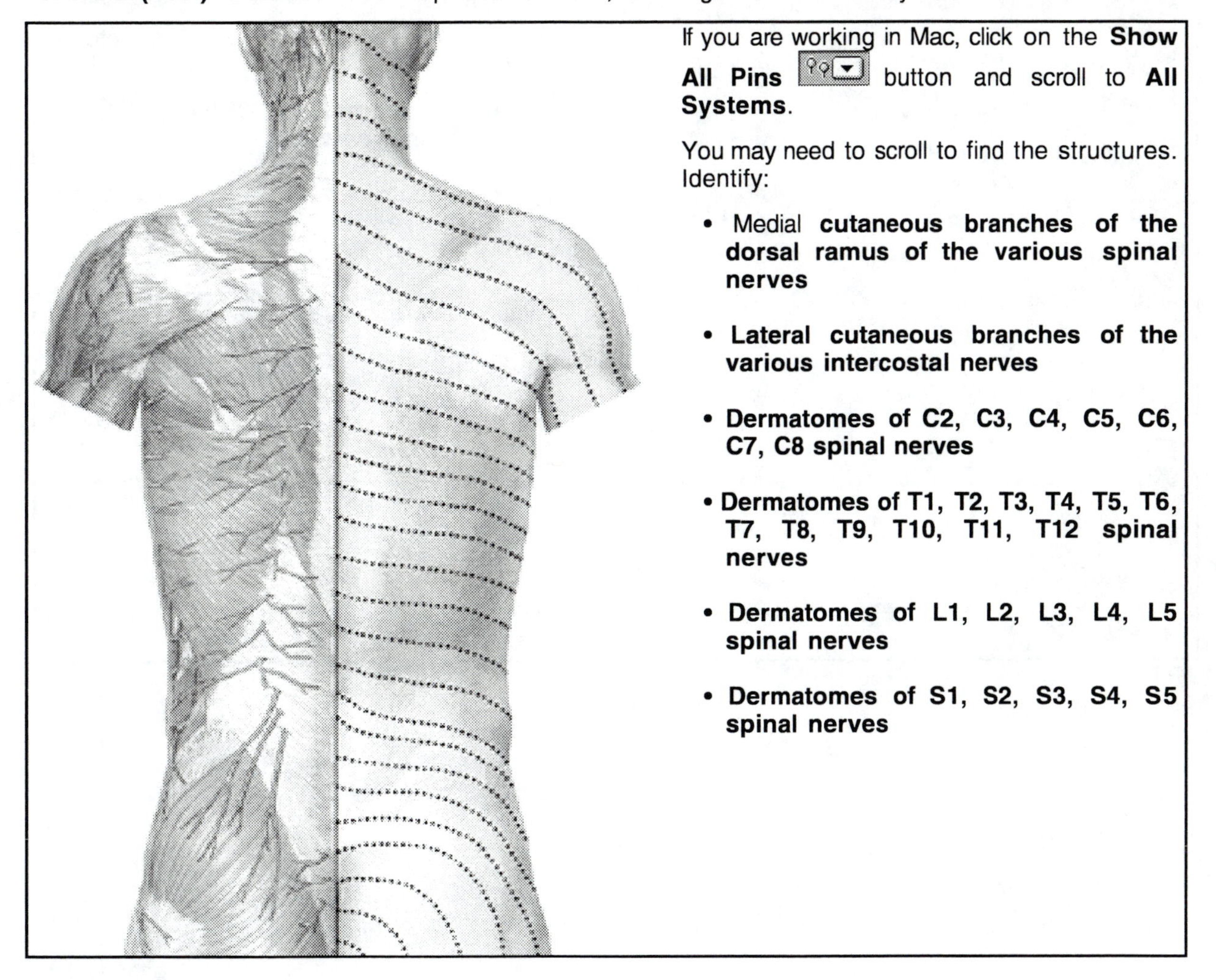

If you are working in Mac, click on the **Show All Pins** button and scroll to **All Systems**.

You may need to scroll to find the structures. Identify:

- Medial **cutaneous branches of the dorsal ramus of the various spinal nerves**
- **Lateral cutaneous branches of the various intercostal nerves**
- **Dermatomes of C2, C3, C4, C5, C6, C7, C8 spinal nerves**
- **Dermatomes of T1, T2, T3, T4, T5, T6, T7, T8, T9, T10, T11, T12 spinal nerves**
- **Dermatomes of L1, L2, L3, L4, L5 spinal nerves**
- **Dermatomes of S1, S2, S3, S4, S5 spinal nerves**

If the spinal cord is severed at C8, would there be feeling in the arms?

---

When you are finished, **close** the window.

Under the **File** menu, drag down to **Open** (then to **Content** if you are working in windows). **Atlas Anatomy** should still be selected. Choose **Region** and **Upper Limb.** Choose **Dermatomes of Upper Limb (Ant).** If the list is not in alphabetical order, the image is close to the top of the list.

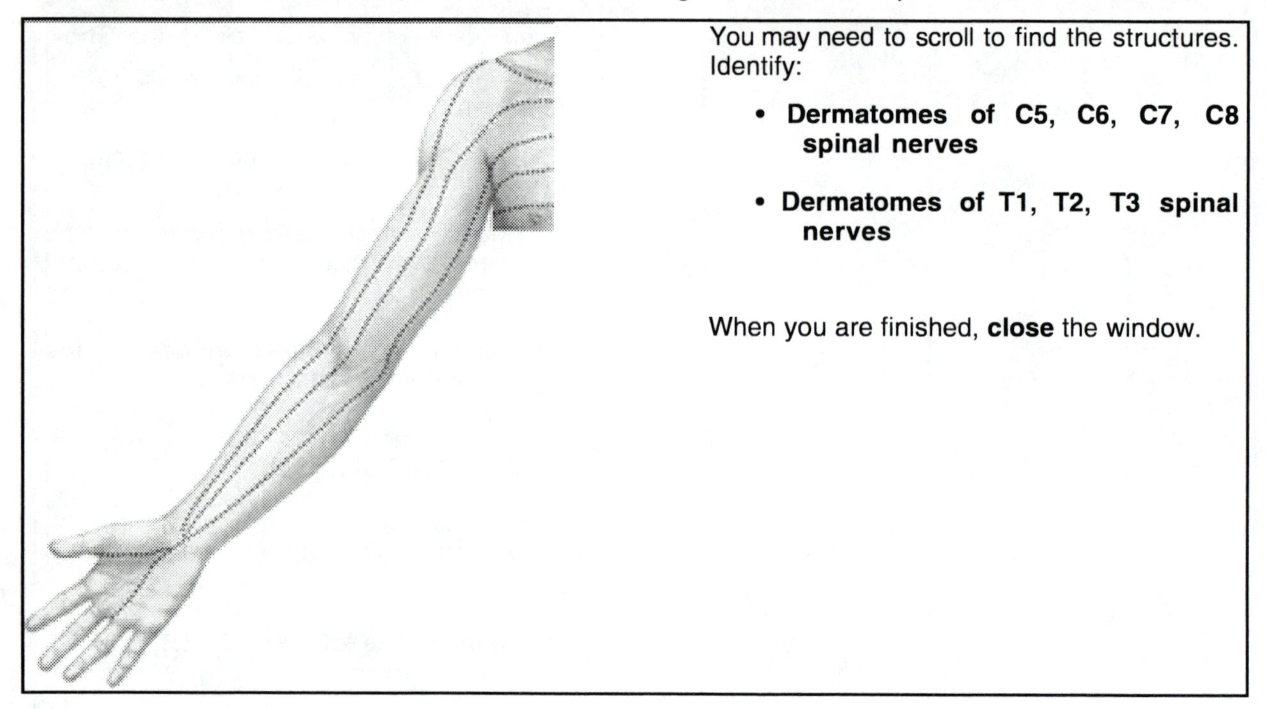

You may need to scroll to find the structures. Identify:

- **Dermatomes of C5, C6, C7, C8 spinal nerves**
- **Dermatomes of T1, T2, T3 spinal nerves**

When you are finished, **close** the window.

Under the **File** menu, drag down to **Open** (then to **Content** if you are working in windows). **Atlas Anatomy** should still be selected. Choose **Region** and **Lower Limb.** Choose **Dermatomes of Lower Limb (Ant).** If the list is not in alphabetical order, the image is close to the top of the list.

Under the **File** menu, drag down to **Open** (then to **Content** if you are working in windows). **Atlas Anatomy** should still be selected. Choose **Region** and **Lower Limb.** Choose **Dermatomes of Lower Limb (Post).** If the list is not in alphabetical order, the image is close to the top of the list.

Position the anterior and posterior images so they can both be seen at the same time on the screen. Do this by dragging the gray bar above the images:

If you are working in Windows, click Window, click Tile, and center the images in their respective windows.

You may need to scroll to find the structures. Identify the following on the anterior image to the left:

- **Dermatomes of L1, L2, L3, L4, L5 spinal nerves**
- **Dermatomes of T10, T11, T12 spinal nerves**
- **Dermatomes of S1, S2, S3 spinal nerves**

You may need to scroll to find the structures. Identify the following on the posterior image to the right:

- **Dermatomes of L1, L2, L3, L4, L5 spinal nerves**
- **Dermatomes of S1, S2, S3, S4, S5 spinal nerves**

When you are finished, **close** the windows.

## Objectives: Spinal Cord & Spinal Nerves:

Student will learn the anatomy of the anatomy of the spinal cord and spinal nerves.

**Structures Associated with Spinal Cord**

- Arachnoid mater
- Cauda equina
- Conus medullaris
- Denticulate ligament
- Dura mater
- Epidural fat
- Epidural space
- External filum terminale
- Gray matter
- Internal filum terminale
- Intervertebral disks
- Pia mater
- Spinal cord
- Subarachnoid space
- Subdural space
- White matter

**Other**

- Dermatomes
- Diaphragm muscle
- Medulla oblongata
- Transverse process of vertebrae

**Spinal Nerves**

- Brachial plexus
- Cervical plexus
- Cutaneous branches of various spinal nerves
- Dorsal rami
- Dorsal root ganglia
- Dorsal root of spinal nerve
- Femoral nerve
- Intercostal nerve
- Lumbar plexus
- Median nerve
- Phrenic nerve
- Radial nerve
- Rami communicans
- Sacral plexus
- Sciatic nerve
- Spinal nerve
- Ulnar nerve
- Ventral rami
- Ventral root

# Cranial Nerves

## Opening A.D.A.M Interactive Anatomy

Open A.D.A.M. Interactive Anatomy according to the directions in the Tutorial.

## 3D Anatomy of the Cranial Nerves

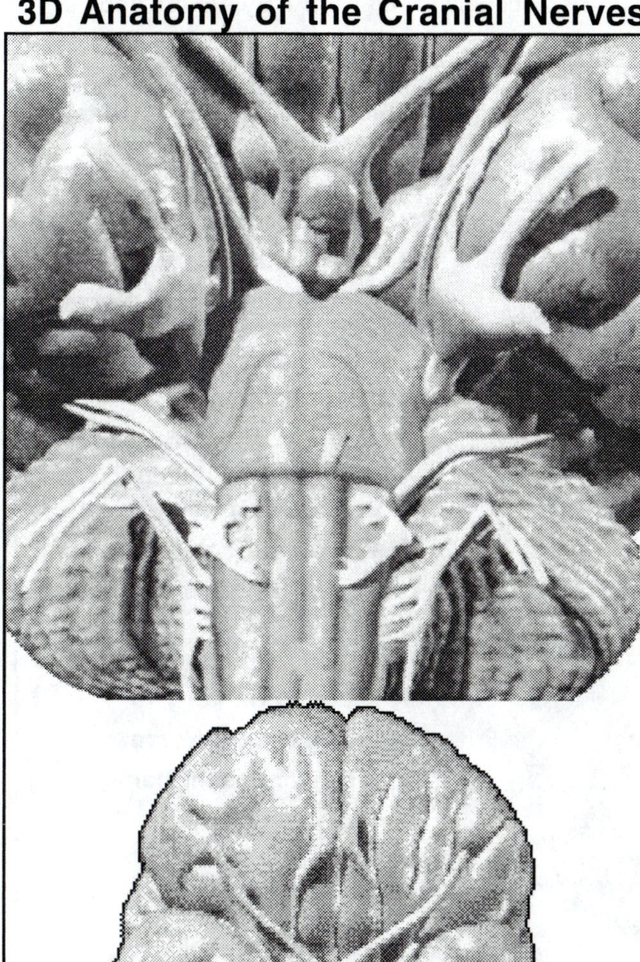

Choose **3D Anatomy**. Click on **3D Brain**. Identify the following by clicking on the **3D Anatomy Structure List**:

- **Abducent {Abducens} nerve {CN VI}**
- **Accessory nerve {CN XI}**
- **Facial nerve {CN VII}**
- **Glossopharyngeal nerve {CN IX}**
- **Hypoglossal nerve {CN XII}**
- **Oculomotor nerve {CN III}**
- **Olfactory nerve {CN I}**
- **Optic chiasma**
- **Optic nerve {CN II}**
- **Trigeminal nerve {CN V}**
- **Trochlear nerve {CN IV}**
- **Vagus nerve {CN X}**
- **Vestibulocochlear Nerve {CN VIII}**

When you are finished, **close** the window.

## Atlas Anatomy of Cranial Nerves

Choose **Atlas Anatomy**. Select **System** and **Nervous.** Choose **Base of Brain (Inf)**. If the list is not in alphabetical order, the image is close to the top of the list.

If you are working in Mac, click on the **Show All Pins** [button icon] button and scroll to **All Systems**. You may need to scroll to find the structures.

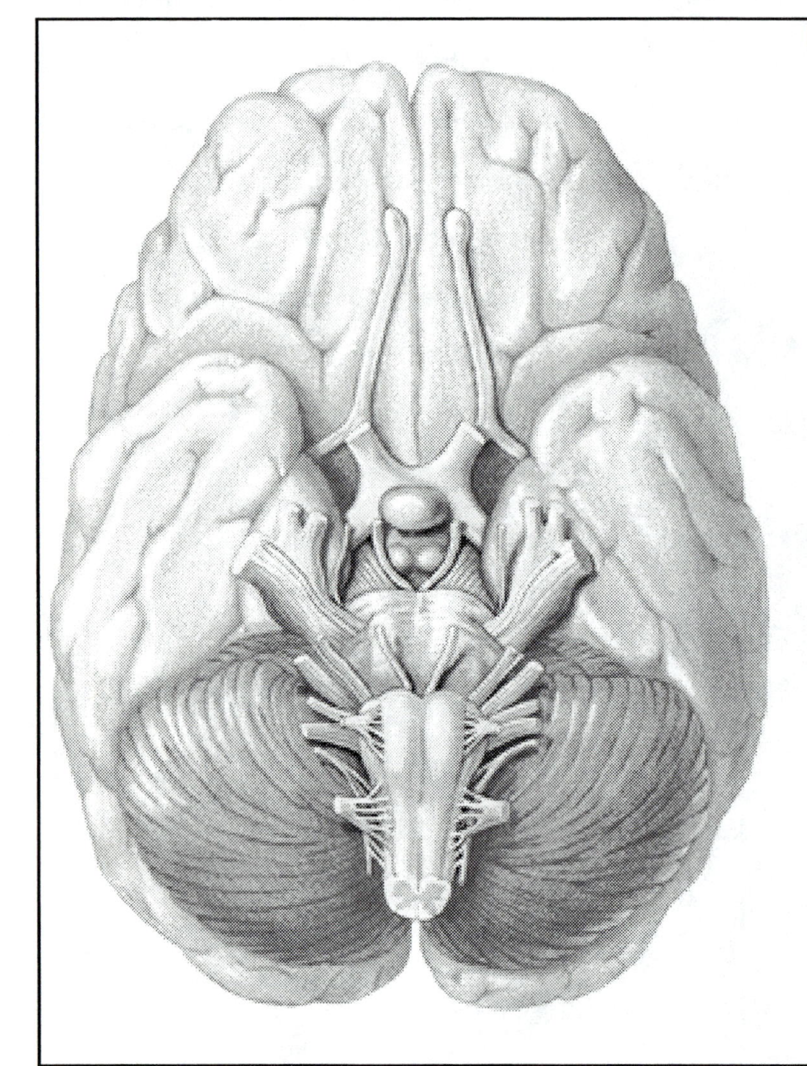

Identify:

- **Olfactory Bulb (CN I Olfactory Nerve ends here)**
- **CN II Optic Nerve**
- **CN III Oculomotor Nerve**
- **CN IV Trochlear Nerve**
- **CN V Trigeminal Nerve - Mandibular division**
- **CN V Trigeminal Nerve - Maxillary Division**
- **CN V Trigeminal Nerve - Ophthalmic Division**
- **CN VI Abducent Nerve**
- **CN VII Facial Nerve**
- **CN VIII Vestibulocochlear Nerve**
- **CN IX Glossopharyngeal Nerve**
- **CN X Vagus Nerve**
- **CN XI Accessory Nerve**
- **CN XII Hypoglossal Nerve**
- **Olfactory tract**
- **Cerebellum**
- **Optic chiasma**
- **Pituitary gland**
- **Pons**
- **Spinal cord**
- **C1 Spinal nerve** (ventral root)

Which cranial nerves emerge from the midbrain?

______________________________________________

Which cranial nerves emerge from the pons?

______________________________________________

Which cranial nerves emerge from the medulla?

______________________________________________

When you are finished, **close** the window.

Under the **File** menu, drag down to **Open** (then to **Content** if you are working in Windows). **Atlas Anatomy** should still be selected. **System** and **Nervous** should still be selected. Choose **Olfactory Nerve in Nasal Cavity**. If the list is not in alphabetical order, the image is about a fourth of the way down.

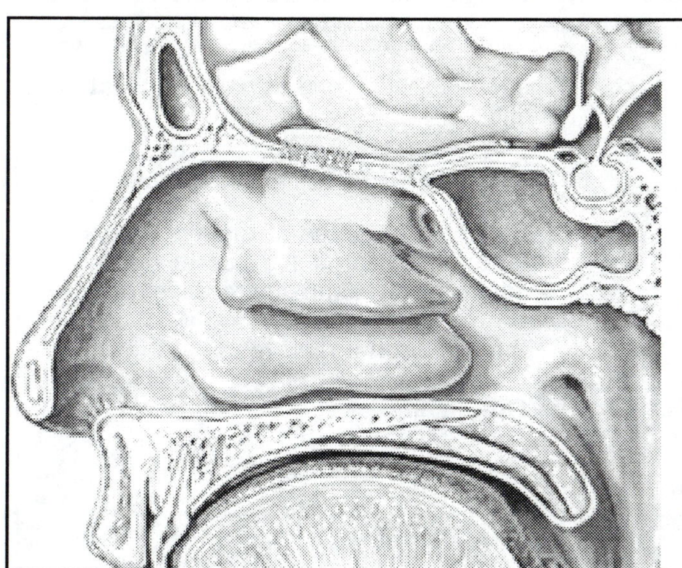

You may need to scroll to find the structures. Identify:

- **Olfactory bulb**
- **CN I Olfactory nerve**
- **Olfactory tract**

Is the olfactory nerve sensory, motor or both?

_______________________________

What is the function of the olfactory nerve?

_______________________________

Where does the olfactory nerve terminate?

_______________________________

When you are finished, **close** the window.

Under the **File** menu, drag down to **Open** (then to **Content** if you are working in Windows). **Atlas Anatomy** should still be selected. **System** and **Nervous** should still be selected. Choose **Optic Nerve in Orbit (Lat)**. If the list is not in alphabetical order, the image is close to the top of the list.

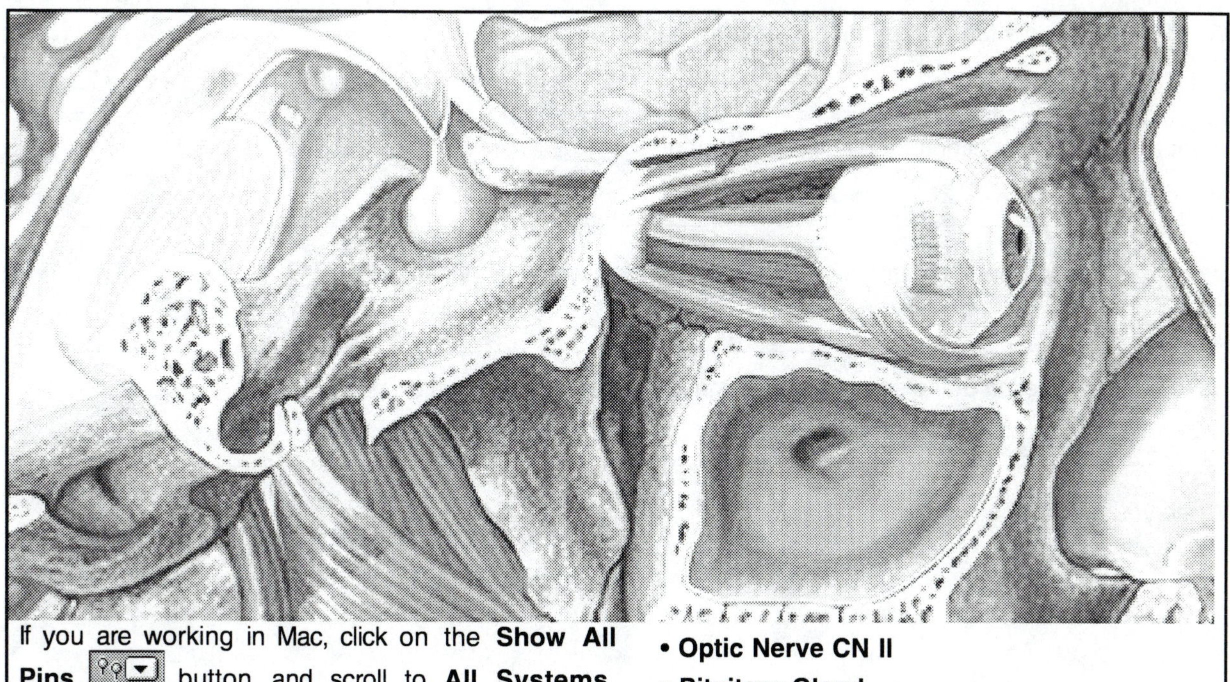

If you are working in Mac, click on the **Show All Pins** button and scroll to **All Systems**. Identify:

- **Maxillary sinus**
- **Midbrain**
- **Optic Nerve CN II**
- **Pituitary Gland**
- **Pons**
- **Superior orbital fissure**

Is the optic nerve sensory, motor or both? _______________________________

What is the function of the optic nerve? _______________________________

When you are finished, **close** the window.

Under the **File** menu, drag down to **Open** (then to **Content** if you are working in Windows). **Atlas Anatomy** should still be selected. **System** and **Nervous** should still be selected. Choose **Oculomotor nerve in Orbit (Ant)**. If the list is not in alphabetical order, the image is close to the top of the list.

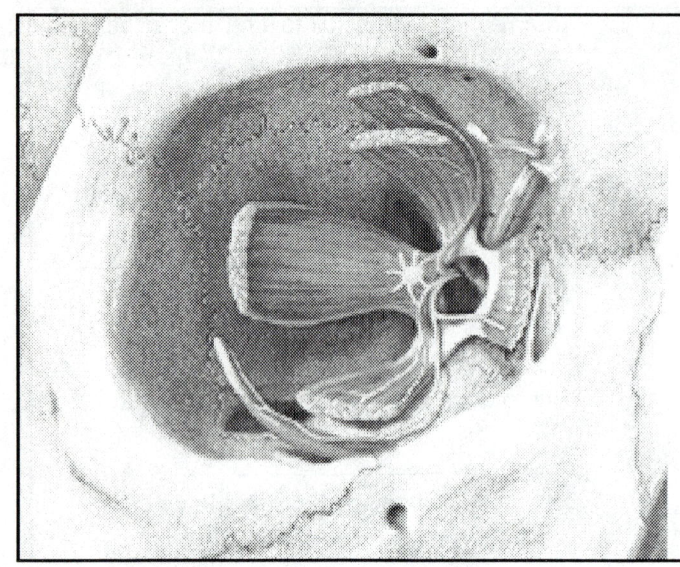

If you are working in Mac, click on the **Show All Pins** button and scroll to **All Systems**. Identify:

- **Oculomotor nerve CN III**
- **Extrinsic muscles of the eye** (inferior oblique, inferior rectus muscle, medial rectus muscle, superior rectus muscle)
- **Inferior orbital fissure**
- **Optic canal**
- **Superior orbital fissure**

Is the oculomotor nerve sensory, motor or both?

___

What is the sensory function of the oculomotor nerve?

___

What is the motor function of the oculomotor nerve?

___

Based on the location of the inferior rectus muscle, what would its action be?

___

Based on the location of the superior rectus muscle, what would its action be?

___

When you are finished, **close** the window.

Under the **File** menu, drag down to **Open** (then to **Content** if you are working in Windows). **Atlas Anatomy** should still be selected. **System** and **Nervous** should still be selected. Choose **Trochlear Nerve (Lat)**. If the list is not in alphabetical order, the image is close to the top of the list.

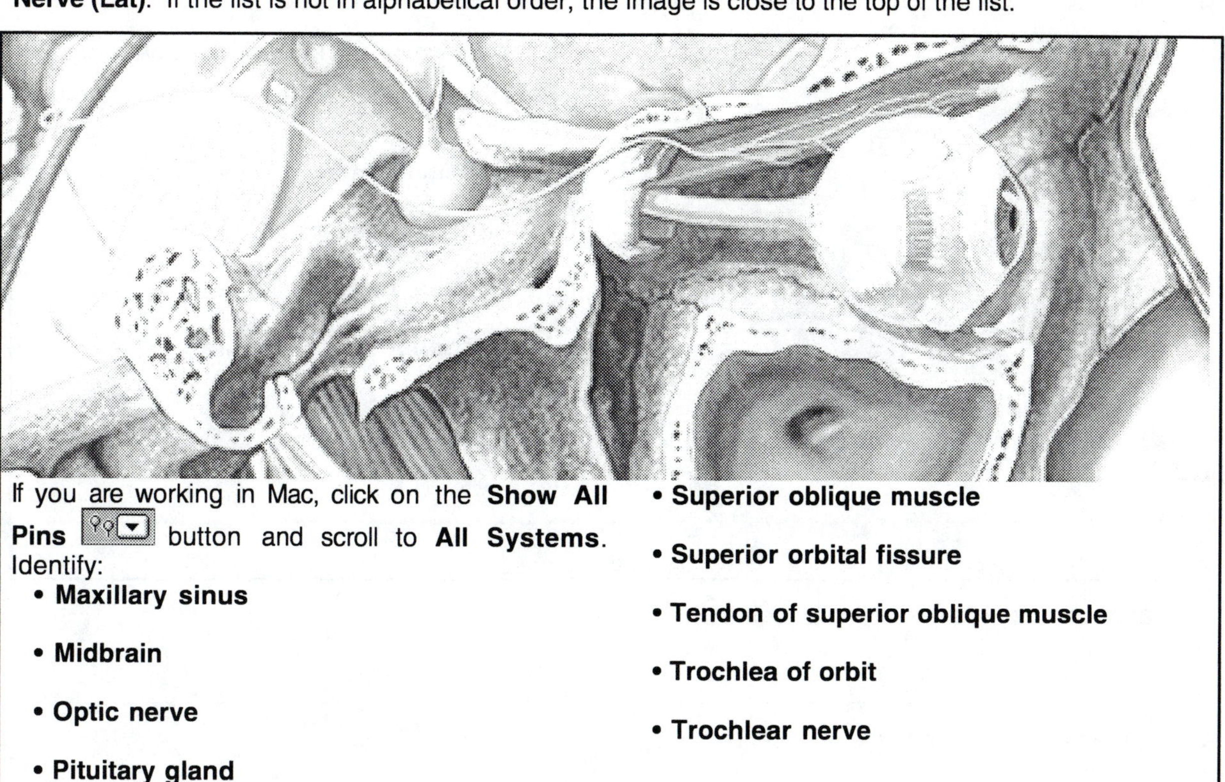

If you are working in Mac, click on the **Show All Pins** button and scroll to **All Systems**. Identify:

- **Maxillary sinus**
- **Midbrain**
- **Optic nerve**
- **Pituitary gland**
- **Pons**
- **Superior oblique muscle**
- **Superior orbital fissure**
- **Tendon of superior oblique muscle**
- **Trochlea of orbit**
- **Trochlear nerve**

Through which foramen does the trochlear nerve leave the skull?

___

What is the motor function of the trochlear nerve?

___

Is the trochlear nerve sensory, motor or both?

___

What is the sensory function of the trochlear nerve?

___

When you are finished, **close** the window.

Under the **File** menu, drag down to **Open** (then to **Content** if you are working in Windows). **Atlas Anatomy** should still be selected. **System** and **Nervous** should still be selected. Choose **Abducent Nerve in Orbit**. If the list is not in alphabetical order, the image is close to the top of the list.

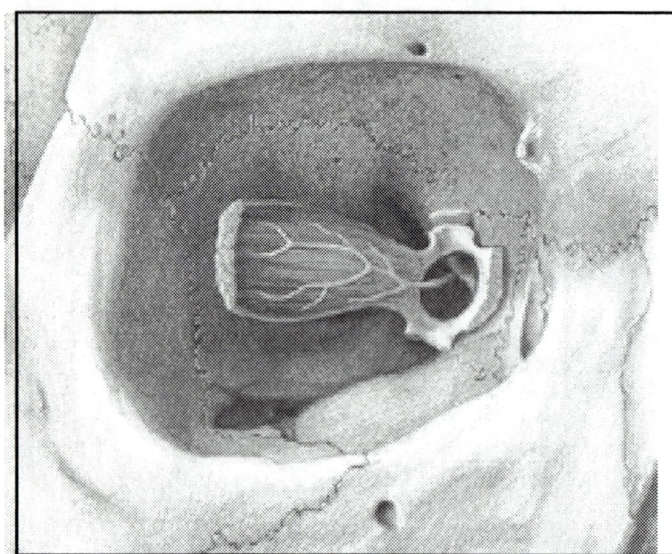

If you are working in Mac, click on the **Show All Pins** button and scroll to **All Systems**. Identify:

- **Abducent nerve CN VI**
- **Inferior orbital fissure**
- **Lateral rectus muscle**
- **Superior orbital fissure**

Is the abducent nerve sensory, motor or both?

---

What is the motor function of the abducent nerve?

---

What is the sensory function of the abducent nerve?

---

Based on the location of the lateral rectus muscle, what would its action be?

---

Through which foramen does the abducent nerve leave the skull?

---

When you are finished, **close** the window.

Under the **File** menu, drag down to **Open** (then to **Content** if you are working in Windows). **Atlas Anatomy** should still be selected. **System** and **Nervous** should still be selected. Choose **Dermatomes of Head/Neck (Lat)**. If the list is not in alphabetical order, the image is close to the top of the list.

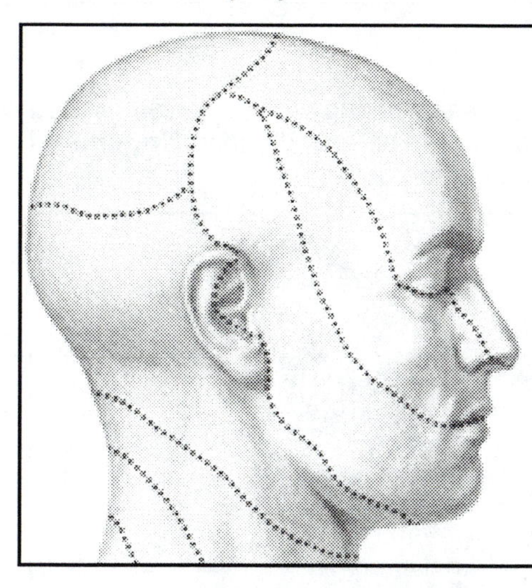

If you are working in Mac, click on the **Show All Pins** button and scroll to **All Systems**. Identify:

- **Dermatome of mandibular division of trigeminal nerve CN V**
- **Dermatome of maxillary division of trigeminal nerve CN V**
- **Dermatome of ophthalmic division of trigeminal nerve CN V**

When you are finished, **close** the window.

Under the **File** menu, drag down to **Open** (then to **Content** if you are working in Windows). **Atlas Anatomy** should still be selected. **System** and **Nervous** should still be selected. Choose **Trigeminal Nerve - V1**. If the list is not in alphabetical order, the image is close to the top of the list.

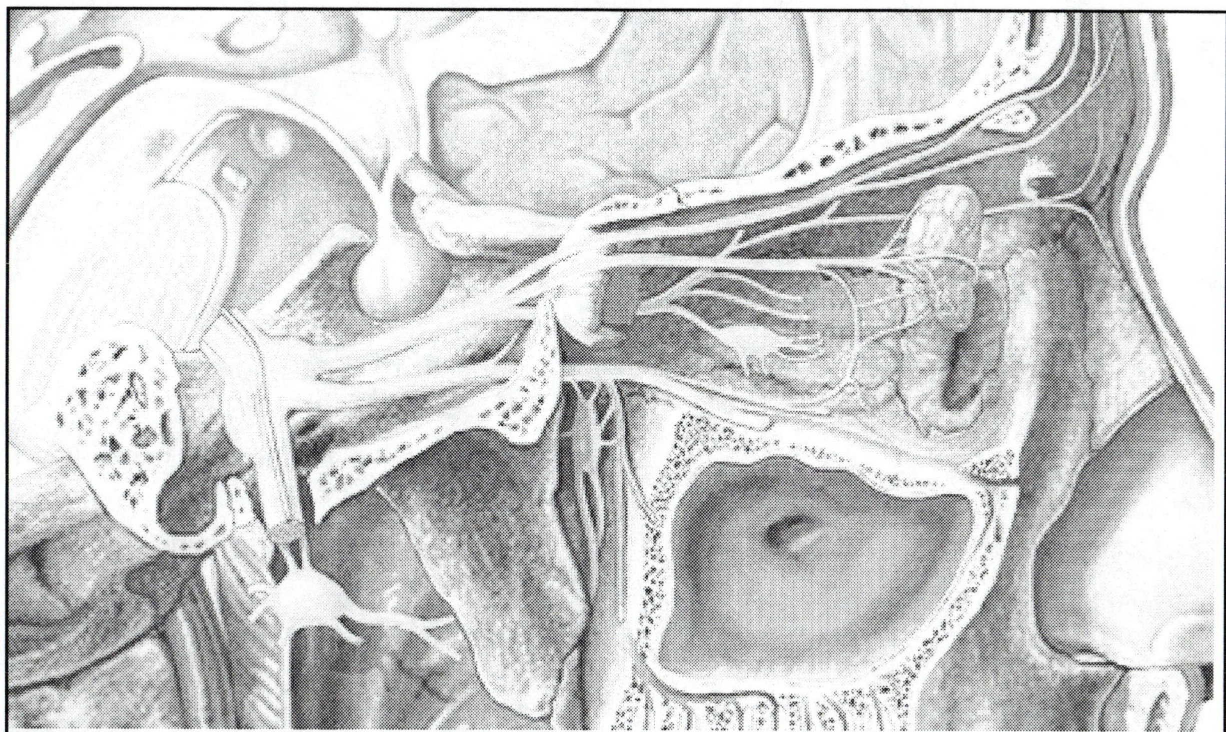

If you are working in Mac, click on the **Show All Pins** button and scroll to **All Systems**. Identify:

- **Superior orbital fissure**
- **Trigeminal nerve CN V**
- **Mandibular division of trigeminal nerve CN V**
- **Maxillary division of trigeminal nerve CN V**
- **Ophthalmic division of trigeminal nerve CN V**
- **Motor root of trigeminal nerve CN V**
- **Trigeminal ganglion**

Through which foramen does the ophthalmic division of trigeminal nerve emerge from the skull?

---

When you are finished, **close** the window.

Under the **File** menu, drag down to **Open** (then to **Content** if you are working in Windows). **Atlas Anatomy** should still be selected. **System** and **Nervous** should still be selected. Choose **Trigeminal Nerve - V2**. If the list is not in alphabetical order, the image is close to the top of the list.

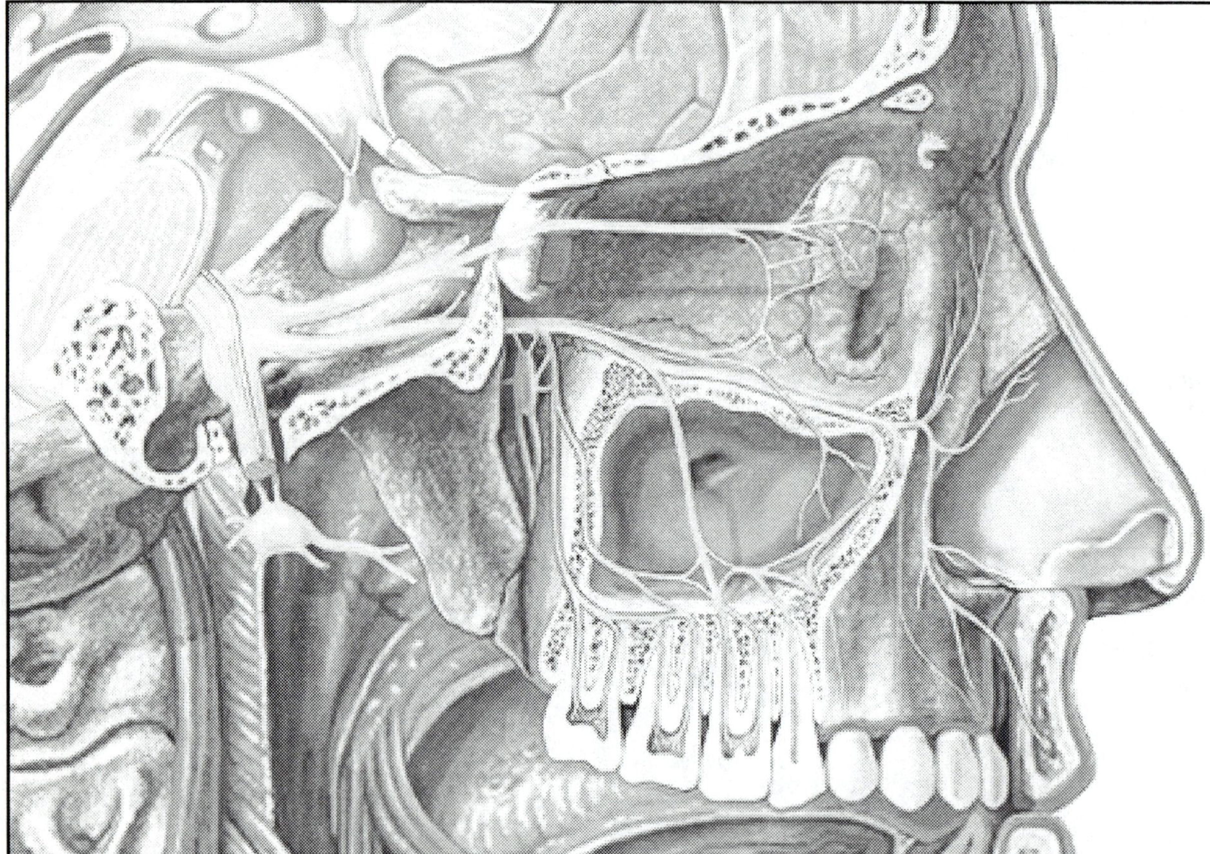

If you are working in Mac, click on the **Show All Pins** button and scroll to **All Systems**. Identify:

- **Trigeminal nerve CN V**
- **Trigeminal ganglion**
- **Mandibular division of trigeminal nerve CN V**
- **Maxillary division of trigeminal nerve CN V**
- **Ophthalmic division of trigeminal nerve CN V**
- **Motor root of trigeminal nerve CN V**

Through which foramen does the maxillary division of the trigeminal nerve emerge from the skull?

---

On the picture above, color in the branches of the maxillary division of the trigeminal nerve. List the names of several of two branches of the maxillary division of the trigeminal nerve that a dentist would inject novocaine near before drilling on the teeth in the upper mouth.

---

When you are finished, **close** the window.

Under the **File** menu, drag down to **Open** (then to **Content** if you are working in Windows). **Atlas Anatomy** should still be selected. **System** and **Nervous** should still be selected. Choose **Trigeminal Nerve - V3**. If the list is not in alphabetical order, the image is close to the top of the list.

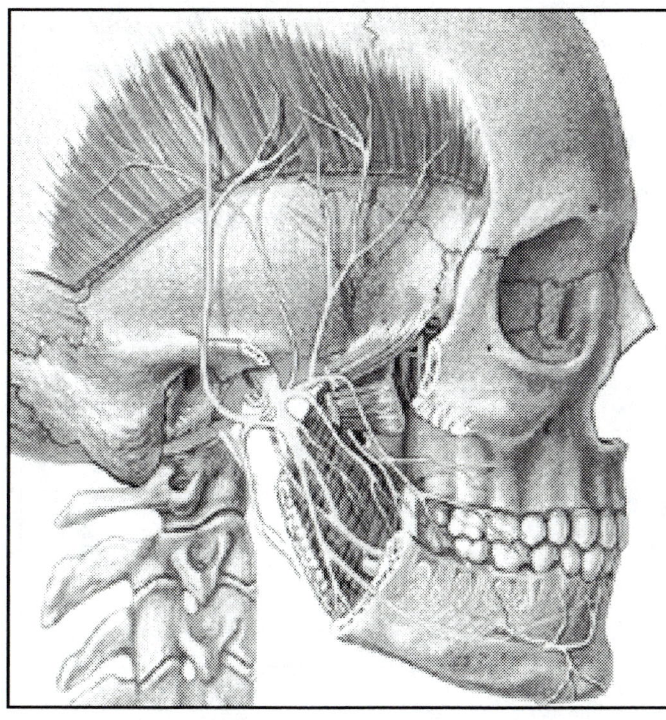

If you are working in Mac, click on the **Show All Pins** button and scroll to **All Systems**. Identify:

- **Mandibular division of trigeminal nerve CN V**
- **Temporalis muscle**
- **Nerve to masseter muscle**

Through which foramen does the mandibular division of the trigeminal nerve emerge from the skull?

On the picture above, color in the branches of the mandibular division of the trigeminal nerve.

What two muscles involved in chewing does the trigeminal nerve innervate?

When you are finished, **close** the window.

 

Under the **File** menu, drag down to **Open** (then to **Content** if you are working in Windows). **Atlas Anatomy** should still be selected. **System** and **Nervous** should still be selected. Choose **Facial Nerve (Lat) 1**. If the list is not in alphabetical order, the image is close to the top of the list.

If you are working in Mac, click on the **Show All Pins** button and scroll to **All Systems**. Identify:

- **Facial nerve CN VII**

On the picture to the left, color in the branches of the facial nerve. List five muscles which are innervated by the facial nerve.

______________________________

______________________________

______________________________

______________________________

______________________________

When you are finished, **close** the window.

Under the **File** menu, drag down to **Open** (then to **Content** if you are working in Windows). **Atlas Anatomy** should still be selected. **System** and **Nervous** should still be selected. Choose **External, Middle, & Inner Ear**. If the list is not in alphabetical order, the image is one-third of the way down from the top of the list.

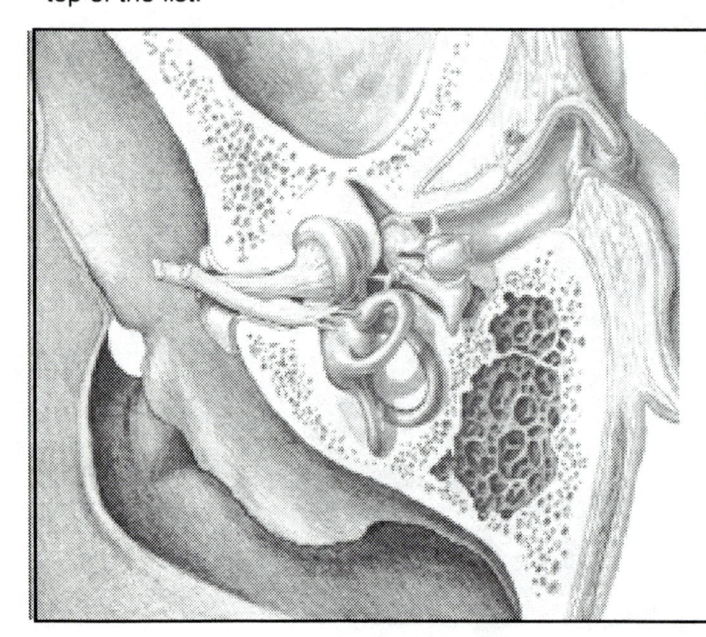

If you are working in Mac, click on the **Show All Pins** button and scroll to **All Systems**. Identify:

- **Bony part of external acoustic meatus**
- **Cochlear division of vestibulocochlear nerve CN VIII**
- **External acoustic meatus**
- **Internal acoustic meatus**
- **Vestibular division of vestibulocochlear nerve CN VIII**

What is the function of the cochlear division of the vestibulocochlear nerve CN VIII?

______________________________

What is the function of the vestibular division of the vestibulocochlear nerve CN VIII?

---

When you are finished, **close** the window.

Under the **File** menu, drag down to **Open** (then to **Content** if you are working in Windows). **Atlas Anatomy** should still be selected. **System** and **Nervous** should still be selected. Choose **Glossopharyngeal Nerve 2**. If the list is not in alphabetical order, the image is close to the top of the list.

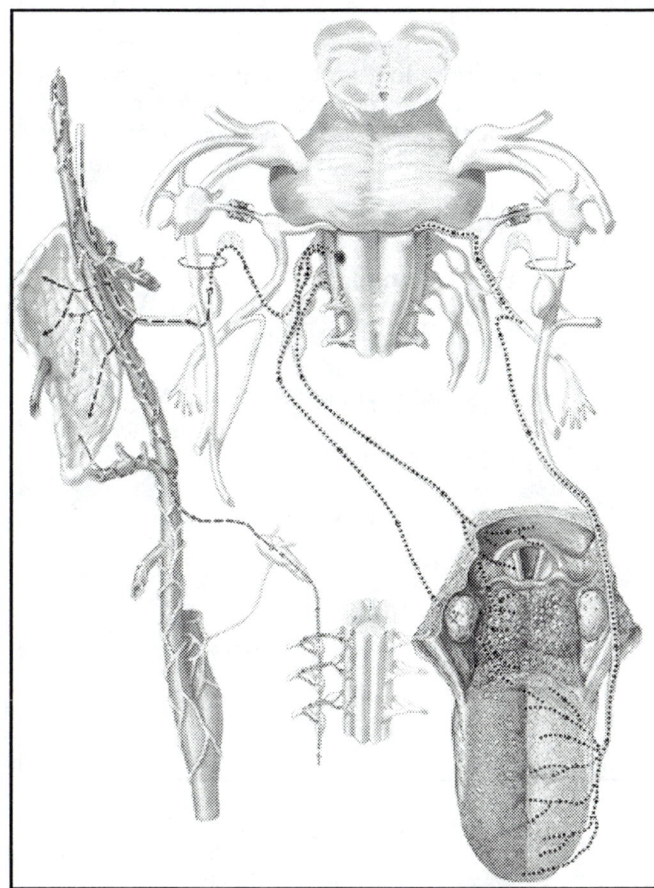

If you are working in Mac, click on the **Show All Pins** button and scroll to **All Systems**.
Identify:

- **Pons**
- **Apex of tongue**
- **Facial nerve CN VII**
- **Glossopharyngeal nerve CN IX**
- **Trigeminal nerve CN V**
- **Vagus nerve CN X**

On the picture to the left, color in the branches of the glossopharyngeal nerve in one color, the branches of the facial nerve in another color, and the branches of the vagus nerve in still another color.

Which nerve is responsible for taste in the anterior tongue?

---

Which nerve is responsible for taste and other sensations in the posterior tongue?

---

Which nerve is responsible for taste in the pharynx (throat)?

---

When you are finished, **close** the window.

Under the **File** menu, drag down to **Open** (then to **Content** if you are working in Windows). **Atlas Anatomy** should still be selected. **System** and **Nervous** should still be selected. Choose **Glossopharyngeal Nerve 1**. If the list is not in alphabetical order, the image is close to the top of the list.

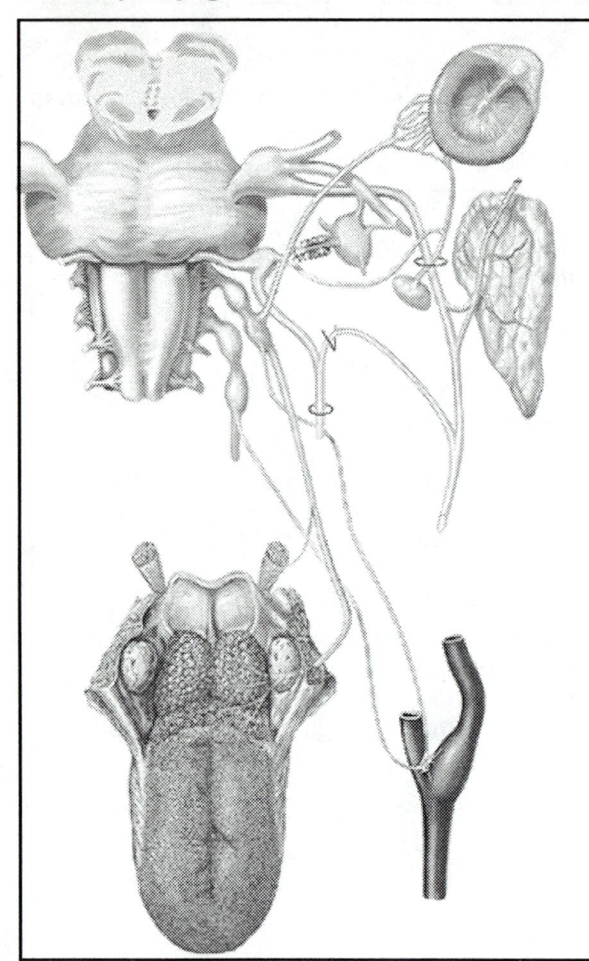

If you are working in Mac, click on the **Show All Pins** button and scroll to **All Systems**. Identify:

- **Carotid branch of glossopharyngeal nerve**
- **Communicating branch of vagus nerve** (carries sensory impulses from major arteries to the brain conveying information about blood pressure and blood gases)
- **Pons**
- **Apex of tongue**
- **Facial nerve CN VII**
- **Glossopharyngeal nerve CN IX**
- **Trigeminal nerve CN V**
- **Vagus nerve CN X**

When you are finished, **close** the window.

Under the **File** menu, drag down to **Open** (then to **Content** if you are working in Windows). **Atlas Anatomy** should still be selected. **System** and **Nervous** should still be selected. Choose **Vagus Nerve (Lat)**. If the list is not in alphabetical order, the image is close to the top of the list.

If you are working in Mac, click on the **Show All Pins** button and scroll to **All Systems**. Identify:

- **Vagus nerve CN X** (all branches)

Name the four branches of the vagus nerve.

________________________________________

________________________________________

________________________________________

________________________________________

When you are finished, **close** the window.

List five facial muscles which are innervated by the vagus nerve.

________________________________________ ________________________________________

________________________________________ ________________________________________

________________________________________

Under the **File** menu, drag down to **Open** (then to **Content** if you are working in Windows). **Atlas Anatomy** should still be selected. **System** and **Nervous** should still be selected. Choose **Autonomic NS-Viscera 2**. If the list is not in alphabetical order, the image is about two-thirds of the way down from the top of the list.

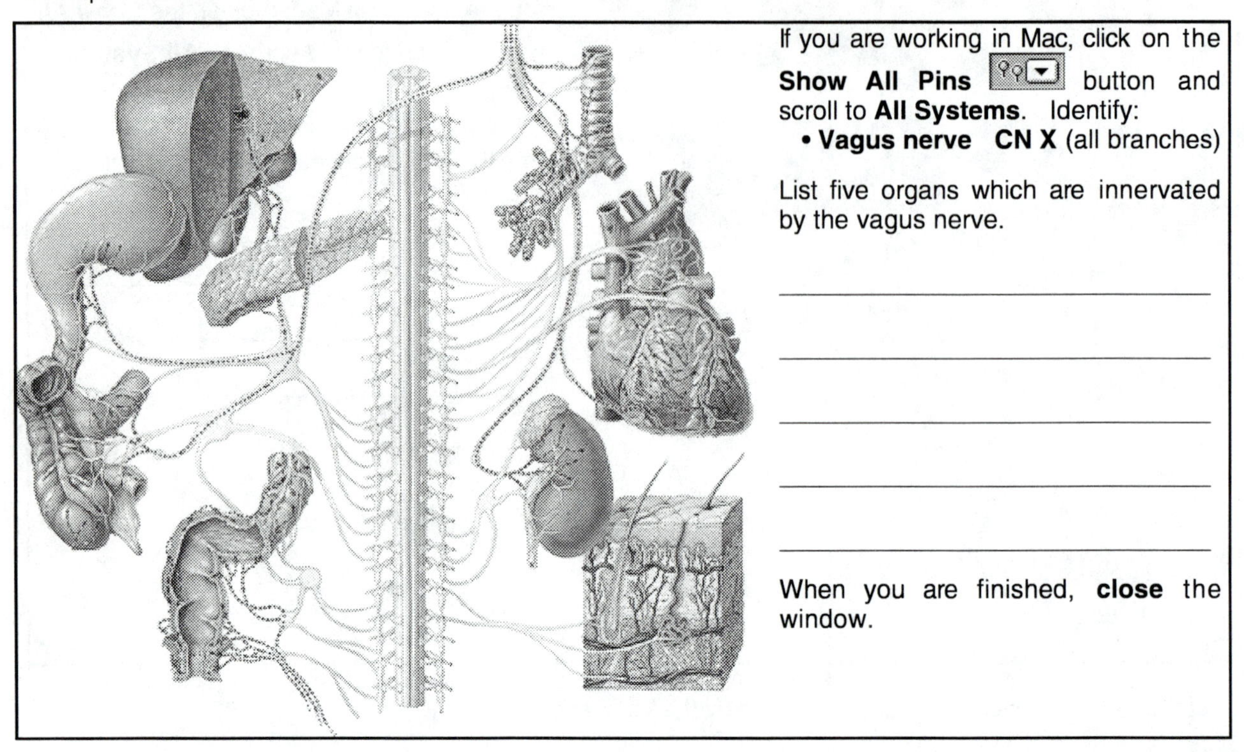

If you are working in Mac, click on the **Show All Pins** button and scroll to **All Systems**. Identify:

- **Vagus nerve CN X** (all branches)

List five organs which are innervated by the vagus nerve.

______________________________

______________________________

______________________________

______________________________

______________________________

When you are finished, **close** the window.

Under the **File** menu, drag down to **Open** (then to **Content** if you are working in Windows). **Atlas Anatomy** should still be selected. **System** and **Nervous** should still be selected. Choose **Accessory Nerve (Lat)**. If the list is not in alphabetical order, the image is close to the top of the list.

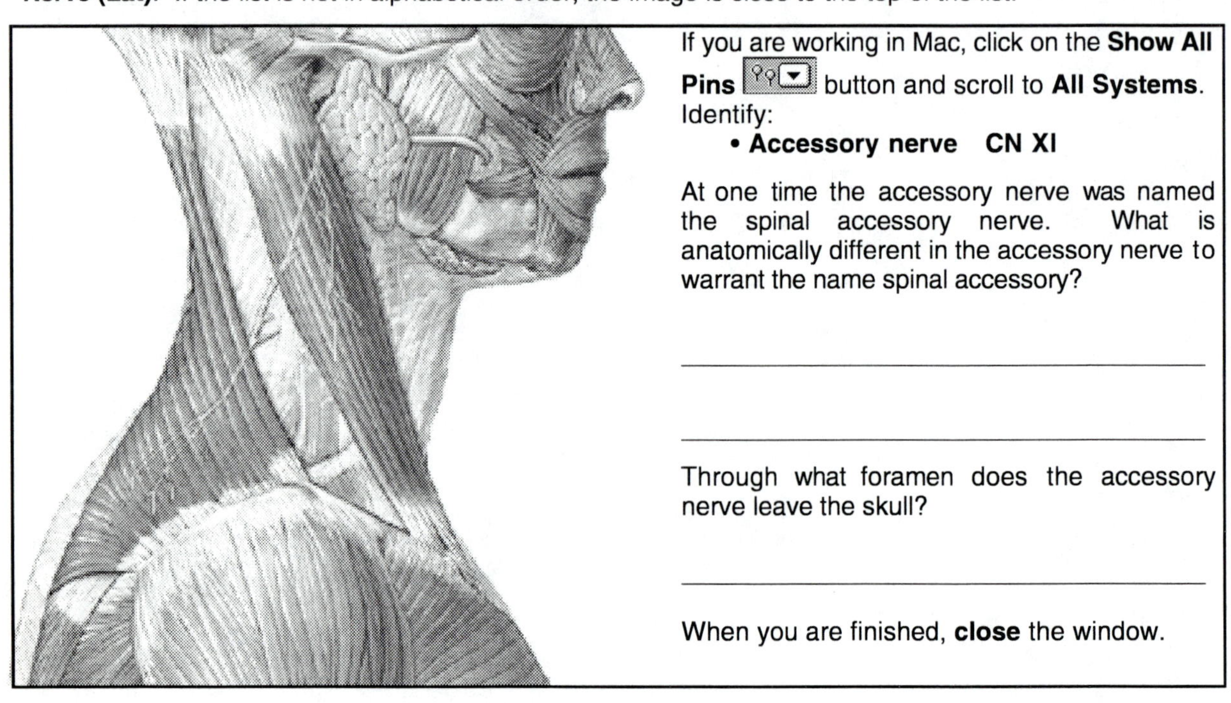

If you are working in Mac, click on the **Show All Pins** button and scroll to **All Systems**. Identify:

- **Accessory nerve CN XI**

At one time the accessory nerve was named the spinal accessory nerve. What is anatomically different in the accessory nerve to warrant the name spinal accessory?

______________________________

______________________________

Through what foramen does the accessory nerve leave the skull?

______________________________

When you are finished, **close** the window.

Under the **File** menu, drag down to **Open** (then to **Content** if you are working in Windows). **Atlas Anatomy** should still be selected. **System** and **Nervous** should still be selected. Choose **Ansa cervicalis (Lat).** If the list is not in alphabetical order, the image is close to the top of the list.

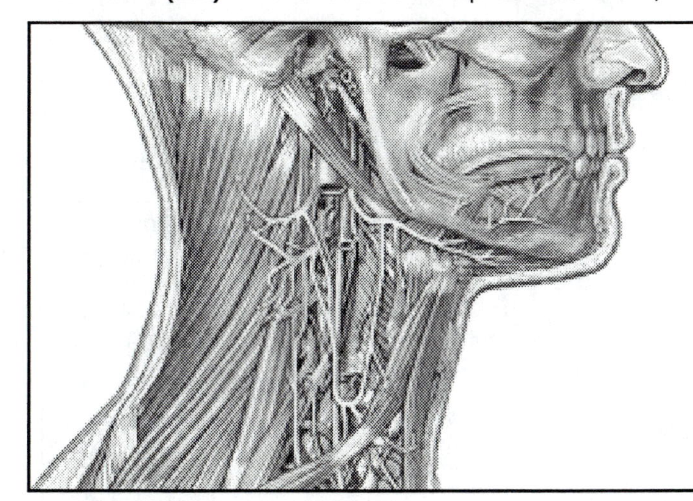

Identify:

- **Hypoglossal nerve CN XII**

When you are finished, **close** the window.

What is the primary function of the hypoglossal nerve?

______________________________________________________________________

## Lateral Dissection of Cranial Nerves

Under the **File** menu, drag down to **Open** (then to **Content** if you are working in Windows).

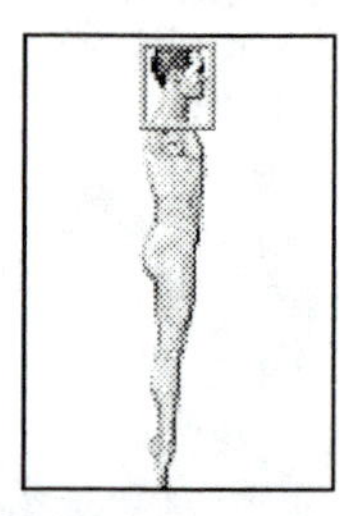

Choose **Dissectible Anatomy.**

Choose either male or female. Select **Lateral.** Click **Open**.

Position the navigator rectangle over the lateral face & neck.

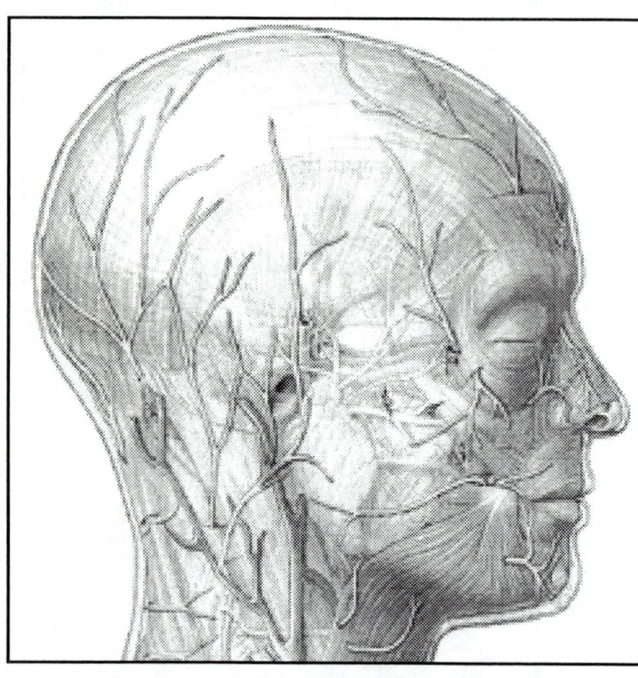

Click on the **Depth Bar** ▼ until you reach **layer 5**. Note the various cutaneous nerves of the face.

From what cranial nerve do these cutaneous nerves arise?

___________________________

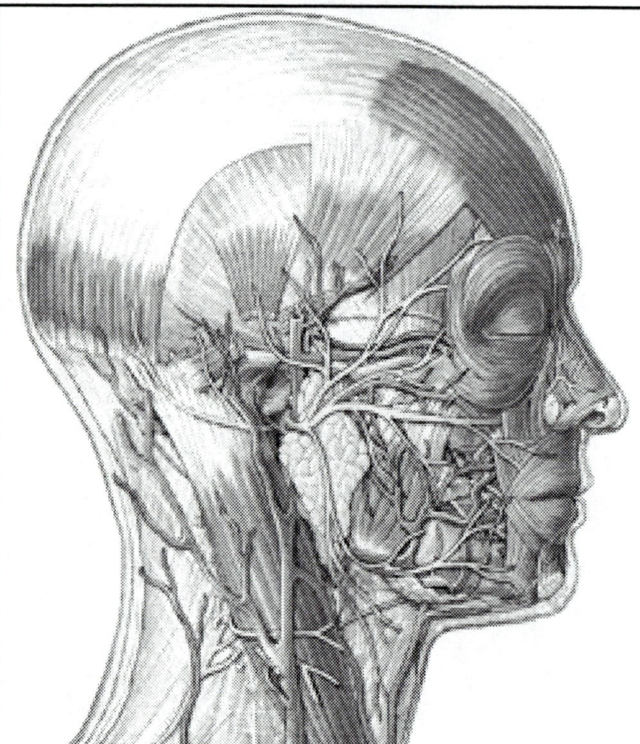

Click on the **Depth Bar** ▼ until you reach **layer 17**. Identify:

- **Facial nerve VII (**including branches)
- **Accessory nerve CN XI**

What is the primary function of the branches of the facial nerve shown here?

___________________________

What is the primary function of the branches of the accessory nerve shown here?

___________________________

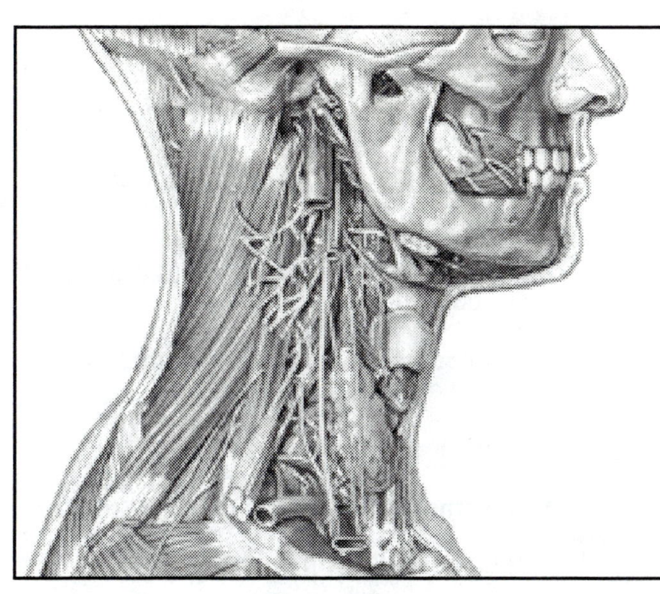

Click on the **Depth Bar** [▼] until you reach **layer 56**. Identify:

- **Vagus nerve X** (including branches)
- **Accessory Nerve XI**
- **Hypoglossal Nerve XII**

What is the primary function of the branches of the vagus nerve shown here?

---

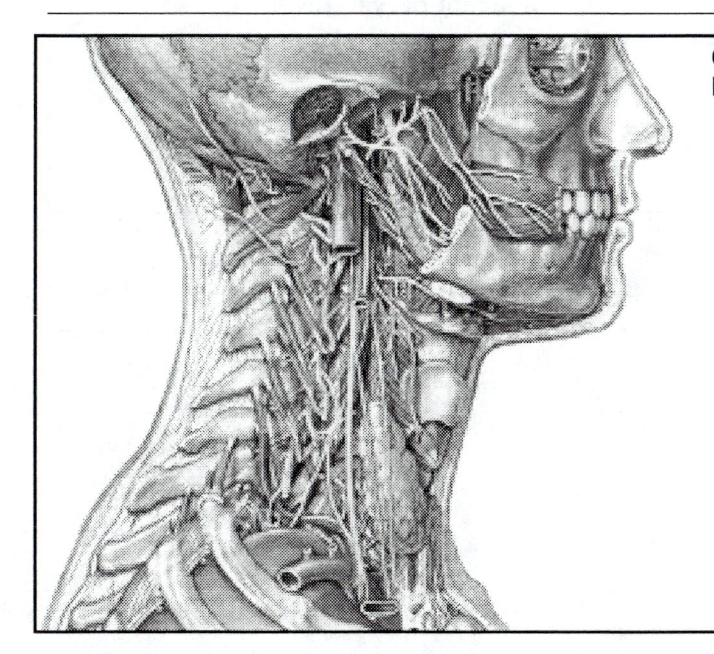

Click on the **Depth Bar** [▼] until you reach **layer 123**. Identify:

- **Vagus nerve X** (including branches)
- **Mandibular division of trigeminal nerve V** (including branches)
- **Hypoglossal Nerve XII**

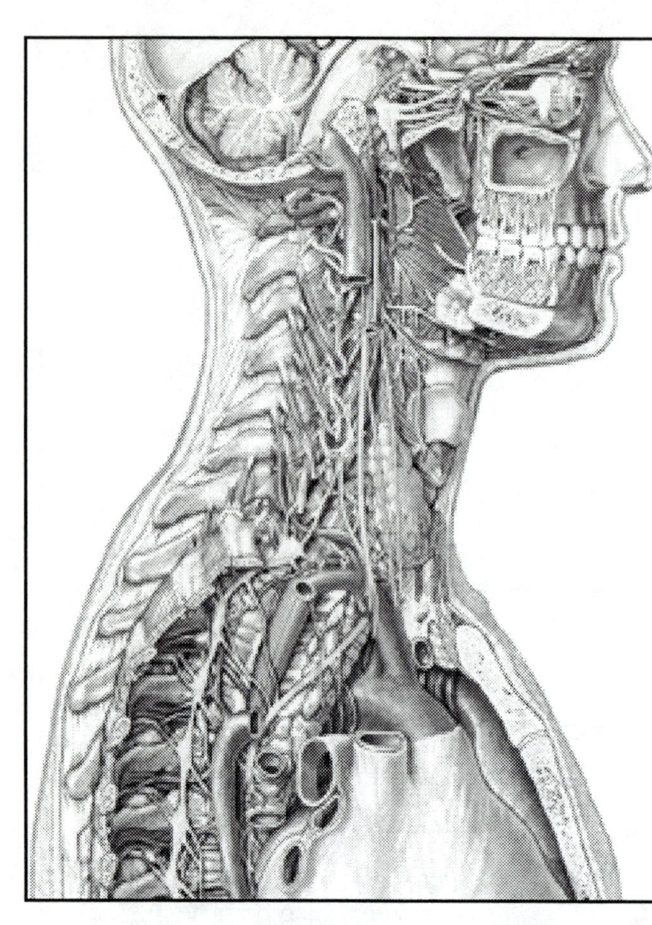

Click on the **Depth Bar** ▼ until you reach **layer 196**. Identify:

- **Vagus nerve X** (including branches)
- **Ophthalmic division of trigeminal nerve V** (including branches)
- **Maxillary division of trigeminal nerve V** (including branches)
- **Mandibular division of trigeminal nerve V** (including branches)
- **Trigeminal ganglion**
- **Oculomotor nerve CN III**
- **Optic nerve CN II**
- **Trochlear nerve CN IV**
- **Abducent nerve CN VI**
- **Hypoglossal Nerve XII**

What are the primary functions of the oculomotor, trochlear, and abducent nerve shown here?

---

## Medial Dissection of Cranial Nerves

Click on the **View Button** and choose **Medial**.

Click on the **Normal** button. Windows— —Mac

Position the navigator rectangle over the face & neck.

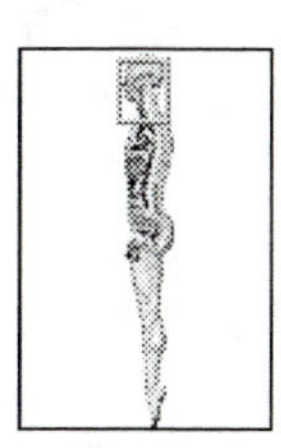

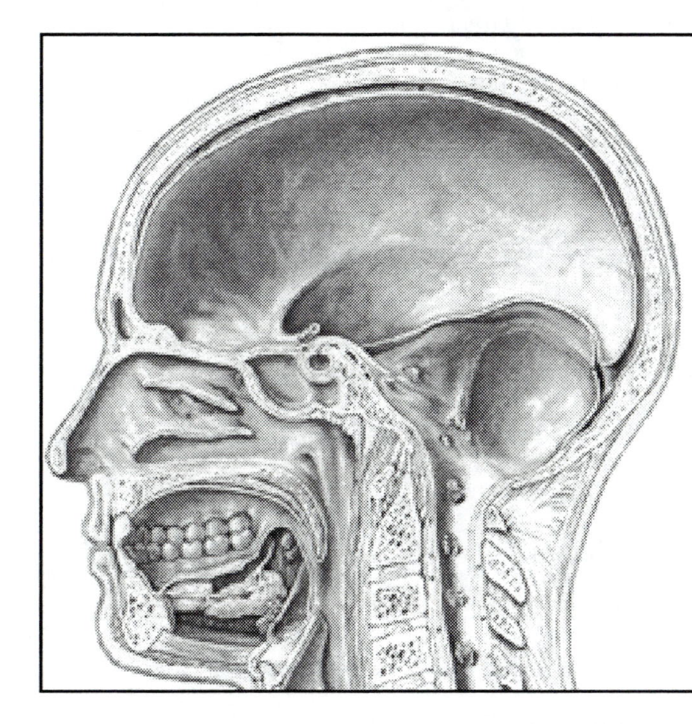

Click on the **Depth Bar** until you reach **layer 101**. Note the exit of the following cranial nerves from the cranium:

- **CN II Optic Nerve**
- **CN VI Abducent Nerve**
- **CN VIII Vestibulocochlear Nerve**
- **CN VII Facial Nerve**
- **CN XII Hypoglossal Nerve**
- **CN XI Accessory Nerve**
- **CN X Vagus Nerve**
- **CN IX Glossopharyngeal Nerve**

When you are finished, **close** the window.

 

## Objectives: Cranial Nerves:

To study the anatomy as well as the motor and sensory functions of the cranial nerves. (Parasympathetic functions of the cranial nerves will be addressed in the lab on the autonomic nervous system.)

**Nerves:**

- CN I Olfactory nerve
- CN II Optic nerve
- CN III Oculomotor nerve
- CN IV Trochlear nerve
- CN V Trigeminal nerve - Mandibular division
- CN V Trigeminal nerve - Maxillary division
- CN V Trigeminal nerve - Ophthalmic division
- Motor root of trigeminal nerve CN V
- Trigeminal ganglion
- CN VI Abducent nerve
- CN VII Facial nerve
- CN VIII Vestibulocochlear nerve
- Cochlear division of vestibulocochlear nerve CN VIII
- Vestibular division of vestibulocochlear nerve CN VIII
- CN IX Glossopharyngeal nerve
- Carotid branch of glossopharyngeal nerve
- CN XI Accessory nerve
- Communicating branch of vagus nerve
- CN X Vagus nerve
- CN XII Hypoglossal nerve
- C1 Spinal nerve (ventral root)
- Nerve to masseter muscle

**Other:**

- Olfactory bulb
- Olfactory tract
- Cerebellum
- Optic chiasma
- Pituitary gland
- Pons
- Spinal cord
- Extrinsic muscles of the eye
- Inferior orbital fissure
- Optic canal
- Superior orbital fissure
- Inferior orbital fissure
- Maxillary sinus
- Midbrain
- Pituitary gland
- Pons
- Tendon of superior oblique muscle
- Trochlea of orbit
- Dermatomes of trigeminal nerve
- Temporalis muscle
- Bony part of external acoustic meatus
- External acoustic meatus
- Internal acoustic meatus
- Apex of tongue

# Autonomic Nervous System

## Opening A.D.A.M Interactive Anatomy

Open A.D.A.M. Interactive Anatomy according to the directions in the Tutorial.

## Anterior Dissection of the Autonomic Nervous System

Choose **Dissectible Anatomy**.

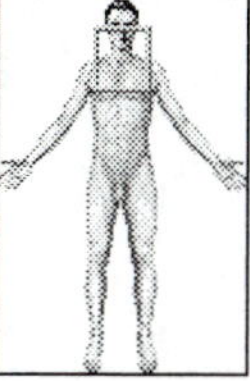

Choose either male or female. Select **Anterior.** Click **Open**.

Enlarge the window and position the navigator rectangle over the neck & chest area.

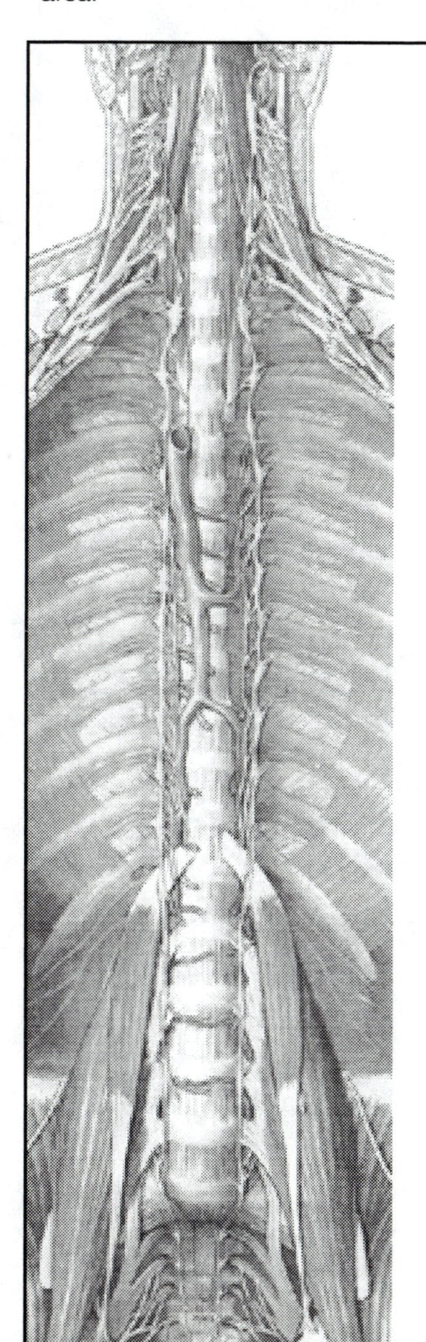

Using the **Depth Bar**, scroll down to **layer 264** and identify:

- **Sympathetic ganglia** (second though sixth thoracic)
- **Sympathetic trunk**
- **White ramus communicans**
- **Gray ramus communicans**
- **Sympathetic ganglia** in the neck area (superior cervical ganglion of sympathetic trunk, middle cervical ganglion of sympathetic trunk, cervicothoracic ganglion of sympathetic trunk)
- **Cardiac branch of sympathetic trunk**
- **Greater thoracic splanchnic nerve**

Why are there no white rami communicans emerging from the sympathetic ganglia in the neck area?

_______________________________________________

Scroll down to the abdomen. You are still at **layer 264**. Identify:

- **Sympathetic ganglia** (seventh though twelfth thoracic)
- **Sympathetic trunk**
- **White ramus communicans**
- **Gray ramus communicans**
- **Ganglia of sympathetic trunk**
- **Thoracic splanchnic nerve** (lesser & greater)

You will now perform a dissection in reverse. Click on the upward **Depth Bar** [▲] until you reach **layer 249**. No graphic has been provided for this layer. Identify:

- **Anterior vagal trunk** (wrapped around the esophagus)

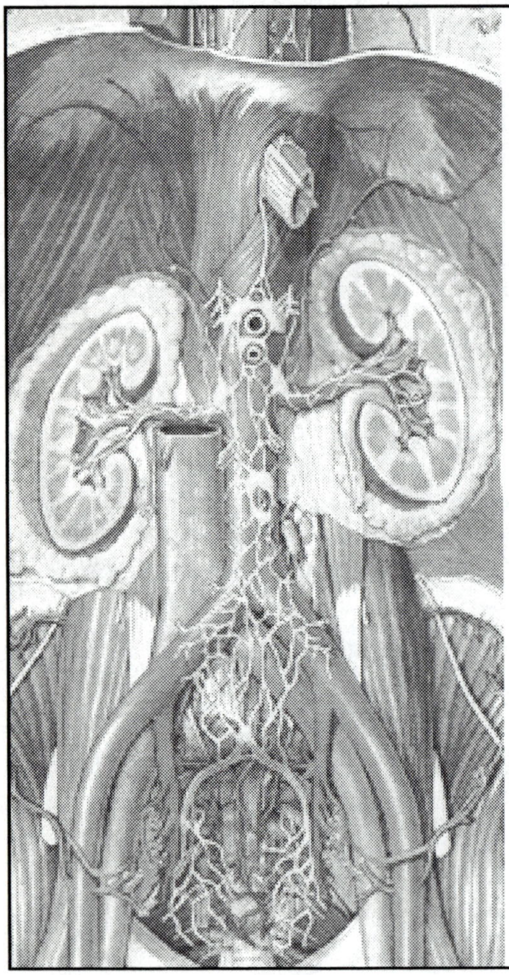

Click on the upward **Depth Bar** [▲] until you reach **layer 239**. Identify:

- **Celiac ganglia** ( left and right)
- **Superior mesenteric ganglia**
- **Inferior mesenteric ganglia**
- **Nerve plexuses** (various)
- **Thoracic splanchnic nerve** (greater)

What organs are innervated by the celiac ganglion?

_______________________________________________

What organs are innervated by the superior mesenteric ganglion?

_______________________________________________

What organs are innervated by the inferior mesenteric ganglion?

_______________________________________________

## Lateral Views of Autonomic Nervous System

Click on the **View Button** and choose **Lateral**.

Click on the **Normal Button**. Windows — Mac

Drag the **Depth Bar** to **layer 0.**

Position the navigator rectangle over the lower face, neck and upper chest area.

Scroll to **layer 233**. Identify:

- **Parasympathetic ganglia** (Submandibular ganglion)
- **Left vagus nerve CN X**

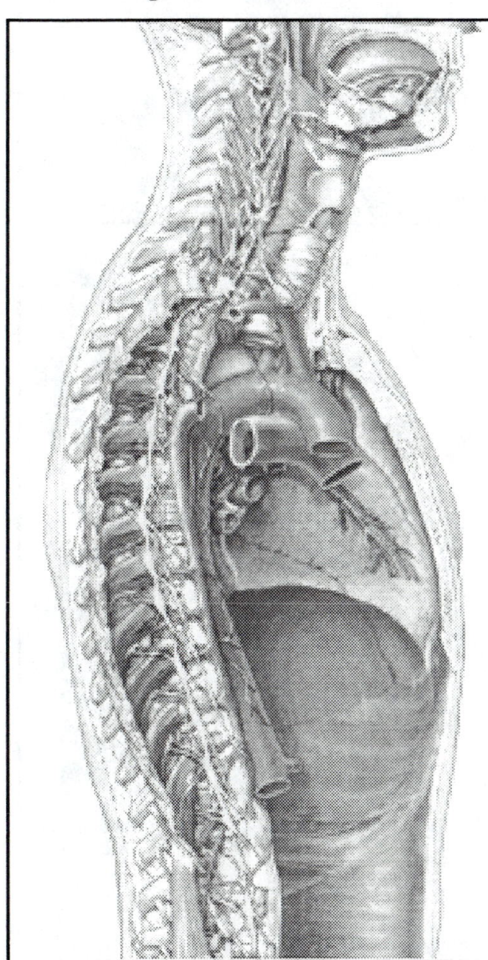

Click on the upward **Depth Bar** until you reach **layer 225**. Identify:

- **Sympathetic ganglia** (second though twelfth thoracic, first and second lumbar)
- **Sympathetic trunk**
- **White ramus communicans**
- **Gray ramus communicans**
- **Sympathetic ganglia** in the neck area (superior cervical ganglion of sympathetic trunk, middle cervical ganglion of sympathetic trunk, cervicothoracic ganglion of sympathetic trunk)
- **Cardiac branch of sympathetic trunk**
- **Greater thoracic splanchnic nerve**

Scroll to the head area. Click on the upward **Depth Bar** until you reach **layer 201**. No graphic has been provided for this layer. Identify:

- **Parasympathetic ganglia** (otic ganglia, pterygopalatine ganglia)
- **Vagus nerve CN X**

When you are finished, **close** the window.

## Atlas Anatomy of Autonomic Nervous System

Under the **File** menu, drag down to **Open** (then to **Content** if you are working in Windows). Click on **Atlas Anatomy**. Select **System** and **Nervous.** Choose **Autonomic NS-Viscera 1**. If the list is not in alphabetical order, the image is about two-thirds from the bottom of the list.

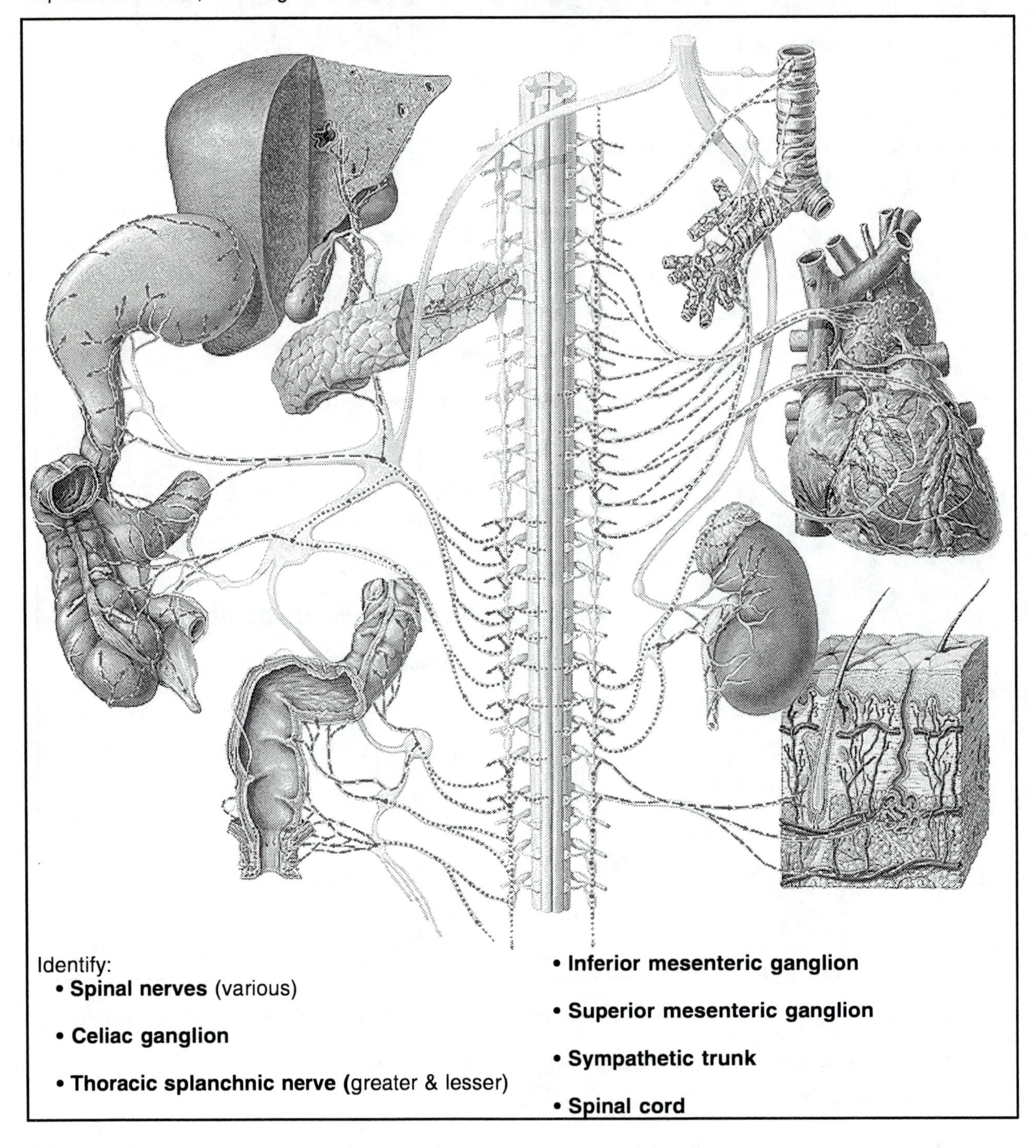

Identify:

- **Spinal nerves** (various)
- **Celiac ganglion**
- **Thoracic splanchnic nerve (**greater & lesser)
- **Inferior mesenteric ganglion**
- **Superior mesenteric ganglion**
- **Sympathetic trunk**
- **Spinal cord**

On the diagram on the previous page, trace the sympathetic pathways to the following organs and indicate the effect of **sympathetic** innervation:

| **Organ** | **Effect of Sympathetic Stimulation:** |
|---|---|
| • Sigmoid colon, rectum, and anal canal | ______________________ |
| • Ascending colon, and ileum (small intestine) | ______________________ |
| • Stomach | ______________________ |
| • Gall bladder and pancreas | ______________________ |
| • Liver (metabolic effects) | ______________________ |
| • Kidney | ______________________ |
| • Adrenal gland | ______________________ |
| • Heart | ______________________ |
| • Bronchi (tubes leading to lungs) | ______________________ |

When you are finished, **close** the window.

Under the **File** menu, drag down to **Open** (then to **Content** if you are working in Windows). **Atlas Anatomy** should still be selected. **System** and **Nervous** should still be selected. Choose **Autonomic NS-Viscera 2**. If the list is not in alphabetical order, the image is about two-thirds of the way down the list.

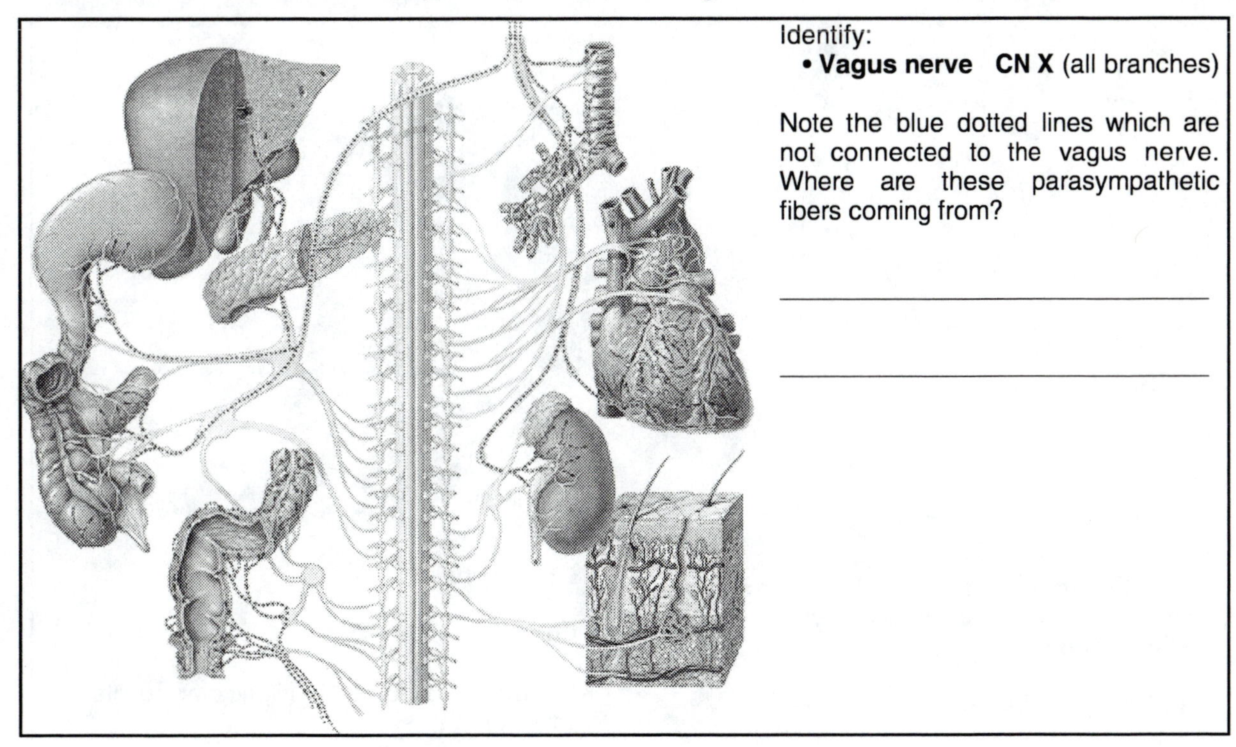

Identify:
- **Vagus nerve CN X** (all branches)

Note the blue dotted lines which are not connected to the vagus nerve. Where are these parasympathetic fibers coming from?

______________________

______________________

 

On the diagram on the previous page, trace the sympathetic pathways to the following organs and indicate the effect of **parasympathetic** innervation:

| Organ | Effect of Parasympathetic Stimulation: |
|---|---|
| • Sigmoid colon, rectum, and anal canal | ______________________ |
| • Ascending colon, and ileum (small intestine) | ______________________ |
| • Stomach | ______________________ |
| • Gall bladder and pancreas | ______________________ |
| • Liver (metabolic effects) | ______________________ |
| • Kidney | ______________________ |
| • Adrenal gland | ______________________ |
| • Heart | ______________________ |
| • Bronchi (tubes leading to lungs) | ______________________ |

When you are finished, **close** the window.

Under the **File** menu, drag down to **Open** (then to **Content** if you are working in Windows). **Atlas Anatomy** should still be selected. **System** and **Nervous** should still be selected. Choose **Ciliary Ganglion - Pathways 1**. If the list is not in alphabetical order, the image is close to the top of the list.

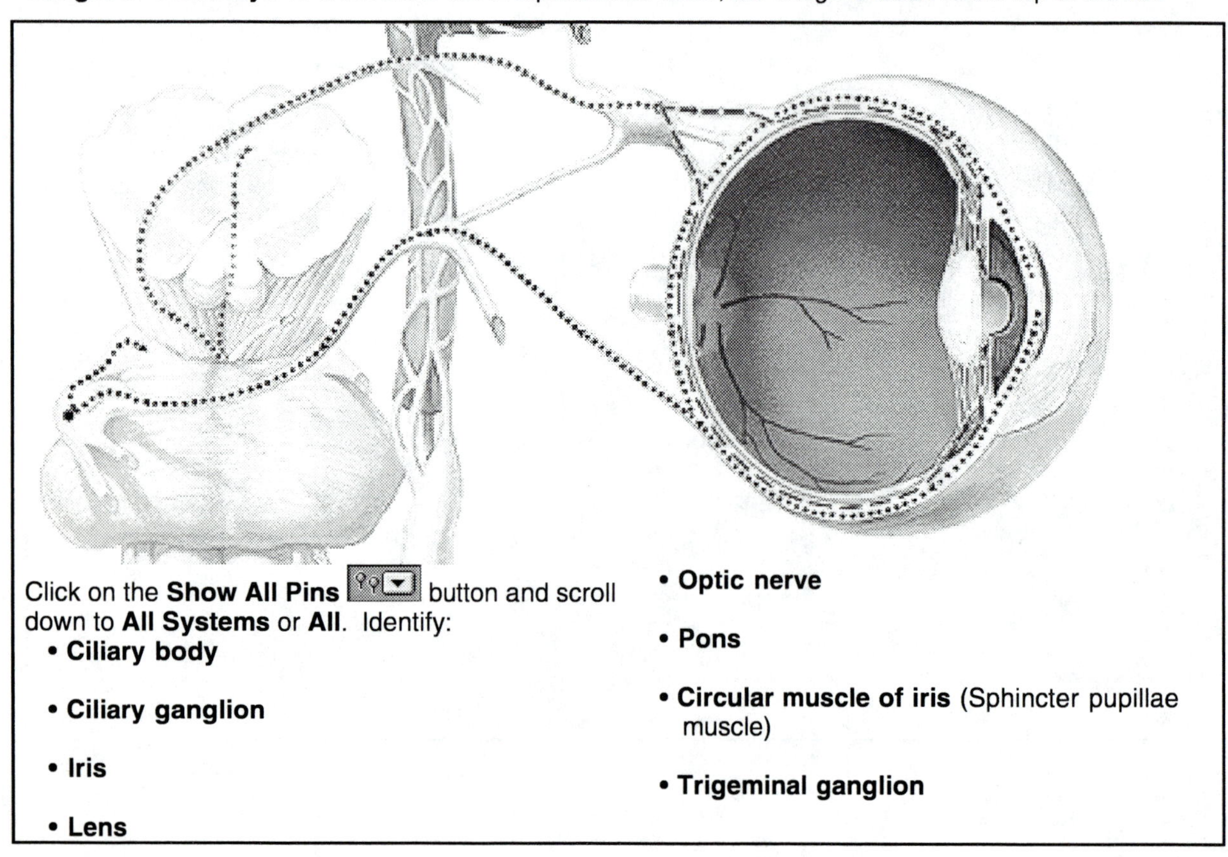

Click on the **Show All Pins** button and scroll down to **All Systems** or **All**. Identify:

- **Ciliary body**
- **Ciliary ganglion**
- **Iris**
- **Lens**
- **Optic nerve**
- **Pons**
- **Circular muscle of iris** (Sphincter pupillae muscle)
- **Trigeminal ganglion**

The diagram on the previous page shows the parasympathetic pathway of the oculomotor nerve to the circular muscle of the iris and the ciliary body (see blue dotted line). What is the effect of the innervation of the circular muscle of the iris?

---

What is the effect of the parasympathetic innervation of the ciliary body?

---

On this diagram, the black dotted line represents the ophthalmic division of the trigeminal nerve. What type of nerve impulses to the eye would it relay?

---

When you are finished, **close** the window.

Under the **File** menu, drag down to **Open** (then to **Content** if you are working in Windows). **Atlas Anatomy** should still be selected. **System** and **Nervous** should still be selected. Choose **Ciliary Ganglion - Pathways 2**. If the list is not in alphabetical order, the image is close to the top of the list.

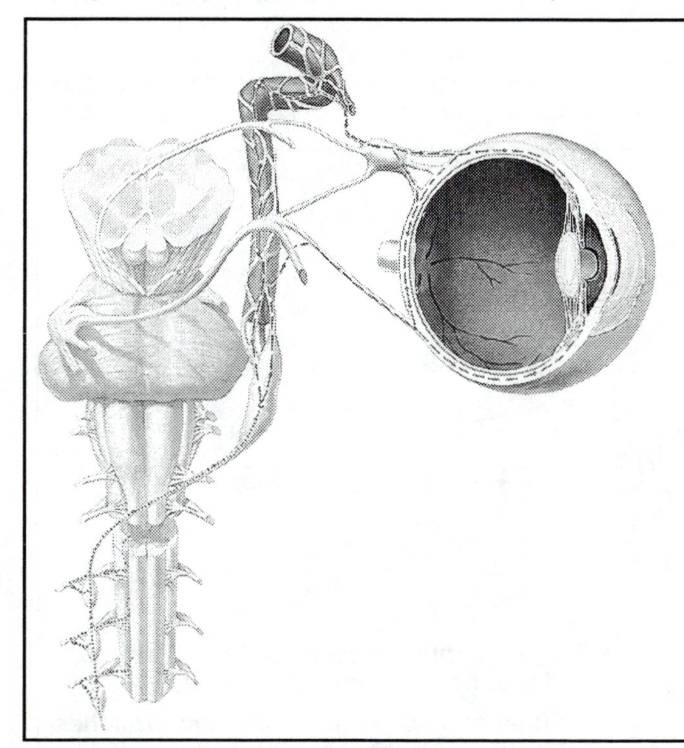

If you are working in Mac, click on the **Show All Pins** button and scroll to **All Systems**. Identify:

- **Ciliary body**
- **Radial muscle of iris** (dilator pupillae muscle)
- **Ganglion of sympathetic trunk**
- **Iris**
- **Lens**
- **Sympathetic trunk**
- **Superior cervical ganglion**
- **Preganglionic sympathetic nerves to eye** (red dotted lines)
- **Postganglionic sympathetic nerves to eye** (red dashed lines)

This diagram shows the sympathetic pathway to the radial muscle of the iris and the ciliary body (see red dotted line). What is the effect of the innervation of the radial muscle of the iris?

---

What is the effect of the sympathetic innervation of the ciliary body?

---

When you are finished, **close** the window.

 

Under the **File** menu, drag down to **Open** (then to **Content** if you are working in Windows). **Atlas Anatomy** should still be selected. **System** and **Nervous** should still be selected. Choose **Oculomotor Nerve in Orbit (Ant)**. If the list is not in alphabetical order, the image is close to the top of the list.

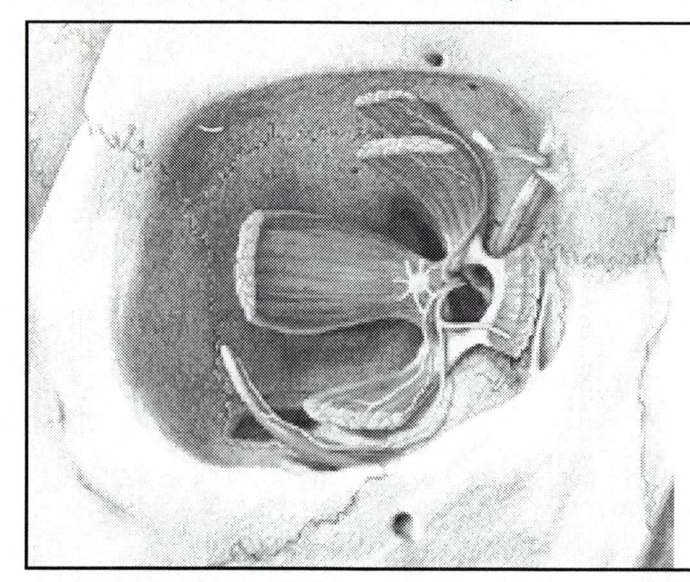

Identify:

- **Oculomotor nerve CN III**
- **Ciliary ganglion**

When you are finished, **close** the window.

Under the **File** menu, drag down to **Open** (then to **Content** if you are working in Windows). **Atlas Anatomy** should still be selected. **System** and **Nervous** should still be selected. Choose **Pterygopalatine Ganglion 2**. If the list is not in alphabetical order, the image is close to the top of the list.

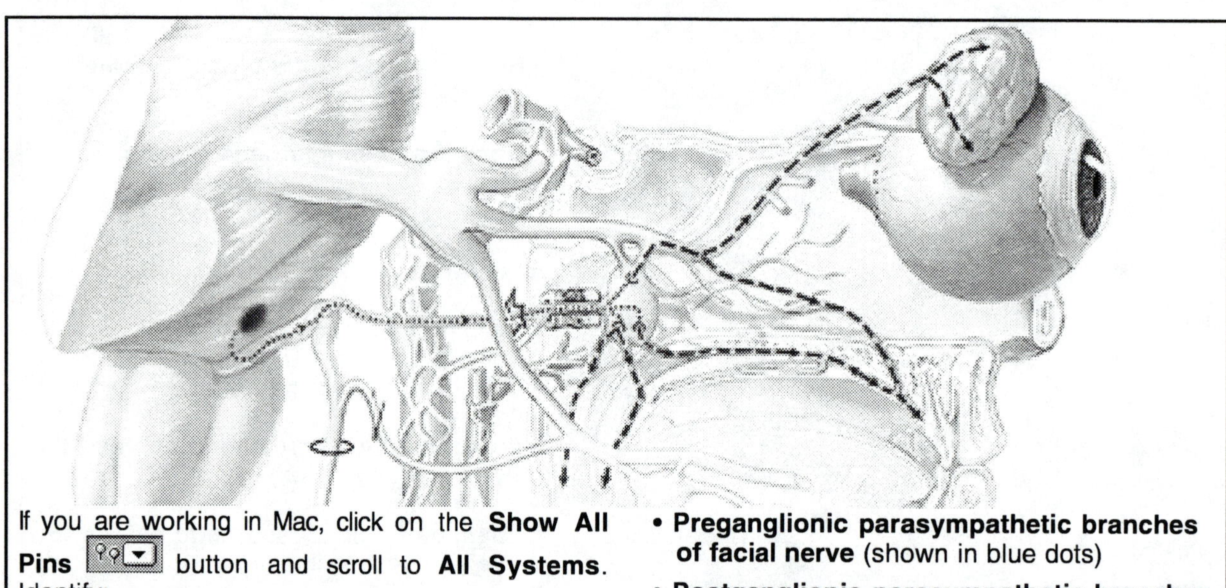

If you are working in Mac, click on the **Show All Pins** button and scroll to **All Systems**. Identify:

- **Location of cell bodies of facial nerve** (superior salivary nucleus)
- **Parasympathetic ganglion** (Pterygopalatine Ganglion)
- **Preganglionic parasympathetic branches of facial nerve** (shown in blue dots)
- **Postganglionic parasympathetic branches of facial nerve** (shown in blue dashes)
- **Lacrimal gland**

This diagram on the previous page shows the parasympathetic pathway of the facial nerve to the lacrimal gland and the nasal mucosa (see blue dashed line). What is the effect of the parasympathetic innervation of the lacrimal gland?

______________________________________________________________________

What is the effect of the parasympathetic innervation of the nasal mucosa?

______________________________________________________________________

When you are finished, **close** the window.

Under the **File** menu, drag down to **Open** (then to **Content** if you are working in Windows). **Atlas Anatomy** should still be selected. **System** and **Nervous** should still be selected. Choose **Pterygopalatine Ganglion 1**. If the list is not in alphabetical order, image is close to the top of the list.

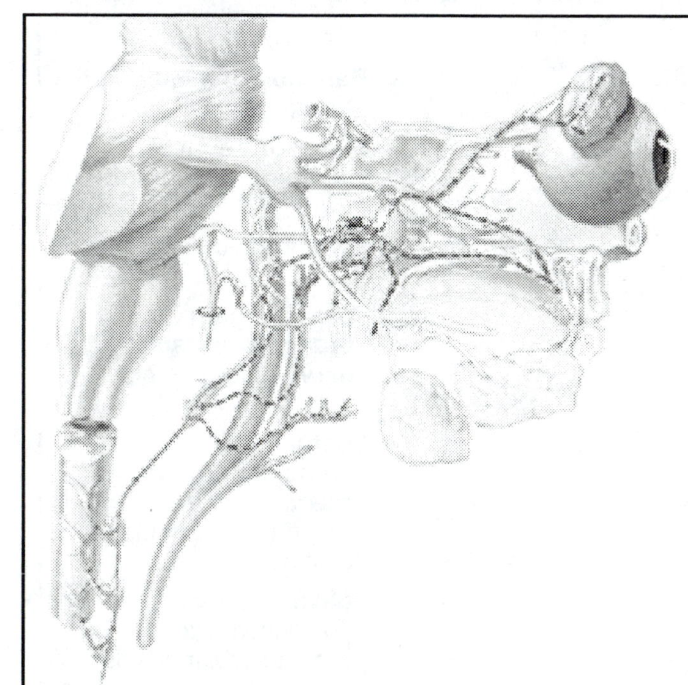

If you are working in Mac, click on the **Show All Pins** button and scroll to **All Systems**. Identify:

- **Ganglion of sympathetic trunk**
- **Lacrimal gland**
- **Superior cervical ganglion of sympathetic trunk**
- **Sympathetic trunk**
- **Preganglionic sympathetic nerve** (shown in red dots)
- **Postganglionic sympathetic nerve** (shown in red dashes)

This diagram shows the sympathetic pathway to the lacrimal gland and the nasal mucosa (see red dashed line). What is the effect of the sympathetic innervation of the lacrimal gland ?

______________________________________________________________________

What is the effect of the sympathetic innervation of the nasal mucosa?

______________________________________________________________________

When you are finished, **close** the window.

Under the **File** menu, drag down to **Open** (then to **Content** if you are working in Windows). **Atlas Anatomy** should still be selected. **System** and **Nervous** should still be selected. Choose **Submandibular Ganglion**. If the list is not in alphabetical order, the image is close to the top of the list.

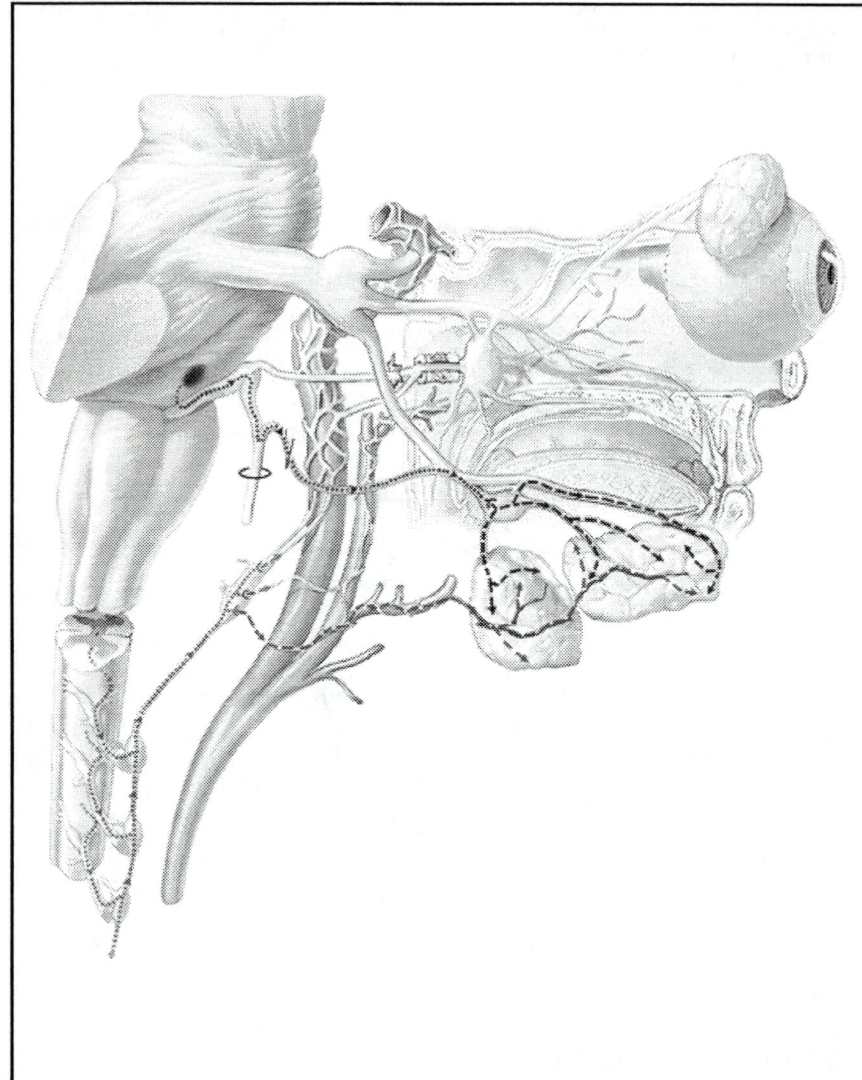

If you are working in Mac, click on the **Show All Pins** button and scroll to **All Systems**. Identify:

- **Location of cell bodies of facial nerve** (superior salivary nucleus)
- **Parasympathetic ganglion** (Pterygopalatine Ganglion)
- **Ganglion of sympathetic trunk**
- **Superior cervical ganglion of sympathetic trunk**
- **Sympathetic trunk**
- **Preganglionic parasympathetic branches of facial nerve** (shown in blue dots)
- **Postganglionic parasympathetic branches of facial nerve** (shown in blue dashes)
- **Salivary glands** (Sublingual and submandibular glands)
- **Preganglionic sympathetic nerve** (shown in red dots)
- **Postganglionic sympathetic nerve** (shown in red dashes)

This diagram shows the sympathetic pathway (see red dotted & dashed line) and the parasympathetic pathway of the facial nerve to the submandibular and sublingual salivary glands. What is the effect of the parasympathetic innervation of the submandibular and sublingual salivary glands?

______________________________________________________________________

What is the effect of the sympathetic innervation of the submandibular and sublingual salivary glands?

______________________________________________________________________

When you are finished, **close** the window.

Under the **File** menu, drag down to **Open** (then to **Content** if you are working in Windows). **Atlas Anatomy** should still be selected. **System** and **Nervous** should still be selected. Choose **Trigeminal Nerve - V1**. If the list is not in alphabetical order, the image is close to the top of the list.

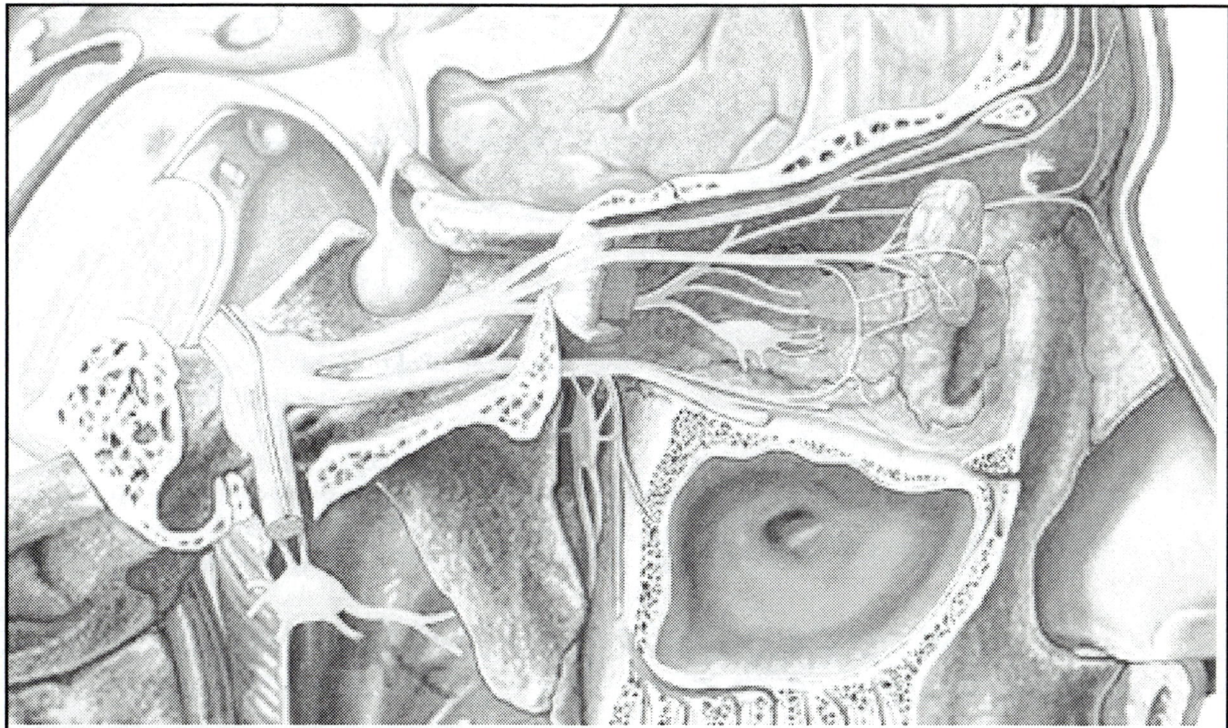

Identify the following as **Parasympathetic ganglia:**

- Ciliary ganglion
- Pterygopalatine ganglion
- Otic ganglion

When you are finished, **close** the window.

 

Under the **File** menu, drag down to **Open** (then to **Content** if you are working in Windows). **Atlas Anatomy** should still be selected. **System** and **Nervous** should still be selected. Choose **Glossopharyngeal Nerve 2.** If the list is not in alphabetical order, the image is close to the top of the list.

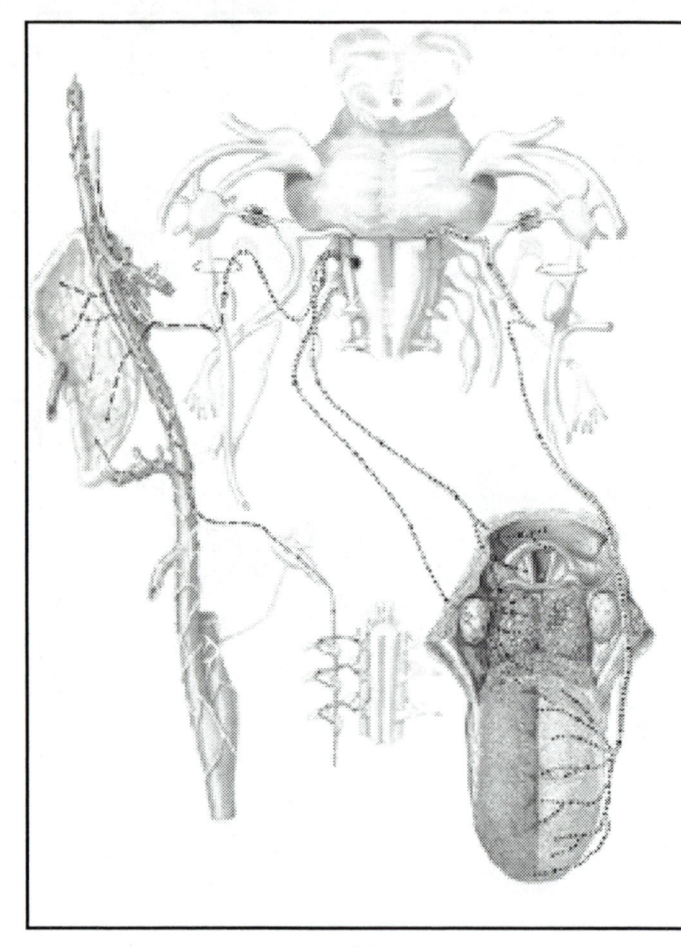

If you are working in Mac, click on the **Show All Pins** button and scroll to **All Systems**. Identify:

- **Location of cell body of glossopharyngeal nerve** (inferior salivary nucleus)
- **Salivary gland** (Parotid gland)
- **Facial nerve CN VII**
- **Glossopharyngeal nerve CN IX**
- **Preganglionic parasympathetic branches of glossopharyngeal nerve** (shown in blue dots)
- **Postganglionic parasympathetic branches of glossopharyngeal nerve** (shown in blue dashes)
- **Preganglionic sympathetic nerve** (shown in red dots)
- **Postganglionic sympathetic nerve** (shown in red dashes)
- **Parasympathetic ganglion** (Otic Ganglion)
- **Ganglion of sympathetic trunk**
- **Superior cervical ganglion of sympathetic trunk**
- **Sympathetic trunk**

This diagram shows the sympathetic pathway (see red dotted & dashed line) and the parasympathetic pathway of the glossopharyngeal nerve (see blue dotted & dashed line) to the parotid salivary glands. What is the effect of the parasympathetic innervation of these glands?

---

What is the effect of the sympathetic innervation of the parotid salivary glands?

---

When you are finished, **close** the window.

## Objectives: Autonomic Nervous System:

To study the anatomy of the sympathetic and parasympathetic nervous systems.

- Anterior vagal trunk
- Cardiac branch of sympathetic trunk
- Celiac ganglia
- Ciliary body
- Ciliary ganglion
- Circular muscle of iris
- Gray ramus communicans
- Cardiac branch of sympathetic trunk
- Facial nerve CN VII
- Glossopharyngeal nerve CN IX
- Inferior mesenteric ganglia
- Iris
- Lens
- Lacrimal gland
- Location of cell bodies of facial nerve
- Location of cell body of glossopharyngeal nerve
- Nerve plexuses
- Oculomotor nerve CN III
- Optic nerve
- Otic ganglion
- Parasympathetic ganglia
- Pons
- Preganglionic parasympathetic branches of facial nerve
- Preganglionic sympathetic nerve
- Preganglionic parasympathetic branches of glossopharyngeal nerve
- Preganglionic sympathetic nerves to eye
- Postganglionic parasympathetic branches of facial nerve
- Postganglionic parasympathetic branches of glossopharyngeal nerve
- Postganglionic sympathetic nerve
- Postganglionic sympathetic nerves to eye
- Pterygopalatine ganglion
- Radial muscle of iris
- Salivary glands
- Spinal nerves
- Spinal cord
- Superior cervical ganglion
- Superior mesenteric ganglia
- Sympathetic ganglia
- Sympathetic trunk
- Thoracic splanchnic nerve (greater & lesser)
- Trigeminal ganglion
- Vagus nerve CN X
- White ramus communicans

# Meninges and Ventricles

## Opening A.D.A.M Interactive Anatomy

Open A.D.A.M. Interactive Anatomy according to the directions in the Tutorial.

## 3D Anatomy of the Meninges & Ventricles of the Brain

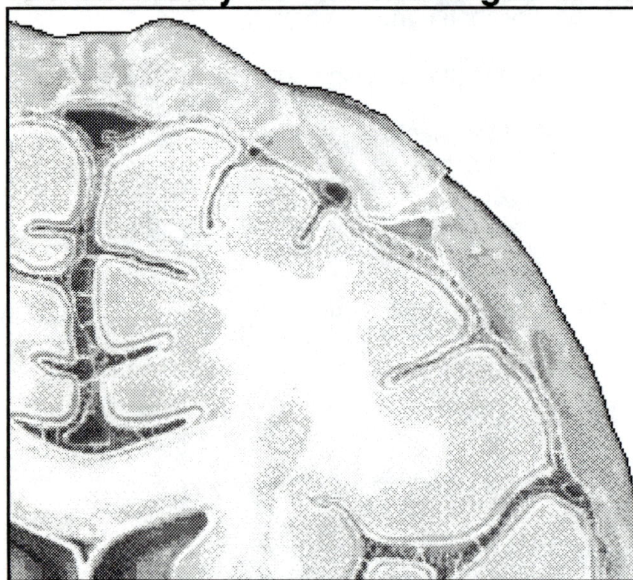

Choose **3D Anatomy**. Click on **3D Brain**. Identify the following by clicking on the **3D Anatomy Structure List**:

- **Arachnoid granulation {Villus}**
- **Arachnoid mater**
- **Dura mater**
- **Pia mater**
- **Subarachnoid space**

What is the function of the arachnoid granulation?

______________________________

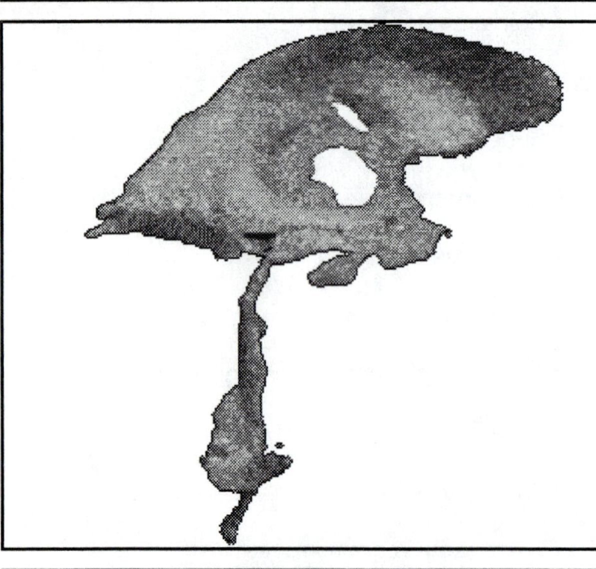

Identify:

- **Cerebral aqueduct**
- **Fourth ventricle**
- **Interventricular foramen**
- **Lateral aperture of fourth ventricle**
- **Lateral ventricle**
- **Median aperture**
- **Third ventricle**

The cerebral aqueduct connects what two ventricles?

______________________________

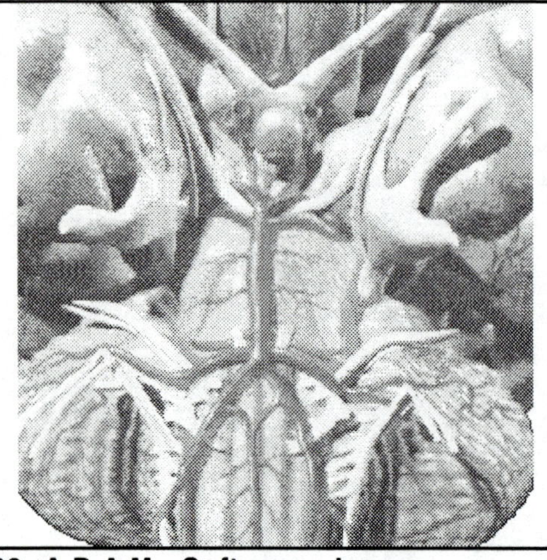

Identify:

- **Cerebral arterial circle**

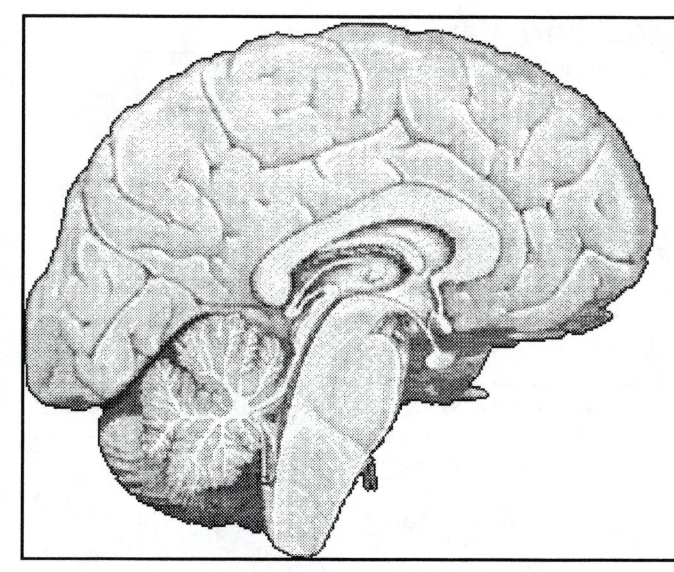

Identify:

- **Choroid plexus**
- **Septum pellucidum**

What is the function of the choroid plexus?

______________________________

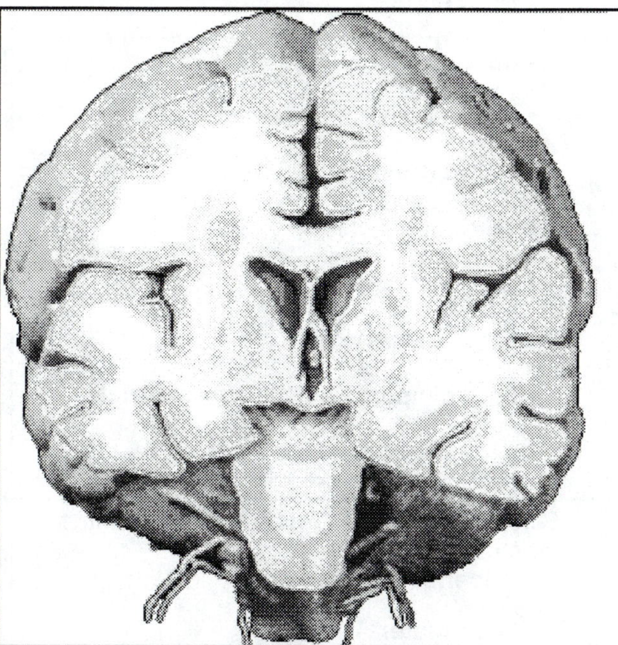

Identify:

- **Lateral ventricle - Coronal cut**

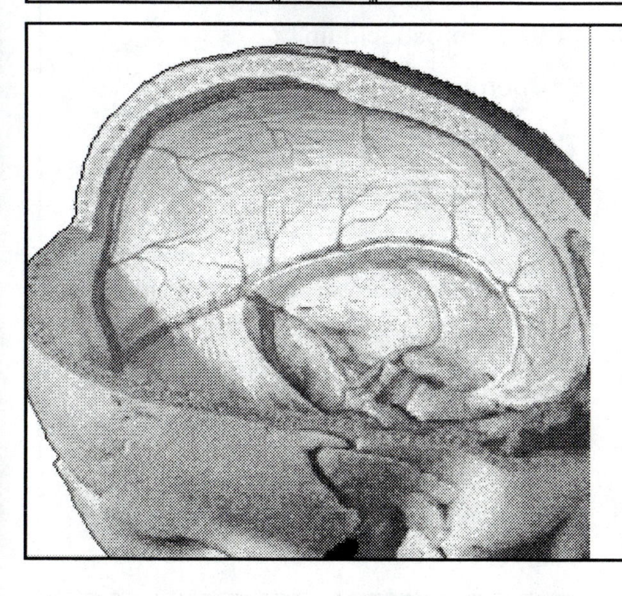

Identify:

- **Cerebral falx**
- **Straight sinus**
- **Superior sagittal sinus**
- **Tentorium cerebelli**

When you are finished, **close** the window.

The inferior and superior sagittal sinuses lie within what structure?

______________________________

## Lateral Dissection of the Meninges and Ventricles

Under the **File** menu, drag down to **Open** (then to **Content** if you are working in Windows).

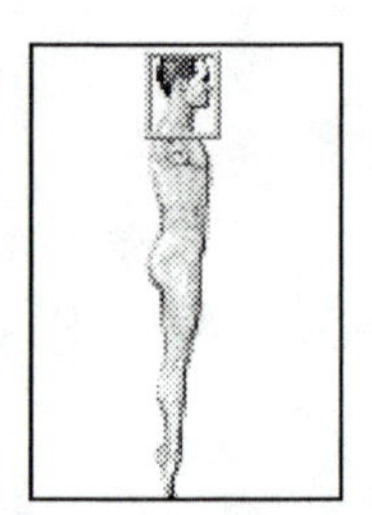

Choose **Dissectible Anatomy**.

Choose either male or female. Select **Lateral.** Click **Open**.

Position the navigator rectangle over the lateral face & neck.

Scroll down on the **Depth Bar** ▼ until you reach **layer 187**. Then click to **layer 188**. No graphic has been provided for this layer. Identify the **dura mater**.

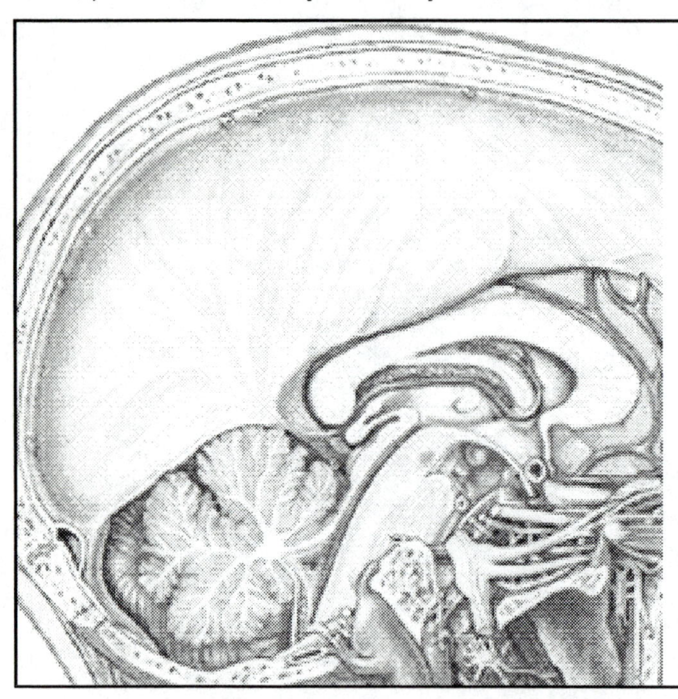

Click on the **Depth Bar** ▼ until you reach **layer 192**. Identify:

- **Superior sagittal sinus**
- **Cerebral falx**
- **Dura mater**
- **Septum pellucidum**
- **Interventricular foramen**
- **Choroid plexus**
- **Third ventricle**
- **Cerebral aqueduct**
- **Fourth ventricle**
- **Lateral aperture of fourth ventricle**

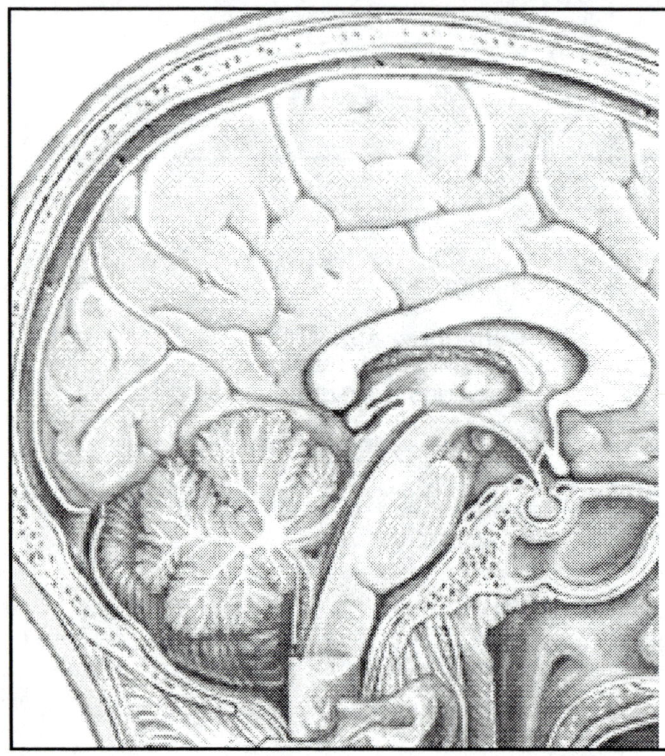

Scroll down on the **Depth Bar** ▼ until you reach **layer 288**. Identify:

- **Superior sagittal sinus**
- **Dura mater**
- **Septum pellucidum**
- **Choroid plexus**
- **Third ventricle**
- **Cerebral aqueduct**
- **Fourth ventricle**
- **Lateral aperture of fourth ventricle**
- **Straight sinus** - cut

The lateral aperture of the fourth ventricle allows cerebrospinal fluid to flow into what structure?

________________________________________

## Anterior Dissection of the Meninges and Ventricles

Click on the **View** button and choose **Anterior**.

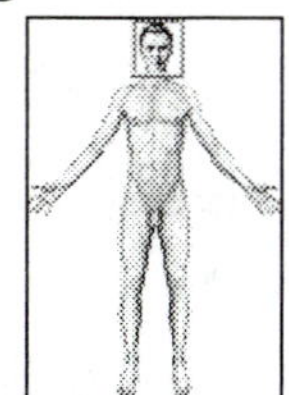

Click on the **Normal** button. Windows— —Mac

Drag the **Depth Bar** to **layer 0.**

Position the navigator rectangle over the head.

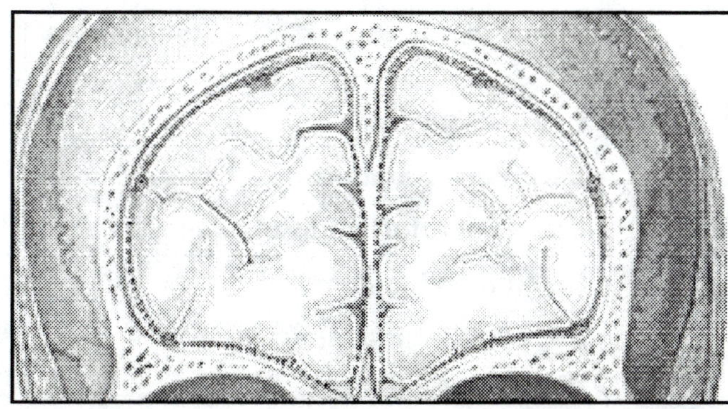

Drag the **Depth Bar** to **layer 48**. Click once to **layer 49** and identify:

- **Falx cerebri** (Cerebral falx)
- **Dura mater**

The falx cerebri is attached inferiorly to what structure of the skull?

______________________________

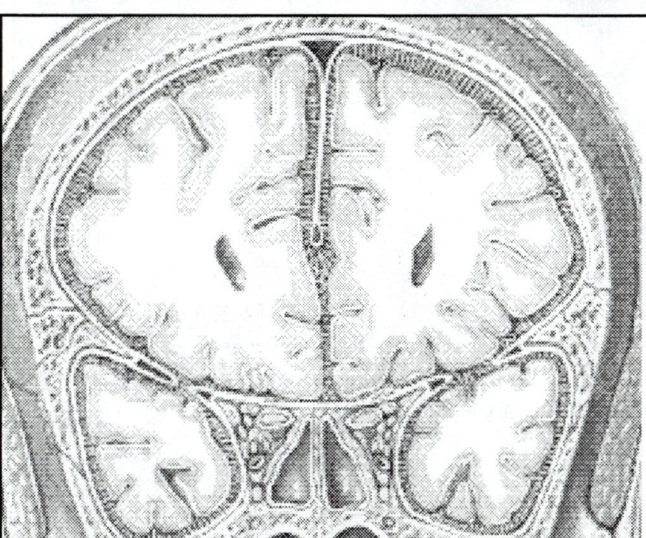

Scroll down on the **Depth Bar** to **layer 274**. Identify:

- **Falx cerebri** (Cerebral falx)
- **Dura mater**
- **Lateral ventricle**
- **Subarachnoid space**
- **Superior sagittal sinus**

The subarachnoid space is located between what two meninges?

______________________________

______________________________

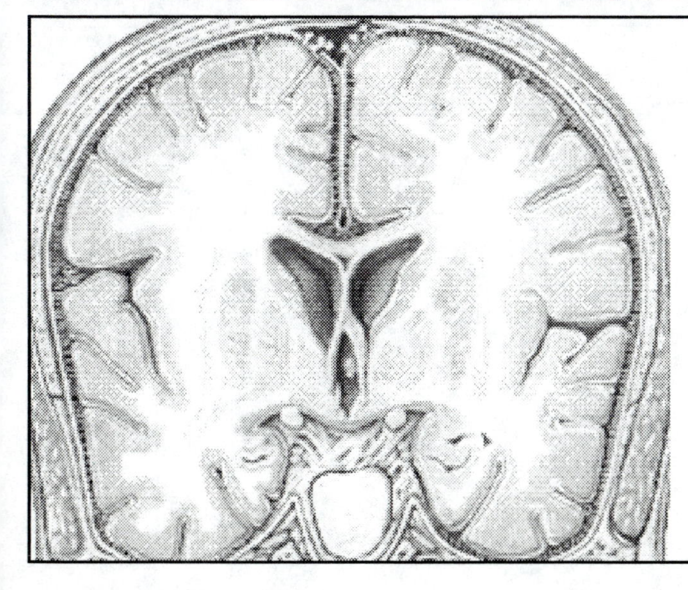

Click a few times on the **Depth Bar** to **layer 276**. Identify:

- **Falx cerebri** (Cerebral falx)
- **Pia mater**
- **Lateral ventricle**
- **Subarachnoid space**
- **Superior sagittal sinus**
- **Third ventricle**

## Medial Dissection of the Meninges and Ventricles

Click on the **View** button and choose **Medial**.

Click on the **Normal** button. Windows— —Mac

Drag the **Depth Bar** to **layer 0.**

Position the navigator rectangle over the head.

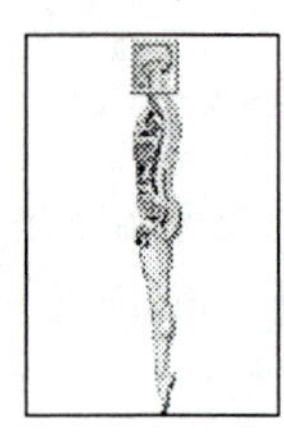

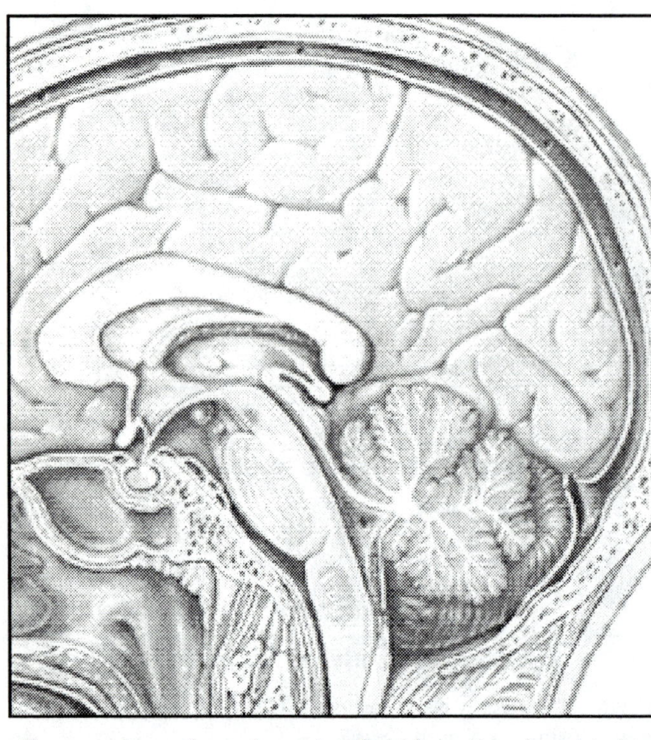

Drag the **Depth Bar** up to **layer 0.** Identify:

- **Choroid plexus**
- **Cerebral aqueduct**
- **Lateral aperture of fourth ventricle**
- **Septum pellucidum**
- **Superior sagittal sinus**
- **Third ventricle**
- **Fourth ventricle**

When you are finished, **close** the window.

Trace a drop of cerebrospinal fluid from the choroid plexus of a lateral ventricle to the lateral aperture of the fourth ventricle naming all the structures through which the drop would pass.

______________________________________________

______________________________________________

______________________________________________

# Atlas Anatomy of the Meninges and Ventricles

Under the **File** menu, drag down to **Open** (then to **Content** if you are working in Windows). Click on **Atlas Anatomy**. Select **Region** and **Head & Neck**. Choose **Sagittal Section of Brain**. If the list is not in alphabetical order, the image is about one-third of the way down the list.

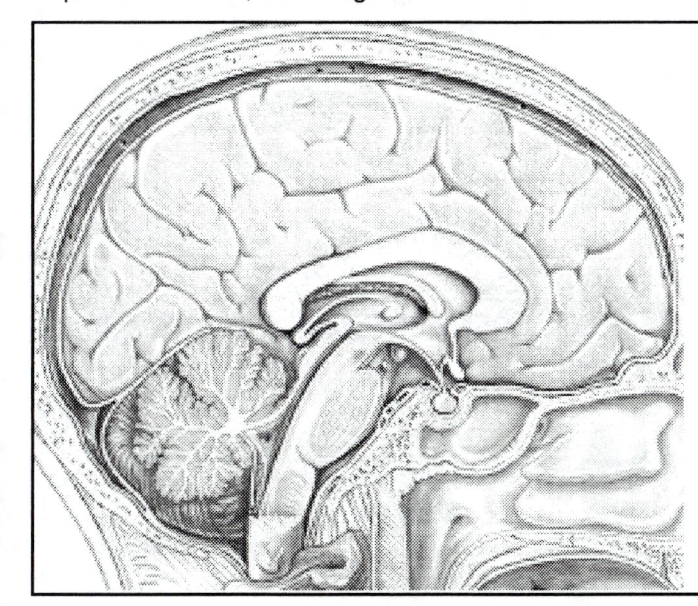

Identify by clicking on the heads of the pins:

- **Cerebral aqueduct**
- **Choroid plexus**
- **Fourth ventricle**
- **Septum pellucidum**
- **Superior sagittal sinus**
- **Tentorium cerebelli**
- **Third ventricle**

When you are finished, **close** the window.

The tentorium cerebelli separates what two parts of the brain?

---

Under the **File** menu, drag down to **Open** (then to **Content** if you are working in Windows). Click on **Atlas Anatomy**. **Region** and **Head & Neck** should still be selected. Choose **Sagittal Section of Head & Neck**. If the list is not in alphabetical order, the image is about one-third of the way down the list.

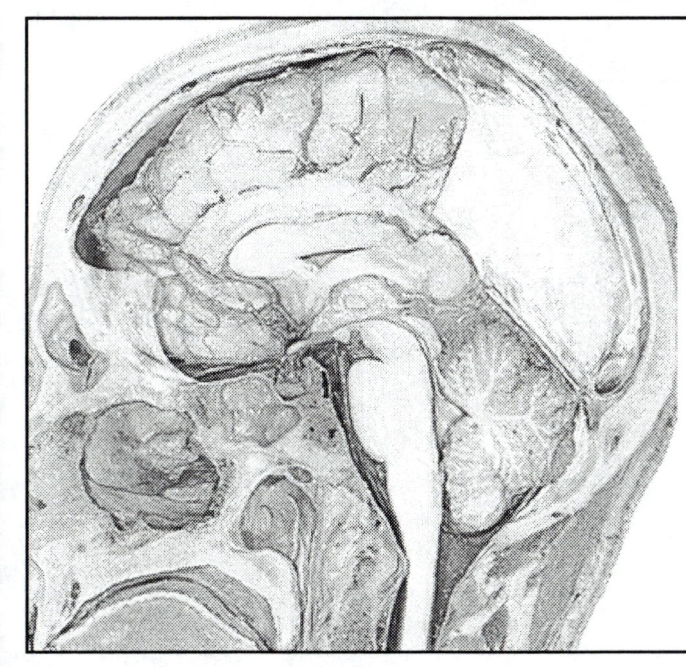

Identify:

- **Cerebral aqueduct**
- **Cerebral falx** (Falx cerebri)
- **Fourth ventricle**
- **Straight dural sinus**
- **Superior sagittal sinus**
- **Third ventricle**

When you are finished, **close** the window.

The fourth ventricle lies between what two structures of the brain?

---

Under the **File** menu, drag down to **Open** (then to **Content** if you are working in Windows). Click on **Atlas Anatomy**. **Region** and **Head & Neck** should still be selected. Choose **Basal Ganglia (Sup)**. If the list is not in alphabetical order, the image is about one-fourth of the way down the list.

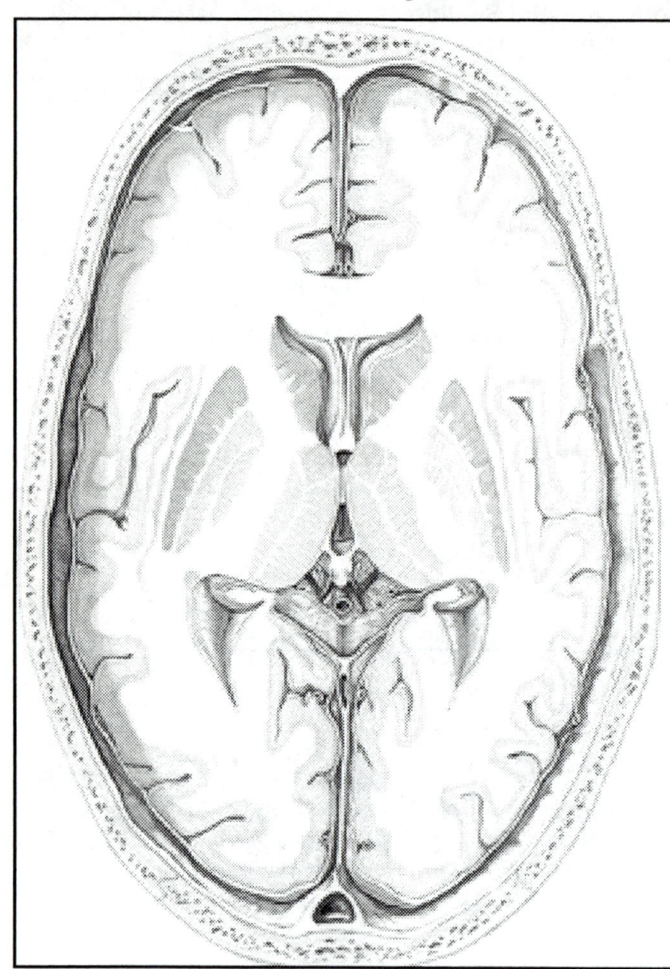

Identify:

- **Arachnoid mater**
- **Dura mater**
- **Cerebral falx** (Falx cerebri)
- **Lateral ventricle**
- **Septum pellucidum**
- **Superior sagittal sinus**
- **Third ventricle**

When you are finished, **close** the window.

The lateral ventricles are separated by what structure?

______________________________________________________________________

Under the **File** menu, drag down to **Open** (then to **Content** if you are working in Windows). Click on **Atlas Anatomy**. **Region** and **Head & Neck** should still be selected. Choose **Hippocampus & Fornix (Sup)**. If the list is not in alphabetical order, the image is about one-third of the way down the list.

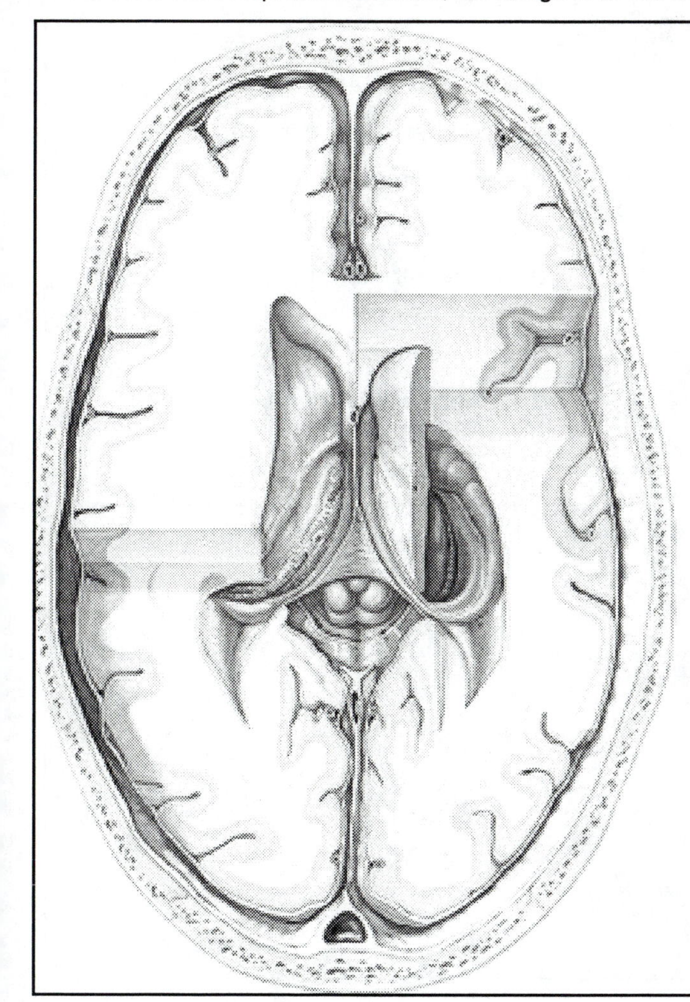

Identify:

- **Arachnoid mater**
- **Dura mater**
- **Cerebral falx** (Falx cerebri)
- **Choroid plexus**
- **Septum pellucidum**
- **Superior sagittal sinus**
- **Lateral ventricle** (frontal horn, occipital horn, temporal horn)

When you are finished, **close** the window.

Under the **File** menu, drag down to **Open** (then to **Content** if you are working in Windows). Click on **Atlas Anatomy**. **Region** and **Head & Neck** should still be selected. Choose **Ventricles of Brain (Sup)**. If the list is not in alphabetical order, the image is about one-third of the way down the list.

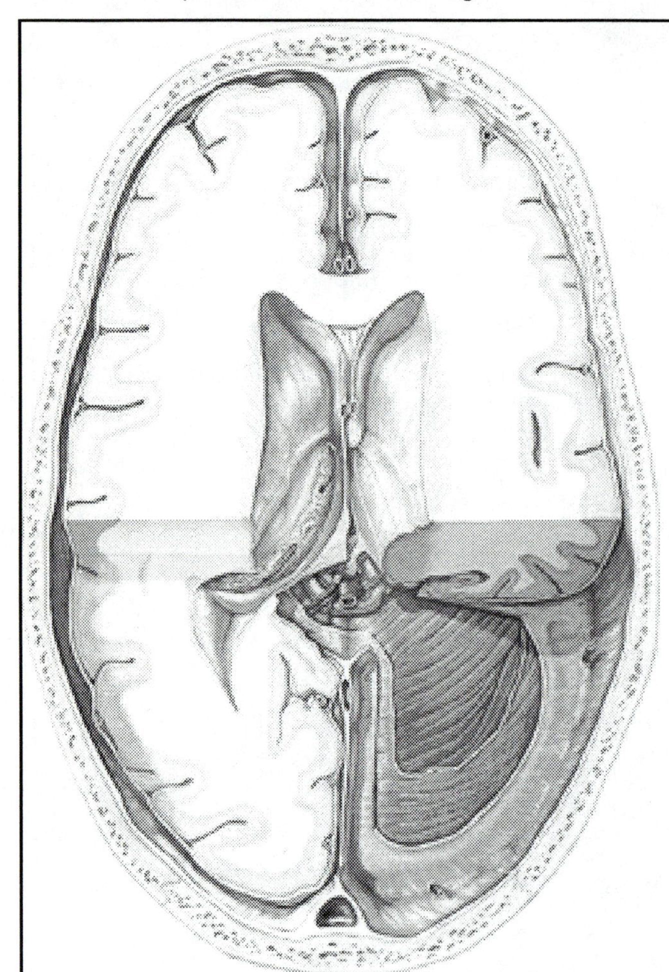

Identify:

- **Arachnoid mater**
- **Cerebral falx** (Falx cerebri)
- **Choroid plexus**
- **Dura mater**
- **Lateral ventricles** (frontal horn, occipital horn)
- **Septum pellucidum**
- **Superior sagittal sinus**
- **Tentorium cerebelli**

When you are finished, **close** the window.

Under the **File** menu, drag down to **Open** (then to **Content** if you are working in Windows). Click on **Atlas Anatomy**. **Region** and **Head & Neck** should still be selected. Choose **Cerebral Arterial Circle (Inf)**. If the list is not in alphabetical order, the image is about one-third of the way down the list.

Because of its high metabolic rate, the brain cannot function without an adequate supply of nutrients and oxygen. The anterior portion of the brain receives blood from the **internal carotid arteries** and the posterior portion receives blood from the **vertebral arteries**. Should either of these supplies be interrupted, the brain will still receive blood from the other due to the structure of the **cerebral arterial circle (circle of Willis)**.

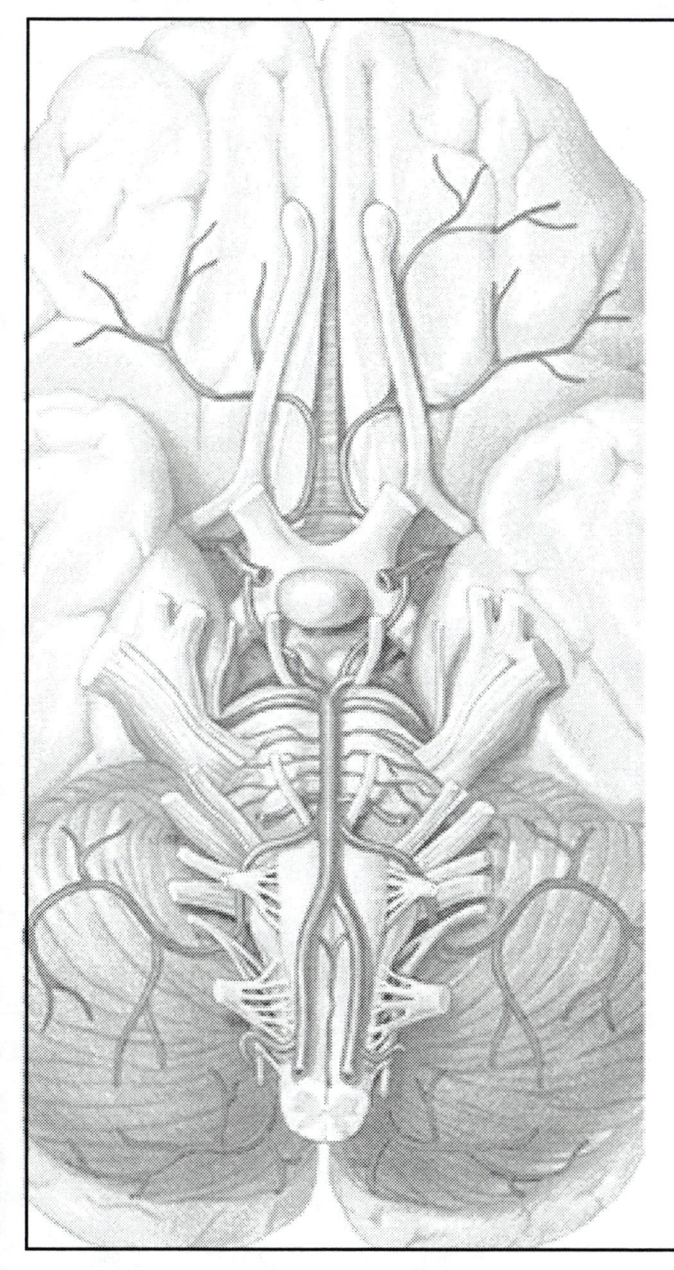

Identify:

- **Basilar artery**
- **Internal carotid artery**
- **Middle cerebral artery**
- **Vertebral artery**
- **Anterior cerebral artery**
- **Anterior communicating artery**
- **Posterior cerebral artery**
- **Anterior inferior cerebellar artery**
- **Posterior inferior cerebellar artery**
- **Posterior communicating artery**

When you are finished, **close** the window.

Blood reaches the brain through what two arteries?

______________________________________________

What four arteries form the cerebral arterial circle (circle of Willis)?

______________________________________________

 

## Objectives: Meninges and Ventricles

To study the meninges, ventricles, and several blood vessels of the brain.

**Ventricles & Meninges:**

- Arachnoid granulation {Villus}
- Arachnoid mater
- Cerebral aqueduct
- Cerebral falx
- Choroid plexus
- Dura mater
- Falx cerebri
- Fourth ventricle
- Interventricular foramen
- Lateral aperture of fourth ventricle
- Lateral ventricle
- Median aperture
- Pia mater
- Septum pellucidum
- Subarachnoid space
- Tentorium cerebelli
- Third ventricle

**Blood Vessels:**

- Basilar artery
- Cerebral arterial circle
- Internal carotid artery
- Middle cerebral artery
- Vertebral artery
- Anterior cerebral artery
- Anterior communicating artery
- Posterior cerebral artery
- Anterior inferior cerebellar artery
- Posterior inferior cerebellar artery
- Posterior communicating artery
- Straight sinus
- Superior sagittal sinus

# Anatomy of the Brain

## Opening A.D.A.M Interactive Anatomy

Open A.D.A.M. Interactive Anatomy according to the directions in the Tutorial.

## 3D Anatomy of the Brain

Choose **3D Anatomy**. Click on **3D Brain**. Identify the following by clicking on the **3D Anatomy Structure List**:

- **Longitudinal fissure of cerebrum**
- **Right hemisphere of cerebrum**
- **Skull**

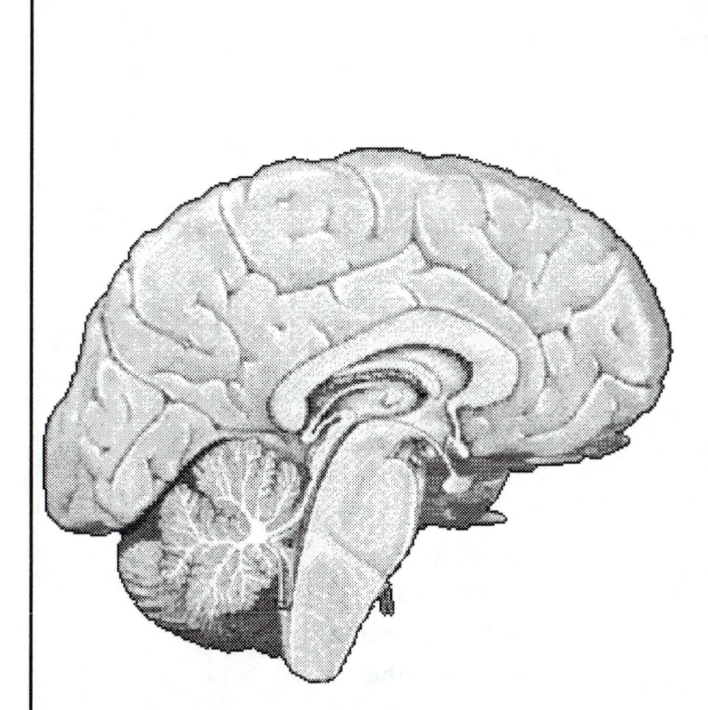

- **Adenohypophysis** (anterior pituitary)
- **Gray matter** - Midsagittal cut (of cerebellum)
- **Corpus callosum -** Midsagittal cut
- **Fornix of brain -** Midsagittal cut
- **Inferior colliculus -** Midsagittal cut
- **Infundibulum of pituitary gland**
- **Interthalamic adhesion -** Midsagittal cut
- **Medulla oblongata** - Midsagittal cut
- **Midbrain**
- **Neurohypophysis** (posterior pituitary)
- **Pineal body** - Midsagittal cut
- **Pituitary gland**
- **Parieto-occipital sulcus**
- **Pons** - Midsagittal cut
- **Superior colliculus** - Midsagittal cut
- **Thalamus - Medial**
- **White matter** - Midsagittal cut (of cerebellum)

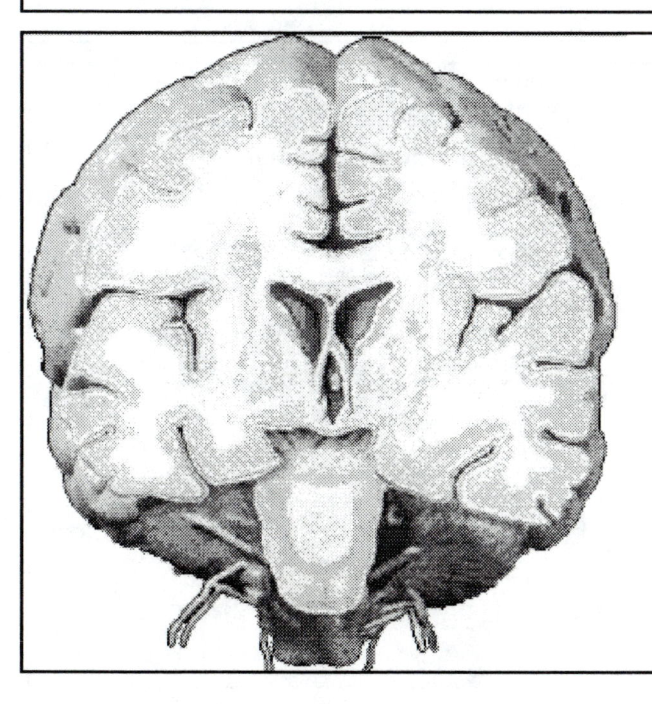

Identify:

- **Corpus callosum**
- **Internal capsule**
- **Pons - Coronal cut**

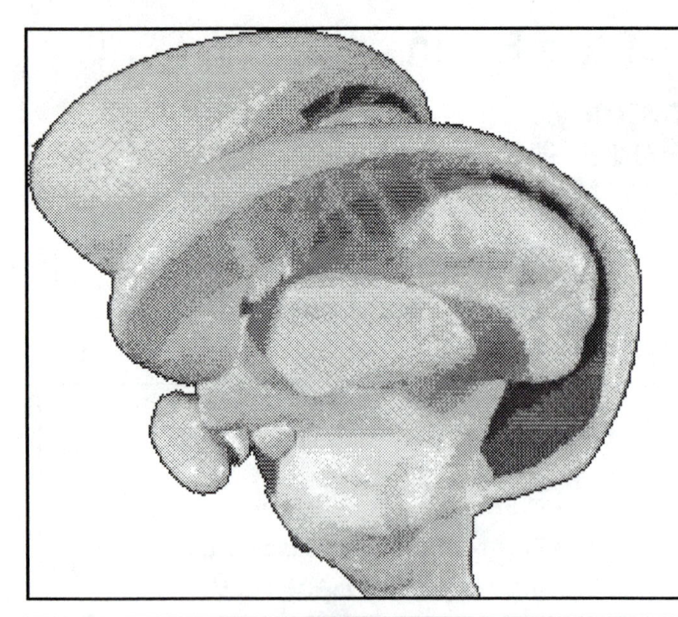

Identify:
- **Caudate nucleus** (part of basal ganglia)
- **Globus pallidus** (part of basal ganglia)
- **Putamen** (part of basal ganglia)
- **Lentiform nucleus** (part of basal ganglia)
- **Thalamus**

Give two functions of the basal ganglia.

_______________________________________________

_______________________________________________

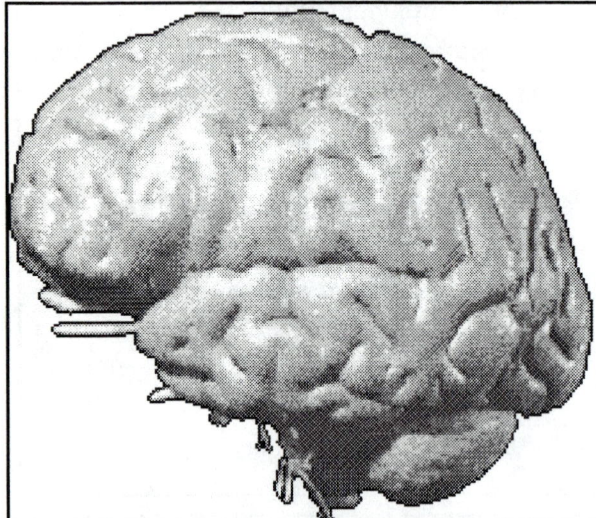

Identify:
- **Occipital lobe**
- **Parietal lobe**
- **Postcentral gyrus**
- **Precentral gyrus**
- **Central sulcus**
- **Cerebellum**
- **Cerebrum**
- **Frontal lobe**
- **Lateral sulcus**
- **Left hemisphere of cerebrum**
- **Temporal lobe**

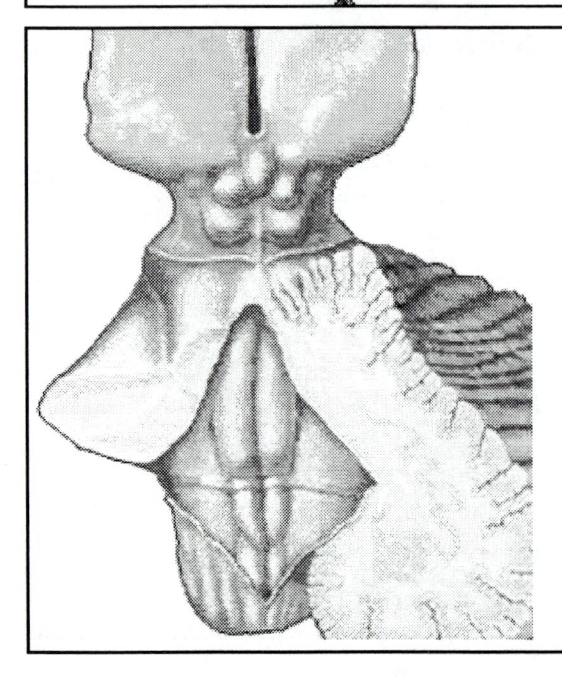

Identify:
- **Inferior colliculus**
- **Corpora quadrigemina**
- **Superior colliculus**
- **Pineal body**

3
What four nuclei are found in the corpora quadrigemina?

_______________________________________________

_______________________________________________

_______________________________________________

_______________________________________________

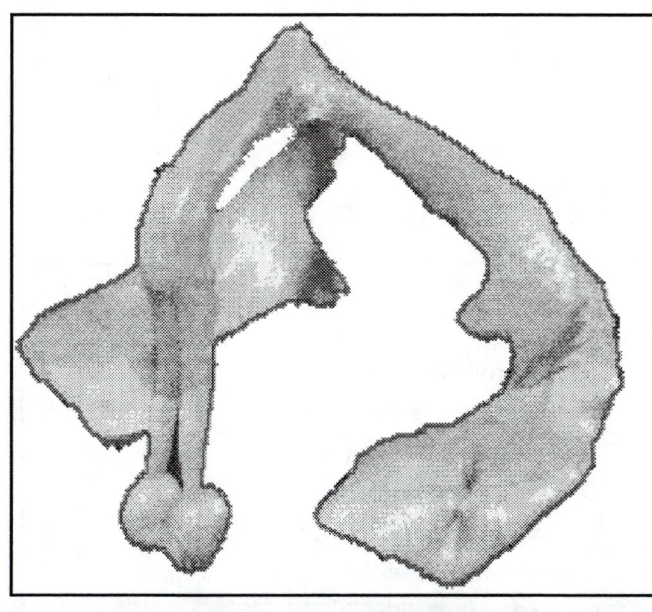

Identify:
- **Hippocampus**
- **Amygdaloid body**

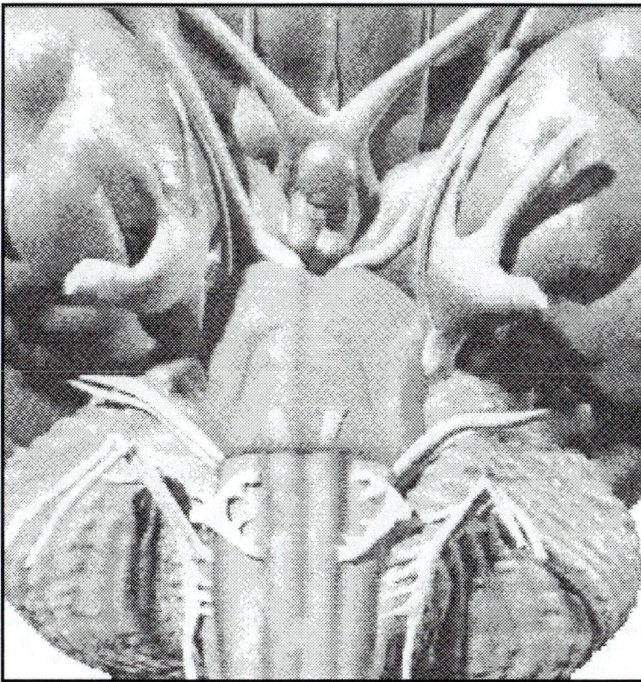

Identify:
- **Mamillary body**
- **Medulla oblongata**
- **Olive**
- **Pons**
- **Pyramid**

What is meant by the "decussation of the pyramids"?

___

___

___

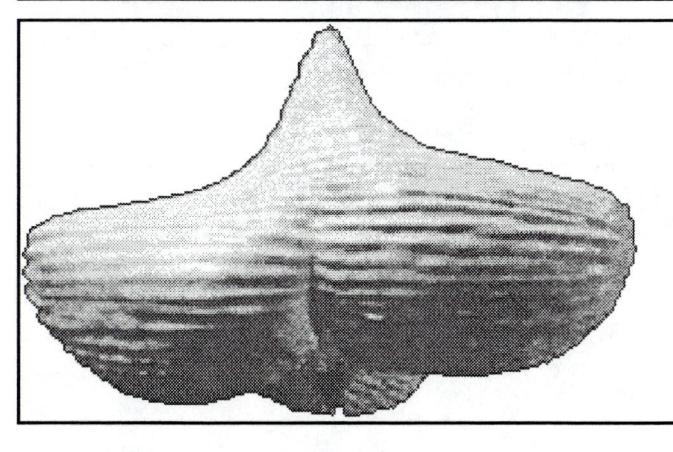

Identify:
- **Right hemisphere of cerebellum**
- **Left hemisphere of cerebellum**
- **Vermis of cerebellum**

When you are finished, **close** the window.

## Lateral Dissection of the Brain

Under the **File** menu, drag down to **Open** (then to **Content** if you are working in windows).

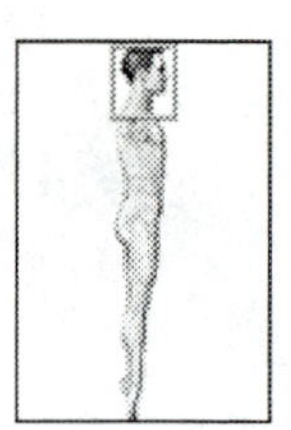

Choose **Dissectible Anatomy**.

Choose male or female. Select **Lateral.** Click **Open**.

Enlarge the window and position the navigation rectangle over the head.

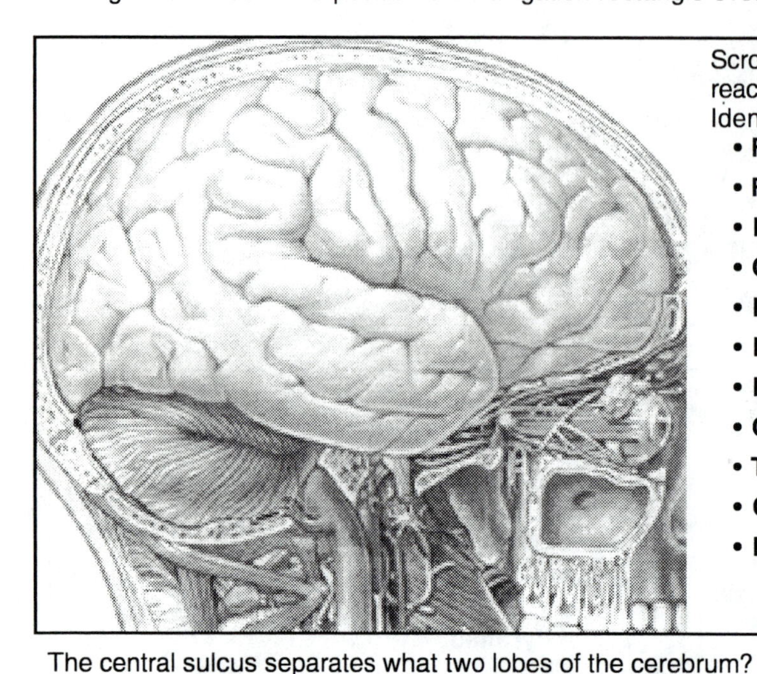

Scroll down on the **Depth Bar** until you reach **layer 187**. Then click to **layer 191**. Identify:

- **Frontal lobe**
- **Frontal gyri**
- **Precentral gyrus**
- **Central sulcus**
- **Postcentral gyrus**
- **Lateral sulcus**
- **Parietal lobe**
- **Occipital lobe**
- **Temporal lobe**
- **Cerebellum**
- **Parieto-occipital sulcus**

The central sulcus separates what two lobes of the cerebrum?

___

The lateral sulcus separates what two lobes of the cerebrum?

___

What is the primary function of the postcentral gyrus?

___

What is the primary function of the precentral gyrus?

___

What structures separate the frontal lobe of the cerebrum from the other lobes?

___

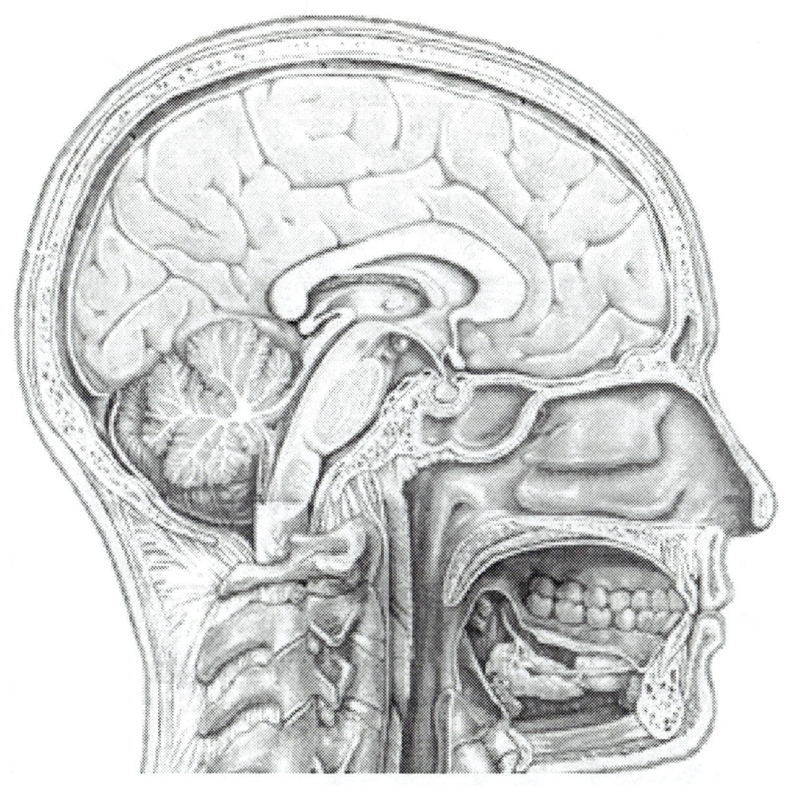

Scroll down on the **Depth Bar** ▼ until you reach **layer 288**. Identify:

- **Optic chiasma**
- **Fornix of brain**
- **Parietal lobe**
- **Corpus callosum**
- **Midbrain**
- **Occipital lobe**
- **Adenohypophysis**
- **Neurohypophysis**
- **Infundibulum**
- **Parieto-occipital sulcus**
- **Superior colliculus**
- **Inferior colliculus**
- **Pineal body**
- **Mamillary body**
- **Pons**
- **Medulla oblongata**
- **Frontal lobe**

What is the name of the hormone secreted by the pineal body?

______________________________________________________________________

The inferior colliculus receives input from which of the special senses?

______________________________________________________________________

The superior and inferior colliculi collectively make up what structure?

______________________________________________________________________

The superior colliculus receives input from which of the special senses?

______________________________________________________________________

What structure acts as a bridge between the cerebellum, cerebrum, mesencephalon, diencephalon, and spinal cord?

______________________________________________________________________

What is the function of the commissural fibers of the corpus callosum?

______________________________________________________________________

## Anterior Dissection

Click on the **View** button and choose **Anterior**.

Click on the **Normal** button. Windows— —Mac

Drag the **Depth Bar** to **layer 0.**

Position the navigator rectangle over the head.

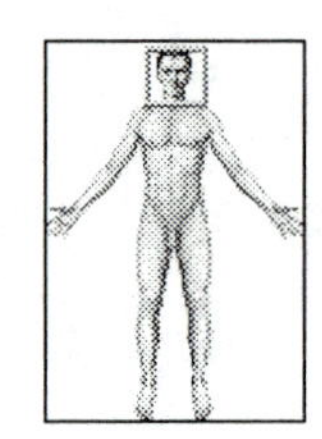

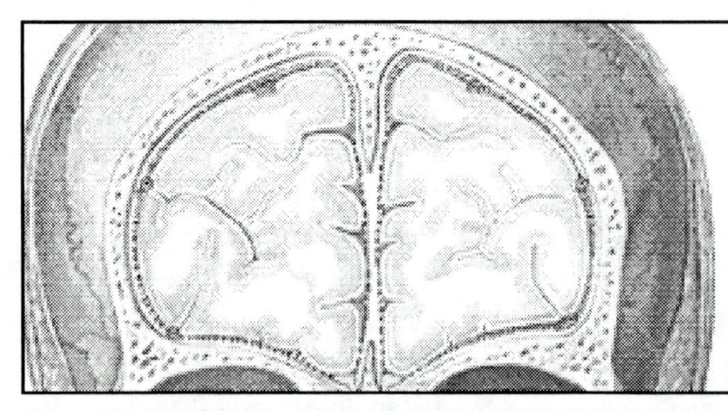

Drag the **Depth Bar** to **layer 48**. Click once to **layer 49** and identify:

- **Frontal lobe**
- **Olfactory bulb**

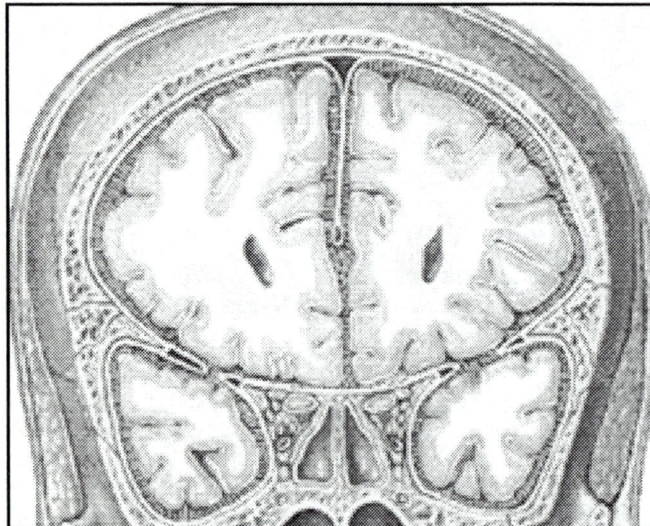

Scroll down on the **Depth Bar** to **layer 64**. Identify:

- **Frontal lobe**
- **Temporal lobe**

The white matter of the brain is composed of what structures?

______________________________

The gray matter of the brain is composed of what structures?

______________________________

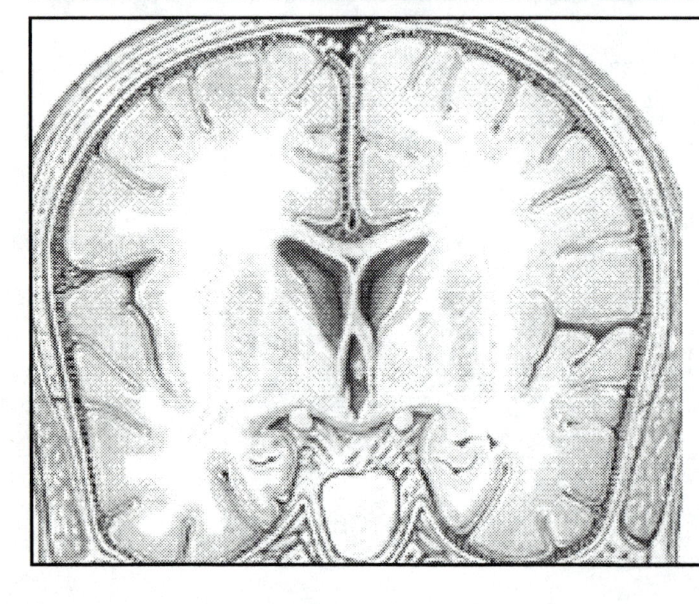

Click on the **Depth Bar** to **layer 330**. Identify:

- **Cingulate gyrus**
- **Caudate nucleus**
- **Globus pallidus**
- **Putamen**
- **Thalamus**
- **Hypothalamus**
- **Corpus callosum**
- **Septum pellucidum**
- **Fornix**

When you are finished, **close** the window.

## Atlas Anatomy of the Brain

Under the **File** menu, drag down to **Open** (then to **Content** if you are working in windows). Choose **Atlas Anatomy**. Select **Region** and **Head & Neck.** Choose **Brain (Lat)**. If the list is not in alphabetical order, the image is about one-third of the way down the list.

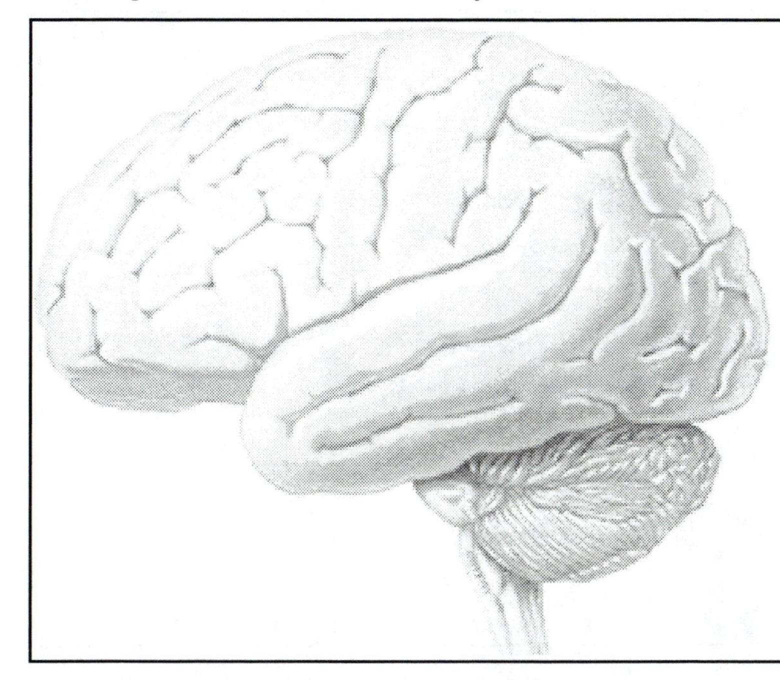

Identify:

- **Central sulcus**
- **Cerebellum**
- **Frontal lobe**
- **Medulla oblongata**
- **Lateral sulcus**
- **Occipital lobe**
- **Parietal lobe**
- **Parieto-occipital sulcus**
- **Pons**
- **Postcentral gyrus**
- **Precentral gyrus**
- **Spinal cord**
- **Temporal lobe**

When you are finished, **close** the window.

Under the **File** menu, drag down to **Open** (then to **Content** if you are working in windows). **Atlas Anatomy** should still be selected. **Region** and **Head & Neck** should still be selected. Choose **Sagittal Section of Brain**. If the list is not in alphabetical order, the image is about one-third of the way down the list.

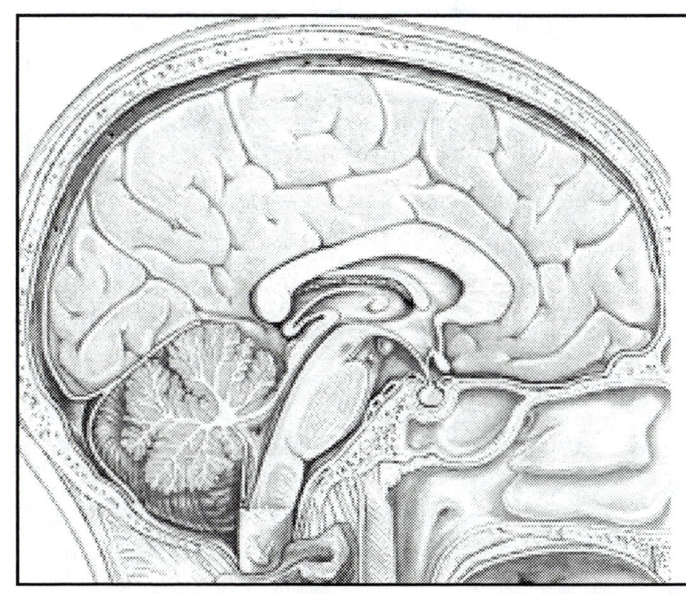

Identify by clicking on the heads of the pins:

- **Central sulcus**
- **Cerebellum**
- **Fornix of brain**
- **Corpus callosum** (genu of, rostrum of)
- **Medulla oblongata**
- **Optic chiasma**
- **Parieto-occipital sulcus**
- **Pineal body**
- **Pituitary gland**
- **Pons**

When you are finished, **close** the window.

The central nervous system and endocrine system are connected by what structure?

_______________________________________________

What structure of the hypothalamus communicates with the endocrine system?

_______________________________________________

Under the **File** menu, drag down to **Open** (then to **Content** if you are working in windows). **Atlas Anatomy** should still be selected. **Region** and **Head & Neck** should still be selected. Choose **Sagittal Section of Head & Neck**. If the list is not in alphabetical order, the image is about one-third of the way down the list.

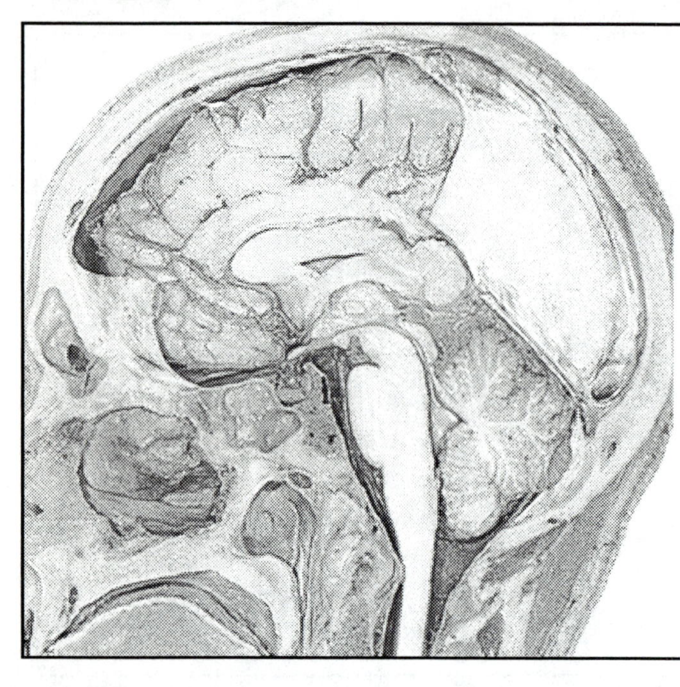

Identify:

- **Cerebellum**
- **Corpus callosum** (genu of, body of, rostrum of)
- **Medulla oblongata**
- **Optic chiasma**
- **Cerebrum**
- **Thalamus**
- **Pituitary gland**
- **Pons**
- **Midbrain**
- **Spinal cord**

When you are finished, **close** the window.

Under the **File** menu, drag down to **Open** (then to **Content** if you are working in windows). **Atlas Anatomy** should still be selected. **Region** and **Head & Neck** should still be selected. Choose **Basal Ganglia (Sup)**. If the list is not in alphabetical order, the image is about one-fourth of the way down the list.

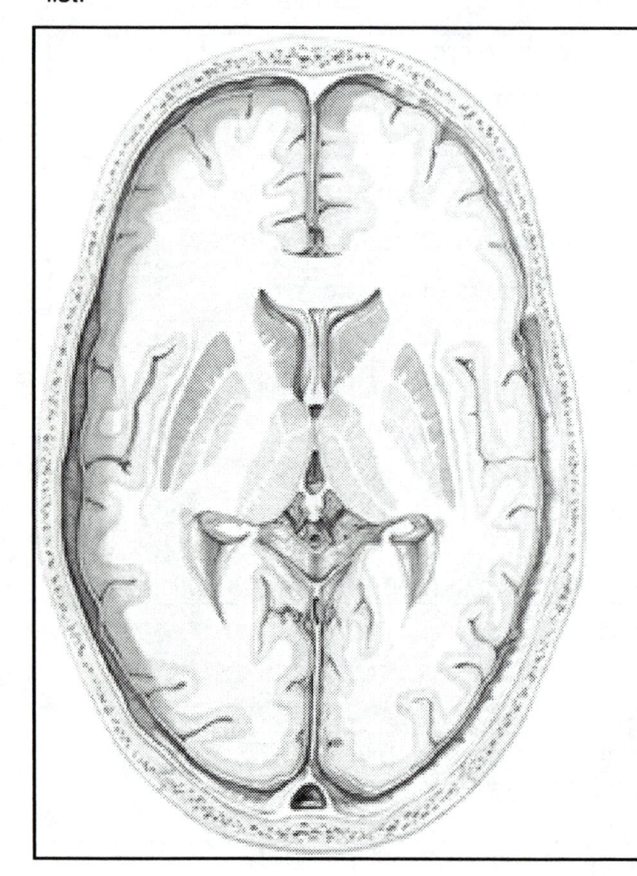

Identify:

- **Internal capsule**
- **Basal ganglia** (caudate nucleus, globus pallidus, putamen, tail of caudate nucleus)
- **Corpus callosum** (genu of)
- **Insula**
- **Pineal body**
- **Thalamus**

When you are finished, **close** the window.

Under the **File** menu, drag down to **Open** (then to **Content** if you are working in windows). Choose **Atlas Anatomy** should still be selected. Select **Region** and **Head & Neck** should still be selected. Choose **Hippocampus & Fornix (Sup)**. If list is not in alphabetical order, the image is about one-fourth of the way down the list.

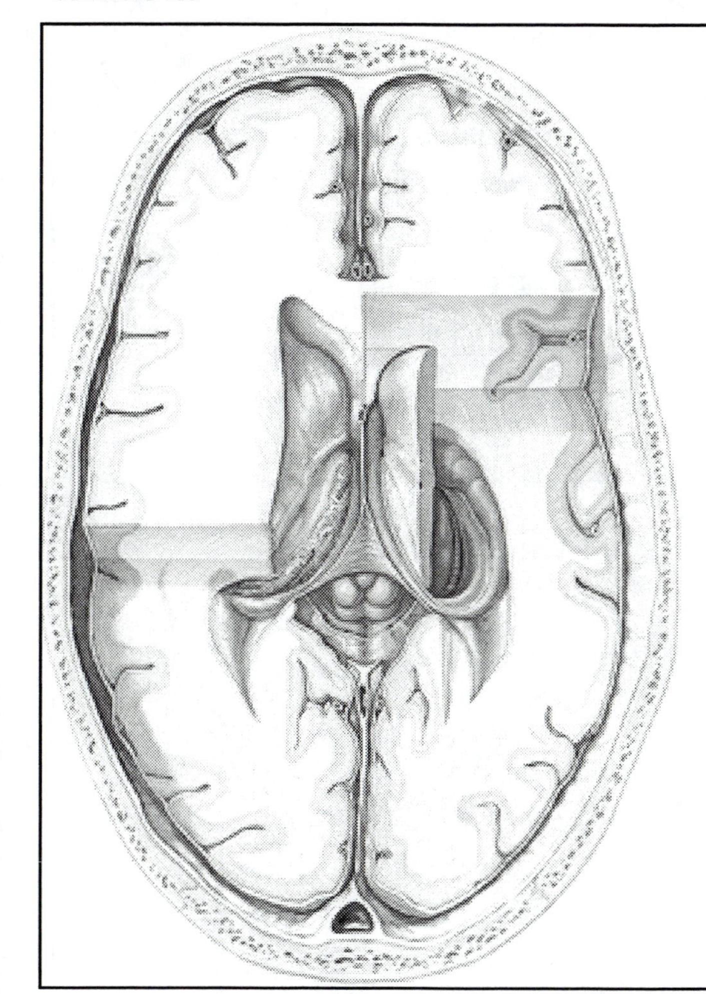

Identify:

- **Basal ganglia** (caudate nucleus)
- **Corpus callosum** (genu of)
- **Corpora quadrigemina** (inferior colliculus, superior colliculus)
- **Pineal body**
- **Thalamus**

When you are finished, **close** the window.

Under the **File** menu, drag down to **Open** (then to **Content** if you are working in windows). **Atlas Anatomy** should still be selected. **Region** and **Head & Neck** should still be selected. Choose **Cerebral Arterial Circle (Inf)**. If the list is not in alphabetical order, the image is about one-third of the way down the list.

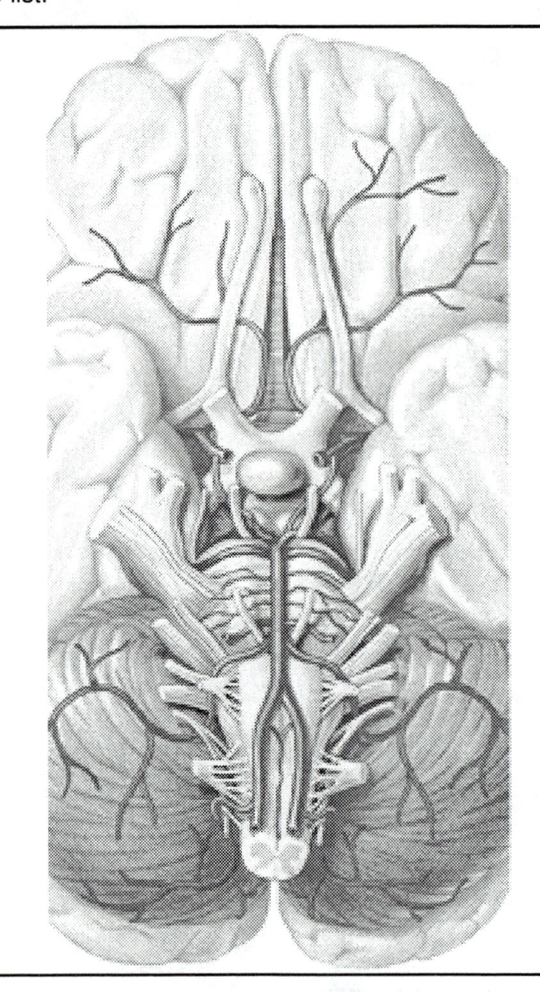

Identify:

- **Cerebellum**
- **Medulla oblongata**
- **Olfactory bulb**
- **Olfactory tract**
- **Optic chiasma**
- **Pituitary gland**
- **Pons**

When you are finished, **close** the window.

Under the **File** menu, drag down to **Open** (then to **Content** if you are working in windows). **Atlas Anatomy** should still be selected. **Region** and **Head & Neck** should still be selected. Choose **Base of Brain (Inf)**. If the list is not in alphabetical order, the image is about one-third of the way down the list.

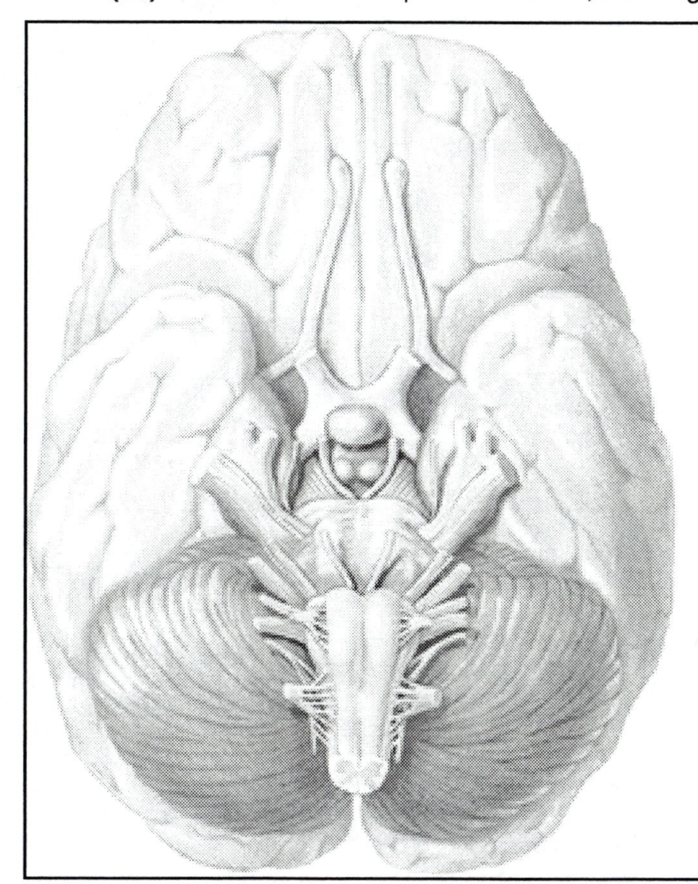

Identify:

- **Cerebellum**
- **Frontal lobe**
- **Mamillary body**
- **Occipital lobe**
- **Olfactory bulb**
- **Optic chiasma**
- **Pons**
- **Pituitary gland**
- **Spinal cord**
- **Temporal lobe**

When you are finished, **close** the window.

 

## Objectives: Brain Anatomy

To study the anatomy of the brain.

- Adenohypophysis (anterior pituitary)
- Amygdaloid body
- Basal ganglia
  - Caudate nucleus
  - Globus pallidus
  - Lentiform nucleus
  - Putamen
- Central sulcus
- Cerebellum
- Cerebrum
- Central sulcus
- Cingulate gyrus
- Corpus callosum (genu of, rostrum of)
- Corpora quadrigemina
  - Inferior colliculus
  - Superior colliculus
- Fornix of brain
- Frontal gyri
- Frontal lobe
- Hippocampus
- Hypothalamus
- Gray matter of cerebellum
- Internal capsule
- Infundibulum of pituitary gland
- Insula
- Internal capsule
- Interthalamic adhesion
- Lateral sulcus
- Left hemisphere of cerebellum
- Longitudinal fissure of cerebrum
- Mamillary body
- Medulla oblongata
- Midbrain
- Neurohypophysis (posterior pituitary)
- Occipital lobe
- Olfactory bulb
- Olive
- Optic chiasma
- Olfactory tract
- Optic chiasma
- Parietal lobe
- Parieto-occipital sulcus
- Pineal body
- Pituitary gland
- Pons
- Postcentral gyrus
- Precentral gyrus
- Pyramid
- Right hemisphere of cerebellum
- Septum pellucidum
- Skull
- Spinal cord
- Temporal lobe
- Thalamus
- Vermis of cerebellum
- White matter

# Eye

## Opening A.D.A.M Interactive Anatomy

Open A.D.A.M. Interactive Anatomy according to the directions in the Tutorial.

## 3D Anatomy of the Eye

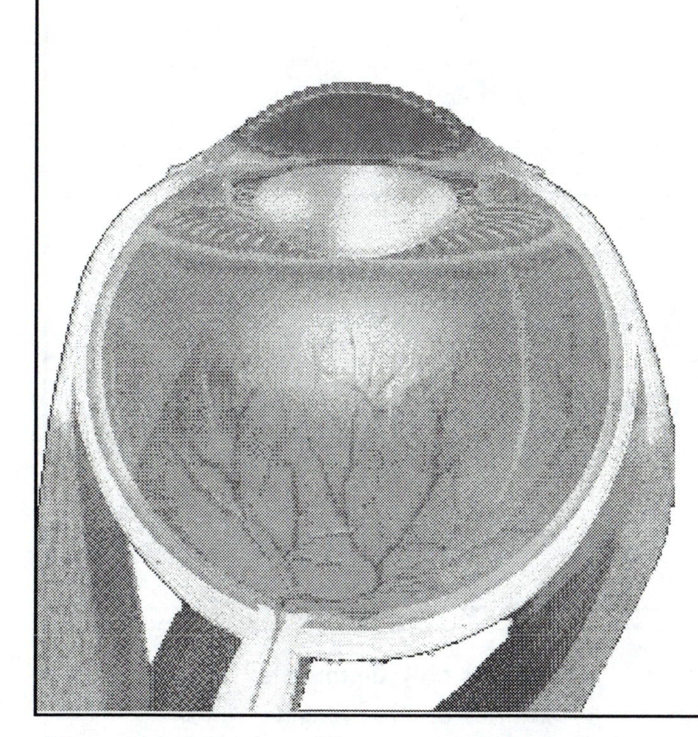

Choose **3D Anatomy**. Click on **3D Eye**. Identify the following by clicking on the **3D Anatomy Structure List**:

- **Anterior chamber of eye**
- **Capsule of lens**
- **Central artery of retina**
- **Central fovea of retina**
- **Central vein of retina**
- **Choroidea** (Choroid)
- **Ciliary body**
- **Cornea** - cut
- **Dilator pupillae muscle** (radial iris smooth muscle)
- **Iris**- cut
- **Lens**
- **Posterior chamber of eye**
- **Retina**
- **Sclera** - cut
- **Scleral venous sinus**
- **Sphincter pupillae muscle** (circular iris smooth muscle)
- **Vitreous humor**

The anterior chamber of the eye contains what substance?

______________________________________________

The posterior chamber of the eye contains what substance?

______________________________________________

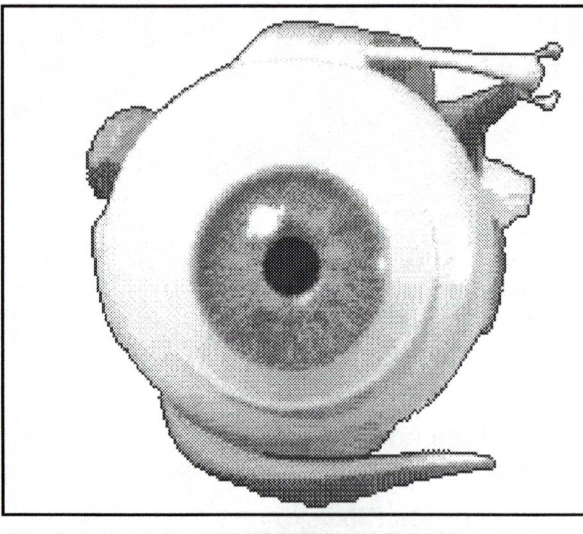

Identify:
- **Conjunctiva**
- **Iris**
- **Pupil**

Medial view:

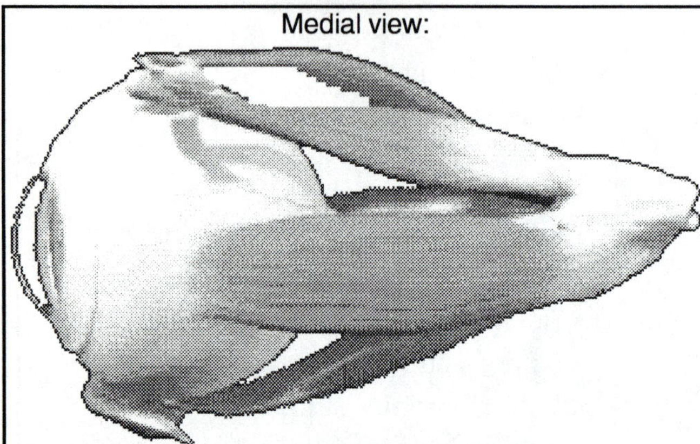

Identify:
- **Cornea**
- **Medial rectus muscle**
- **Medial rectus muscle** - Action
- **Sclera**

The medial rectus muscle moves the eye in what direction?

_______________________________________

Inferior view:

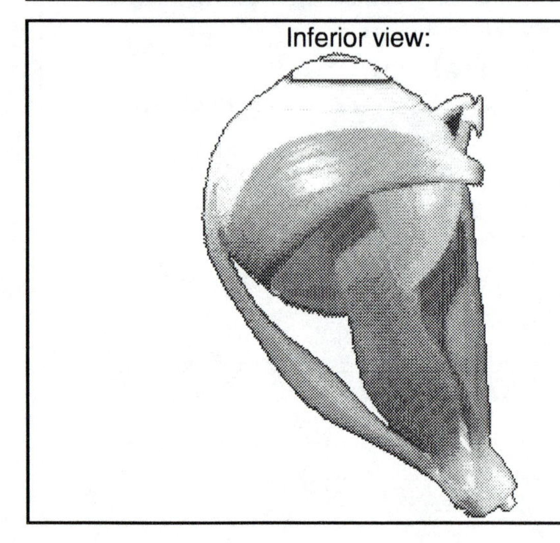

Identify:
- **Inferior oblique muscle**
- **Inferior oblique muscle**- Action
- **Inferior rectus muscle**
- **Inferior rectus muscle** - Action

The inferior oblique muscle moves the eye in what direction?

_______________________________________

The inferior rectus muscle moves the eye in what direction?

_______________________________________

Lateral view:

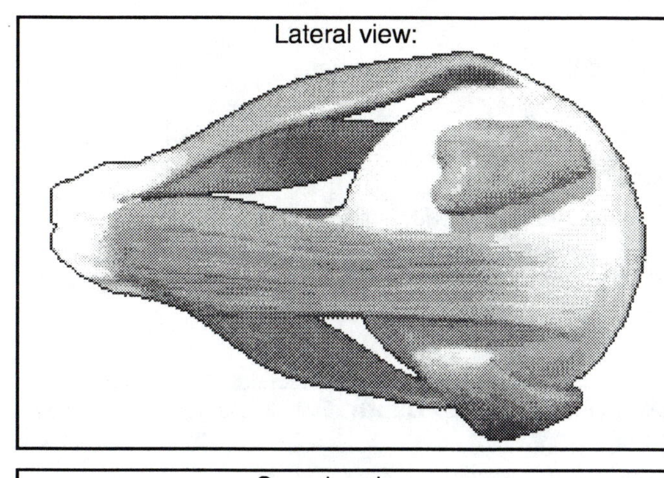

Identify:

- **Lacrimal gland**
- **Lateral rectus muscle**
- **Lateral rectus muscle**- Action
- **Optic nerve (CN II)**

The lateral rectus muscle moves the eye in what direction?

___________________________________

Superior view:

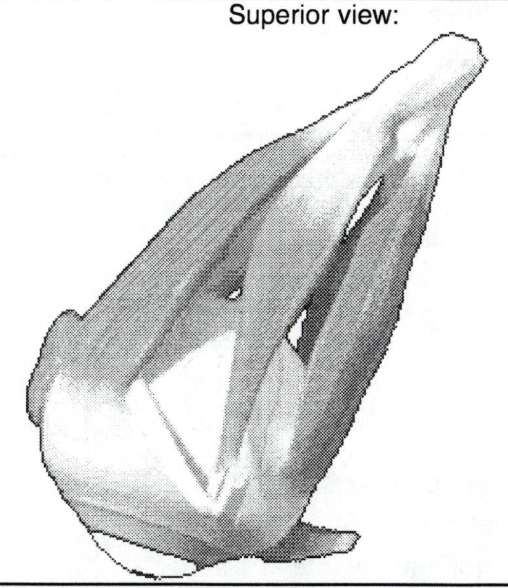

- **Superior oblique muscle**
- **Superior oblique muscle**- Action
- **Superior rectus muscle**
- **Superior rectus muscle**- Action

The superior oblique muscle moves the eye in what direction?

___________________________________

The superior rectus muscle moves the eye in what direction?

___________________________________

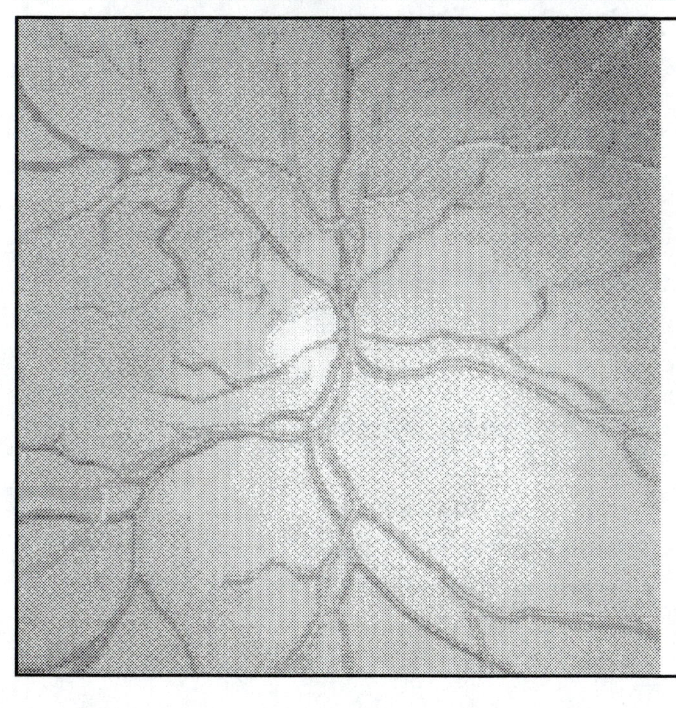

Identify:

- **Macula**
- **Optic disk**
- **Orbital animation**

The macula contains (1) no rods (2) no cones (3) no rods or cones (4) both rods and cones.

__________

The optic disk contains (1) no rods (2) no cones (3) no rods or cones (4) both rods and cones.

__________

When you are finished, **close** the window.

## Anterior Dissection of the Eye

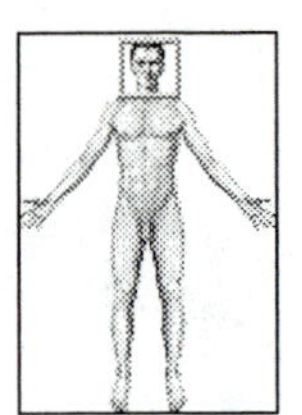

Under the **File** menu, drag down to **Open** (then to **Content** if you are working in Windows).

Choose **Dissectible Anatomy**.

Choose male or female. Select **Anterior.** Click **Open**.

Enlarge the window and position the navigation rectangle over the head.

Click on the **Depth Bar** ▼ until you reach **layer 12**. No graphic has been provided for this layer. Identify:

- **Orbicularis oculi**

Click once on the **Depth Bar** ▼ to **layer 13**. Identify:

- **Superior tarsal plate**
- **Inferior tarsal plate**

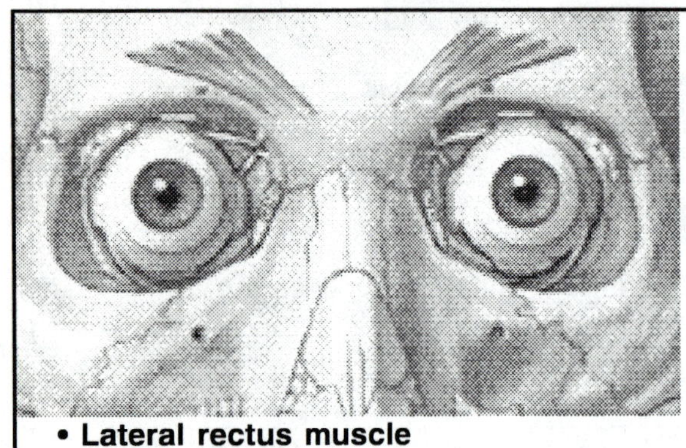

Click a few times on the **Depth Bar** ▼ until you reach **layer 38**. Identify:

- **Cornea covering pupil**
- **Cornea covering iris**
- **Conjunctiva**
- **Lacrimal gland**
- **Lacrimal sac**
- **Sclera**
- **Superior rectus muscle**
- **Inferior oblique muscle**
- **Superior oblique muscle**
- **Medial rectus muscle**
- **Lateral rectus muscle**
- **Superior levator palpebrae muscle**
- **Orbital fat**

What is the function of the superior levator palpebrae muscle?

___________________________________________________________________________

The cornea is a part of the ______________________________ (conjunctiva, sclera)?

Click a few times on the **Depth Bar** ▼ to **layer 40**. No graphic has been provided for this layer. Identify:

- **Orbital fat**

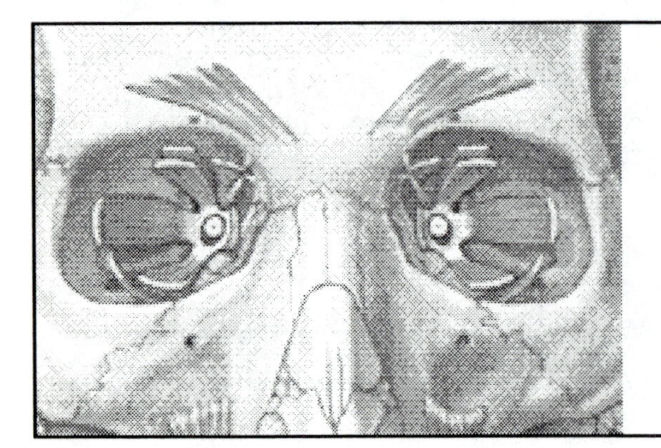

Click once on the **Depth Bar** ▼ to **layer 41**. Identify:

- **Superior rectus muscle**
- **Optic nerve (CN II)**
- **Inferior oblique muscle**
- **Medial rectus muscle**
- **Lateral rectus muscle**
- **Superior oblique muscle**
- **Superior levator palpebrae muscle**

## Lateral Dissection of the Eye

Click on the **View** button and choose **Lateral**.

Click on the **Normal** button. Windows— —Mac

Drag the **Depth Bar** to **layer 0.**

Position the navigator rectangle over the head.

Click on the **Depth Bar** until you reach **layer 15**. No graphic has been provided for this layer. Identify:

- **Orbicularis oculi**

Click on the **Depth Bar** to **layer 34**. No graphic has been provided for this layer. Identify:

- **Tarsal plate.**

Click on the **Depth Bar** to **layer 116**. No graphic has been provided for this layer. Identify:

- **Orbital fat**

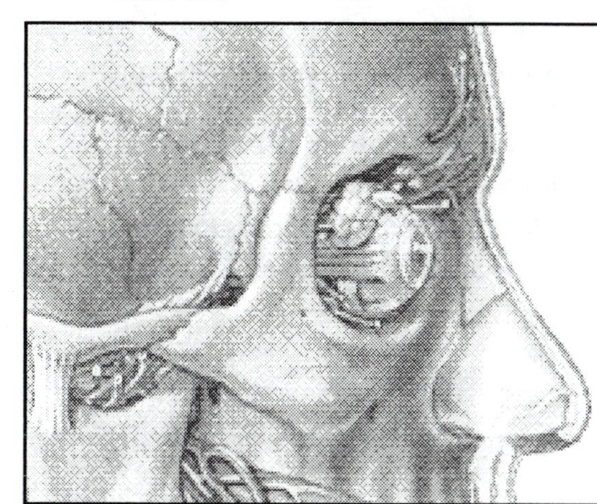

Click on the **Depth Bar** until you reach **layer 117**. Identify:

- **Cornea**
- **Lacrimal gland**
- **Sclera**
- **Lateral rectus muscle**

What is the function of the lacrimal gland?

_______________________________________________

Drag the **Depth Bar** to **level 194**. No graphic has been provided for this layer. Identify the **lacrimal gland**.

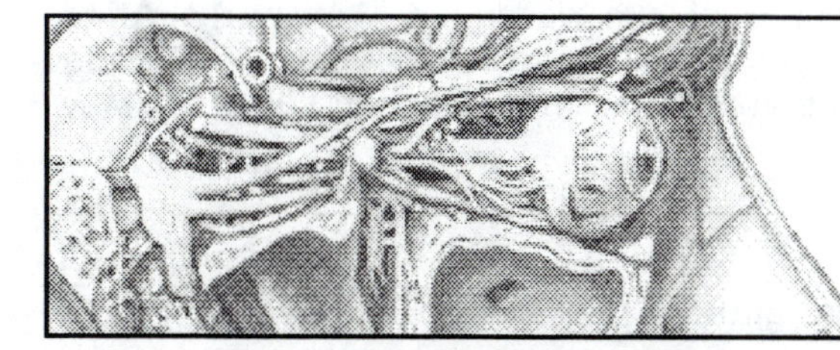

Click a few times on the **Depth Bar** to **layer 196**. Identify:

- **Cornea**
- **Sclera**
- **Optic nerve (CN II)**

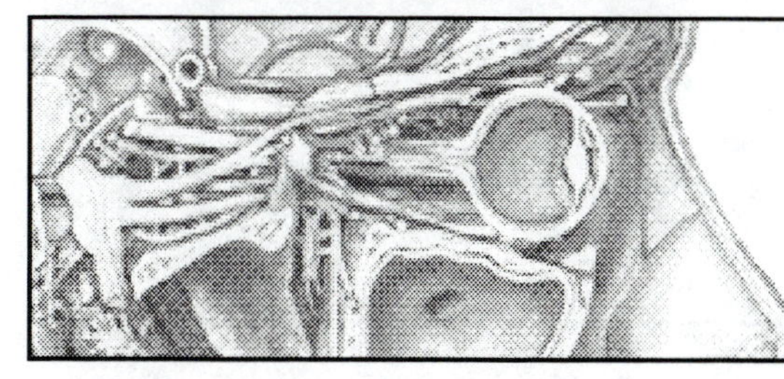

Click once on the **Depth Bar** to **layer 197**. Identify:

- **Retina**
- **Lens**
- **Optic nerve (CN II)**
- **Anterior chamber of eye**

Click once on the **Depth Bar** to **layer 198**. No graphic has been provided for this layer. Identify the **lacrimal sac**.

When you are finished, **close** the window.

 

## Atlas Anatomy of the Eye

Under the **File** menu, drag down to **Open** (then to **Content** if you are working in Windows). Choose **Atlas Anatomy**. Select **Region** and **Head & Neck.** Choose **Extrinsic Eye Muscles (Ant)**. If the list is not in alphabetical order, the image is slightly over half of the way down the list.

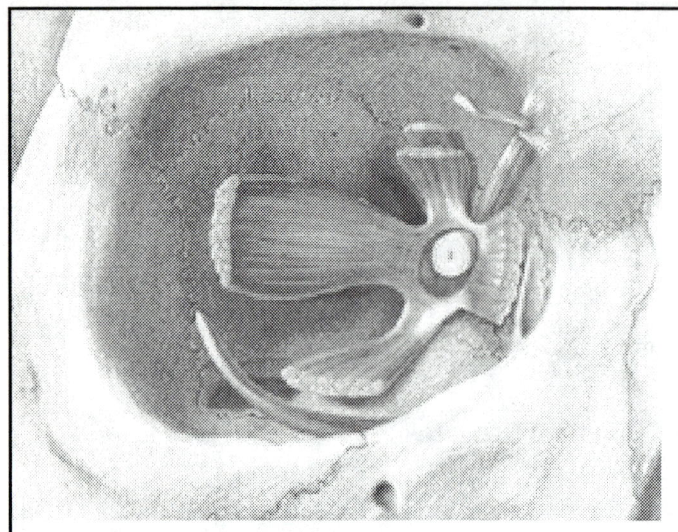

Identify:

- **Fossa of lacrimal sac**
- **Frontal process of maxilla**
- **Frontal process of zygomatic bone**
- **Inferior oblique muscle**
- **Inferior orbital fissure**
- **Inferior rectus muscle**
- **Lateral rectus muscle**
- **Levator palpebrae superioris muscle**
- **Medial rectus muscle**
- **Optic nerve (CN II)**
- **Orbital part of greater wing of sphenoid bone**
- **Orbital plate of frontal bone**
- **Orbital surface of maxilla**
- **Orbital surface of zygomatic bone**
- **Superior oblique fissure**
- **Superior orbital fissure**
- **Superior rectus muscle**
- **Zygomatic process of frontal bone**

When you are finished, **close** the window.

The oculomotor nerve controls what extrinsic eye muscles?

______________________________________________________________

The trochlear nerve controls what extrinsic eye muscles?

______________________________________________________________

Under the **File** menu, drag down to **Open** (then to **Content** if you are working in Windows). **Atlas Anatomy** should still be selected. **Region** and **Head & Neck** should still be selected. Choose **Glands of Head & Neck (Lat)**. If the list is not in alphabetical order, the image is slightly less than half of the way down the list.

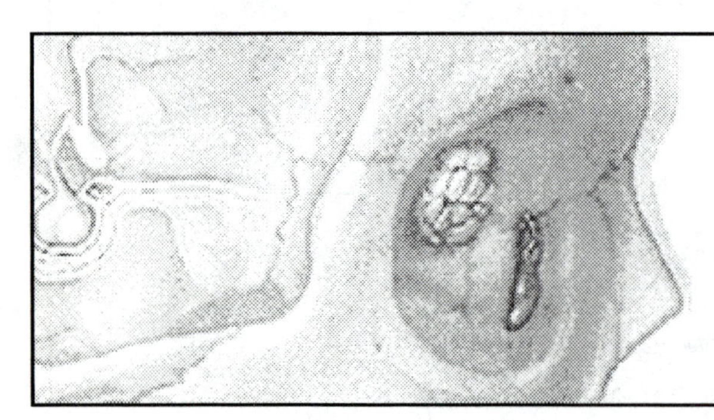

Identify:

- **Lacrimal gland**
- **Lacrimal sac**

When you are finished, **close** the window.

Under the **File** menu, drag down to **Open** (then to **Content** if you are working in windows). **Atlas Anatomy** should still be selected. **Region** and **Head & Neck** should still be selected. Choose **Optic Nerve in Orbit (Lat)**. If the list is not in alphabetical order, the image is almost half of the way down the list.

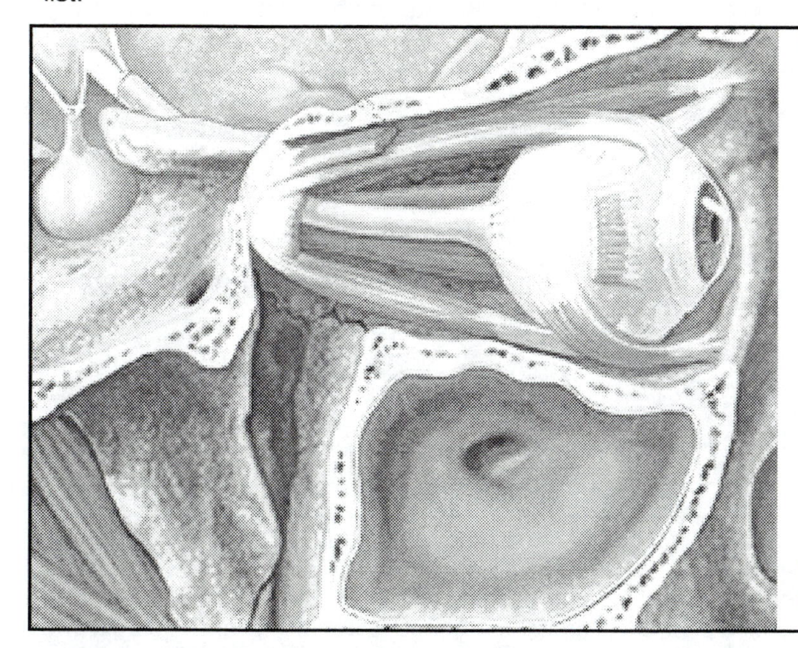

Identify:

- **Inferior oblique muscle**
- **Inferior rectus muscle**
- **Lateral rectus muscle**
- **Levator palpebrae superioris muscle**
- **Maxillary sinus**
- **Optic nerve (CN II)**
- **Superior oblique muscle**
- **Superior orbital fissure**
- **Superior rectus muscle**

When you are finished, **close** the window.

Under the **File** menu, drag down to **Open** (then to **Content** if you are working in Windows). **Atlas Anatomy** should still be selected. **Region** and **Head & Neck** should still be selected. Choose **Ciliary Ganglion - Pathways 1**. If the list is not in alphabetical order, the image is about one-third of the way down the list.

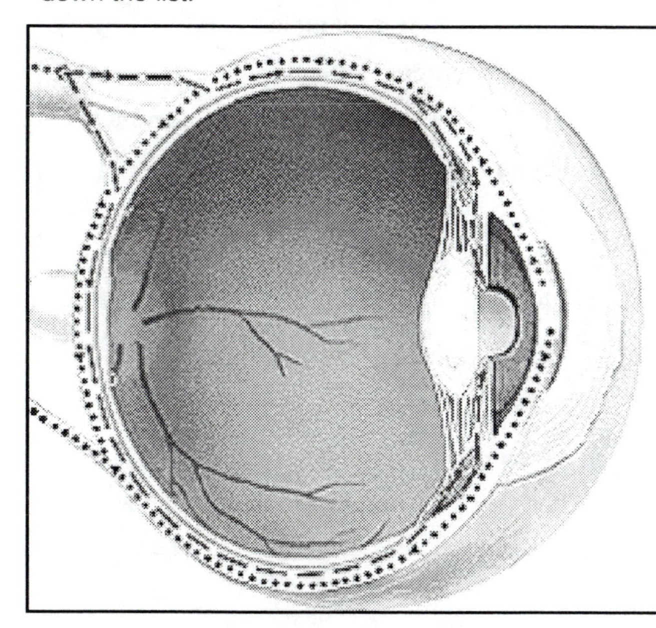

Identify:

- **Ciliary body**
- **Cornea**
- **Iris**
- **Lens**
- **Retina**
- **Sclera**

When you are finished, **close** the window.

What is the function of the ciliary body?

---

Trace the path of light from the anterior surface of the eye to the retina naming the structures (in order) that the light would pass through.

---

 

## Cadaver Images of the Eye

Under the **File** menu, drag down to **Open** (then to **Content** if you are working in Windows). **Atlas Anatomy** should still be selected. **Region** and **Head & Neck** should still be selected. Choose **Dissection of Orbit (Lat)**. If the list is not in alphabetical order, the image is about half of the way down the list.

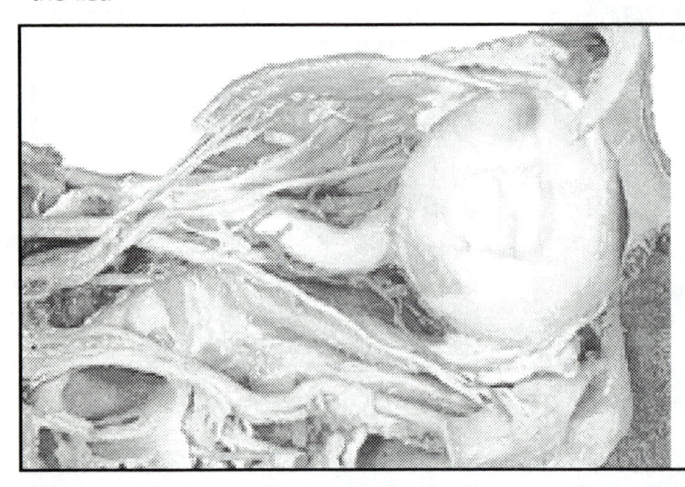

Identify:

- **Inferior oblique muscle**
- **Inferior rectus muscle**
- **Lateral rectus muscle**
- **Levator palpebrae superioris muscle**
- **Optic nerve (CN II)**
- **Sclera**
- **Superior rectus muscle**

When you are finished, **close** the window.

Under the **File** menu, drag down to **Open** (then to **Content** if you are working in Windows). **Atlas Anatomy** should still be selected. **Region** and **Head & Neck** should still be selected. Choose **Eyeball & Extrinsic Eye Muscles**. If the list is not in alphabetical order, the image is about half of the way down the list.

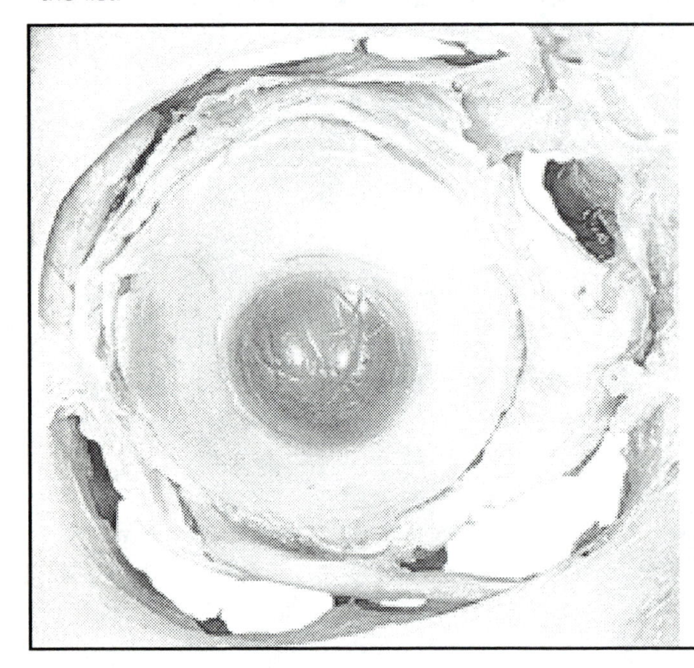

Identify:

- **Inferior oblique muscle**
- **Inferior rectus muscle**
- **Lateral rectus muscle**
- **Levator palpebrae superioris muscle**
- **Lacrimal gland**
- **Conjunctiva**
- **Superior oblique muscle**
- **Cornea**
- **Superior rectus muscle**
- **Medial rectus muscle**
- **Sclera**

When you are finished, **close** the window.

Under the **File** menu, drag down to **Open** (then to **Content** if you are working in Windows). Choose **Atlas Anatomy** should still be selected. **Region** and **Head & Neck** should still be selected. Choose **Sagittal Section of Eyeball**. If the list is not in alphabetical order, the image is slightly over half of the way down the list.

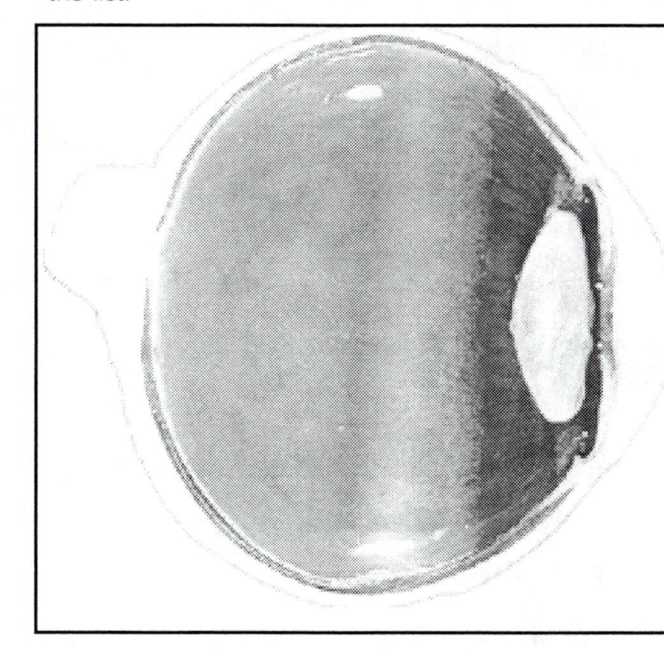

Identify:

- **Anterior chamber of eye**
- **Ciliary body**
- **Cornea**
- **Iris**
- **Lens**
- **Optic disk**
- **Optic nerve (CN II)**
- **Posterior chamber of eye**
- **Pupil**
- **Retina**
- **Sclera**

When you are finished, **close** the window.

## Objectives: Eye

To learn the anatomy of the eye and surrounding structures.

- Anterior chamber of eye
- Capsule of lens
- Central artery of retina
- Central fovea of retina
- Central vein of retina
- Choroidea (Choroid)
- Ciliary body
- Conjunctiva
- Cornea
- Dilator pupillae muscle (radial iris smooth muscle)
- Fossa of lacrimal sac
- Frontal process of maxilla
- Frontal process of zygomatic bone
- Inferior oblique muscle
- Inferior orbital fissure
- Inferior rectus muscle
- Inferior tarsal plate
- Iris
- Lacrimal gland
- Lacrimal sac
- Lateral rectus muscle
- Lens
- Levator palpebrae superioris muscle
- Macula
- Maxillary sinus
- Medial rectus muscle
- Optic disk
- Optic nerve (CN II)
- Orbicularis oculi muscle
- Orbital fat
- Orbital part of greater wing of sphenoid bone
- Orbital plate of frontal bone
- Orbital surface of maxilla
- Orbital surface of zygomatic bone
- Posterior chamber of eye
- Pupil
- Retina
- Sclera
- Scleral venous sinus
- Sphincter pupillae muscle (circular iris smooth muscle)
- Superior oblique fissure
- Superior oblique muscle
- Superior orbital fissure
- Superior rectus muscle
- Superior tarsal plate
- Tarsal plate.
- Vitreous humor
- Zygomatic process of frontal bone

# Ear

## Opening A.D.A.M Interactive Anatomy

Open A.D.A.M. Interactive Anatomy according to the directions in the Tutorial.

## 3D Anatomy of the Ear

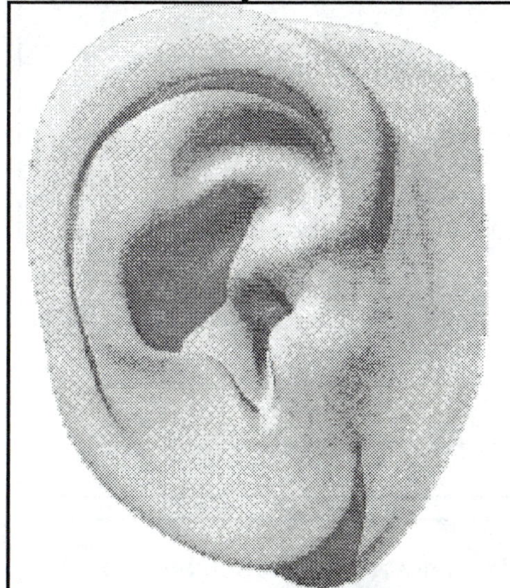

Choose **3D Anatomy**. Click on **3D Ear**. Identify the following by clicking on the **3D Anatomy Structure List**:

- **Auricle (Pinna)**
- **Helix of auricle**
- **Lobule of Auricle**
- **Tympanic membrane**

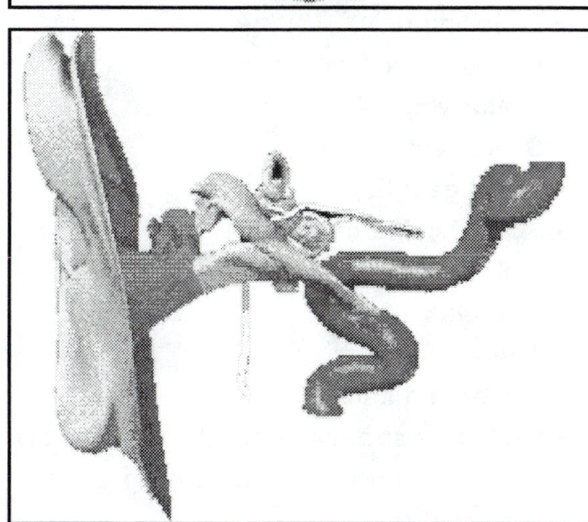

Identify :

- **Auditory tube**
- **Stapes - cut**
- **Ampulla of inner ear**
- **Vestibulocochlear nerve (CN VIII)**

The vestibulocochlear nerve carries the impulses of what two senses?

______________________________

______________________________

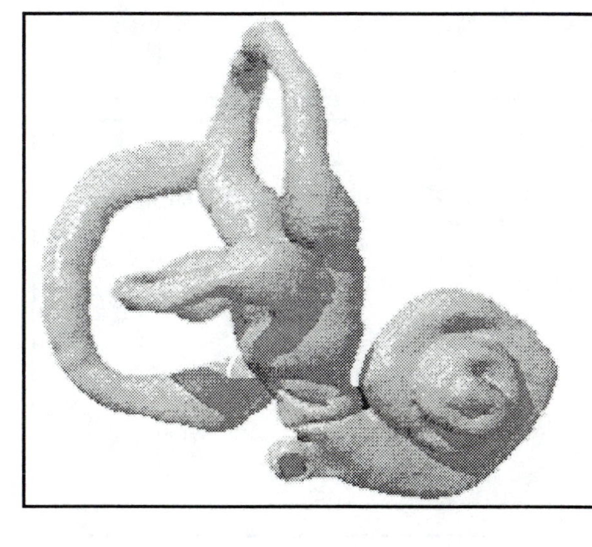

Identify :

- **Anterior semicircular canal**
- **Cochlea**
- **Lateral semicircular canal**
- **Oval window**
- **Posterior semicircular canal**
- **Round window**
- **Saccule of membranous labyrinth of inner ear**
- **Vestibule of inner ear**

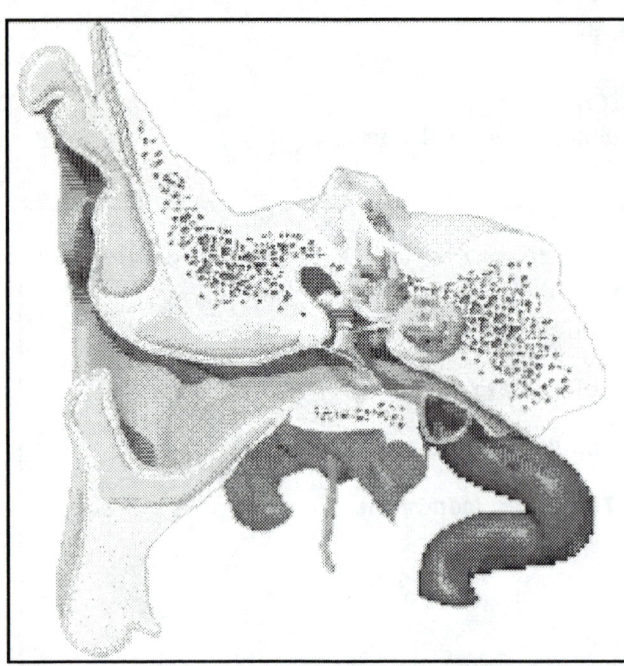

Identify :

- **Auditory tube** - cut
- **Incus** - cut
- **Malleus** - cut
- **Tympanic cavity** - cut
- **Tympanic membrane** - cut

What is the function of the auditory tube?

______________________________

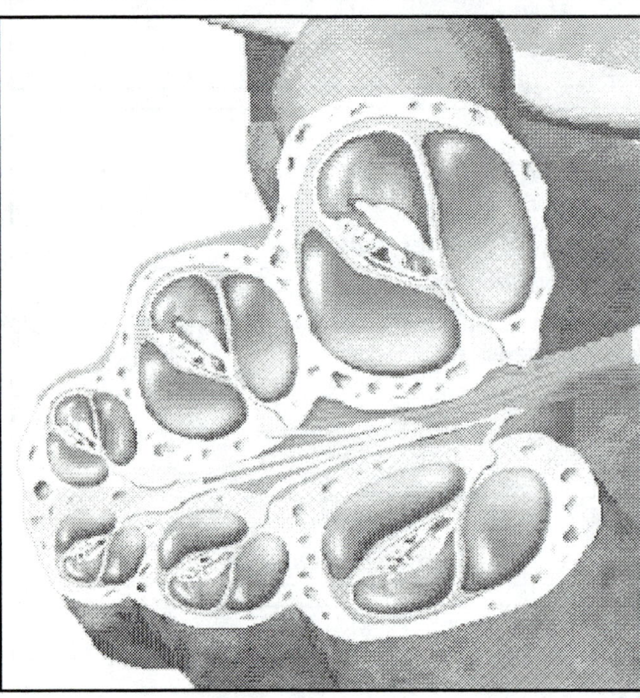

Identify :

- **Basilar membrane of cochlea**
- **Endolymph**
- **Helicotrema**
- **Scala tympani**
- **Scala vestibuli**
- **Spiral ganglion**
- **Spiral lamina of cochlea**
- **Spiral organ**
- **Modiolus**
- **Perilymph**
- **Tectorial membrane**
- **Vestibular surface** (vestibular membrane)

Which of the structures listed belong to the skeletal system?

______________________________

Identify :

- **Utricle of membranous labyrinth of inner ear**
- **Cochlear duct**

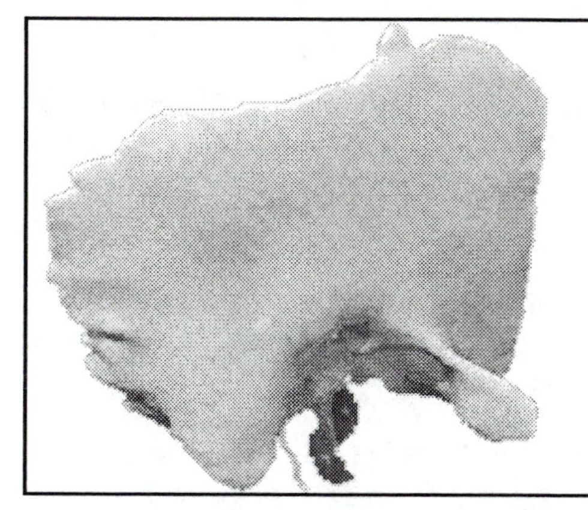

Identify :

- **External auditory (acoustic) meatus**
- **Temporal bone**

What glands of the integumentary system are located in the external auditory meatus? What do these glands produce?

______________________________

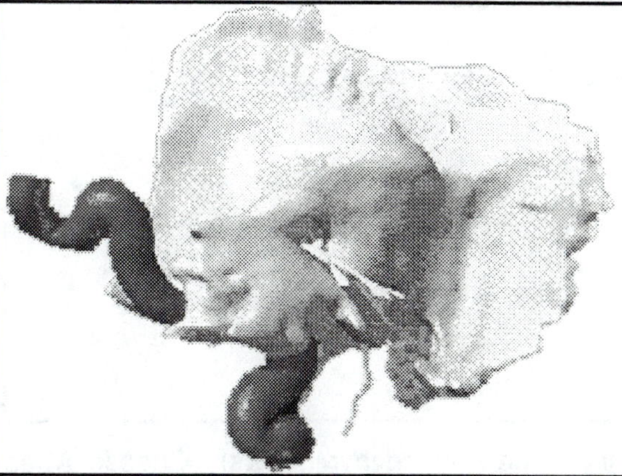

Identify :

- **Internal auditory (acoustic) meatus**

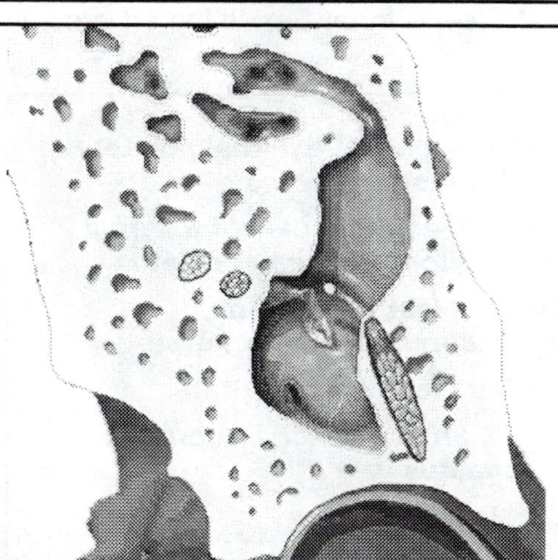

Identify :

- **Mastoid antrum**
- **Round window** - midsagittal cut
- **Stapedius muscle** - midsagittal cut
- **Stapes** - midsagittal cut
- **Tensor tympani muscle**

What is the function of the tensor tympani and stapedius muscles?

______________________________

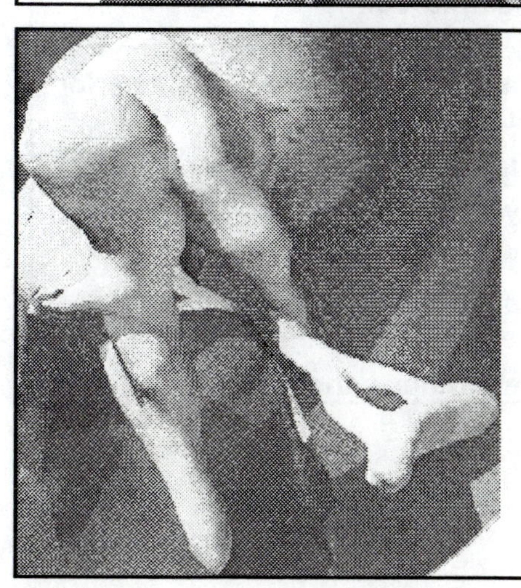

Identify :

- **Incus**
- **Malleus**
- **Stapes**

When you are finished, **close** the window.

## Atlas Anatomy of Ear

Under the **File** menu, drag down to **Open** (then to **Content** if you are working in Windows). Choose **Atlas Anatomy**. Select **Region** and **Head & Neck.** Choose **Auricle of Ear**. If the list is not in alphabetical order, the image is about close to the bottom of the list.

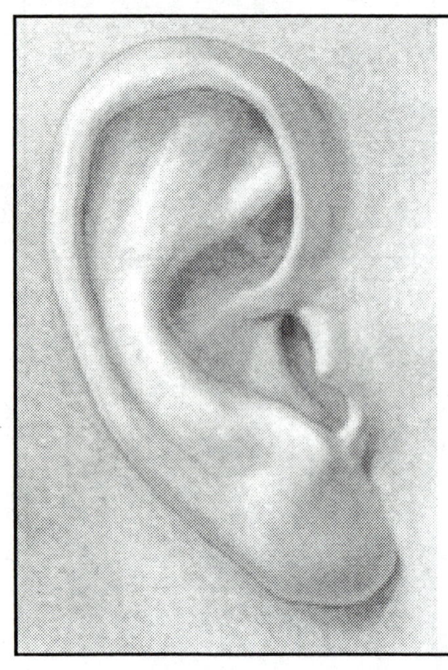

Identify :

- **Helix of auricle**
- **Auricle (Pinna)**
- **External acoustic meatus**
- **Lobule of auricle**

When you are finished, **close** the window.

Under the **File** menu, drag down to **Open** (then to **Content** if you are working in windows). Choose **Atlas Anatomy**. Select **Region** and **Head & Neck.** Choose **External, Middle, & Inner Ear**. If the list is not in alphabetical order, the image is close to the bottom of the list.

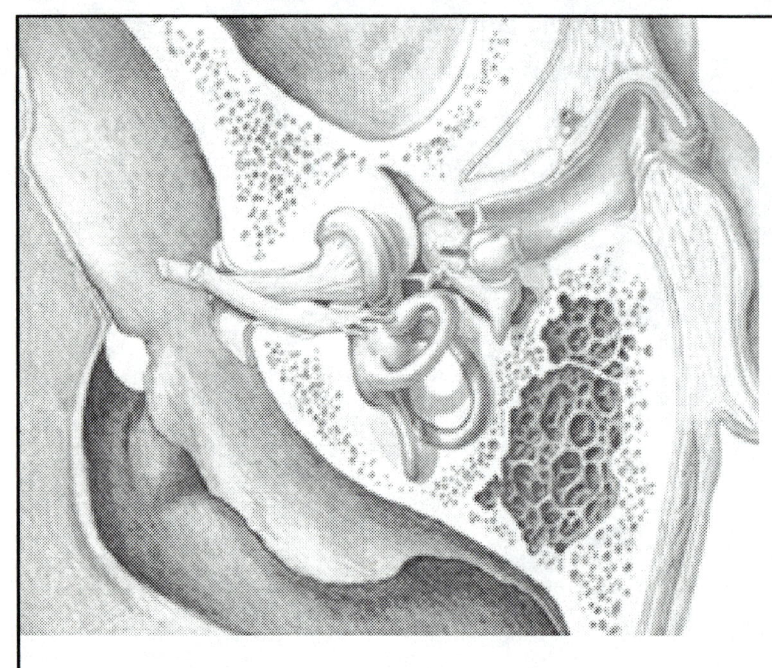

Identify:

- **Bony part of external auditory (acoustic) meatus**
- **Cartilaginous part of external auditory (acoustic) meatus**
- **Cochlear division of vestibulocochlear nerve**
- **Cochlear duct**
- **External auditory (acoustic) meatus**
- **Malleus**
- **Mastoid cell**
- **Membranous labyrinth of inner ear (Anterior, lateral, & posterior ducts)**
- **Stapes**
- **Tympanic cavity**
- **Tympanic membrane**
- **Vestibular division of vestibulocochlear nerve**
- **Vestibular ganglion**

When you are finished, **close** the window.

## Objectives: Ear

Student will learn the following parts of the ear:

- Anterior semicircular canal
- Ampulla of inner ear
- Auricle (Pinna)
- Auditory tube
- Basilar membrane of cochlea
- Bony part of external auditory (acoustic) meatus
- Cartilaginous part of external auditory (acoustic) meatus
- Cochlea
- Cochlear division of vestibulocochlear nerve
- Cochlear duct
- Endolymph
- External auditory (acoustic) meatus
- Helix of auricle
- Helicotrema
- Incus
- Internal auditory (acoustic) meatus
- Lateral semicircular canal
- Lobule of auricle
- Malleus
- Mastoid cell
- Mastoid antrum
- Membranous labyrinth of inner ear (anterior, lateral, & posterior ducts)
- Modiolus
- Oval window
- Perilymph
- Posterior semicircular canal
- Round window
- Saccule of membranous labyrinth of inner ear
- Scala tympani
- Scala vestibuli
- Spiral ganglion
- Spiral lamina of cochlea
- Spiral organ
- Stapes
- Stapedius muscle
- Tectorial membrane
- Temporal bone
- Tensor tympani muscle
- Tympanic cavity
- Tympanic membrane
- Utricle of membranous labyrinth of inner ear
- Vestibule of inner ear
- Vestibular division of vestibulocochlear nerve
- Vestibulocochlear nerve (CN VIII)
- Vestibular ganglion

# Heart

## Opening A.D.A.M Interactive Anatomy

Open A.D.A.M. Interactive Anatomy according to the directions in the Tutorial.

## Anterior Dissection of the Heart

Choose **Dissectible Anatomy**.

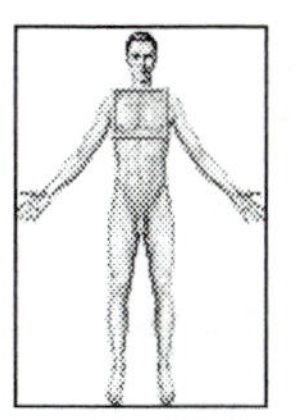

Choose either male or female. Select **Anterior.** Click **Open**.

Enlarge the window and position the image over the chest.

Click on the **Depth Bar** ▼ until you reach **layer 157**. No graphic has been provided for this layer.

What bones protect the heart anteriorly? ____________________

Click on the **Depth Bar** ▼ to **layer 168**. No graphic has been provided for this layer. Identify:

- **Pericardial sac**

Click twice on the **Depth Bar** ▼ to **layer 170**. No graphic has been provided for this layer. Identify:

- **Epicardium**

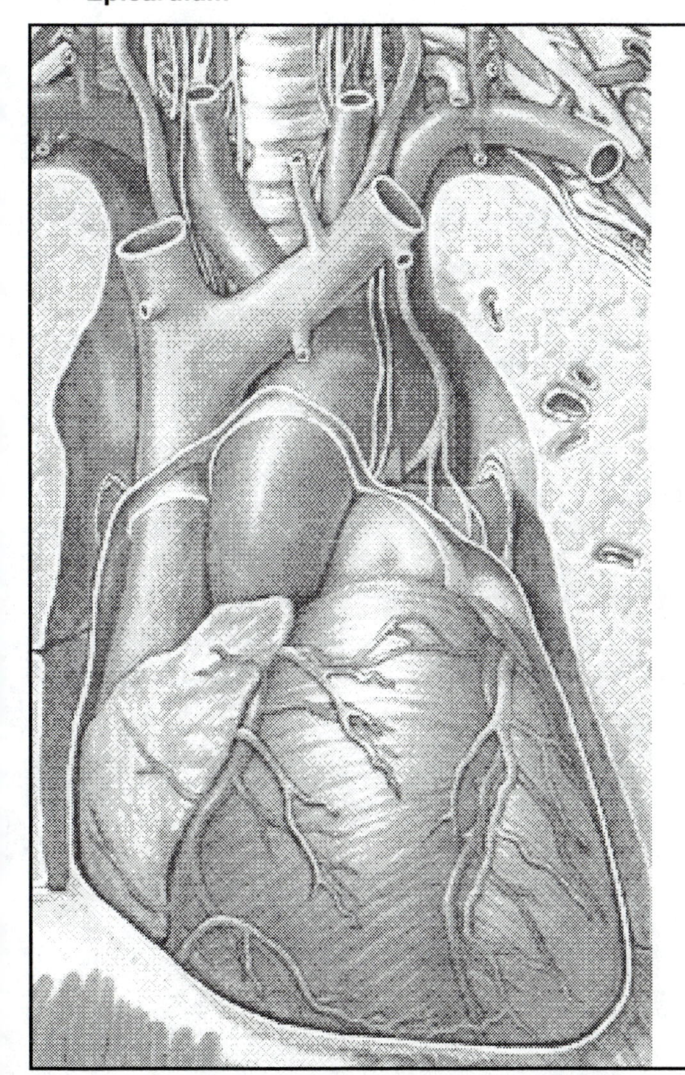

Click once on the **Depth Bar** ▼ until you reach **layer 171**. Identify:

- **Superior vena cava**
- **Right brachiocephalic vein**
- **Left brachiocephalic vein**
- **Pericardial sac** (anterior of)
- **Coronary arteries** (right coronary artery, anterior interventricular artery)
- **Coronary veins** (great cardiac vein, anterior cardiac vein, right marginal vein)
- **Apex of heart**
- **Ascending aorta**
- **Arch of aorta**
- **Pulmonary trunk**
- **Left pulmonary artery**
- **Right atrium**
- **Right ventricle**
- **Left ventricle**

What is the name of the cavity in which the heart is located?

---

The **mediastinum**, which separates the two pleural cavities, contains the heart and three other organs. Name these three. (Hint: Check layers 164 and 256.)

---

Click on the **Depth Bar** ▼ until you reach **layer 173**. No graphic has been provided for this layer. Identify the **coronary sulcus** and the **anterior interventricular sulcus**.

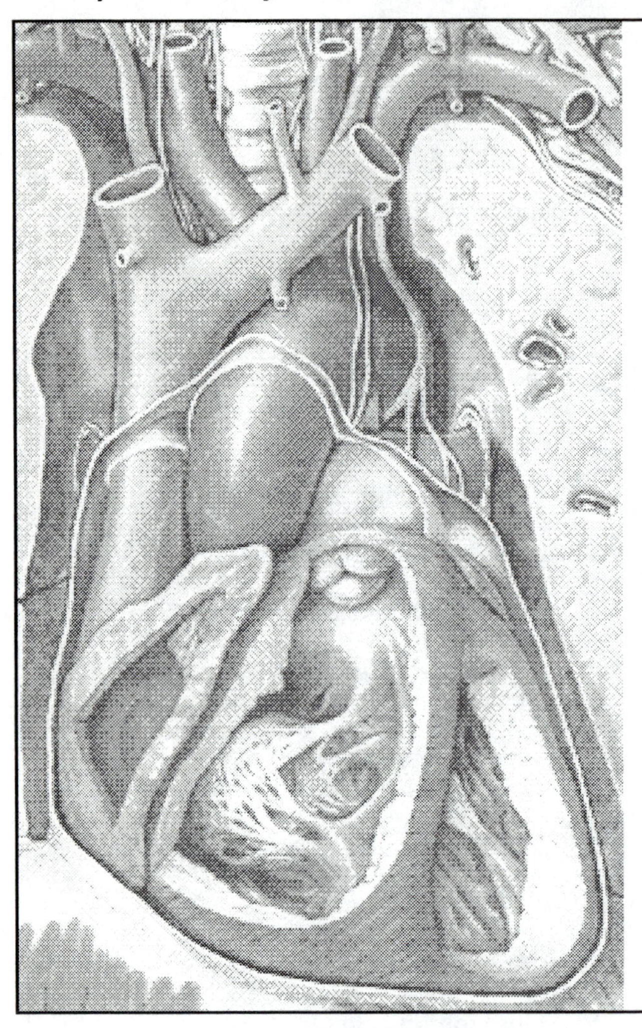

Click on the **Depth Bar** ▼ until you reach **layer 174**. Identify:

- **Chordeae tendineae** (Tendinous chords)
- **Superior vena cava**
- **Fossa ovalis** (Oval fossa)
- **Right atrium**
- **Right ventricle**
- **Tricuspid valve** (Right AV valve)
- **Papillary muscles**
- **Pulmonary valve**
- **Left ventricle**
- **Coronary sulcus**
- **Right brachiocephalic vein**
- **Left brachiocephalic vein**
- **Pulmonary trunk**
- **Ascending aorta**
- **Arch of aorta**
- **Interventricular septum**
- **Pericardial sac** (interior of)
- **Left pulmonary artery**

Click once on the **Depth Bar** ▼ to **layer 175**. No graphic has been provided for this layer. Identify:
- **Pericardial sac**

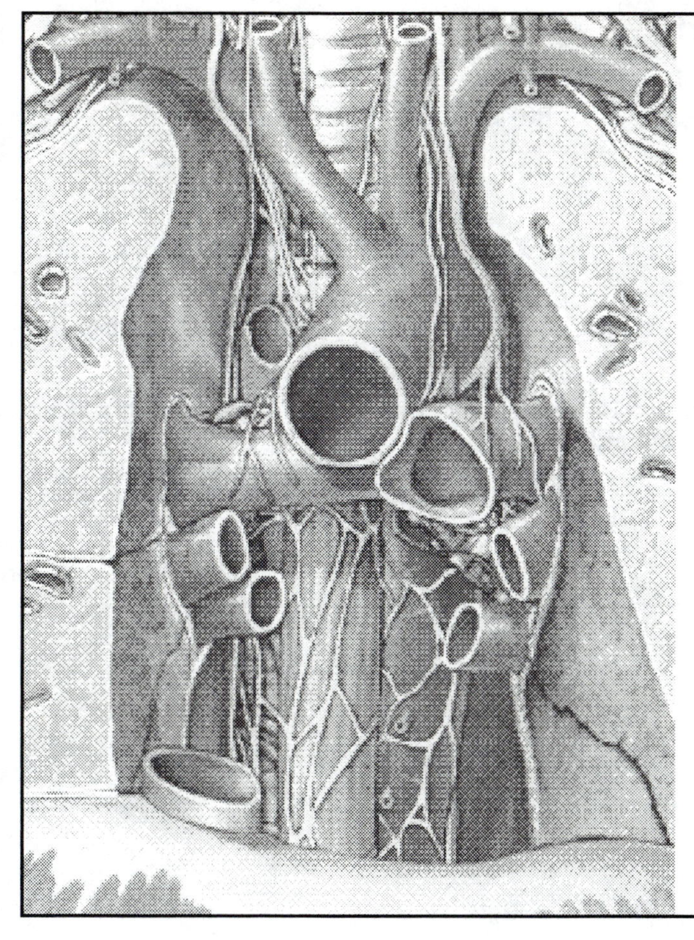

Click on the **Depth Bar** [▼] to **layer 177**. Identify:

- **Inferior vena cava**
- **Right inferior pulmonary vein**
- **Left inferior pulmonary vein**
- **Right superior pulmonary vein**
- **Left superior pulmonary vein**
- **Pulmonary trunk**
- **Arch of aorta**
- **Brachiocephalic artery**
- **Right common carotid artery**
- **Left common carotid artery**
- **Descending thoracic aorta**
- **Left pulmonary artery**
- **Right pulmonary artery**
- **Right subclavian artery**
- **Left subclavian artery**
- **Esophagus**
- **Trachea**

The pulmonary veins bring ____________________ (oxygenated, deoxygenated) blood from the ________________ (heart, lungs, body) to the ________________ (heart, lungs, body).

The pulmonary arteries bring ________________________________ (oxygenated, deoxygenated) blood from the ________________ (heart, lungs, body) to the ________________ (heart, lungs, body).

The aorta brings ________________ (oxygenated, deoxygenated) blood from the ________________ (heart, lungs, body) to the ________________ (heart, lungs, body).

## Lateral Dissection of the Heart

Click on the **View** button and choose **Lateral**.

Click on the **Normal** button.

Drag the **Depth Bar** to **layer 0.**

Position the navigator rectangle over the chest.

Scroll to **layer 139**. Click on the **Depth Bar** until you reach **layer 147** and then click again until you reach **layer 155**. Scroll to **layer 210**. No graphics have been provided for these layers. Identify the **pericardial sac**.

Click a few times on the **Depth Bar** to **layer 212**. No graphic has been provided for this layer. Identify the **epicardium** and the **adipose tissue** around the heart.

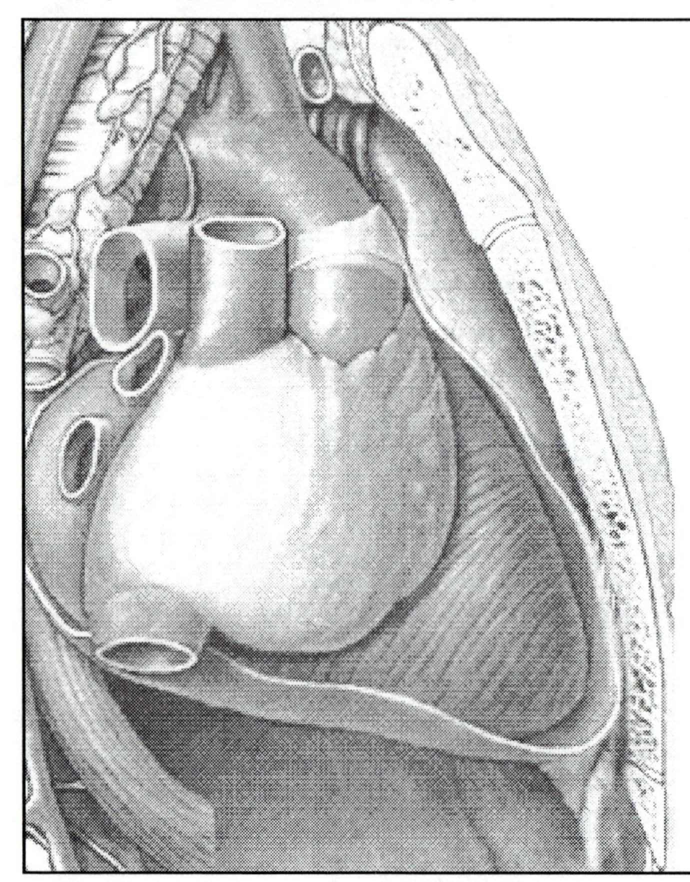

Click on the **Depth Bar** until you reach **layer 219**. Identify:

- **Body of sternum, manubrium, xiphoid process**
- **Right superior pulmonary vein**
- **Thymus gland**
- **Pericardial sac**
- **Ascending aorta**
- **Arch of aorta**
- **Right ventricle**
- **Right atrium**
- **Superior vena cava**
- **Inferior vena cava**
- **Right inferior pulmonary vein**
- **Right pulmonary artery**
- **Left atrium**
- **Auricle of right atrium**
- **Esophagus**
- **Trachea**

The inferior vena cava brings blood to the heart from which region(s) of the body?

---

The superior vena cava brings blood to the heart from which region(s) of the body?

---

Trace a drop of blood through the heart from the inferior vena cava to the pulmonary trunk, naming all of the structures (in order) through which the blood would pass.

---

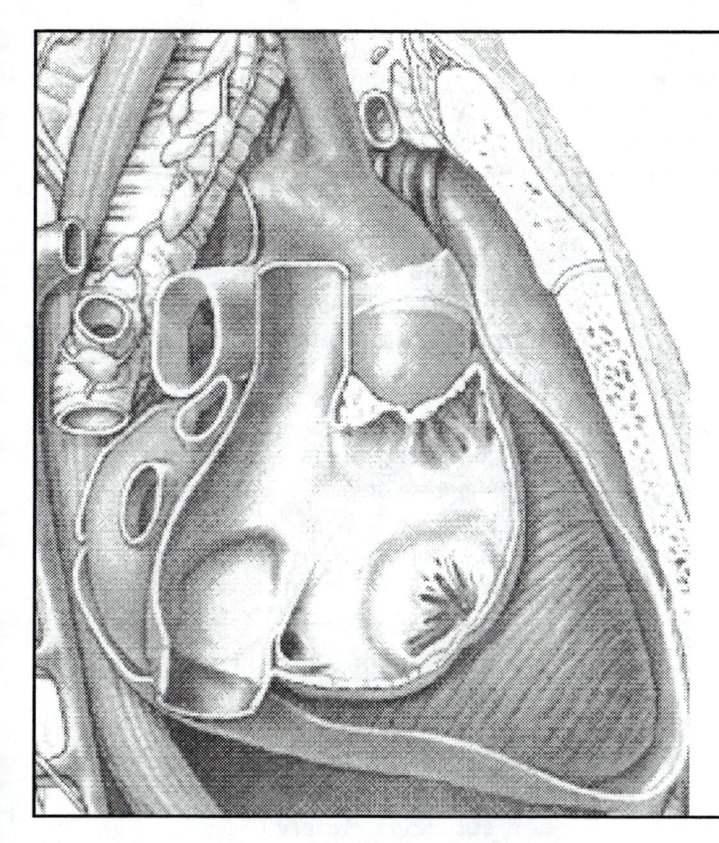

Click once on the **Depth Bar** [▼] to **layer 220**. Identify:

- **Left atrium**
- **Right inferior pulmonary vein**
- **Right superior pulmonary vein**
- **Pericardial sac**
- **Ascending aorta**
- **Tricuspid valve**
- **Right ventricle**
- **Right atrium**
- **Auricle of right atrium**
- **Superior vena cava**
- **Inferior vena cava**
- **Right pulmonary artery**
- **Fossa ovalis**
- **Chordae tendineae**
- **Esophagus**
- **Trachea**

When you are finished, **close** dissectible anatomy.

What is the function of the chordae tendineae?

______________________________________________________________________

 

## Atlas Anatomy of the Heart

Under the **File** menu, drag down to **Open** (then to **Content** if you are working in Windows). Click on **Atlas Anatomy**. Select **Region** and **Thorax.** Choose **Heart & Great Vessels (Ant)**. If the list is not in alphabetical order, the image is about two-thirds of the way down the list.

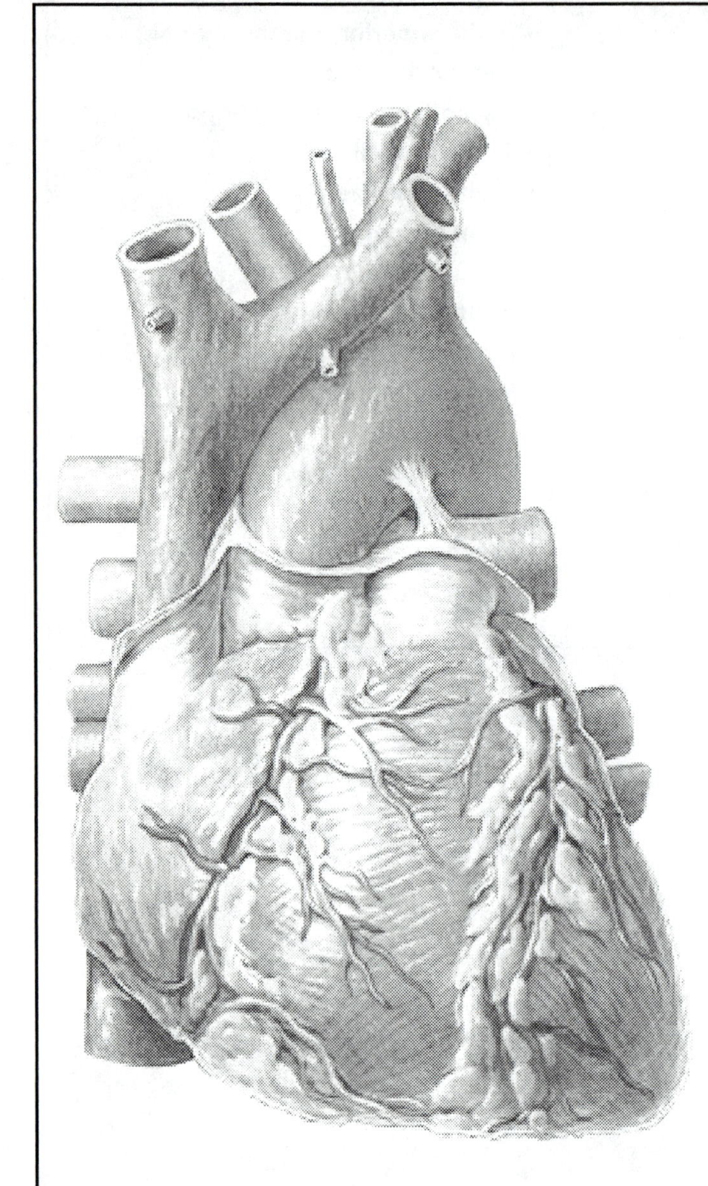

Identify:

- **Apex of heart**
- **Arch of aorta**
- **Ascending aorta**
- **Auricle of left atrium**
- **Auricle of right atrium**
- **Brachiocephalic trunk**
- **Coronary arteries** (right coronary artery, anterior interventricular artery, right marginal artery)
- **Coronary veins** (great cardiac vein, anterior cardiac vein, small cardiac vein)
- **Inferior vena cava**
- **Left brachiocephalic vein**
- **Left common carotid artery**
- **Left inferior pulmonary vein**
- **Left pulmonary artery**
- **Left subclavian artery**
- **Left superior pulmonary vein**
- **Left ventricle**
- **Ligamentum arteriosum**
- **Pericardium**
- **Pulmonary trunk**
- **Right atrium**
- **Right brachiocephalic vein**
- **Right inferior pulmonary vein**
- **Right pulmonary artery**
- **Right superior pulmonary vein**
- **Right ventricle**
- **Superior vena cava**

What blood vessels join to form the superior vena cava?

___

What fetal structure becomes the ligamentum arteriosum in the adult?

___

When you are finished, **close** the window.

Under the **File** menu, drag down to **Open** (then to **Content** if you are working in Windows). Click on **Atlas Anatomy**. **Region** and **Thorax** should still be selected. Choose **Heart & Great Vessels (Post)**. If the list is not in alphabetical order, the image is about two-thirds of the way down the list.

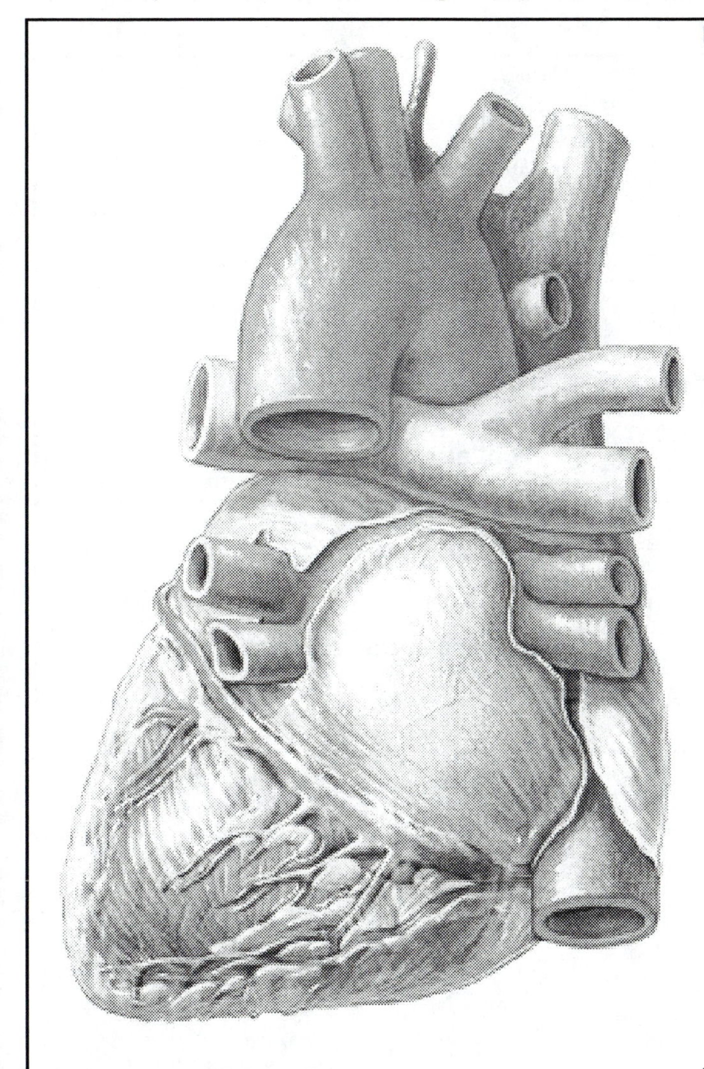

Identify:

- **Apex of heart**
- **Arch of aorta**
- **Brachiocephalic trunk**
- **Coronary arteries** (posterior interventricular artery, circumflex branch of left coronary artery)
- **Coronary veins** (great cardiac vein, middle cardiac vein)
- **Coronary sinus**
- **Inferior vena cava**
- **Left atrium**
- **Left brachiocephalic vein**
- **Left common carotid artery**
- **Left inferior pulmonary vein**
- **Left pulmonary artery**
- **Left subclavian artery**
- **Left superior pulmonary vein**
- **Left ventricle**
- **Pericardial sac/Pericardium**
- **Right atrium**
- **Right brachiocephalic vein**
- **Right inferior pulmonary vein**
- **Right pulmonary artery**
- **Right superior pulmonary vein**
- **Right ventricle**
- **Superior vena cava**

When you are finished, **close** the window.

What is the function of the coronary arteries?

______________________________________________________________________

## Cross Section of the Heart

Under the **File** menu, drag down to **Open** (then to **Content** if you are working in Windows). Click on **Atlas Anatomy**. **Region** and **Thorax** should still be selected. Choose **T8 Vertebra (Inf)**. If the list is not in alphabetical order, the image is about half way down the list.

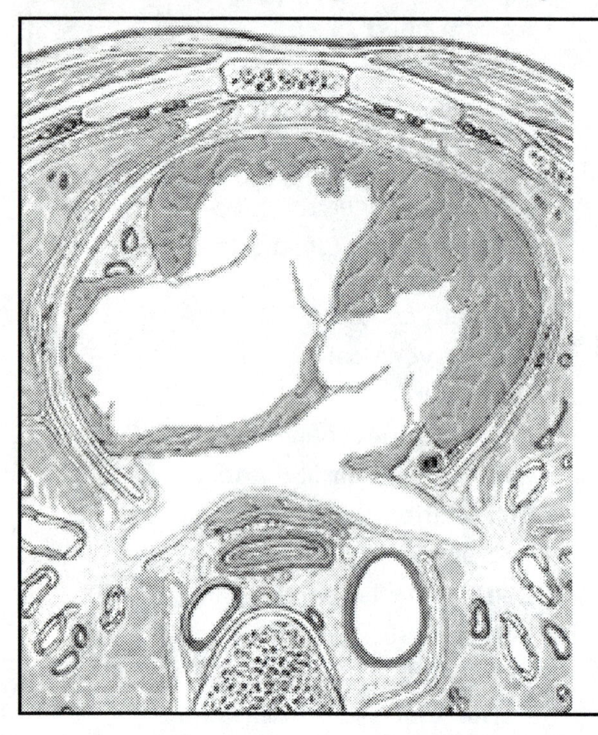

Identify:

- **Sternum**
- **Myocardium**
- **Epicardium**
- **Parietal pericardium** (Pericardium)
- **Pericardial cavity**
- **Left atrium**
- **Left ventricle**
- **Right atrium**
- **Right ventricle**
- **Endocardium**
- **Esophagus**
- **Thoracic aorta** (descending aorta)

When you are finished, **close** the window.

## Cadaver Images of the Heart

Under the **File** menu, drag down to **Open** (then to **Content** if you are working in Windows). Click on **Atlas Anatomy**. **Region** and **Thorax** should still be selected. Choose **Thoracic Viscera (Ant)**. If the list is not in alphabetical order, the image is about half way down the list.

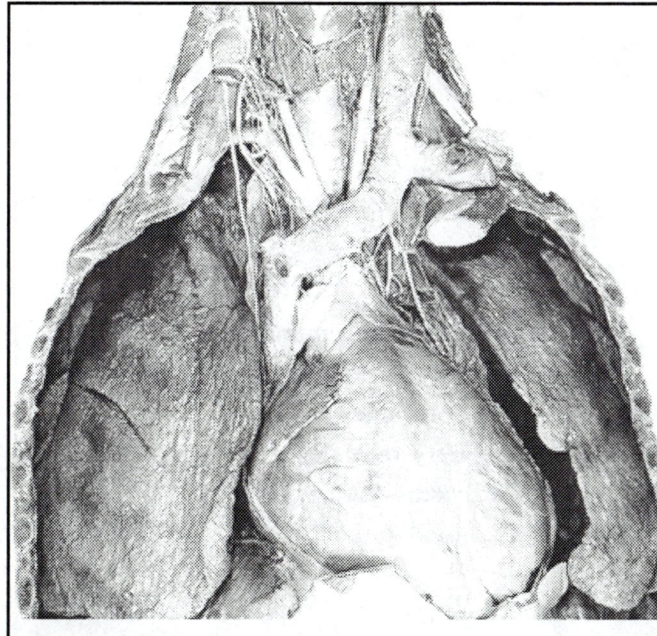

Identify:

- **Anterior interventricular sulcus**
- **Apex of heart**
- **Ascending aorta**
- **Auricle of right atrium**
- **Brachiocephalic trunk**
- **Left brachiocephalic vein**
- **Left common carotid artery**
- **Left internal jugular vein**
- **Left ventricle**
- **Pericardium** (parietal)
- **Right brachiocephalic vein**
- **Right common carotid artery**
- **Right subclavian artery**
- **Right ventricle**
- **Superior vena cava**
- **Trachea**

When you are finished, **close** the window.

Under the **File** menu, drag down to **Open** (then to **Content** if you are working in Windows). Click on **Atlas Anatomy**. **Region** and **Thorax** should still be selected. Choose **Dissection of Left Mediastinum**. If the list is not in alphabetical order, the image is about one-third of the way down the list.

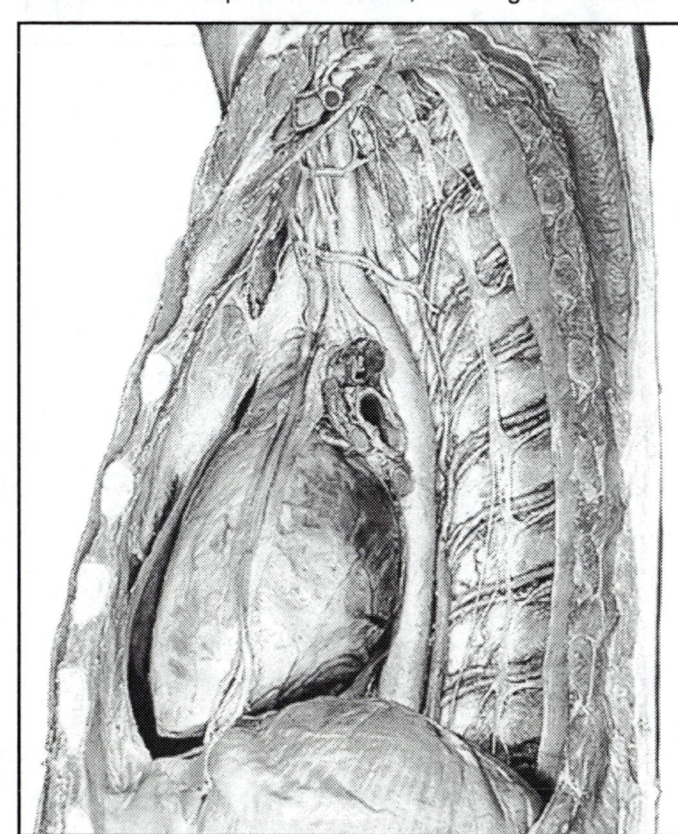

Identify:

- **Apex of heart**
- **Arch of aorta**
- **Diaphragm**
- **Epicardium**
- **Left inferior pulmonary vein**
- **Left pulmonary artery**
- **Pericardial cavity**
- **Pericardium** (parietal)
- **Thoracic aorta** (descending aorta)

When you are finished, **close** the window.

Under the **File** menu, drag down to **Open** (then to **Content** if you are working in Windows). Click on **Atlas Anatomy**. **Region** and **Thorax** should still be selected. Choose **Dissection of Heart Chambers**. If the list is not in alphabetical order, the image is about three-fourths of the way down.

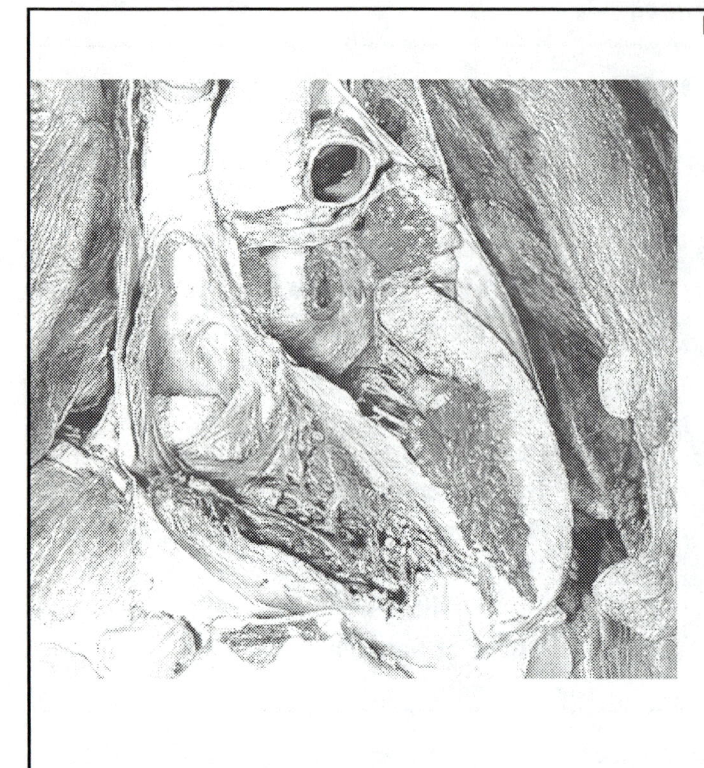

Identify:

- **Apex of heart**
- **Arch of aorta**
- **Ascending aorta**
- **Diaphragm**
- **Fossa ovalis**
- **Inferior vena cava**
- **Interventricular septum**
- **Left atrium**
- **Left inferior pulmonary vein**
- **Left lung**
- **Left ventricle**
- **Papillary muscle**
- **Pericardial sac**
- **Mitral valve** (posterior cusp)
- **Pulmonary trunk**
- **Right atrium**
- **Right lung**
- **Right ventricle**
- **Superior vena cava**
- **Tendinous cord** (chorda tendinea)

When you are finished, **close** the window.

## 3D Anatomy of the Heart

Choose **3D Anatomy**. Choose **3D Heart.** Identify the following. It is suggested you hold a heart model in your hand and identify:

- **Cardiac veins**
- **Coronary arteries**
- **Valves of heart**

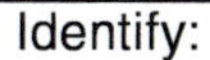

Identify:

- **Pulmonary valve** (anterior cusp, left cusp, right cusp)

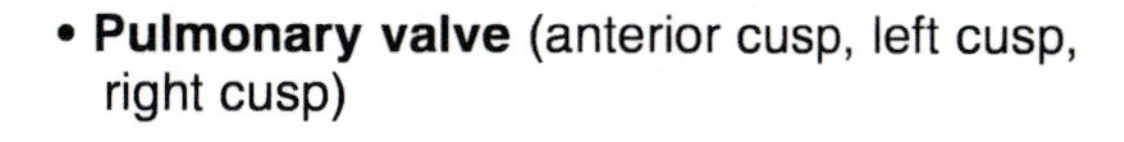

- **Tricuspid valve** (anterior cusp, cusp anterior, posterolateral, posterior cusp)
- **Papillary muscle** (anterior, right ventricle)

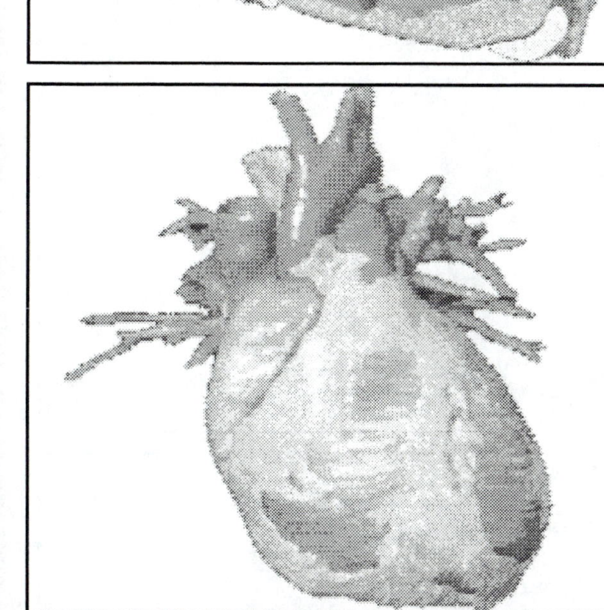

Identify:

- **Aorta**
- **Auricle of right atrium**
- **Pulmonary trunk**
- **Right ventricle**

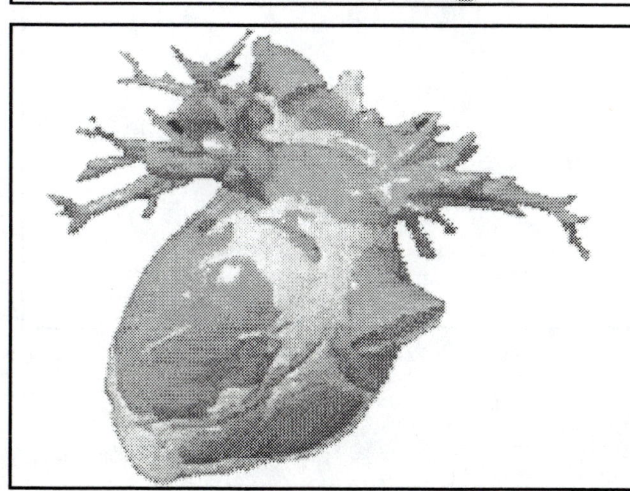

Identify:

- **Coronary sinus**
- **Inferior vena cava**
- **Left atrium**
- **Pulmonary arteries**
- **Pulmonary vein**

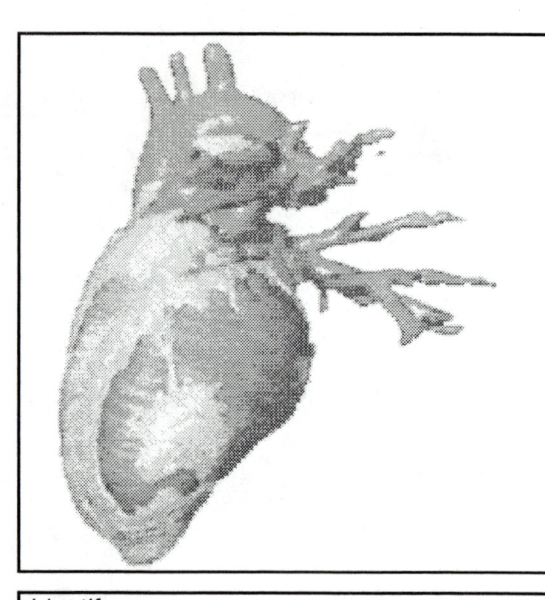

Identify:

- **Auricle of left atrium**
- **Left ventricle**

Identify:
- **Fossa ovalis**

Identify:
- **Mitral valve** (cusp, posterior cusp)

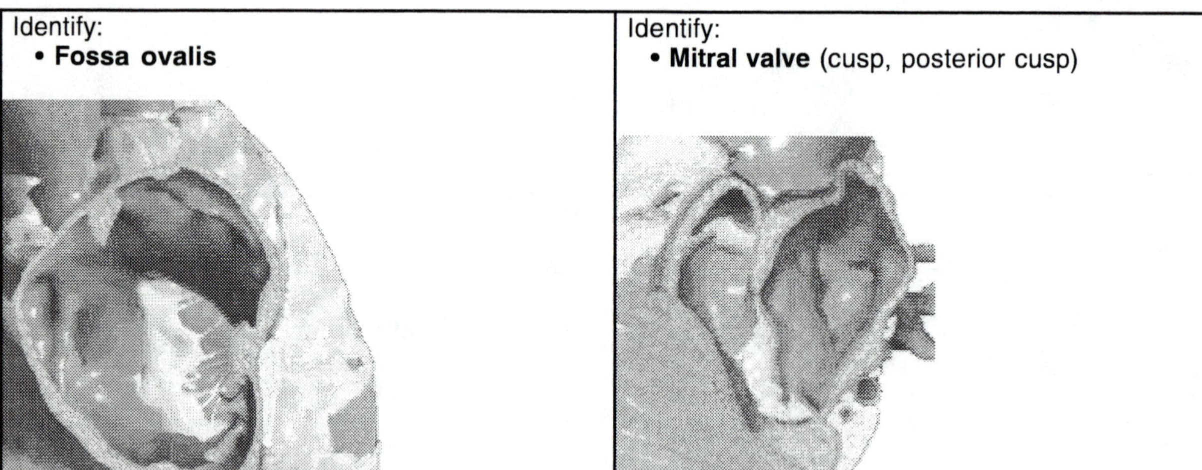

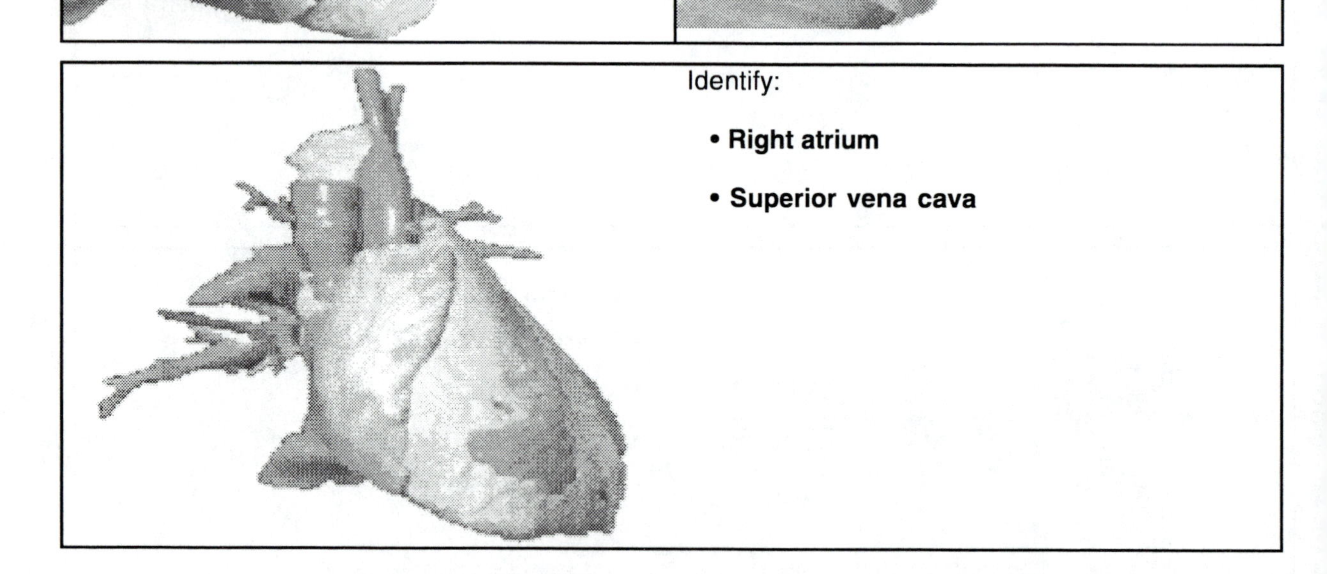

Identify:

- **Right atrium**
- **Superior vena cava**

## Objectives: Heart

To learn the anatomy of the heart and its surrounding structures.

**Blood Vessels**

- Aorta
- Arch of aorta
- Ascending aorta
- Descending aorta
- Brachiocephalic artery
- Brachiocephalic veins
- Cardiac veins
- Common carotid arteries
- Coronary arteries
- Coronary sinus
- Coronary veins
- Inferior vena cava
- Superior vena cava
- Internal jugular veins
- Pulmonary arteries
- Pulmonary trunk
- Pulmonary veins
- Subclavian arteries

**Other**

- Mediastinum
- Coronary sulcus
- Xiphoid process
- Manubrium
- Sternum
- Esophagus
- Trachea
- Thymus gland
- Ligamentum arteriosum
- Diaphragm
- Lungs

**Heart**

- Anterior atrioventricular sulcus
- Interventricular septum
- Chordae tendineae
- Apex of heart
- Auricle of left atrium
- Auricle of right atrium
- Left atrium
- Right atrium
- Left ventricle
- Right ventricle
- Endocardium
- Myocardium
- Epicardium
- Fossa ovalis
- Papillary muscle
- Pericardial sac

**Valves**

- Pulmonary valve
- Tricuspid valve
- Mitral valve

# Arteries and Veins

## Opening A.D.A.M Interactive Anatomy

Open A.D.A.M. Interactive Anatomy according to the directions in the Tutorial.

## Atlas Anatomy of the Arteries and Veins of the Head and Neck

Under the **File** menu, drag down to **Open** (then to **Content** if you are working in Windows). Click on **Atlas Anatomy**. Select **System** and **Cardiovascular**. Choose **Arteries of Head and Neck**. If the list is not in alphabetical order, the image is a little way down from the top of the list.

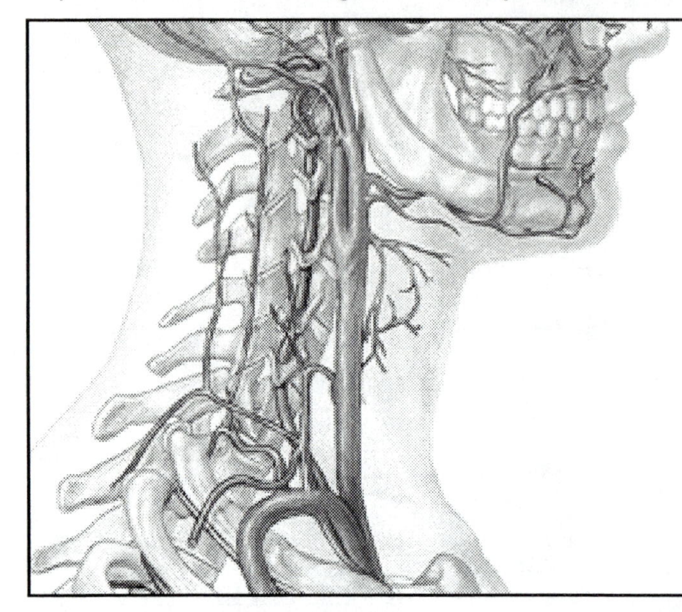

Identify the following by clicking on the heads of the pins:

- **Common carotid artery**
- **External carotid artery**
- **Internal carotid artery**
- **Subclavian artery**
- **Vertebral artery**

When you are finished, **close** the window.

What foramen does the vertebral artery go through?

---

Under the **File** menu, drag down to **Open** (then to **Content** if you are working in Windows). Click on **Atlas Anatomy**. **System** and **Cardiovascular** should still be selected. Choose **Deep Veins of Head (Lat)**. If the list is not in alphabetical order, the image is about one-third of the way down the list.

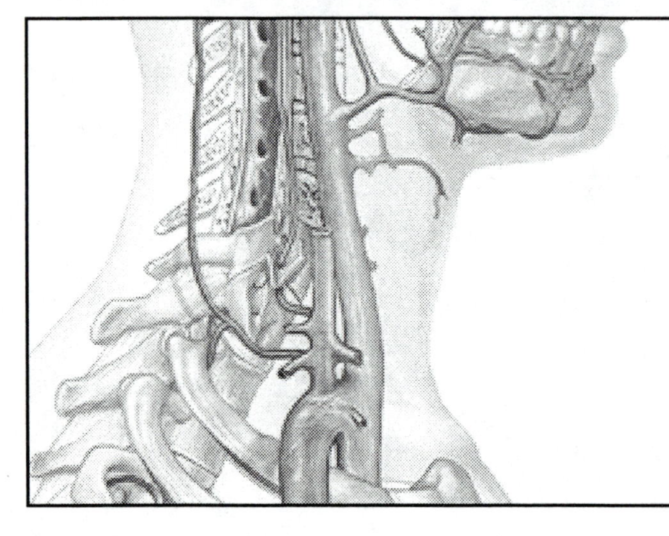

Identify :

- **External jugular vein**
- **Internal jugular vein**
- **Right brachiocephalic vein**
- **Right subclavian vein**

When you are finished, **close** the window.

# Anterior Dissection of the Arteries and Veins of the Head and Neck

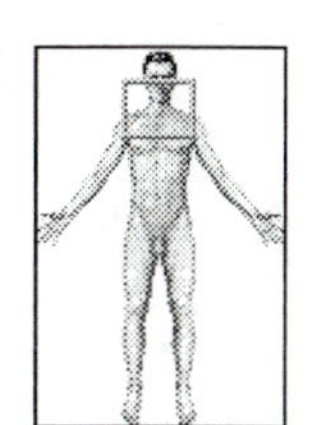

Under the **File** menu, drag down to **Open** (then to **Content** if you are working in Windows).

Choose **Dissectible Anatomy**.

Choose either male or female. Select **Anterior.** Click **Open**.

Enlarge the window and position the image over the neck and upper chest area.

Click several times on the **Depth Bar** ▼ to remove the skin and fascia until you reach **layer 30.** No graphic has been provided for this layer. Identify the **external jugular vein**.

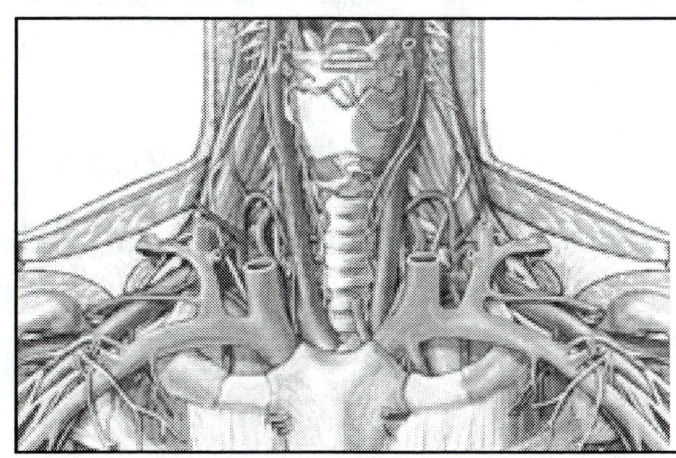

Drag down on the **Depth Bar** ▼ until you reach **layer 79**. Identify:

- **Internal jugular vein**
- **External jugular vein**
- **Brachiocephalic vein**
- **Subclavian vein**
- **Axillary vein**
- **Common carotid arteries**
- **Internal carotid artery**
- **External carotid artery**

When you are finished, **close** the window.

Which is larger in diameter, the external or internal jugular vein?

______________________________________________

## Atlas Anatomy of the Arteries and Veins of the Arm

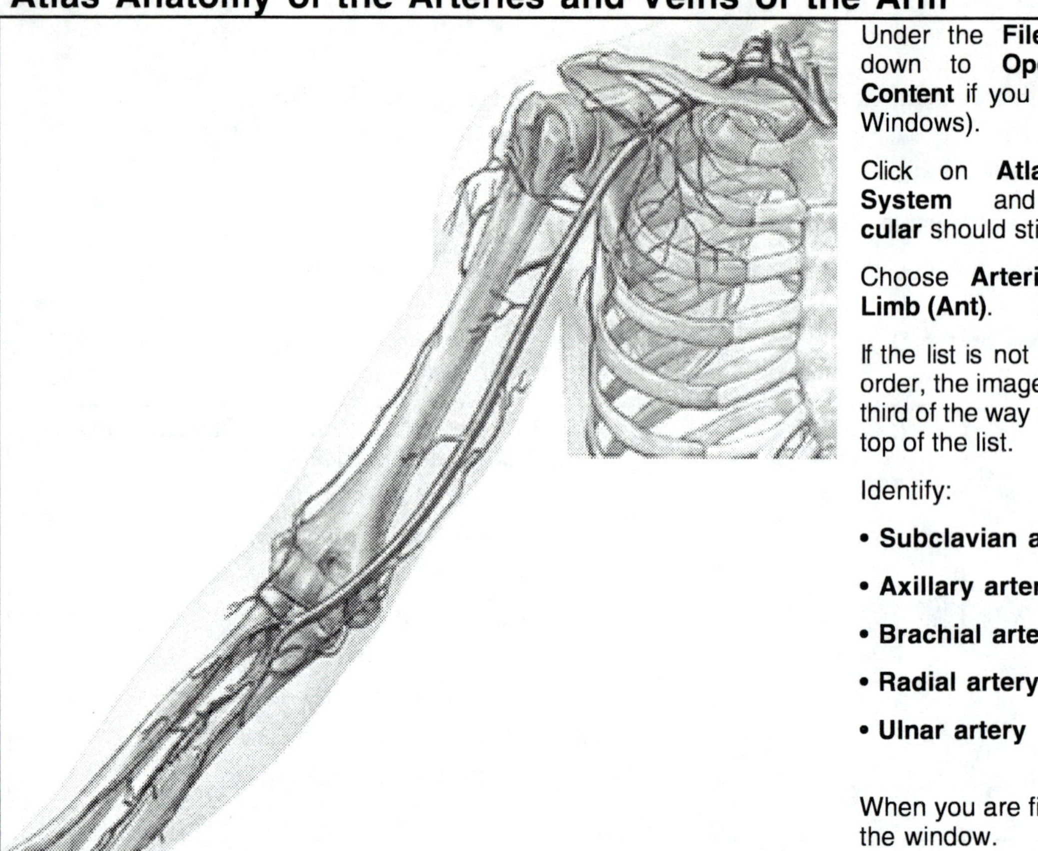

Under the **File** menu, drag down to **Open** (then to **Content** if you are working in Windows).

Click on **Atlas Anatomy**. **System** and **Cardiovascular** should still be selected.

Choose **Arteries of Upper Limb (Ant)**.

If the list is not in alphabetical order, the image is about one-third of the way down from the top of the list.

Identify:

- **Subclavian artery**
- **Axillary artery**
- **Brachial artery**
- **Radial artery**
- **Ulnar artery**

When you are finished, **close** the window.

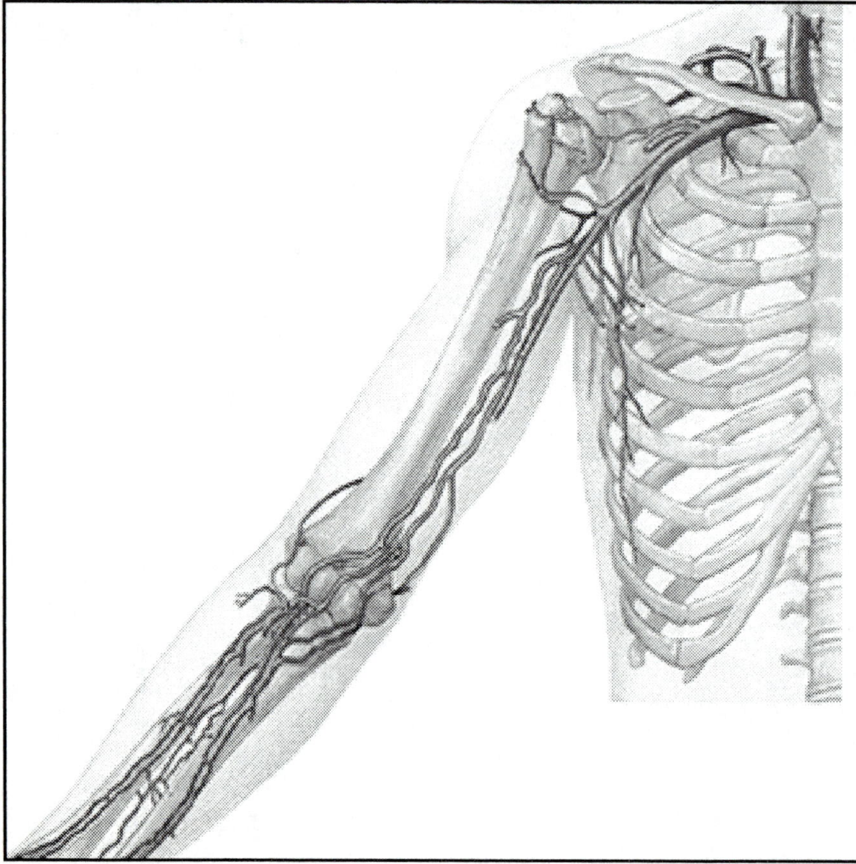

Under the **File** menu, drag down to **Open** (then to **Content** if you are working in Windows).

Click on **Atlas Anatomy**. **System** and **Cardiovascular** should still be selected.

Choose **Deep Veins of Upper Limb (Ant).**

If the list is not in alphabetical order, the image is about one-third of the way down from the top of the list.

Identify:

- **Axillary vein**
- **Basilic vein**
- **Brachial vein**
- **External jugular vein**
- **Internal jugular vein**
- **Radial vein**
- **Ulnar vein**

When you are finished, **close** the window.

## Anterior Dissection of the Arteries and Veins of the Arm

Under the **File** menu, drag down to **Open** (then to **Content** if you are working in Windows).

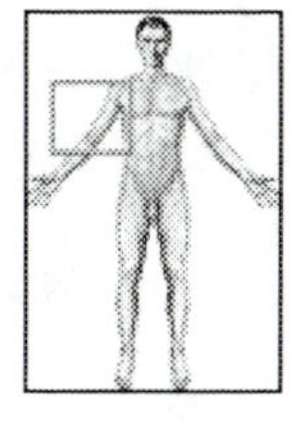

Choose **Dissectible Anatomy**.

Choose either male or female. Select **Anterior.** Click **Open**.

Drag the depth bar to **layer 0.**

Enlarge the window and position the image over the neck and upper arm area.

Click on the **Depth Bar** ▼ to **layer 5.** Identify:

- **Cephalic vein**
- **Median cubital vein**
- **Basilic vein**

Which vein is typically used for venipuncture?

---

Is the cephalic vein superficial or deep?

---

Is the basilic vein superficial or deep?

---

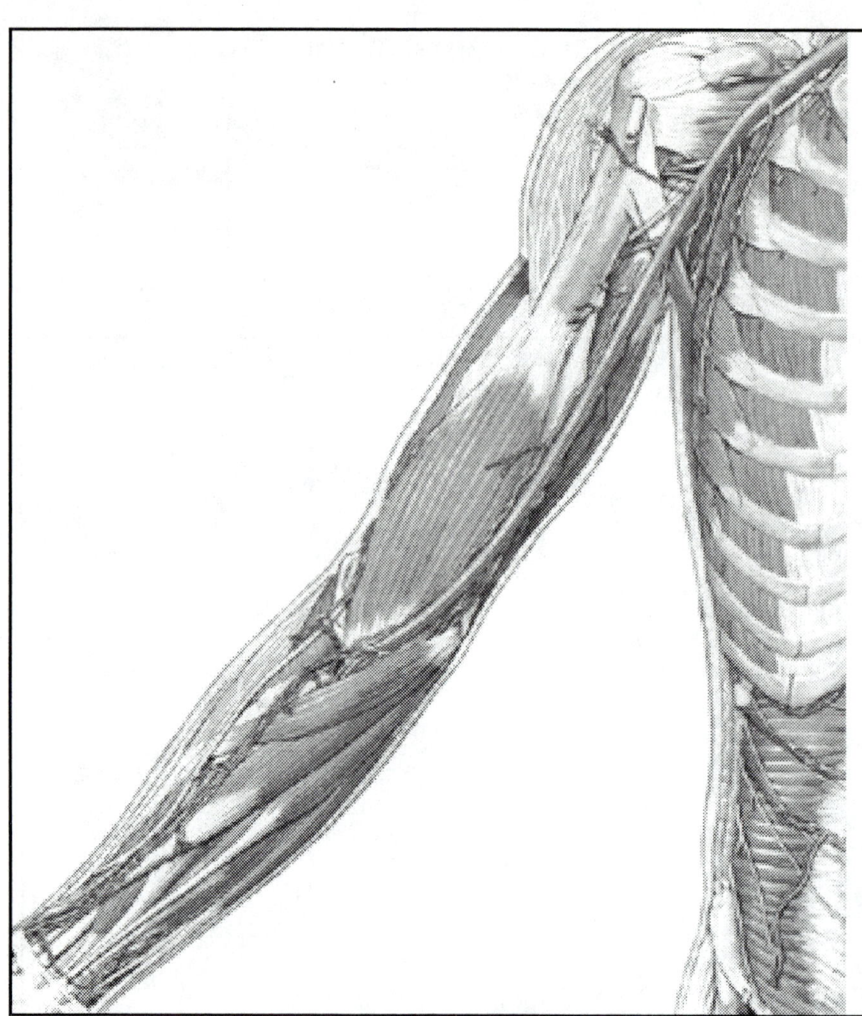

Click on the **Depth Bar** [▼] to **layer 88**. You will need to scroll to see all of these structures. Identify:

- **Axillary artery**
- **Brachial artery**
- **Brachial vein**
- **Radial artery**
- **Radial vein**
- **Ulnar artery**
- **Ulnar vein**

Is the radial vein deep or superficial?

______________________________

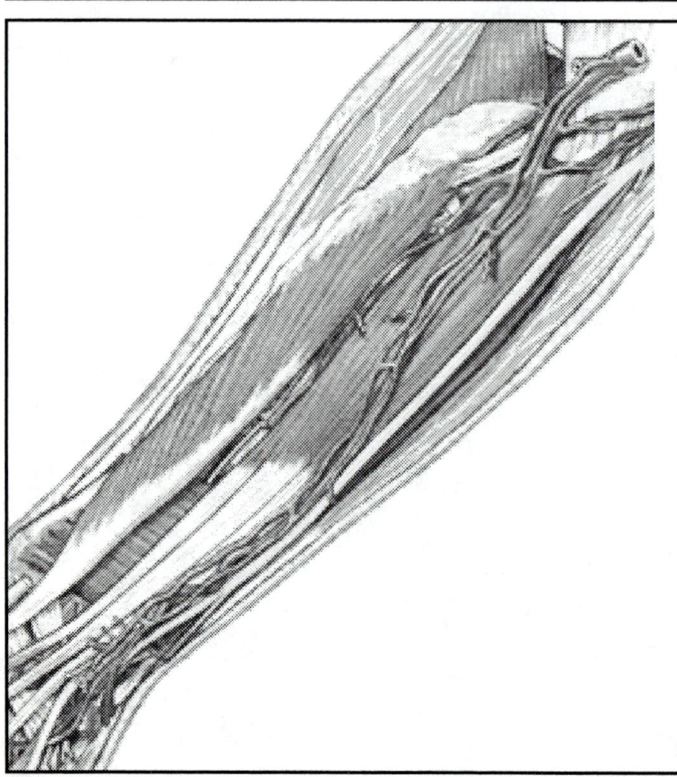

Position the image over the lower arm. Click on the **Depth Bar** [▼] to **layer 122**. Identify:

- **Ulnar artery**
- **Ulnar vein**

Is the ulnar vein deep or superficial?

______________________________

When you are finished, **close** the window.

## Atlas Anatomy of the Arteries and Veins of the Leg

Under the **File** menu, drag down to **Open** (then to **Content** if you are working in Windows). Click on **Atlas Anatomy**. **System** and **Cardiovascular** should still be selected. Choose **Arteries of Lower Limb (Ant)**. If the list is not in alphabetical order, the image is close to the bottom of the list.

Under the **File** menu, drag down to **Open** (then to **Content** if you are working in Windows). Click on **Atlas Anatomy**. **System** and **Cardiovascular** should still be selected. Choose **Deep Veins of Lower Limb (Ant)**. If the list is not in alphabetical order, the image is close to the bottom of the list.

Position the anterior and posterior images so they can both be seen at the same time on the screen. Do this by dragging the gray bar above the images:

If you are working in Windows, click Window, click Tile, and center the images in their respective windows.

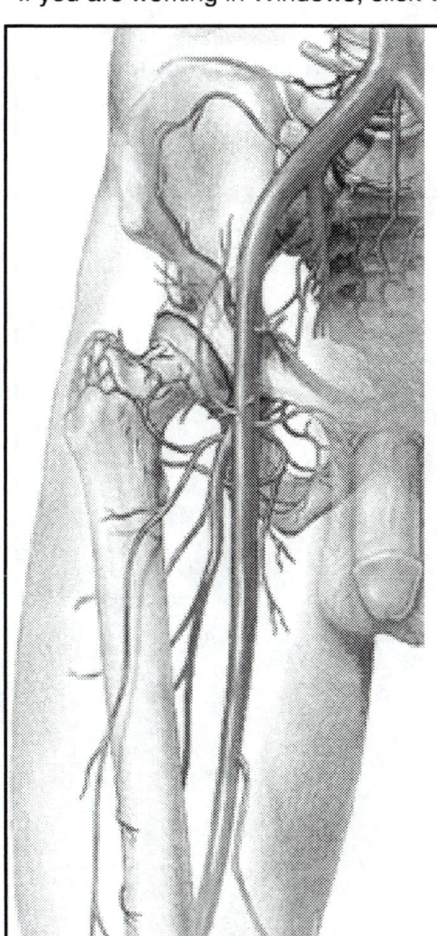

Identify the arteries to the left:

- **Common iliac artery**
- **External iliac artery**
- **Internal iliac artery**
- **Femoral artery**

Identify the veins to the right:

- **Common iliac vein**
- **External iliac vein**
- **Femoral vein**
- **Inferior vena cava**
- **Great saphenous vein**

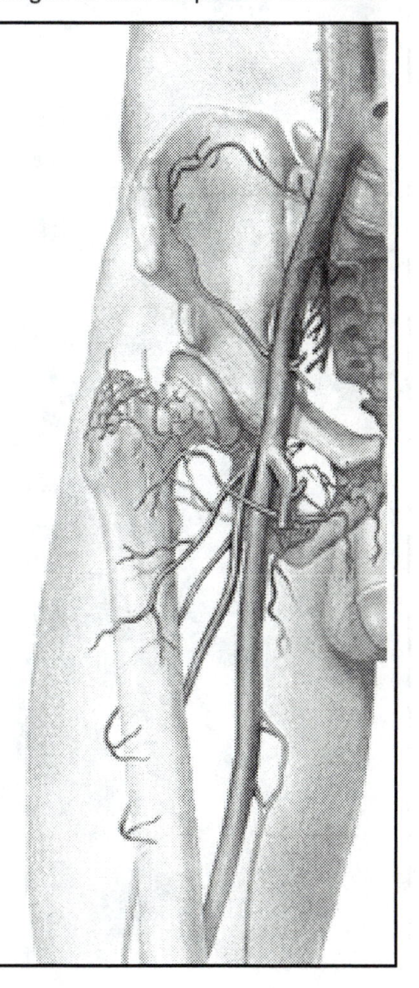

Is the femoral vein deep or superficial?

---

When you are finished, **close** the windows.

## Medial Dissection of the Arteries and Veins of the Leg

Under the **File** menu, drag down to **Open** (then to **Content** if you are working in Windows).

Choose **Dissectible Anatomy**.

Choose either male or female. Select **Medial.** Click **Open**.

Enlarge the window and position the image over the upper leg area.

Drag the **Depth Bar** ▼ to **layer 0**.

Click on the **Normal** button.

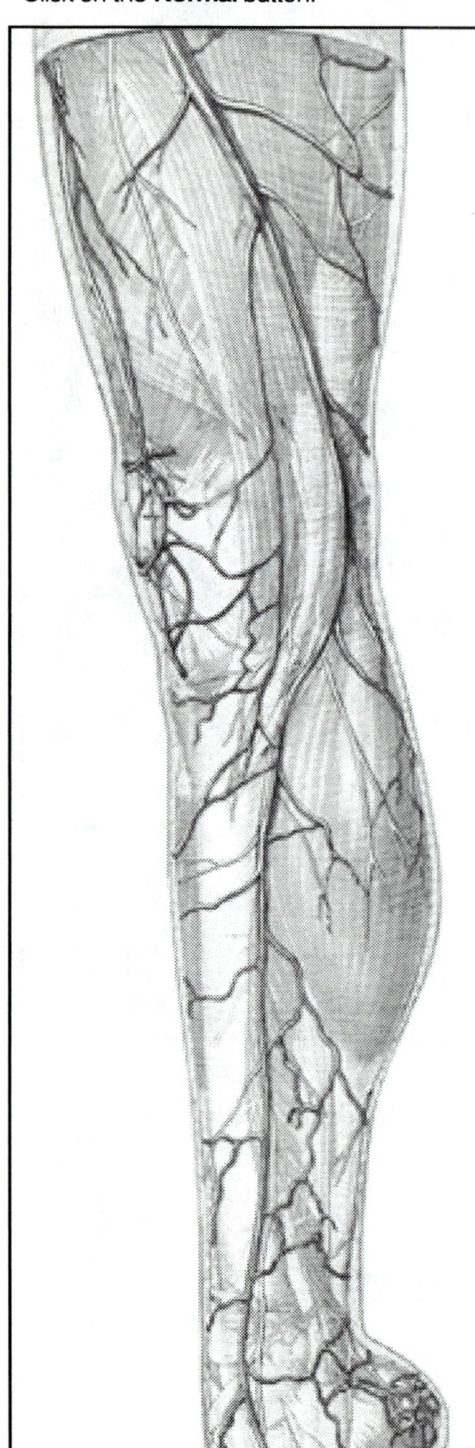

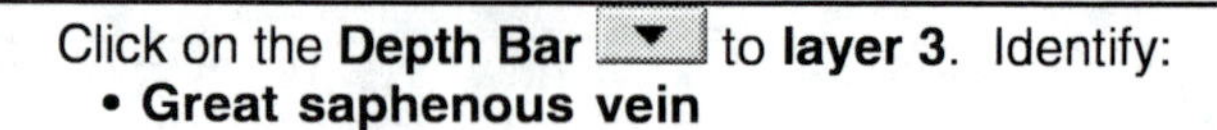

Click on the **Depth Bar** ▼ to **layer 3**. Identify:

- **Great saphenous vein**

Follow the great saphenous vein all the way down the leg. This is the vein that is often used for coronary bypass surgery.

## Atlas Anatomy of the Arteries and Veins of the Trunk

Under the **File** menu, drag down to **Open** (then to **Content** if you are working in Windows). Click on **Atlas Anatomy**. **System** and **Cardiovascular** should still be selected. Choose **Arteries of Trunk (Ant)**. If the list is not in alphabetical order, the image is almost half of the way down the list.

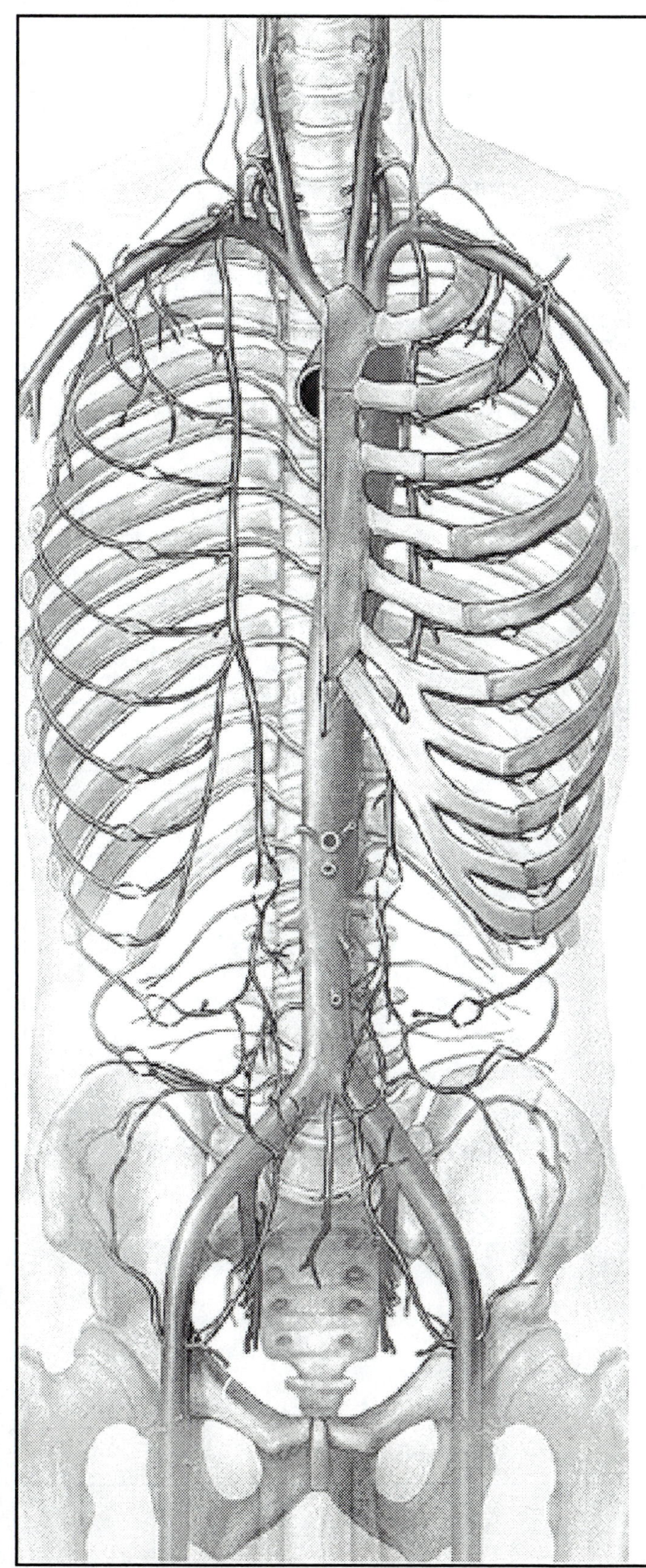

Identify :

- **Abdominal aorta**
- **Arch of aorta**
- **Ascending aorta**
- **Axillary artery**
- **Brachiocephalic trunk**
- **Celiac trunk**
- **External iliac artery**
- **Femoral artery**
- **Inferior mesenteric artery**
- **Internal iliac artery**
- **Common carotid artery**
- **Common iliac artery**
- **Renal artery**
- **Subclavian artery**
- **Vertebral artery**
- **Superior mesenteric artery**
- **Thoracic aorta**

Trace the flow of blood from the ascending aorta to the femoral artery.

________________________________

________________________________

________________________________

________________________________

________________________________

When you are finished, **close** the window.

Under the **File** menu, drag down to **Open** (then to **Content** if you are working in Windows). Click on **Atlas Anatomy**. **System** and **Cardiovascular** should still be selected. Choose **Parietal Veins of Trunk (Ant)**. If the list is not in alphabetical order, the image is almost half of the way down the list.

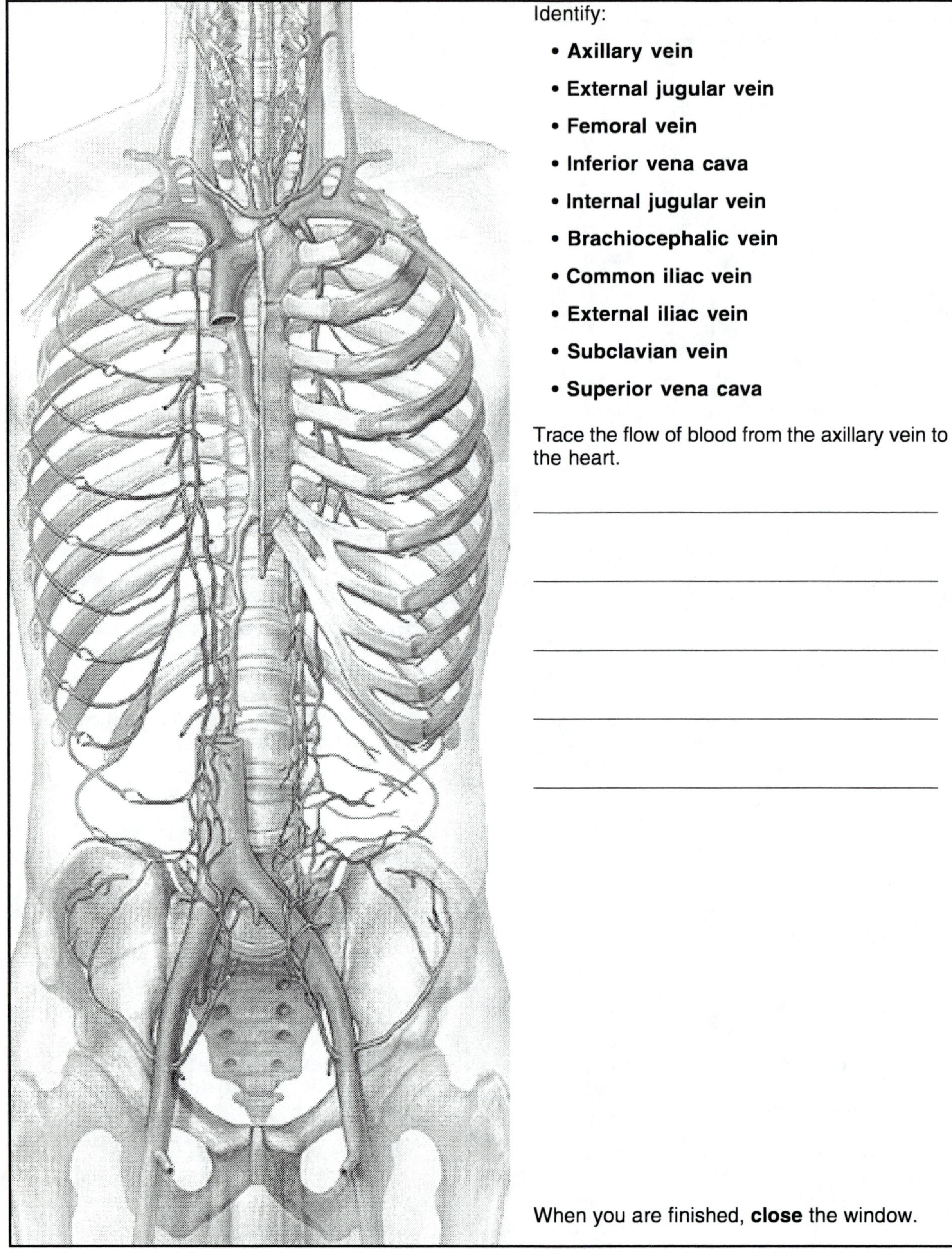

Identify:

- **Axillary vein**
- **External jugular vein**
- **Femoral vein**
- **Inferior vena cava**
- **Internal jugular vein**
- **Brachiocephalic vein**
- **Common iliac vein**
- **External iliac vein**
- **Subclavian vein**
- **Superior vena cava**

Trace the flow of blood from the axillary vein to the heart.

When you are finished, **close** the window.

Under the **File** menu, drag down to **Open** (then to **Content** if you are working in Windows). Click on **Atlas Anatomy**. **System** and **Cardiovascular** should still be selected. Choose **Mesenteric Veins**. If the list is not in alphabetical order, the image is about two-thirds of the way down the list.

Identify:

- **Celiac trunk**
- **External iliac vein**
- **Femoral artery**
- **Femoral vein**
- **Inferior mesenteric vein**
- **Inferior vena cava**
- **Common iliac vein**
- **Portal vein**
- **Splenic vein**
- **Superior mesenteric artery**
- **Superior mesenteric vein**

What three veins drain into the portal vein?

_______________________________________

_______________________________________

_______________________________________

When you are finished, **close** the window.

Under the **File** menu, drag down to **Open** (then to **Content** if you are working in Windows). Click on **Atlas Anatomy**. **System** and **Cardiovascular** should still be selected. Choose **Superior Mes-enteric Artery 1**. If the list is not in alphabetical order, the image is about two thirds of the way down.

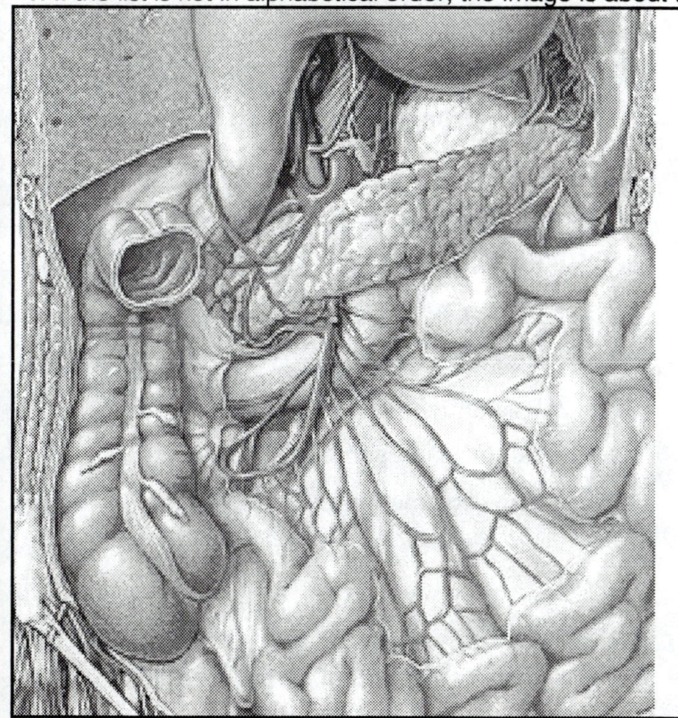

Identify:

- **Celiac trunk**
- **Common hepatic artery**
- **Portal vein**
- **Proper hepatic artery**
- **Splenic artery**
- **Splenic vein**
- **Superior mesenteric artery**

When you are finished, **close** the window.

Under the **File** menu, drag down to **Open** (then to **Content** if you are working in Windows). Click on **Atlas Anatomy**. **System** and **Cardiovascular** should still be selected. Choose **Blood Supply to Stomach**. If the list is not in alphabetical order, the image is about two thirds of the way down the list.

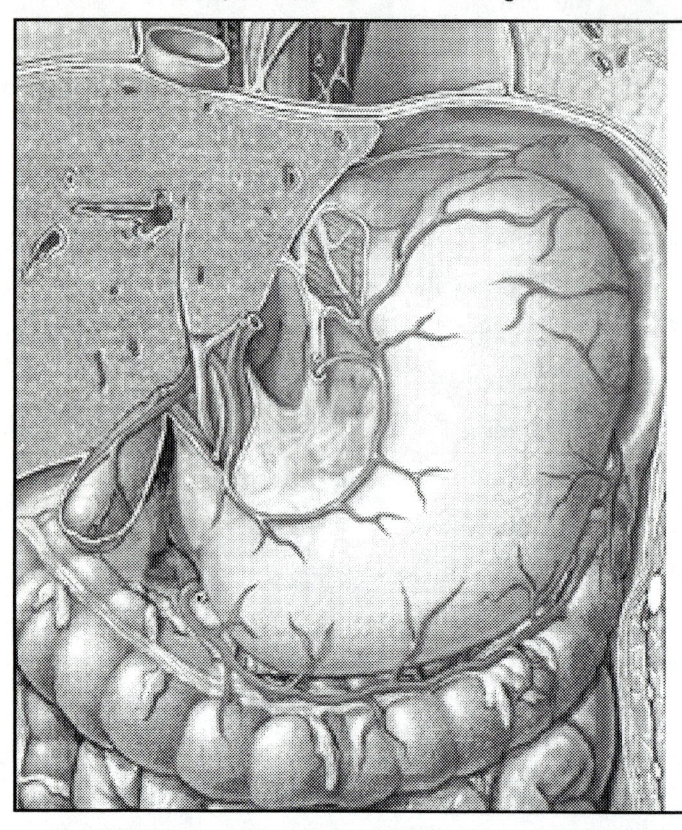

Identify:

- **Left gastric artery**
- **Inferior vena cava**
- **Portal vein**
- **Proper hepatic artery**
- **Thoracic aorta**

When you are finished, **close** the window.

Under the **File** menu, drag down to **Open** (then to **Content** if you are working in Windows). Click on **Atlas Anatomy**. **System** and **Cardiovascular** should still be selected. Choose **Blood Supply to Pancreas/Spleen**. If the list is not in alphabetical order, the image is a little over half way down the list.

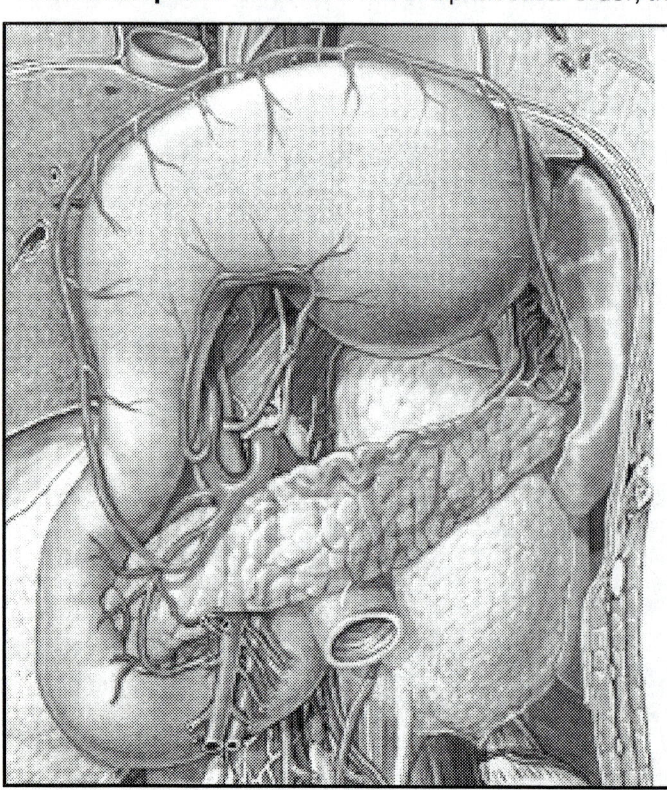

Identify:

- **Celiac trunk**
- **Common hepatic artery**
- **Inferior mesenteric vein**
- **Inferior vena cava**
- **Left gastric artery**
- **Portal vein**
- **Proper hepatic artery**
- **Splenic artery**
- **Splenic vein**
- **Superior mesenteric artery**
- **Superior mesenteric vein**

When you are finished, **close** the window.

## Anterior Dissection of the Arteries and Veins of the Trunk

Under the **File** menu, drag down to **Open** (then to **Content** if you are working in Windows).

Choose **Dissectible Anatomy**.

Choose either male or female. Select **Anterior.** Click **Open**.

Enlarge the window and position the image over the abdomen.

Drag the **Depth Bar** to **layer 0**.

Click on the **Normal** button.

Scroll down to **layer 201**. Identify the **left gastric artery**. No graphic has been provided for this layer.

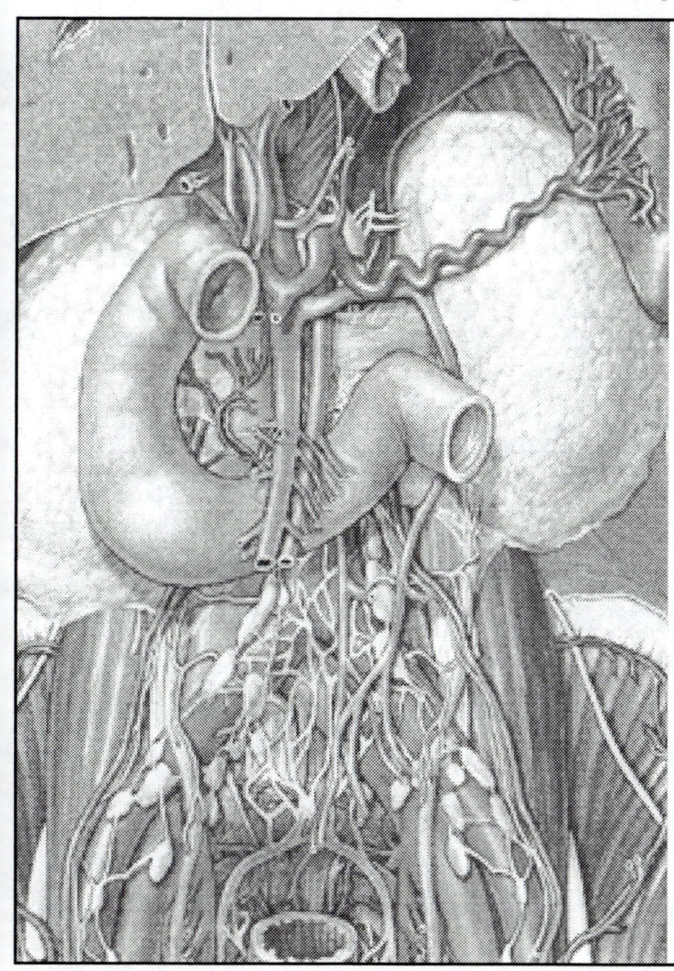

Click on the **Depth Bar** until you reach **layer 216**. Identify:

- **Inferior vena cava**
- **Common hepatic artery**
- **Inferior mesenteric vein**
- **Inferior mesenteric artery**
- **Portal vein**
- **Proper hepatic artery**
- **Splenic artery**
- **Splenic vein**
- **Superior mesenteric artery**
- **Superior mesenteric vein**

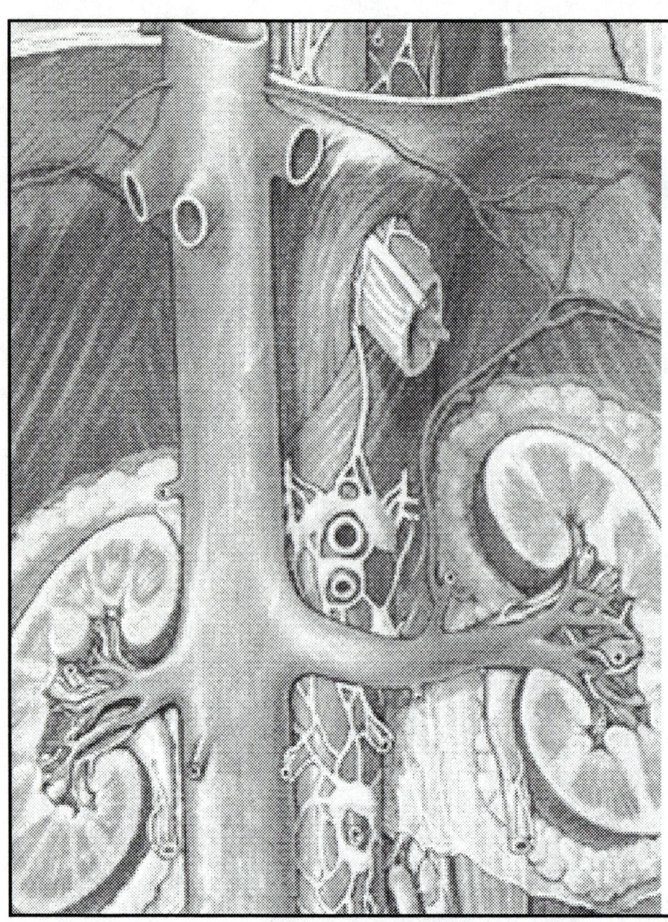

Click on the **Depth Bar** ▼ until you reach **layer 238**. Identify:

- **Inferior vena cava**
- **Hepatic veins**
- **Renal veins**
- **Celiac trunk**
- **Abdominal aorta**

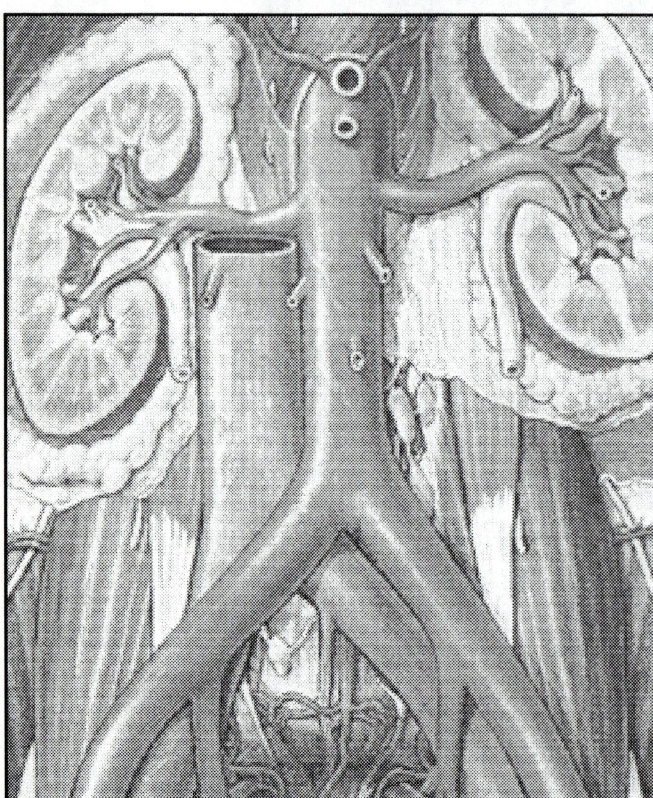

Click on the **Depth Bar** ▼ until you reach **layer 240**. Identify:

- **Inferior vena cava**
- **Common iliac veins**
- **External iliac vein**
- **Abdominal aorta**
- **Celiac trunk**
- **Superior mesenteric artery**
- **Inferior mesenteric artery**
- **Renal arteries**
- **Common Iliac arteries**
- **External iliac artery**
- **Internal Iliac artery**

## Cadaver Images of the Arteries and Veins

Under the **File** menu, drag down to **Open** (then to **Content** if you are working in Windows). Click on **Atlas Anatomy**. **System** and **Cardiovascular** should still be selected. Choose **Vessels in Left Mediastinum**. If the list is not in alphabetical order, the image is about half way down the list.

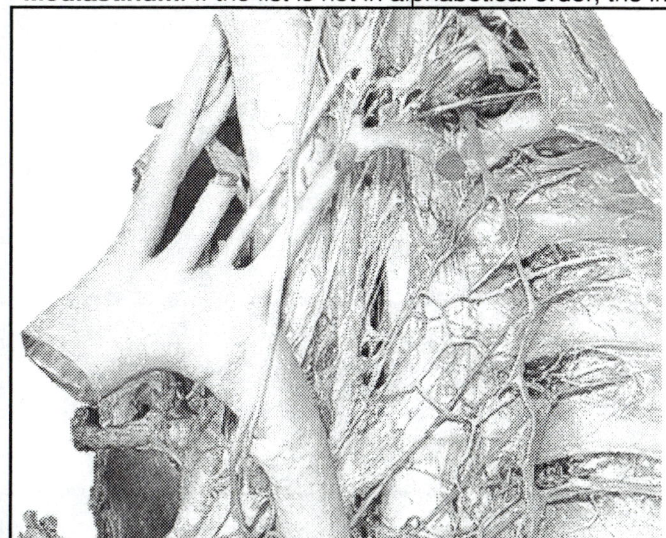

Identify:

- **Arch of aorta**
- **Ascending aorta**
- **Brachiocephalic trunk**
- **Left common carotid artery**
- **Right common carotid artery**
- **Left subclavian artery**
- **Right subclavian artery**
- **Thoracic aorta**

When you are finished, **close** the window.

Under the **File** menu, drag down to **Open** (then to **Content** if you are working in Windows). Click on **Atlas Anatomy**. **System** and **Cardiovascular** should still be selected. Choose **Dissection of Kidneys (Ant)**. If the list is not in alphabetical order, the image is about three fourths of the way down.

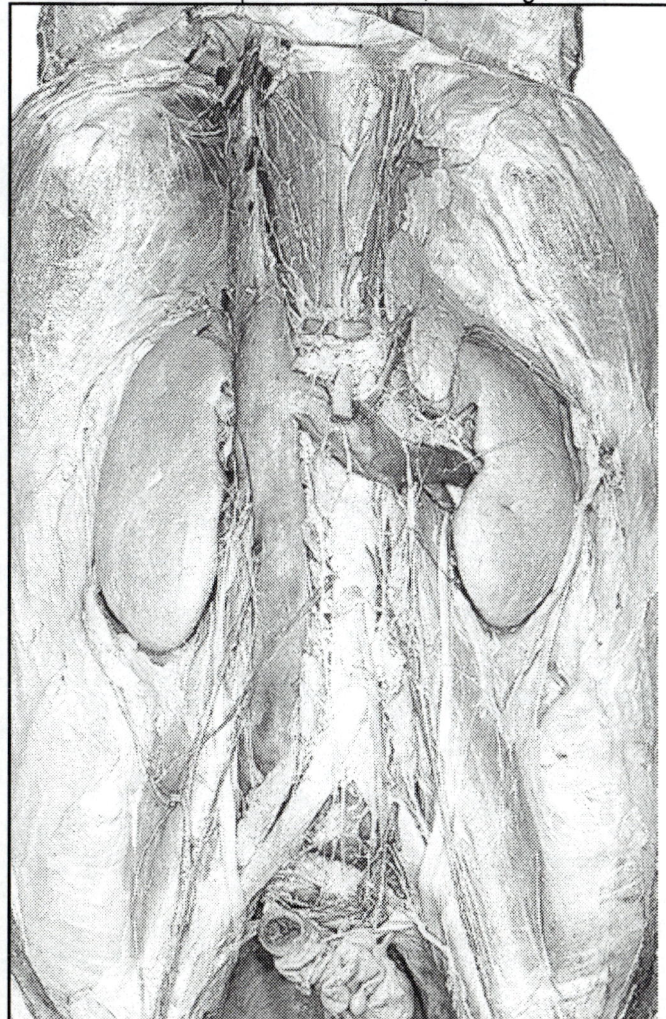

If you are working in Mac, click on the **Show All Pins** button and scroll to **All Systems**. Identify:

- **Abdominal aorta**
- **Inferior mesenteric artery**
- **Inferior vena cava**
- **Common iliac artery**
- **Left & right renal veins**
- **Superior mesenteric artery**
- **Left and right kidney**
- **Left and right ureter**
- **Diaphragm**

When you are finished, **close** the window.

## Objectives: Arteries and Veins

To learn the major arteries and veins of the body.

**Arteries**

- Common carotid arteries
- External carotid arteries
- Internal carotid arteries
- Subclavian arteries
- Vertebral arteries
- Axillary arteries
- Brachial arteries
- Radial arteries
- Ulnar arteries
- Common iliac arteries
- External iliac arteries
- Internal iliac arteries
- Femoral arteries
- Arch of aorta
- Abdominal aorta
- Ascending aorta
- Thoracic aorta
- Brachiocephalic trunk
- Celiac trunk artery
- Inferior mesenteric arteries
- Superior mesenteric arteries
- Renal arteries
- Common hepatic arteries
- Proper hepatic arteries
- Splenic artery
- Gastric artery

**Veins**

- External jugular veins
- Internal jugular veins
- Brachiocephalic veins
- Subclavian veins
- Axillary veins
- Basilic veins
- Brachial veins
- Radial veins
- Ulnar veins
- Cephalic veins
- Median cubital vein
- Common iliac veins
- External iliac veins
- Femoral veins
- Inferior vena cava
- Superior vena cava
- Great saphenous vein
- Portal veins
- Splenic veins
- Inferior mesenteric veins
- Superior mesenteric veins
- Renal veins
- Hepatic veins

**Other**

- Diaphragm
- Kidneys
- Ureter

# Lymphatic System

## Opening A.D.A.M Interactive Anatomy

Open A.D.A.M. Interactive Anatomy according to the directions in the Tutorial.

## Anterior Dissection of Lymphatic System

Choose **Dissectible Anatomy**.

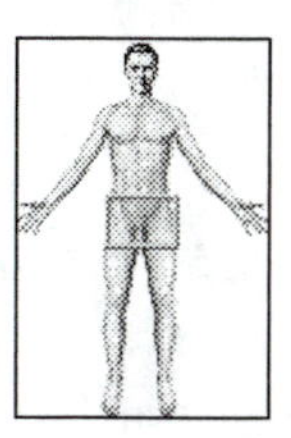

Choose either male or female. Select **Anterior.** Click **Open**.

Enlarge the window and position the navigator rectangle over the groin.

Click on the **Depth Bar** ▼ to **layer 3**. No graphic has been provided for this layer. Identify the **lymph nodes**.

Position the navigator rectangle over the head and neck area.

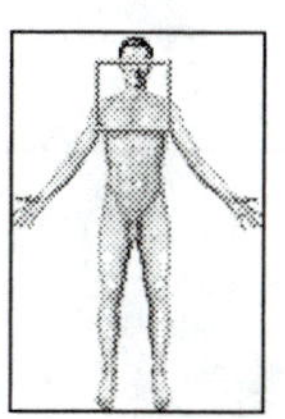

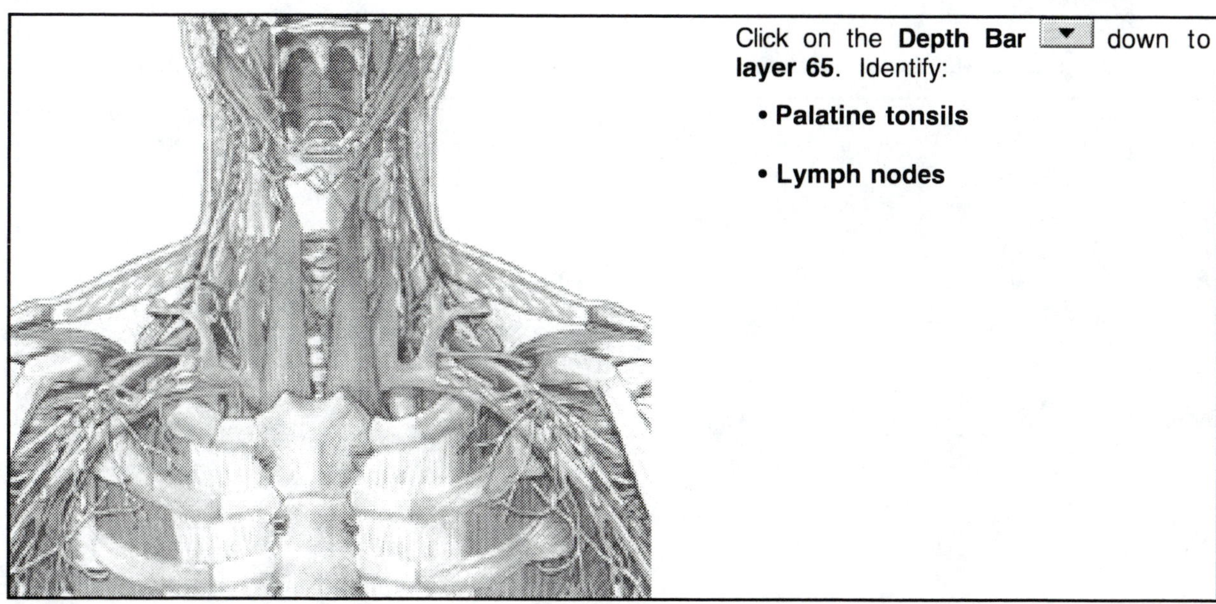

Click on the **Depth Bar** ▼ down to **layer 65**. Identify:

- **Palatine tonsils**
- **Lymph nodes**

What is one function of the lymph nodes?

_______________________________________________

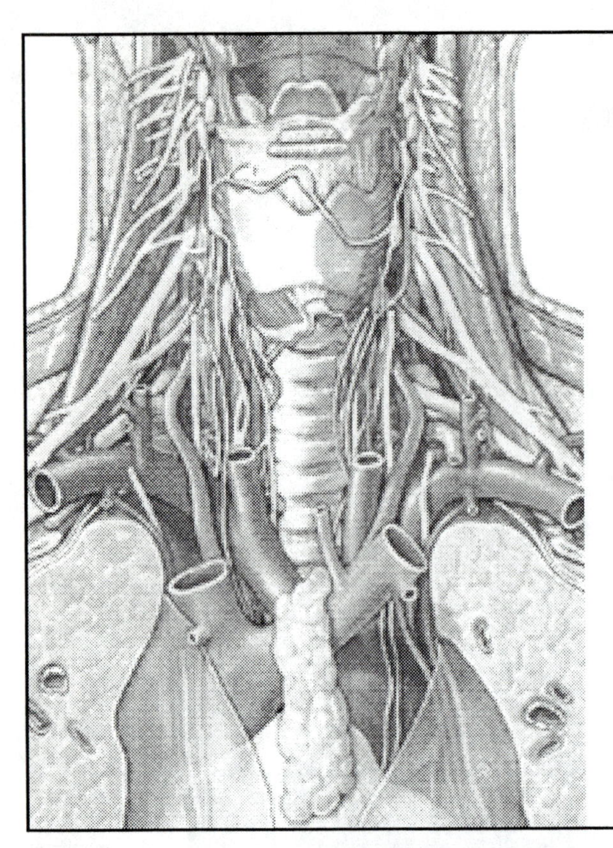

Click on the **Depth Bar** ▼ until you reach **layer 164**. Identify:

- **Thymus gland**
- **Lymph nodes**
- **Thoracic duct**

What is one function of the thymus gland?

________________________________________

________________________________________

Move the image to the abdominal area. Click several times on the **Depth Bar** ▼ until you reach **layer 219**. Identify:

- **Spleen**
- **Lymph nodes in abdomen**

What is one function of the spleen?

________________________________________

________________________________________

Click several times on the **Depth Bar** ▼ to **layer 237**. No graphic has been provided for this layer. Identify:

- **Lymph nodes in abdomen**

Click on the **Depth Bar** ▼ until you reach **layer 251**. No graphic has been provided for this layer. Identify:

- **Lymph nodes in abdomen**

Scroll up to the thorax area. You are still at **layer 251**. Identify:

- **Lymph nodes in thorax**

- **Thoracic duct**

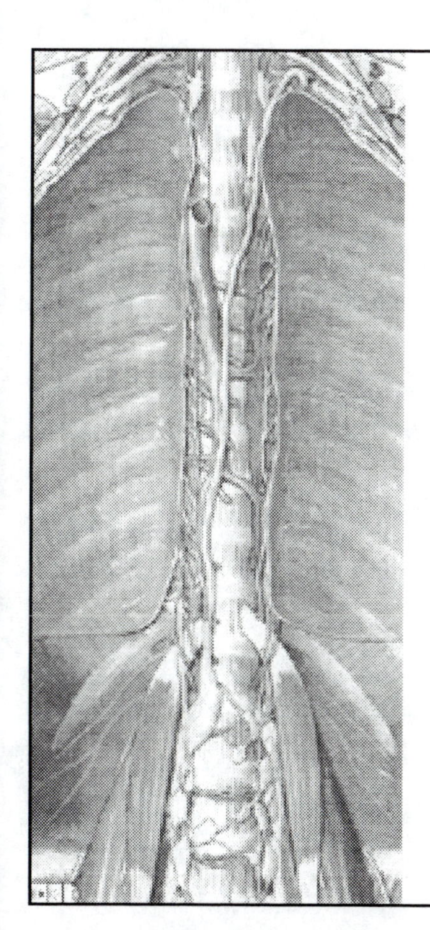

Click on the **Depth Bar** ▼ until you reach **layer 262**. Identify:

- **Thoracic duct**

- **Lymph nodes in abdomen**

- **Chyle cistern (Cisterna chyli)**

Scroll to **layer 330**. No graphic has been provided for this layer. Identify:

- **Bone Marrow**

What is one function of the bone marrow?

_______________________________________________

## Medial Dissection of the Lymphatic System

Click on the **View** button and choose **Medial**.

Click on the **Normal** button. Windows— —Mac

Drag the **Depth Bar** to **layer 0.**

Position the navigator rectangle over the lower head and chest area.

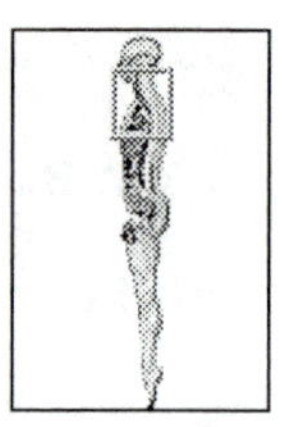

At **layer 0**, identify:

- **Pharyngeal tonsils**
- **Lingual tonsil**
- **Lymph nodes**
- **Thymus**
- **Thoracic duct**

What is one function of the tonsils?

______________________________

______________________________

## Lateral Dissection of the Lymphatic System

Click on the **View** button and choose **Lateral**.

Click on the **Normal** button. Windows— —Mac

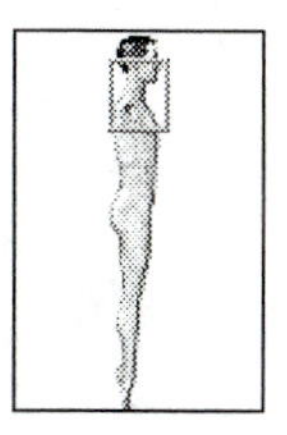

Drag the **Depth Bar** to **layer 0.**

Position the navigator rectangle over the face and upper chest area.

Click on the **Depth Bar** until you reach **layer 21**. No graphic has been provided for this layer. Identify:

- **Lymph nodes in face, neck and axillary area**

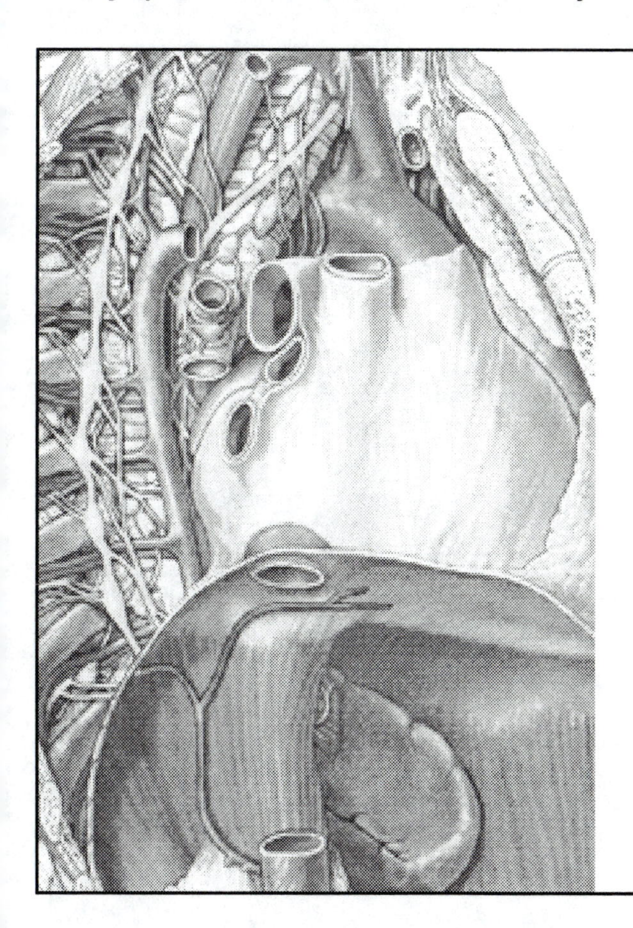

Scroll down to the chest and upper abdomen. Click on the **Depth Bar** until you reach **layer 174**. Identify:

- **Lymph nodes**
- **Spleen**
- **Thymus**

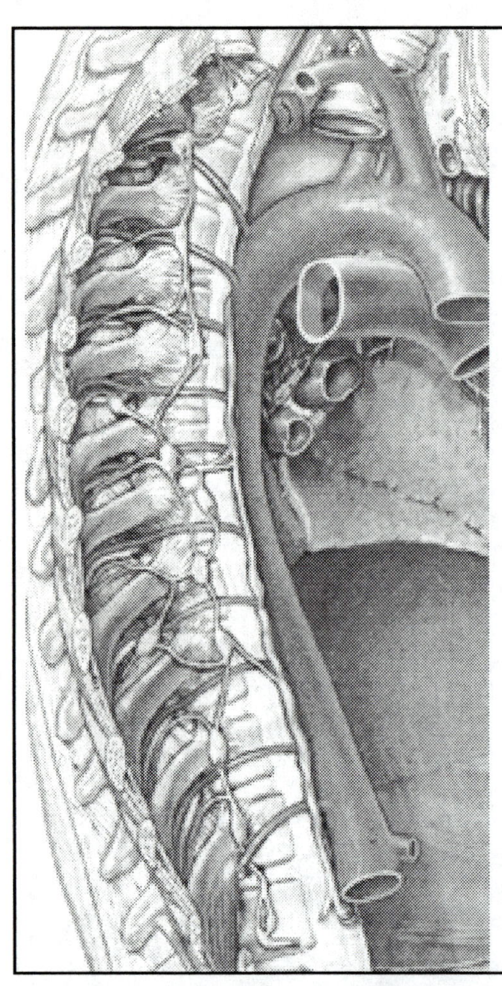

Click on the **Depth Bar** until you reach **layer 228**. Identify:

- **Lymph nodes**
- **Thoracic duct**

What is the function of the thoracic duct?

______________________________________________

______________________________________________

Scroll up to the face area. Click on the **Depth Bar** until you reach **layer 257**. Identify:

- **Pharyngeal tonsils**
- **Lingual tonsils**

When you are finished, **close** dissectible anatomy.

## Atlas Anatomy of the Lymphatic System

Under the **File** menu, drag down to **Open** (then to **Content** if you are working in Windows). Click on **Atlas Anatomy**. Select **System** and **Lymphatic**. Choose **Surface of Tongue (Dorsal)**. If the list is not in alphabetical order, the image is close to the top of the list.

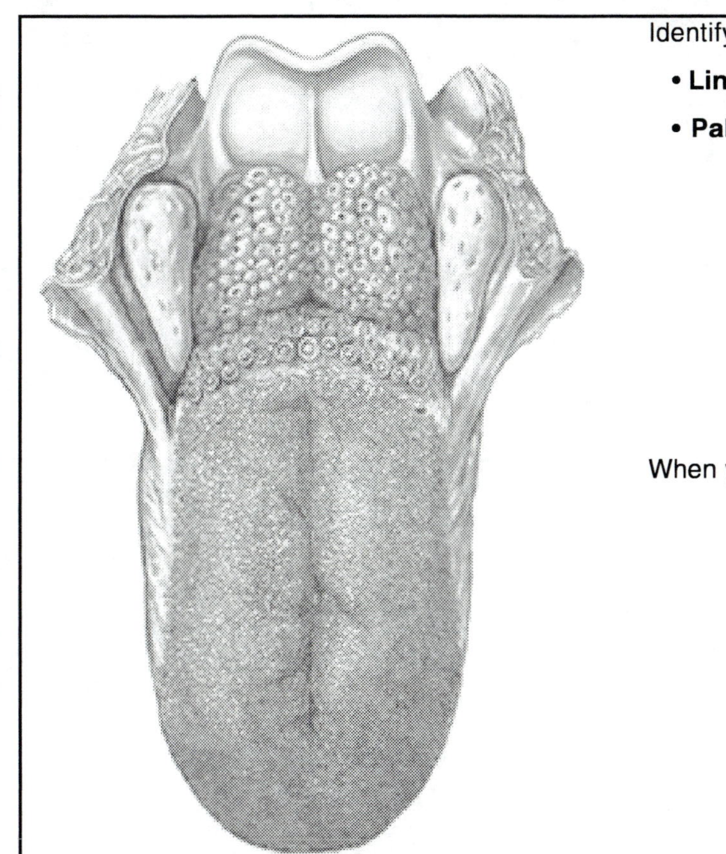

Identify:

- **Lingual tonsils**
- **Palatine tonsils**

When you are finished, **close** the window.

## Objectives: Lymphatic System

To explore the structures of the lymphatic system.

**Tonsils**

- Palatine tonsils
- Pharyngeal tonsils
- Lingual tonsils

**Other**

- Lymph nodes
- Thymus gland
- Thoracic duct
- Spleen
- Bone marrow

# Nose, Pharynx & Larynx

## Opening A.D.A.M Interactive Anatomy

Open A.D.A.M. Interactive Anatomy according to the directions in the Tutorial.

## 3D Anatomy of the Larynx

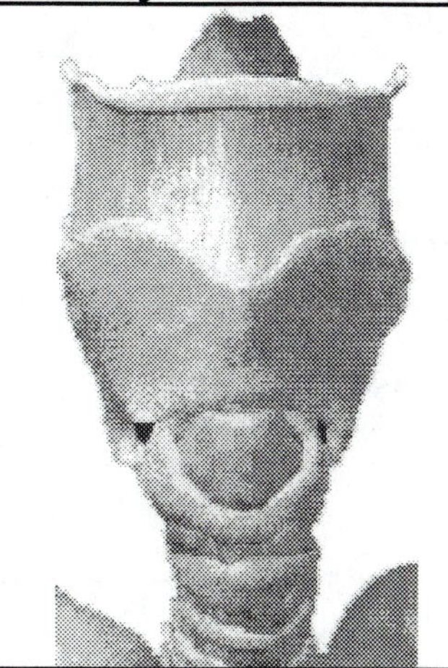

Choose **3D Anatomy**. Click on **3D Lungs**. Identify the following by clicking on the **3D Anatomy Structure List**:

- **Cricoid cartilage**
- **Hyoid bone**
- **Thyroid cartilage**

Identify:
- **Epiglottis**

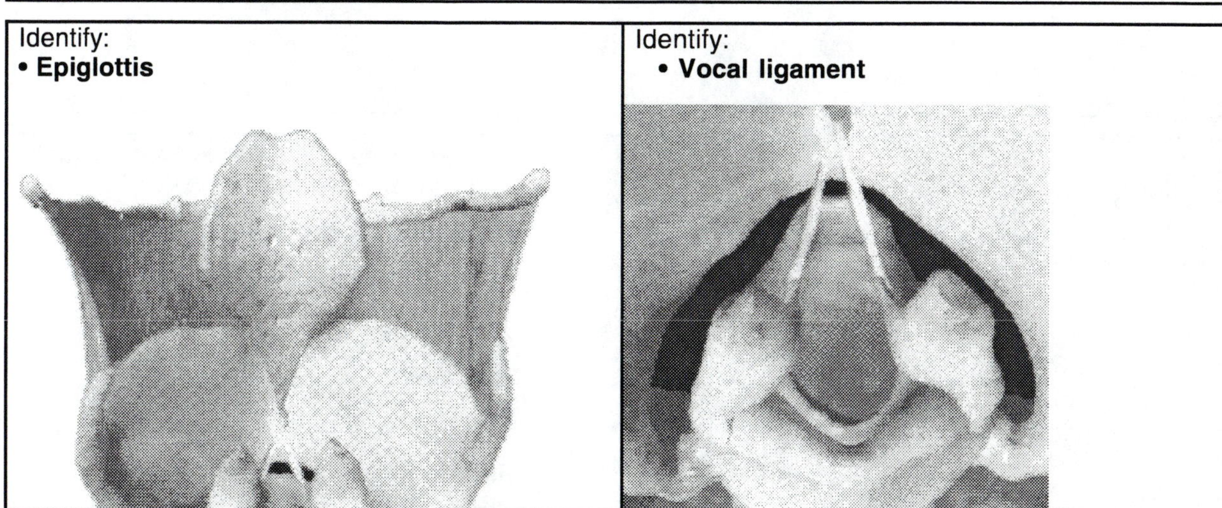

Identify:
- **Vocal ligament**

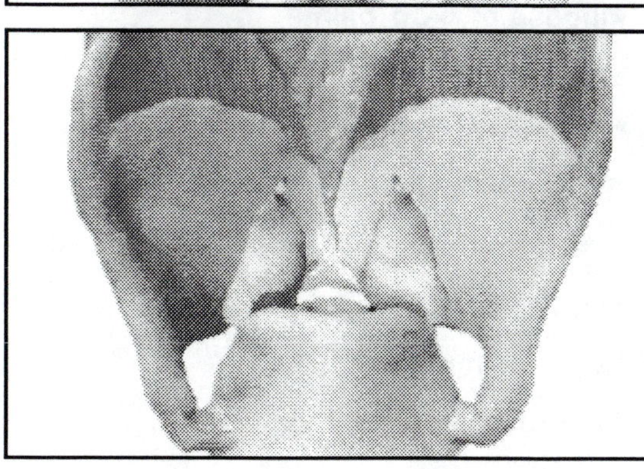

Identify:

- **Arytenoid cartilage**
- **Corniculate cartilage**

When you are finished, **close** the window.

## Anterior Dissection of the Nose, Pharynx, & Larynx

Under the **File** menu, drag down to **Open** (then to **Content** if you are working in Windows).

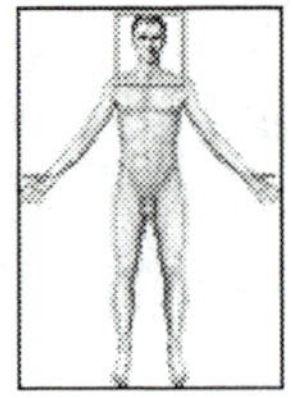

Choose **Dissectible Anatomy**.

Choose either male or female. Select **Anterior.** Click **Open**.

Enlarge the window and position the image over the head and neck.

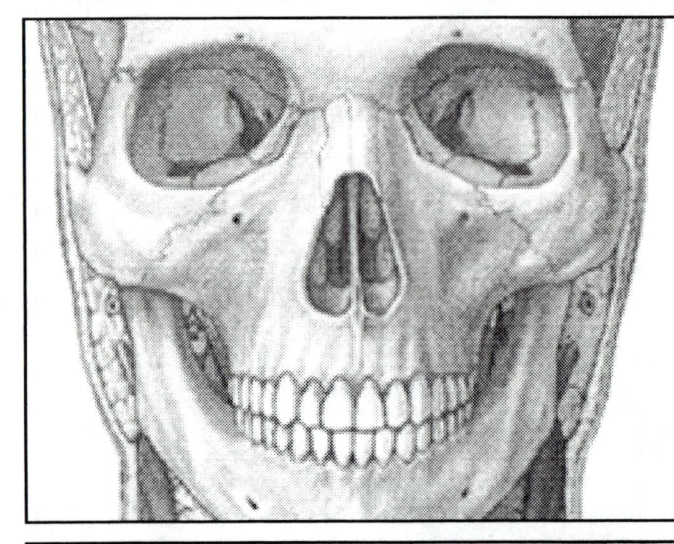

Drag down on the **Depth Bar** ▼ until you reach **layer 48**. Identify:

- **Nasal passage**
- **Inferior nasal conchae**
- **Middle nasal conchae**

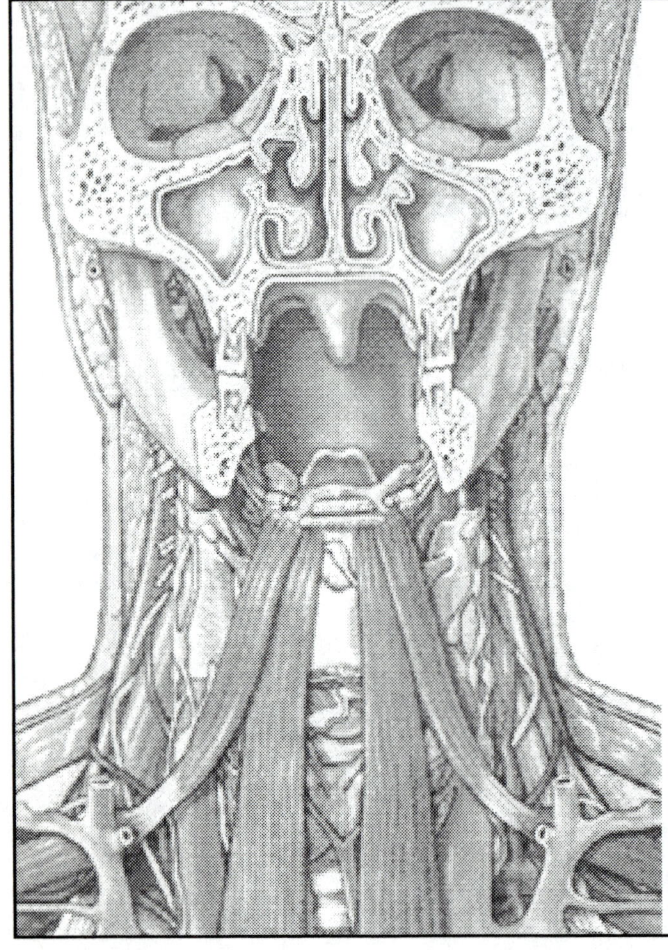

Click several times on the **Depth Bar** ▼ until you reach **layer 59**. Identify:

- **Tracheal ring cartilage**
- **Thyroid cartilage**
- **Hyoid bone**
- **Epiglottis**
- **Posterior wall of oral pharynx**
- **Mucosa of uvula and soft palate**
- **Mucosa of hard palate**
- **Palatine tonsil**
- **Maxillary sinus** (mucosa of)
- **Ethmoidal cells**
- **Inferior meatus**
- **Middle meatus**
- **Superior meatus**
- **Inferior nasal conchae**
- **Middle nasal conchae**
- **Mucosa of nose**
- **Perpendicular plate of ethmoid bone**

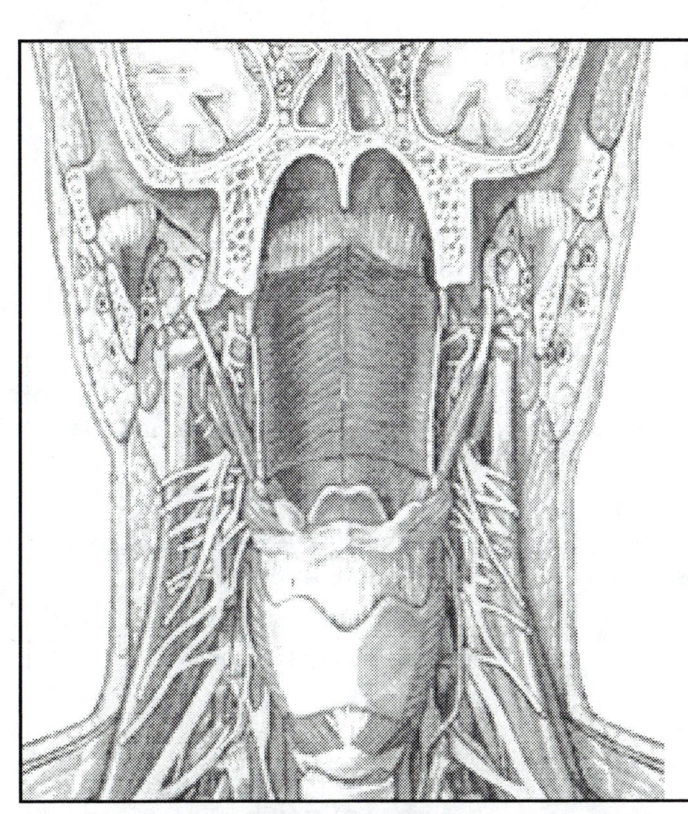

Drag down on the **Depth Bar** [▼] until you reach **layer 253**. Identify:

- **Sphenoidal sinus**
- **Epiglottis**
- **Hyoid bone**
- **Thyrohyoid ligament**
- **Thyroid cartilage**
- **Cricothyroid ligament**
- **Cricoid cartilage**
- **Tracheal ring cartilage**

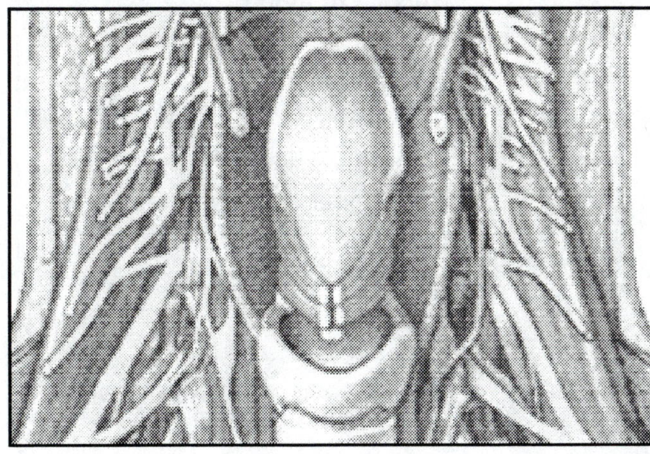

Click a few times on the **Depth Bar** [▼] until you reach **layer 255**. Identify:

- **Epiglottis**
- **Cricoid cartilage**

Click once on the **Depth Bar** [▼] to **layer 256**. No graphic has been provided for this layer. Identify:

- **Arytenoid cartilage**
- **Cricoid cartilage**
- **Corniculate cartilage**

## Medial Dissection of the Upper Respiratory System

Click on the **View** button and choose **Medial**.

Click on the **Normal** button. Windows— —Mac

Drag the **Depth Bar** to **layer 0.**

Position the navigator rectangle over the head and neck.

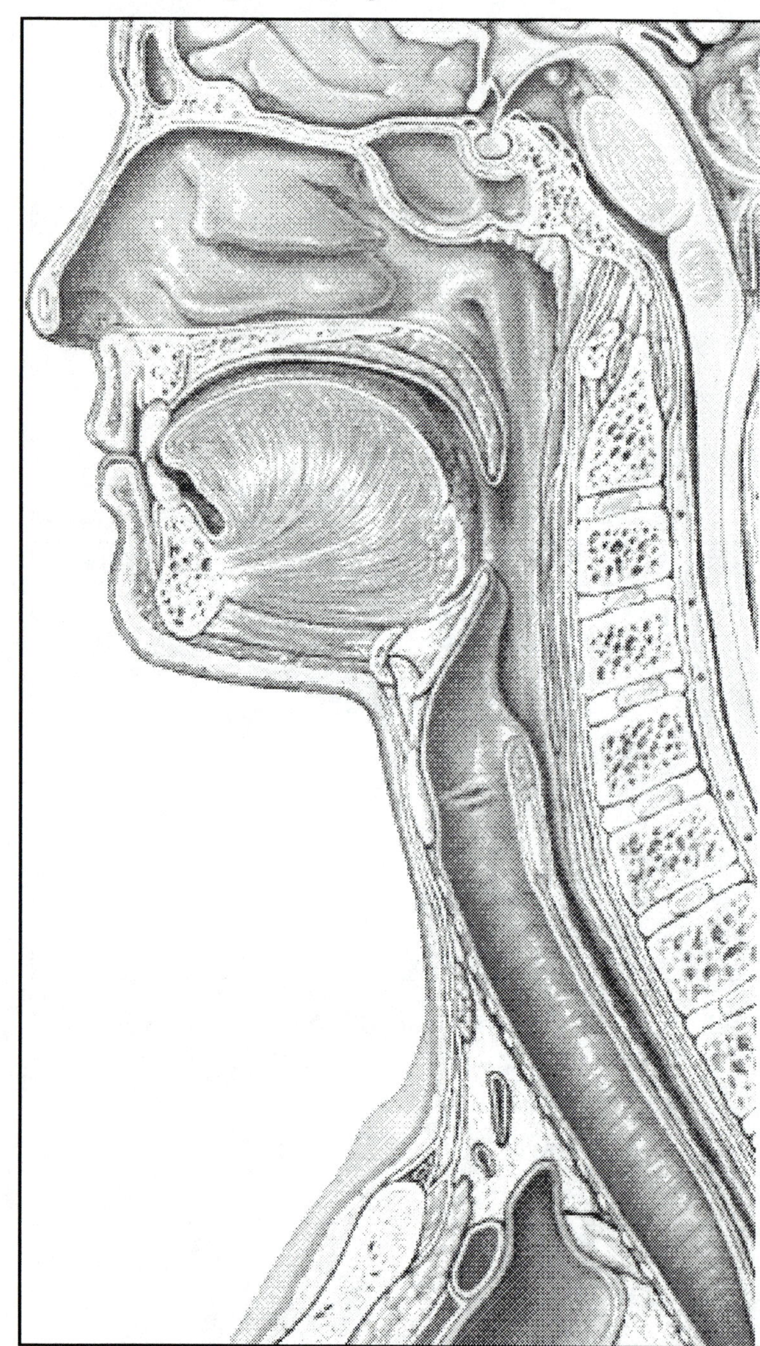

Identify:

- **Nasal cartilage**
- **Frontal sinus**
- **Palatine process of the maxilla** (hard palate)
- **Vestibule of nose**
- **Nasal cavity proper**
- **Inferior nasal conchae**
- **Middle nasal conchae**
- **Superior nasal conchae**
- **Epiglottis**
- **Epiglottic cartilage**
- **Oropharynx**
- **Mucosa of mouth**
- **Pharyngeal tonsils**
- **Lingual tonsils**
- **Hyoid bone**
- **Thyroid cartilage**
- **Pharyngeal opening of auditory tube**
- **Laryngopharynx**
- **Esophagus**
- **Trachea**
- **Larynx** (composed of laryngeal vestibule and infraglottic cavity)
- **Tongue**
- **Nasopharynx**
- **Vocal fold**
- **Cricoid cartilage**
- **Tracheal ring cartilage**
- **Isthmus of thyroid gland**

## Lateral Dissection of the Upper Respiratory System

Click on the **View** button and choose **Lateral**.

Click on the **Normal** button. Windows — Mac

Drag the **Depth Bar** to **layer 0**.

Position the navigator rectangle over the head and neck.

Drag the **Depth Bar** down to **layer 207**. No graphic has been provided for this layer. Identify:

- **Maxillary sinus**

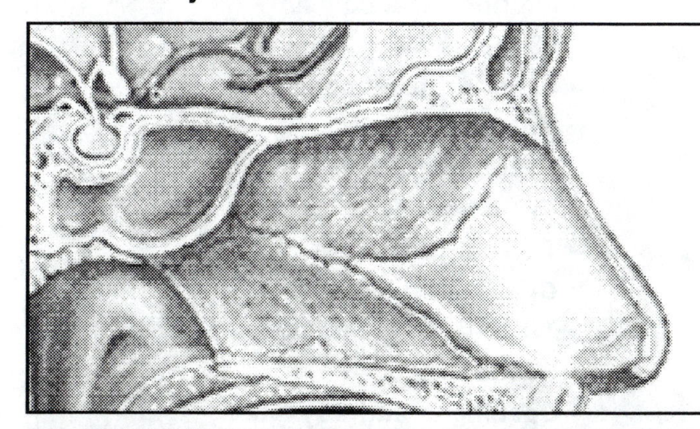

Click on the **Depth Bar** to **layer 209**. Identify:

- **Nasal septal cartilage**
- **Vomer perpendicular plate of the ethmoid bone**

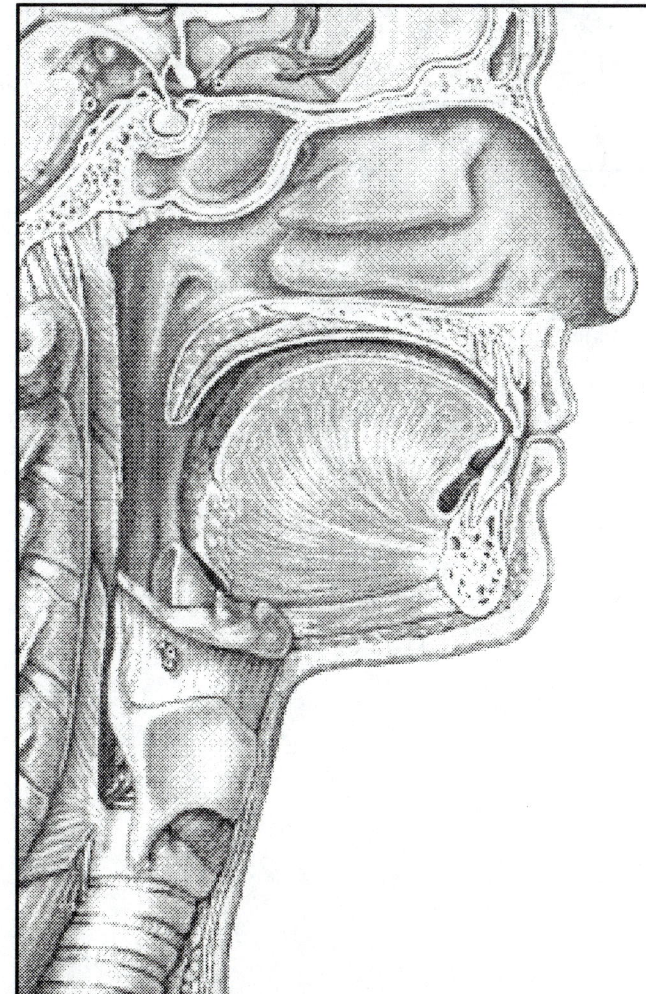

Click on the **Depth Bar** to **layer 256**. Identify:

- **Frontal sinus**
- **Sphenoidal sinus**
- **Palatine process of the maxilla**
- **Palatine bone**
- **Vestibule of nose**
- **Nasal cavity proper**
- **Inferior nasal concha**
- **Middle nasal concha**
- **Superior nasal concha**
- **Mucosa of mouth**
- **Pharyngeal tonsils**
- **Hyoid bone**
- **Thyroid cartilage**
- **Cricoid cartilage**
- **Tracheal ring cartilage**
- **Isthmus of thyroid gland**
- **Tongue**
- **Nasopharynx**
- **Opening of auditory tube**
- **Esophagus**
- **Oropharynx**
- **Epiglottis**
- **Thyrohyoid ligament**
- **Maxillary sinus**

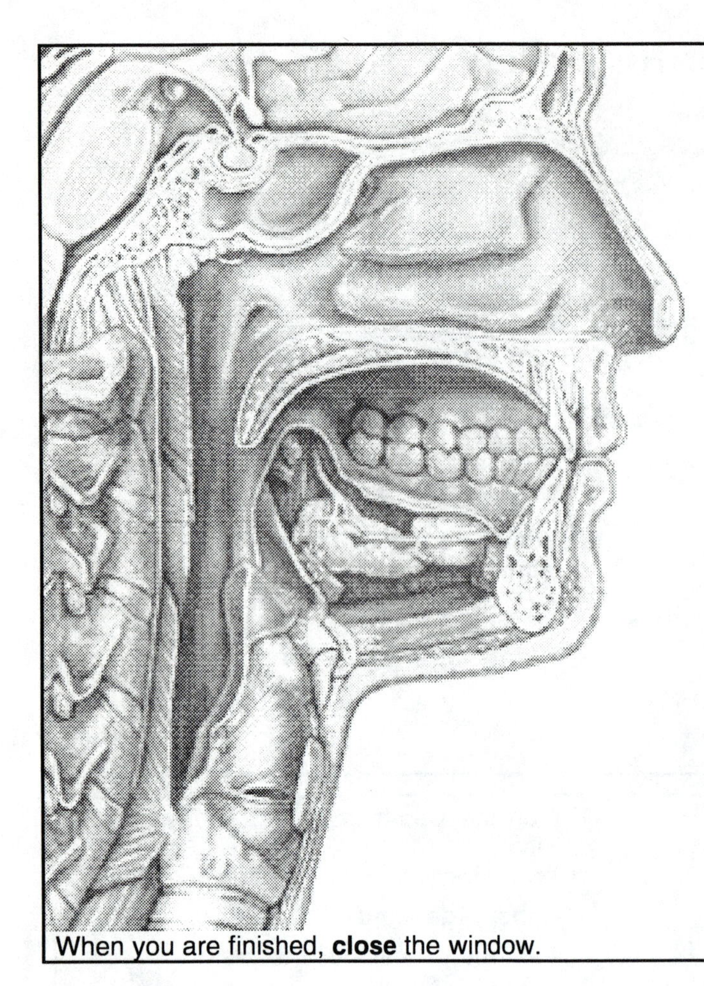

Click on the **Depth Bar** ▼ down to **layer 288**. Identify:

- **Frontal sinus**
- **Sphenoidal sinus**
- **Palatine process of the maxilla**
- **Palatine bone**
- **Vestibule of nose**
- **Nasal cavity proper**
- **Inferior nasal concha**
- **Middle nasal concha**
- **Superior nasal concha**
- **Mucosa of mouth**
- **Pharyngeal tonsils**
- **Hyoid bone**
- **Thyroid cartilage**
- **Cricoid cartilage**
- **Tracheal ring cartilage**
- **Nasopharynx**
- **Opening of auditory tube**
- **Esophagus**
- **Laryngopharynx**
- **Oropharynx**
- **Epiglottis**

When you are finished, **close** the window.

The nasal conchae serve what role in the respiratory system?

______________________________________________________________________

List the regions of the pharynx in order from superior to inferior.

______________________________________________________________________

## Atlas Anatomy of the Upper Respiratory System

Under the **File** menu, drag down to **Open** (then to **Content** if you are working in Windows). Click on **Atlas Anatomy**. Select **System** and **Respiratory**. Choose **Laryngeal Muscles (Ant)**. If the list is not in alphabetical order, the image is almost half of the way down the list.

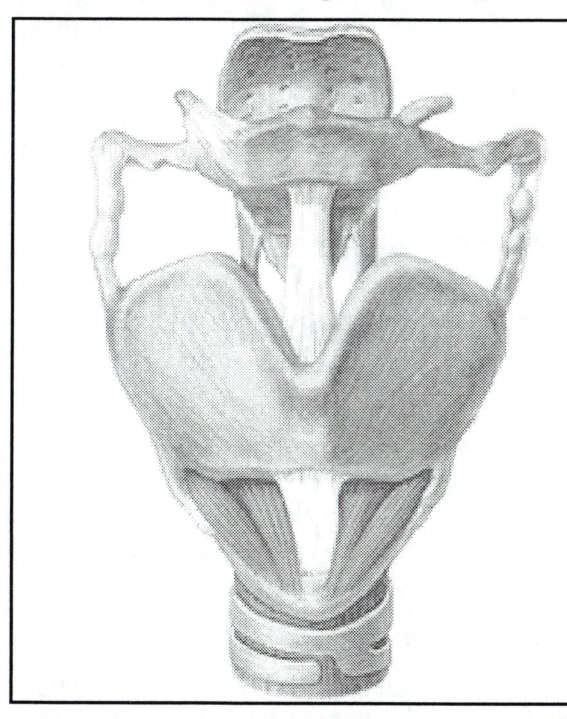

If you are working in Mac, click on the **Show All Pins** button and scroll to **All Systems**. Identify:

- **Body of hyoid bone**
- **Cricoid cartilage**
- **Cricothyroid ligament**
- **Epiglottis**
- **Thyrohyoid ligament**
- **Thyroid cartilage**
- **Tracheal ring cartilage**

When you are finished, **close** the window.

What is the common name given to the thyroid cartilage? ______________________________

Under the **File** menu, drag down to **Open** (then to **Content** if you are working in Windows). Click on **Atlas Anatomy**. **System** and **Respiratory** should still be selected. Choose **Laryngeal Muscles (Post) 1**. If the list is not in alphabetical order, the image is almost half of the way down the list.

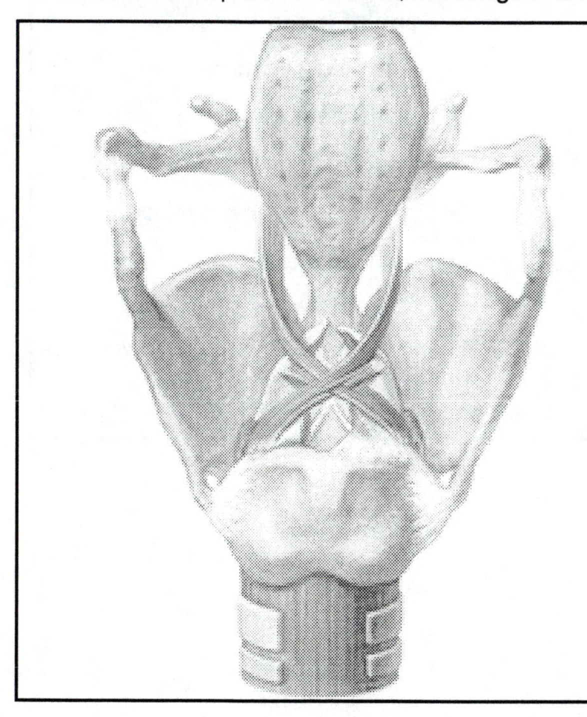

If you are working in Mac, click on the **Show All Pins** button and scroll to **All Systems**. Identify:

- **Body of hyoid bone**
- **Arytenoid cartilage**
- **Corniculate cartilage**
- **Epiglottis**
- **Lamina of cricoid cartilage**
- **Lamina of thyroid cartilage**
- **Tracheal ring cartilage**
- **Vocal ligament**

When you are finished, **close** the window.

Under the **File** menu, drag down to **Open** (then to **Content** if you are working in Windows). Click on **Atlas Anatomy**. **System** and **Respiratory** should still be selected. Choose **Laryngeal Muscles (Post) 2**. If the list is not in alphabetical order, the image is almost half of the way down the list.

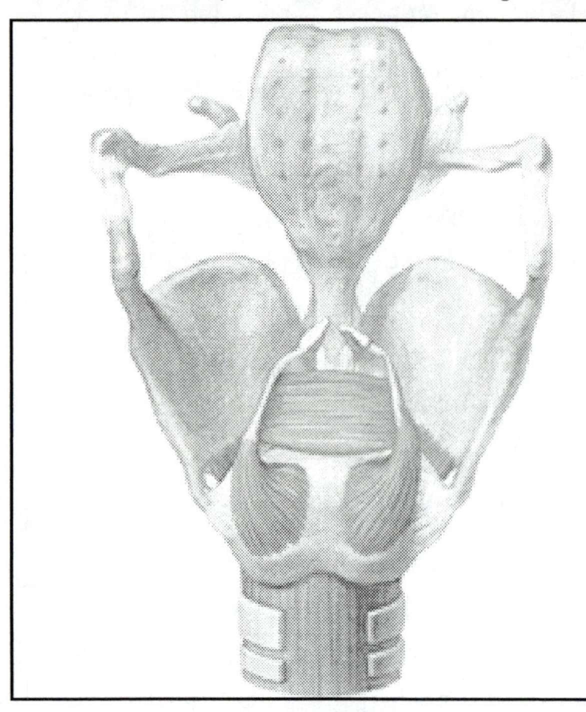

If you are working in Mac, click on the **Show All Pins** button and scroll to **All Systems**. Identify:

- **Body of hyoid bone**
- **Arytenoid cartilage**
- **Corniculate cartilage**
- **Epiglottis**
- **Lamina of cricoid cartilage**
- **Thyroid cartilage**
- **Tracheal ring cartilage**

Give two functions of the arytenoid and corniculate cartilages.

________________________________________

________________________________________

When you are finished, **close** the window.

Under the **File** menu, drag down to **Open** (then to **Content** if you are working in Windows). Click on **Atlas Anatomy**. **System** and **Respiratory** should still be selected. Choose **Laryngeal Muscles (Sup)**. If the list is not in alphabetical order, the image is almost half of the way down the list.

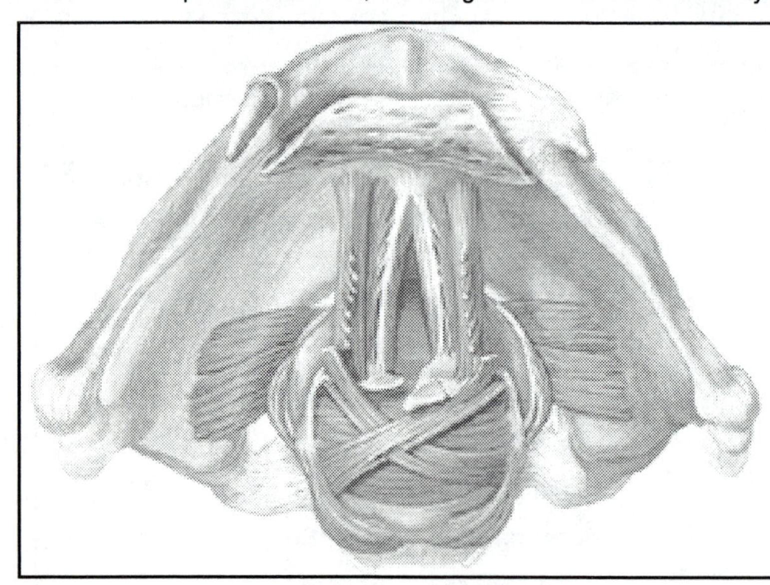

If you are working in Mac, click on the **Show All Pins** button and scroll to **All Systems**. Identify:

- **Body of hyoid bone**
- **Arytenoid cartilage**
- **Corniculate cartilage**
- **Epiglottis**
- **Orifice between vocal cords**
- **Vocal ligament**

When you are finished, **close** the window.

# Cadaver Images of the Upper Respiratory System

Under the **File** menu, drag down to **Open** (then to **Content** if you are working in Windows). Click on **Atlas Anatomy**. **System** and **Respiratory** should still be selected. Choose **Sagittal Section of Head & Neck**. If the list is not in alphabetical order, the image is near the top of the list.

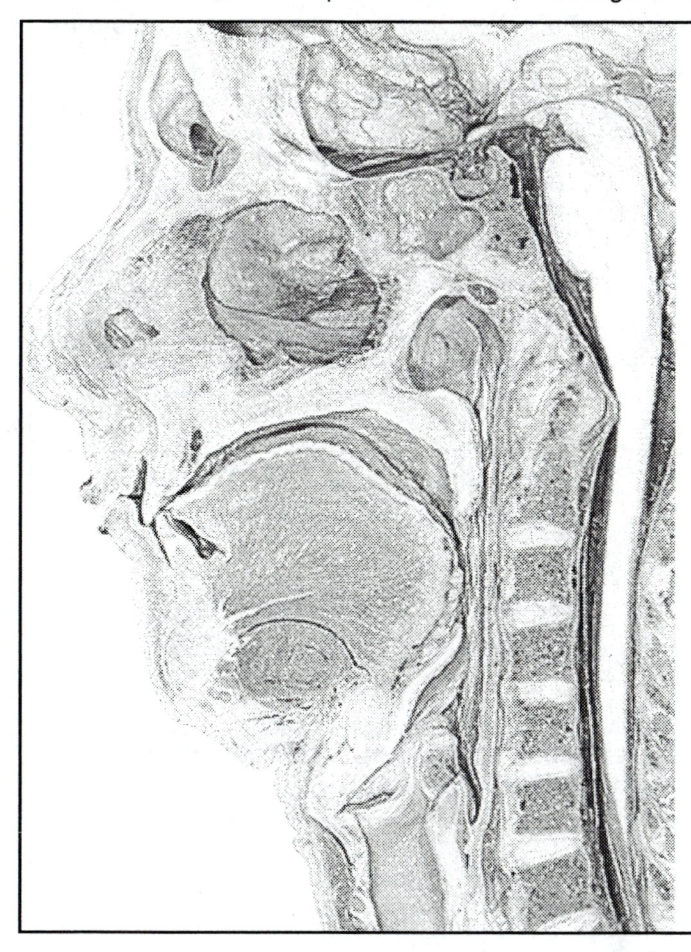

If you are working in Mac, click on the **Show All Pins** button and scroll to **All Systems**. Identify:

- **Lingual tonsil**
- **Laryngopharynx**
- **Inferior nasal concha**
- **Epiglottis**
- **Middle nasal concha**
- **Nasopharynx**
- **Oral cavity**
- **Oropharynx**
- **Pharyngeal orifice of auditory tube**
- **Pharyngeal tonsil**
- **Sphenoidal sinus**
- **Superior nasal concha**
- **Thyroid cartilage**
- **Uvula**
- **Vestibular fold**
- **Vocal fold**

When you are finished, **close** the window.

The pharyngeal tonsils are commonly known as the ________________________________________.

The vestibular folds are commonly known as the ________________________________________.

Under the **File** menu, drag down to **Open** (then to **Content** if you are working in Windows). Click on **Atlas Anatomy**. **System** and **Respiratory** should still be selected. Choose **Sagittal Section of Larynx**. If the list is not in alphabetical order, the image is about half way down the list.

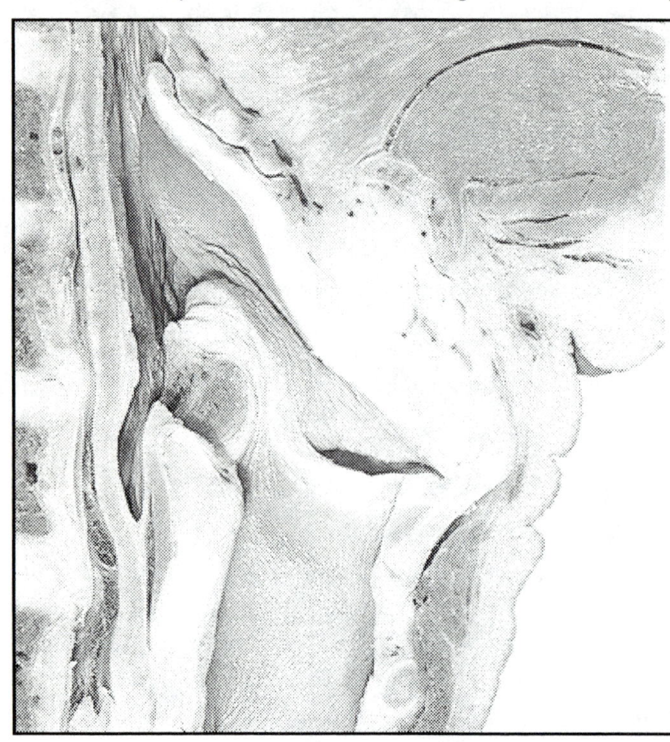

If you are working in Mac, click on the **Show All Pins** button and scroll to **All Systems**. Identify:

- **Lingual tonsil**
- **Laryngopharynx**
- **Cricothyroid ligament**
- **Epiglottis**
- **Cricoid cartilage**
- **Body of hyoid bone**
- **Lamina of cricoid cartilage**
- **Larynx**
- **Trachea**
- **Ventricle of larynx**
- **Thyroid cartilage**
- **Vestibule of larynx**
- **Vestibular fold**
- **Vocal fold**

When you are finished, **close** the window.

Under the **File** menu, drag down to **Open** (then to **Content** if you are working in Windows). Click on **Atlas Anatomy**. **System** and **Respiratory** should still be selected. Choose **Lateral Wall of Palate**. If the list is not in alphabetical order, the image is about one-third of the way down the list.

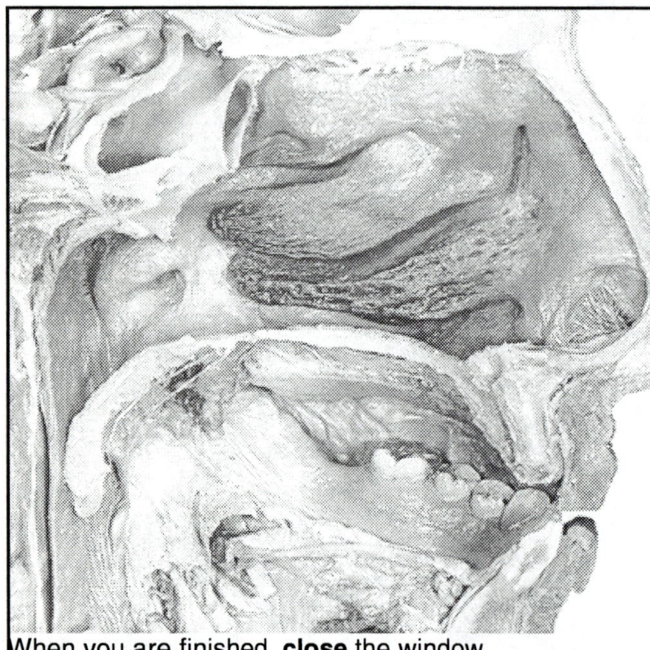

If you are working in Mac, click on the **Show All Pins** button and scroll to **All Systems**. Identify:

- **Hard palate**
- **Inferior nasal concha**
- **Inferior nasal meatus**
- **Middle nasal concha**
- **Middle nasal meatus**
- **Nasal vestibule**
- **Nasopharynx**
- **Oropharynx**
- **Pharyngeal orifice of auditory tube**
- **Pharyngeal tonsil**
- **Soft palate**
- **Sphenoidal sinus**
- **Superior nasal concha**
- **Superior nasal meatus**
- **Uvula**

When you are finished, **close** the window.

What is the primary function of the nasal meatuses?

Under the **File** menu, drag down to **Open** (then to **Content** if you are working in Windows). Click on **Atlas Anatomy**. **System** and **Respiratory** should still be selected. Choose **Dissection of Larynx (Sup)**. If the list is not in alphabetical order, the image is about one-third of the way down the list.

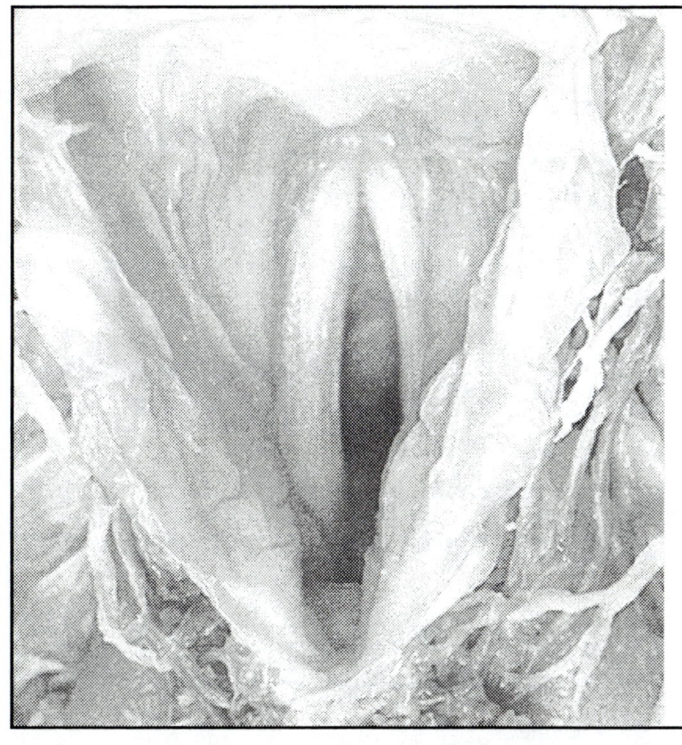

Identify:

- **Epiglottis**
- **Orifice between vocal cords**
- **Ventricle of larynx**
- **Vestibular fold**
- **Vocal fold**

When you are finished, **close** the window.

Under the **File** menu, drag down to **Open** (then to **Content** if you are working in Windows). Click on **Atlas Anatomy**. **System** and **Respiratory** should still be selected. Choose **Lateral Wall of Nasal Cavity**. If the list is not in alphabetical order, the image is about half way down the list.

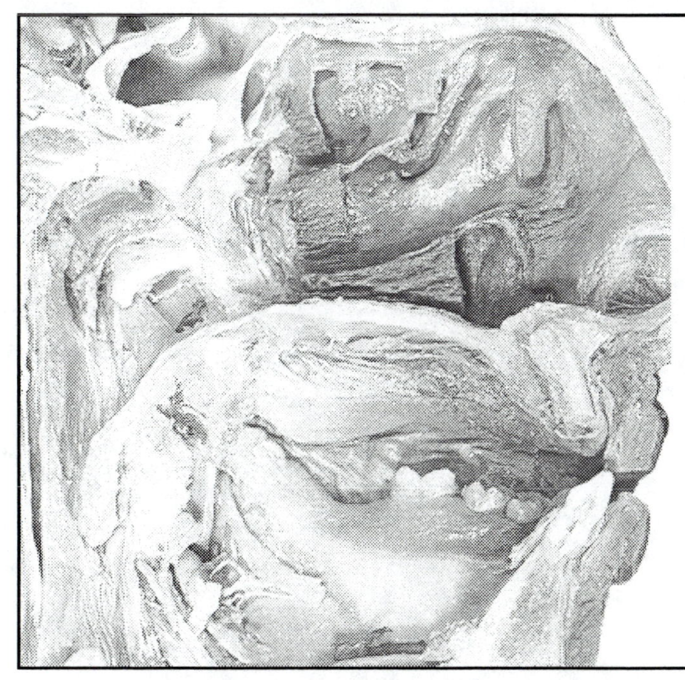

If you are working in Mac, click on the **Show All Pins** button and scroll to **All Systems**. Identify:

- **Hard palate**
- **Inferior nasal concha**
- **Ethmoidal cell**
- **Middle nasal concha**
- **Nasal vestibule**
- **Pharyngeal orifice of auditory tube**
- **Soft palate**
- **Sphenoidal sinus**
- **Superior nasal concha**
- **Uvula**

When you are finished, **close** the window.

The hard palate is made up of portions of what two bones?

______________________________________________

## Objectives: Nose, Pharynx & Larynx:

To study the anatomy of the nose, pharynx, and larynx.

- Arytenoid cartilage
- Corniculate cartilage
- Cricoid cartilage
- Cricothyroid ligament
- Epiglottis
- Esophagus
- Ethmoidal cells
- Frontal sinus
- Hard palate
- Hyoid bone
- Inferior meatus
- Inferior nasal conchae
- Larynx
- Laryngopharynx
- Lingual tonsils
- Maxillary sinus
- Middle meatus
- Middle nasal conchae
- Nasal cartilage
- Nasal passage
- Nasopharynx
- Oropharynx
- Palatine tonsils
- Perpendicular plate of ethmoid bone
- Pharyngeal tonsils
- Soft palate
- Sphenoidal sinus
- Superior meatus
- Superior nasal conchae
- Thyrohyoid ligament
- Thyroid cartilage
- Thyroid gland
- Tongue
- Trachea
- Tracheal ring cartilage
- Uvula
- Vestibular fold
- Vestibule of nose
- Vocal fold
- Vocal ligament

# Lower Respiratory System

## Opening A.D.A.M Interactive Anatomy

Open A.D.A.M. Interactive Anatomy according to the directions in the Tutorial.

## 3D Anatomy of the Lower Respiratory System

Choose **3D Anatomy**. Click on **3D Lungs**. Identify the following by clicking on the **3D Anatomy Structure List**:

- **Carina**
- **Diaphragmatic surface of inferior lobe of left lung**
- **Hilum of left lung**
- **Hilum of right lung**

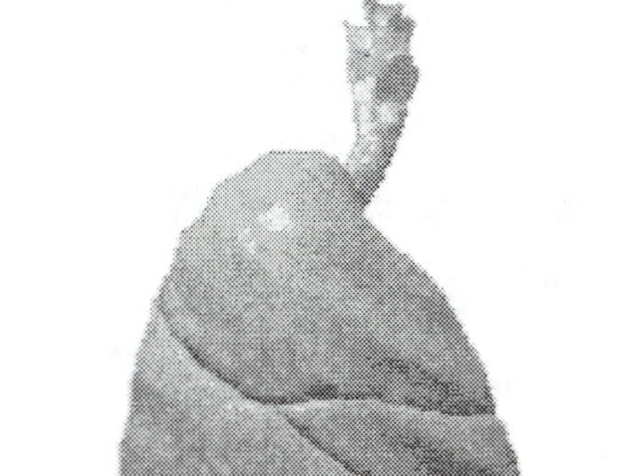

- **Apex of right lung**
- **Horizontal fissure of right lung**
- **Inferior lobe of right lung**
- **Middle lobe of right lung**
- **Oblique fissure of right lung**
- **Superior lobe of right lung**

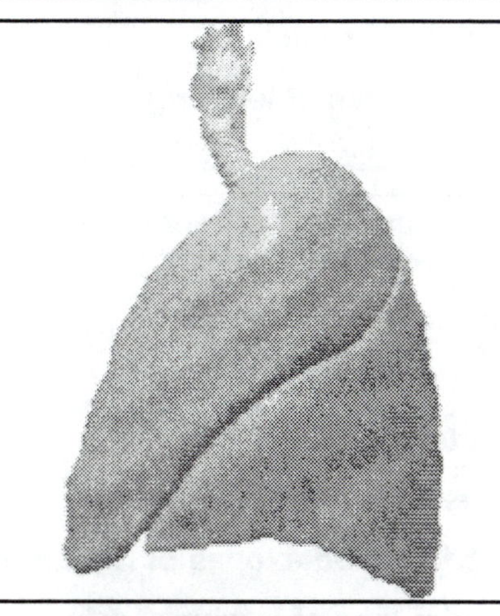

- **Apex of left lung**
- **Oblique fissure of left lung**
- **Inferior lobe of left lung**
- **Superior lobe of left lung**

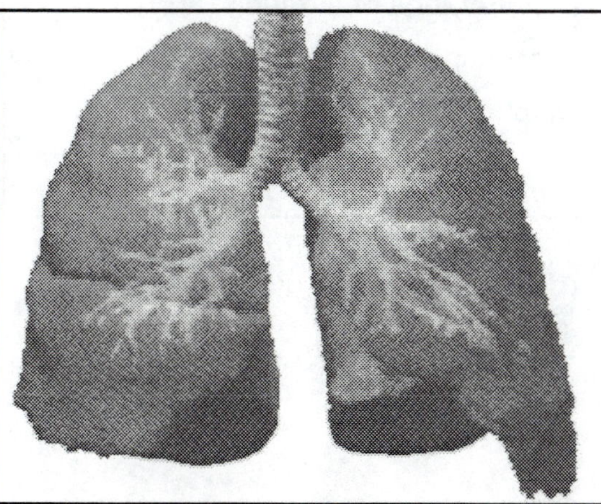

- **Cardiac notch of left lung**
- **Left primary bronchus** - Anterior
- **Right primary bronchus** - Anterior
- **Trachea**
- **Tracheal cartilage ring**

When you are finished, **close** the window.

## Anterior Dissection of the Lower Respiratory System

Under the **File** menu, drag down to **Open** (then to **Content** if you are working in Windows).

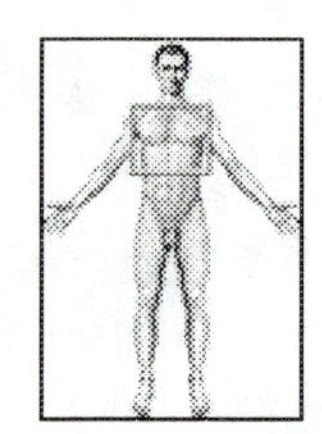

Choose **Dissectible Anatomy**.

Choose either male or female. Select **Anterior.** Click **Open**.

Enlarge the window and position the image over the thoracic region.

Drag down on the **Depth Bar** ▼ until you reach **layer 150**. Then slowly click on the **Depth Bar** ▼ to **layer 161**. No graphic has been provided for this layer. Identify:

- **Parietal pleura**.

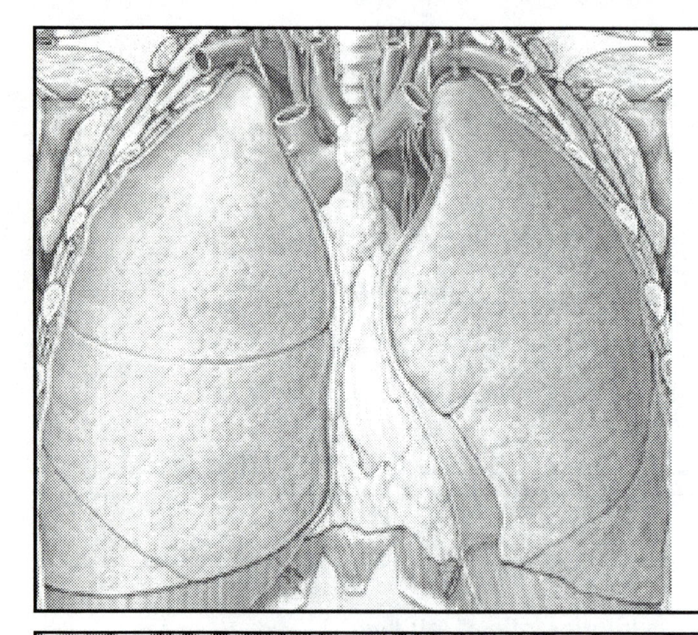

Click once on the **Depth Bar** ▼ to **layer 162**. Identify:

- **Tracheal ring cartilage**
- **Trachea**
- **Superior, middle, & inferior lobes of right lung**
- **Superior & inferior lobes of left lung**
- **Horizontal and oblique fissures of right lung**
- **Oblique fissure of left lung**
- **Parietal pleura**
- **Apex of right & left lung**
- **Cardiac notch of left lung**

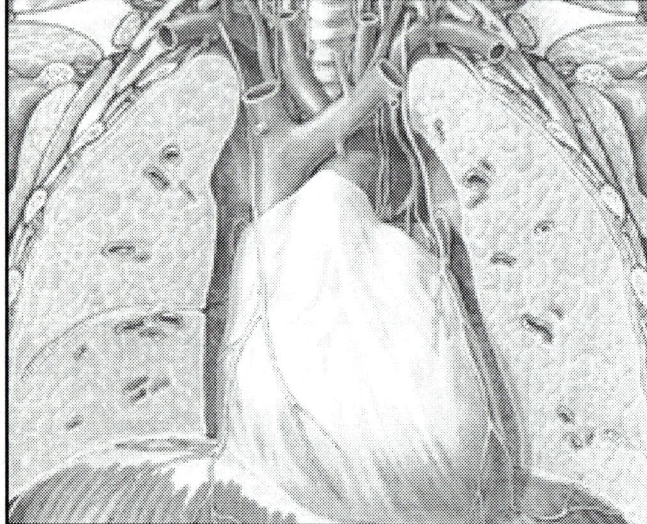

Click a few times on the **Depth Bar** ▼ until you reach **layer 168**. Identify:

- **Tracheal ring cartilage**
- **Trachea**
- **Parietal pleura**
- **Visceral pleura**
- **Superior, middle lobes of right lung**
- **Superior lobe of left lung**
- **Phrenic nerves** (right & left)
- **Diaphragm**

The phrenic nerves originate from which plexus?

___

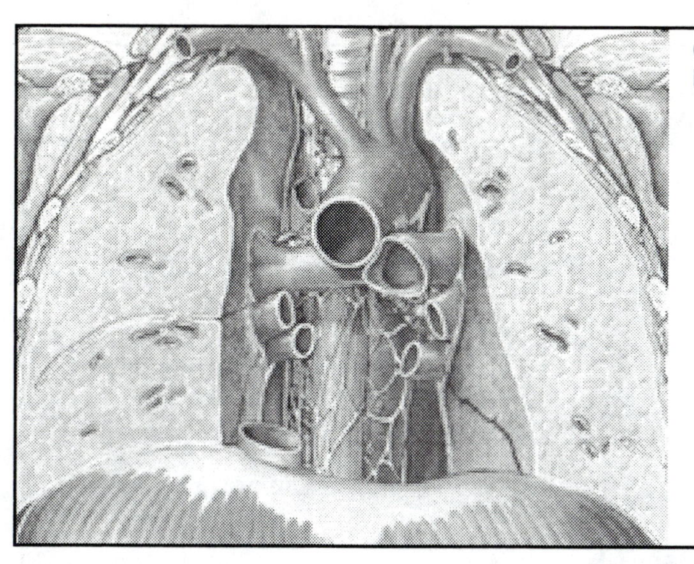

Click on the **Depth Bar** [▼] until you reach **layer 193**. Identify:

- **Superior, middle lobes of right lung**
- **Superior lobe of left lung**
- **Visceral pleura**
- **Parietal pleura**
- **Pulmonary arteries and veins within lungs**
- **Diaphragm**
- **Pulmonary arteries going into the lungs**
- **Pulmonary veins emerging from the lungs**
- **Pulmonary trunk**

______________________________ is a condition which may be caused by excessive secretion of pleural fluid or inflammation of the visceral and parietal pleura.

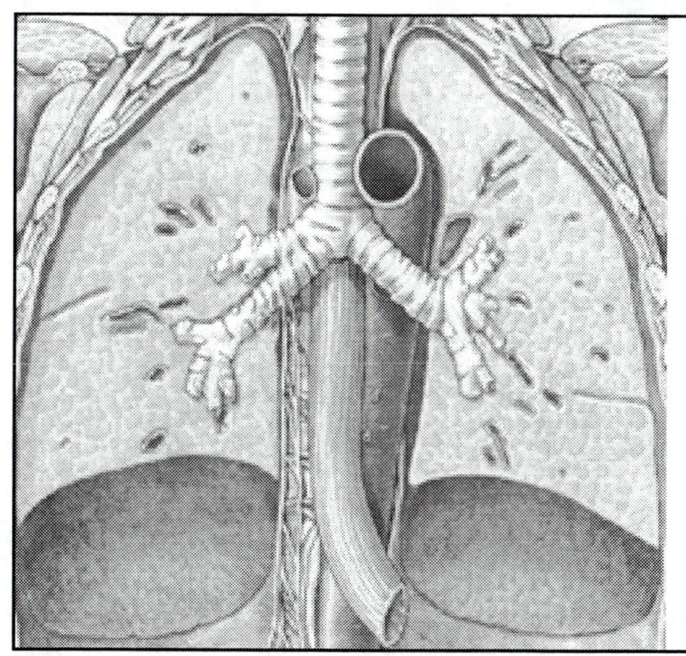

Click on the **Depth Bar** [▼] until you reach **layer 253**. Identify:

- **Tracheal ring cartilage**
- **Trachea**
- **Superior & inferior lobes of right lung**
- **Superior & inferior lobes of left lung**
- **Bronchial cartilage**
- **Left & right primary bronchi**
- **Secondary bronchi**
- **Tertiary bronchi**
- **Pulmonary arteries and veins within lungs**
- **Esophagus**

 

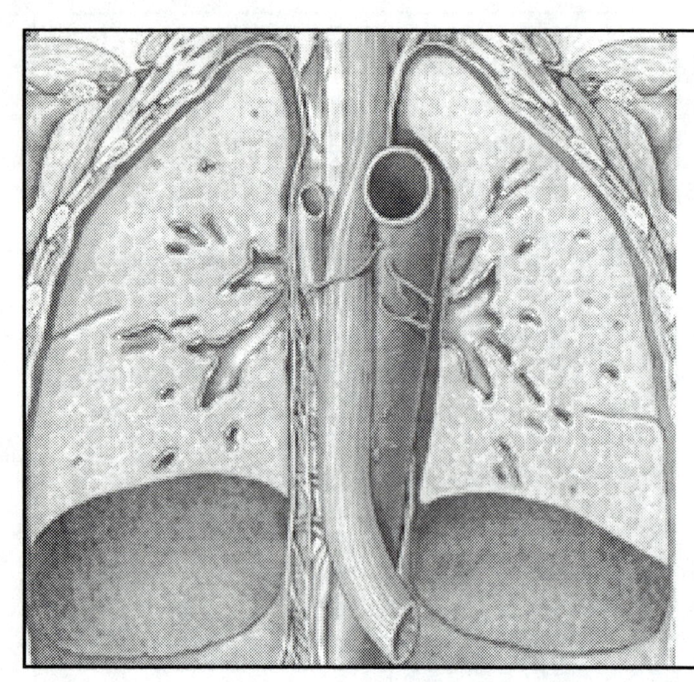

Click a few times on the **Depth Bar** ▼ to **layer 259**. Identify:

- **Superior & inferior lobes of right lung**
- **Superior & inferior lobes of left lung**
- **Parietal pleura**
- **Secondary bronchi**
- **Tertiary bronchi**
- **Pulmonary arteries and veins within lungs**
- **Esophagus**
- **Descending thoracic aorta**

Click a few times on the **Depth Bar** ▼ to **layer 262**. No graphic has been provided for this layer. Identify the **parietal pleura**.

## Medial Dissection of the Lower Respiratory System

Click on the **View** button and choose **Medial**.

Click on the **Normal** button. Windows— —Mac

Drag the **Depth Bar** to **layer 0.**

Position the navigator rectangle over the thorax.

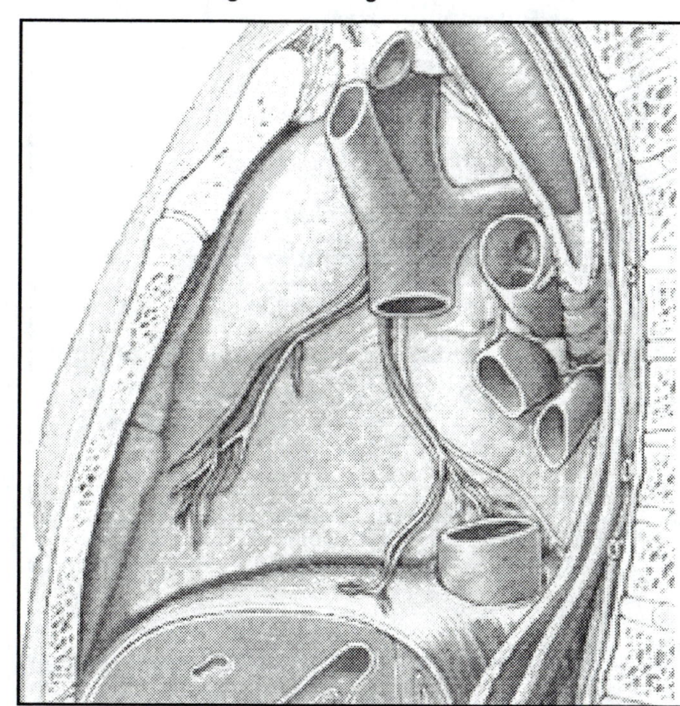

Click on the **Depth Bar** to **layer 67**. Identify:

- **Esophagus**
- **Trachea** (mucosa of)
- **Pulmonary veins**
- **Pulmonary artery**
- **Parietal pleura**
- **Phrenic nerve**
- **Diaphragm**
- **Carina**
- **Bronchial cartilage**

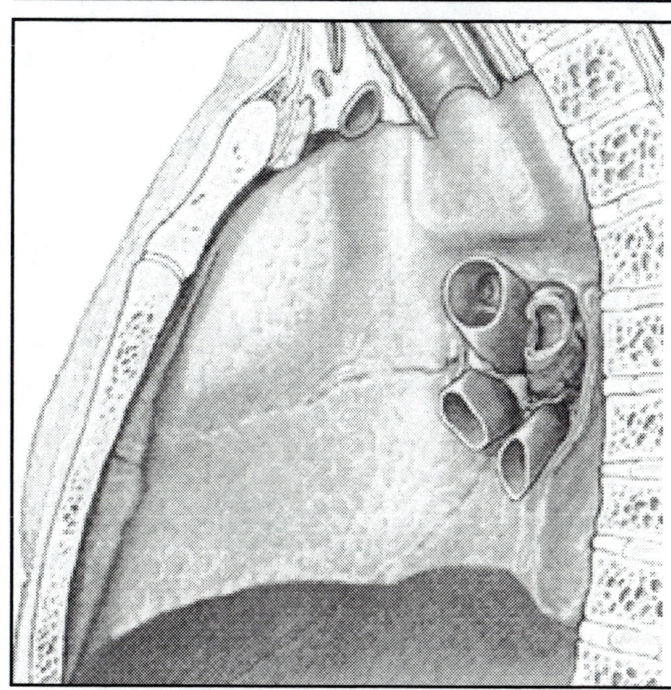

Click on the **Depth Bar** to **layer 94**. Identify:

- **Esophagus**
- **Trachea**
- **Pulmonary veins**
- **Pulmonary artery**
- **Parietal pleura**
- **Right primary bronchus**
- **Lymph nodes**

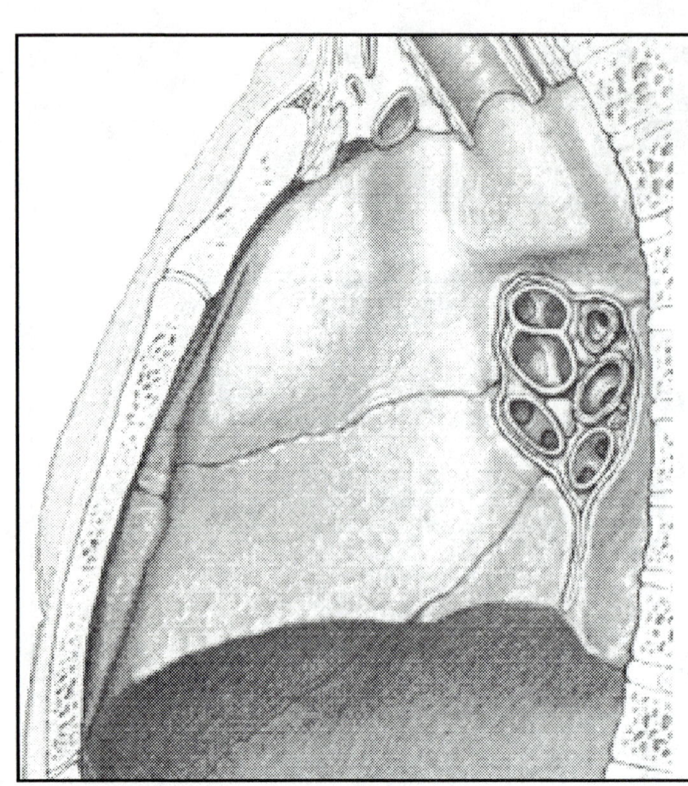

Click a few times on the **Depth Bar** [▼] to **layer 96**. Identify:

- **Esophagus**
- **Trachea**
- **Pulmonary veins**
- **Pulmonary arteries**
- **Parietal pleura**
- **Right secondary bronchus**
- **Lymph nodes**
- **Superior lobe of right lung**
- **Middle lobe of right lung**
- **Inferior lobe of right lung**
- **Horizontal and oblique fissures of right lung**

# Lateral Dissection of the Lower Respiratory System

Click on the **View** button and choose **Lateral**.

Click on the **Normal** button. Windows— —Mac

Drag the **Depth Bar** to **layer 0.**

Position the navigator rectangle over the thorax.

Drag the **Depth Bar** down to **layer 139**. No graphic has been provided for this layer. Identify the **parietal pleura.**

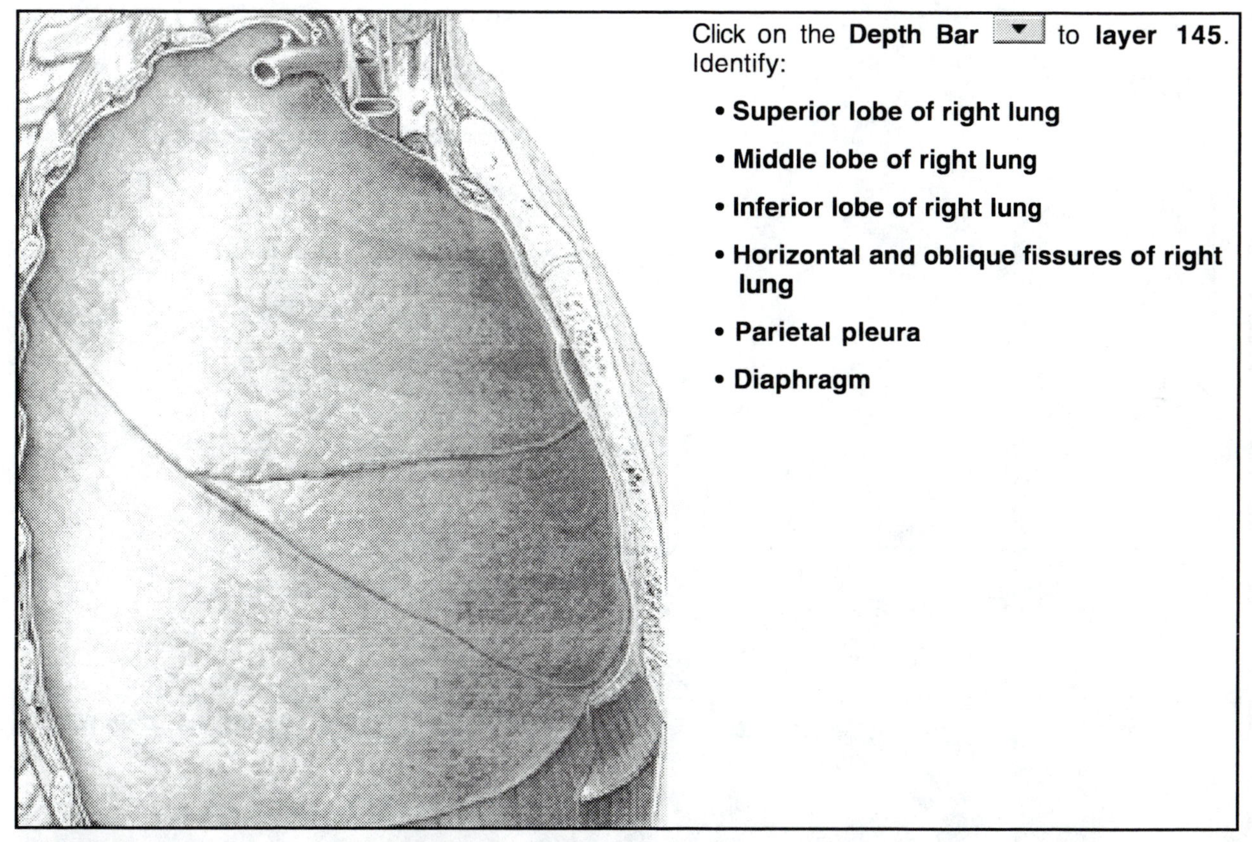

Click on the **Depth Bar** to **layer 145**. Identify:

- **Superior lobe of right lung**
- **Middle lobe of right lung**
- **Inferior lobe of right lung**
- **Horizontal and oblique fissures of right lung**
- **Parietal pleura**
- **Diaphragm**

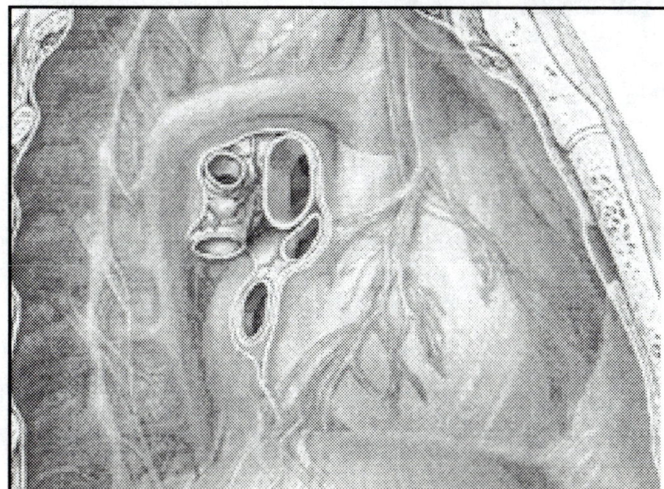

Click once on the **Depth Bar** to **layer 146**. Identify:

- **Bronchi** (secondary & tertiary)
- **Pulmonary arteries and veins**
- **Parietal pleura**

Click on the **Depth Bar** [▼] to **layer 149**. No graphic has been provided for this layer. Identify:

- **Phrenic nerve**
- **Diaphragm**

Click on the **Depth Bar** [▼] to reach **layer 248**. No graphic has been provided for this layer. Identify:

- **Parietal pleura**

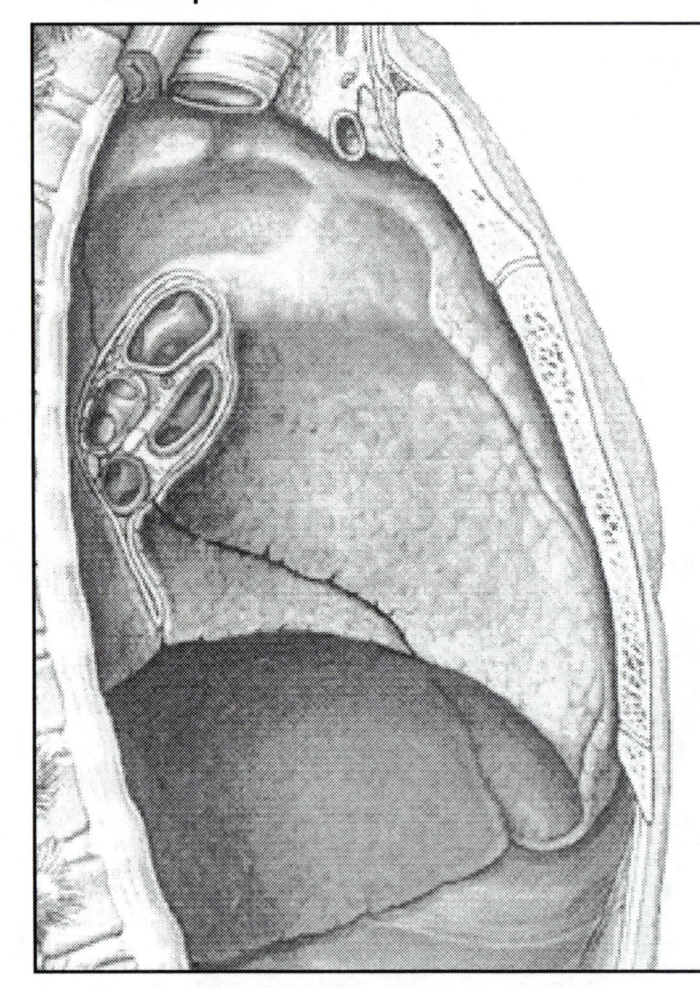

Click once on the **Depth Bar** [▼] to **layer 249**. Identify:

- **Esophagus**
- **Trachea**
- **Pulmonary arteries and veins**
- **Left primary bronchus**
- **Superior lobe of left lung**
- **Inferior lobe of left lung**
- **Oblique fissure of left lung**

When you are finished, **close** the window.

## Atlas Anatomy of the Lower Respiratory System

Under the **File** menu, drag down to **Open** (then to **Content** if you are working in Windows). Click on **Atlas Anatomy**. Select **System** and **Respiratory**. Choose **Landmarks of Thorax (Ant)**. If the list is not in alphabetical order, the image is slightly over half of the way down the list.

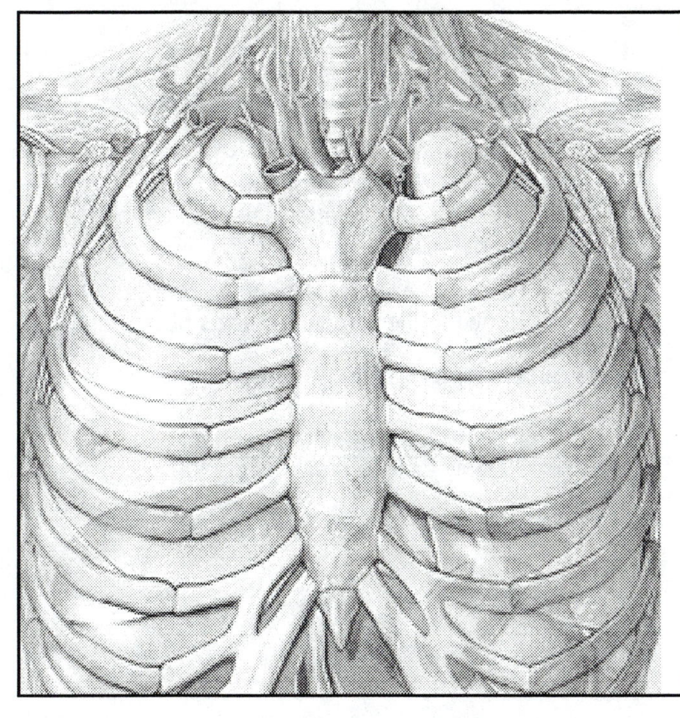

Identify :

- **Apex of left lung**
- **Apex of right lung**
- **Cardiac notch of left lung**
- **Horizontal fissure of right lung**
- **Inferior lobe of left lung**
- **Inferior lobe of right lung**
- **Middle lobe of right lung**
- **Oblique fissure of left lung**
- **Oblique fissure of right lung**
- **Superior lobe of left lung**
- **Superior lobe of right lung**
- **Trachea**

When you are finished, **close** the window.

Under the **File** menu, drag down to **Open** (then to **Content** if you are working in Windows). Click on **Atlas Anatomy**. **System** and **Respiratory** should still be selected. Choose **Bronchial Tree (Ant)**. If the list is not in alphabetical order, the image is a little over half of the way down the list.

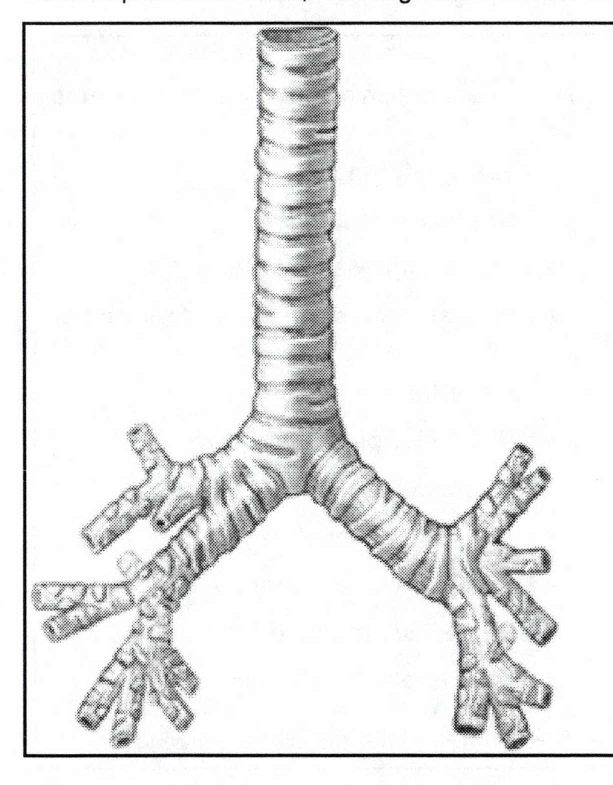

Identify:

- **Bronchopulmonary segment** (tertiary bronchus)
- **Left** (main) **primary bronchus**
- **Right** (main) **primary bronchus**
- **Secondary bronchi (5)** (Superior, middle, and inferior lobe bronchus of right lung, superior, and inferior lobe bronchus of left lung)
- **Tertiary bronchi** (10 shown on right, 8 shown on left) (colored ends of bronchi)
- **Trachea**
- **Tracheal cartilage ring**

When you are finished, **close** the window.

 

Under the **File** menu, drag down to **Open** (then to **Content** if you are working in Windows). Click on **Atlas Anatomy**. **System** and **Respiratory** should still be selected. Choose **Bronchial Tree (Post)**. If the list is not in alphabetical order, the image is a little over half of the way down the list.

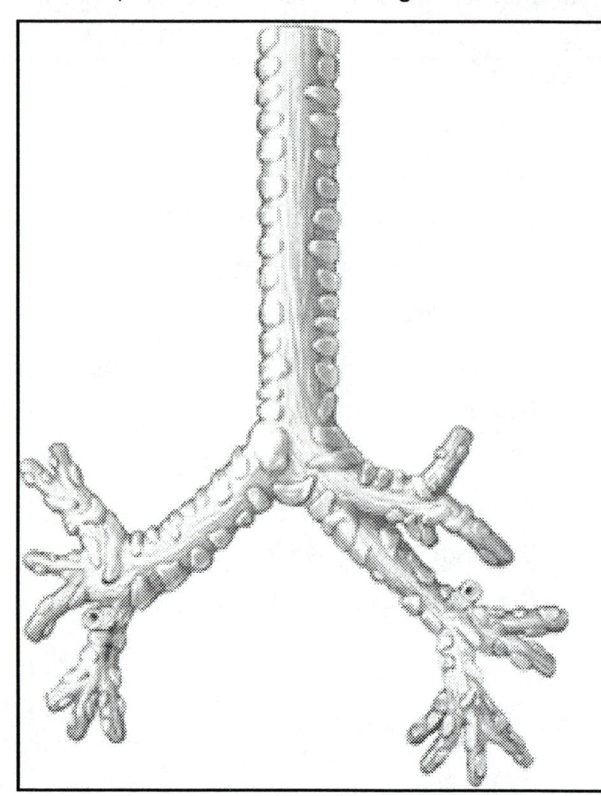

Identify:

- **Left** (main) **primary bronchus**
- **Right** (main) **primary bronchus**
- **Secondary bronchi** (5) (Superior, middle, and inferior lobe bronchus of right lung, superior, and inferior lobe bronchus of left lung)
- **Tertiary bronchi** (10 shown on right, 8 shown on left) (colored ends of bronchi)
- **Tracheal cartilage ring**

What is the function of the tracheal cartilage rings? Why are the rings not complete?

______________________________________________

______________________________________________

When you are finished, **close** the window.

Under the **File** menu, drag down to **Open** (then to **Content** if you are working in Windows). Click on **Atlas Anatomy**. **System** and **Respiratory** should still be selected. Choose **Left Lung (Med)**. If the list is not in alphabetical order, the image is a little over half of the way down the list.

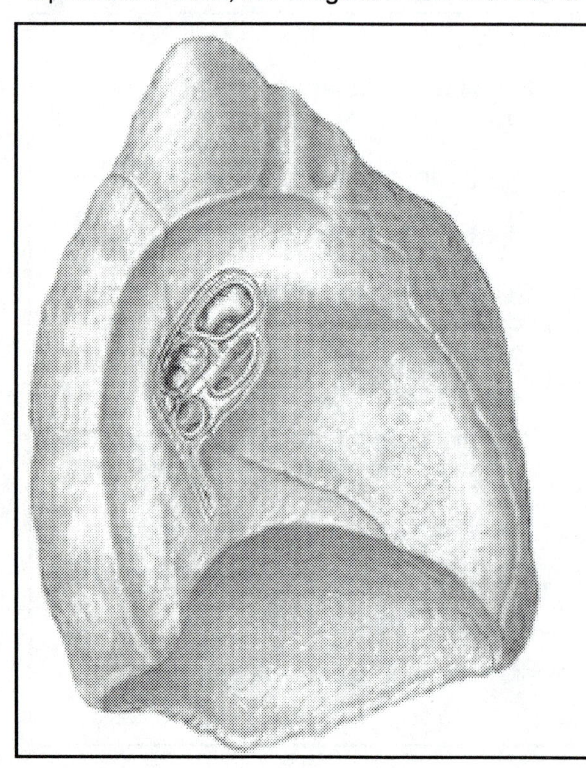

If you are working in Mac, click on the **Show All Pins** button and scroll to **All Systems**. Identify:

- **Apex of left lung**
- **Left primary bronchus**
- **Cardiac impression in lung**
- **Diaphragmatic surface of inferior lobe of lung**
- **Inferior lobe of left lung**
- **Left inferior pulmonary vein**
- **Left pulmonary artery**
- **Left primary bronchus**
- **Left superior pulmonary vein**
- **Oblique fissure of left lung**
- **Superior lobe of left lung**

When you are finished, **close** the window.

Under the **File** menu, drag down to **Open** (then to **Content** if you are working in Windows). Click on **Atlas Anatomy**. **System** and **Respiratory** should still be selected. Choose **Right Lung (Med)**. If the list is not in alphabetical order, the image is a little over half of the way down the list.

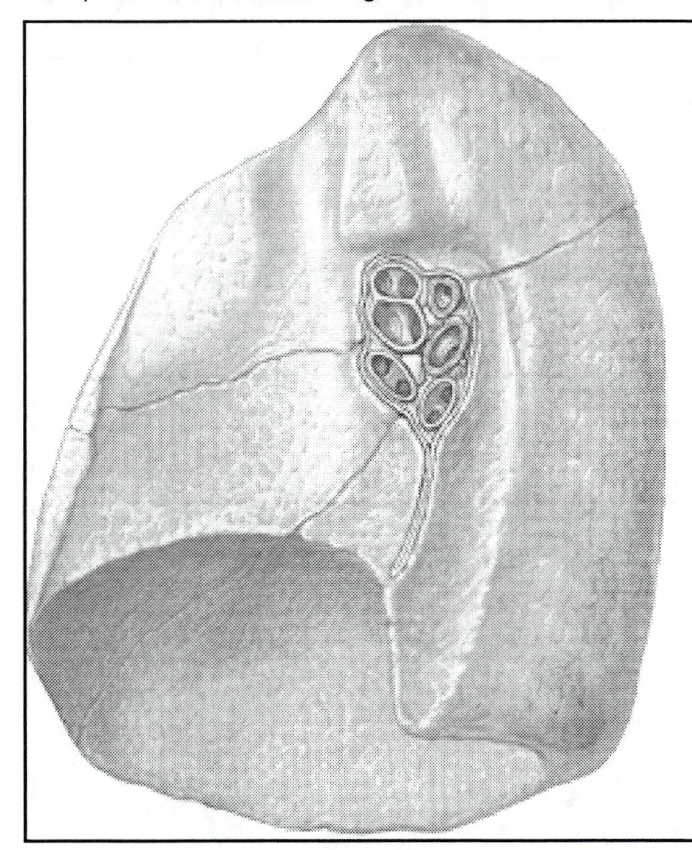

If you are working in Mac, click on the **Show All Pins** button and scroll to **All Systems**. Identify:

- **Apex of right lung**
- **Horizontal fissure of right lung**
- **Cardiac impression in lung**
- **Inferior lobe of right lung**
- **Middle lobe of right lung**
- **Right inferior pulmonary vein**
- **Right pulmonary artery**
- **Secondary bronchus**
- **Tertiary bronchus**
- **Right superior pulmonary vein**
- **Oblique fissure of right lung**
- **Superior lobe of right lung**

When you are finished, **close** the window.

## Cross Section of the Lower Respiratory System

Under the **File** menu, drag down to **Open** (then to **Content** if you are working in Windows). Click on **Atlas Anatomy**. **System** and **Respiratory** should still be selected. Choose **T8 Vertebra (Inf)**. If the list is not in alphabetical order, the image is three quarters of the way down the list.

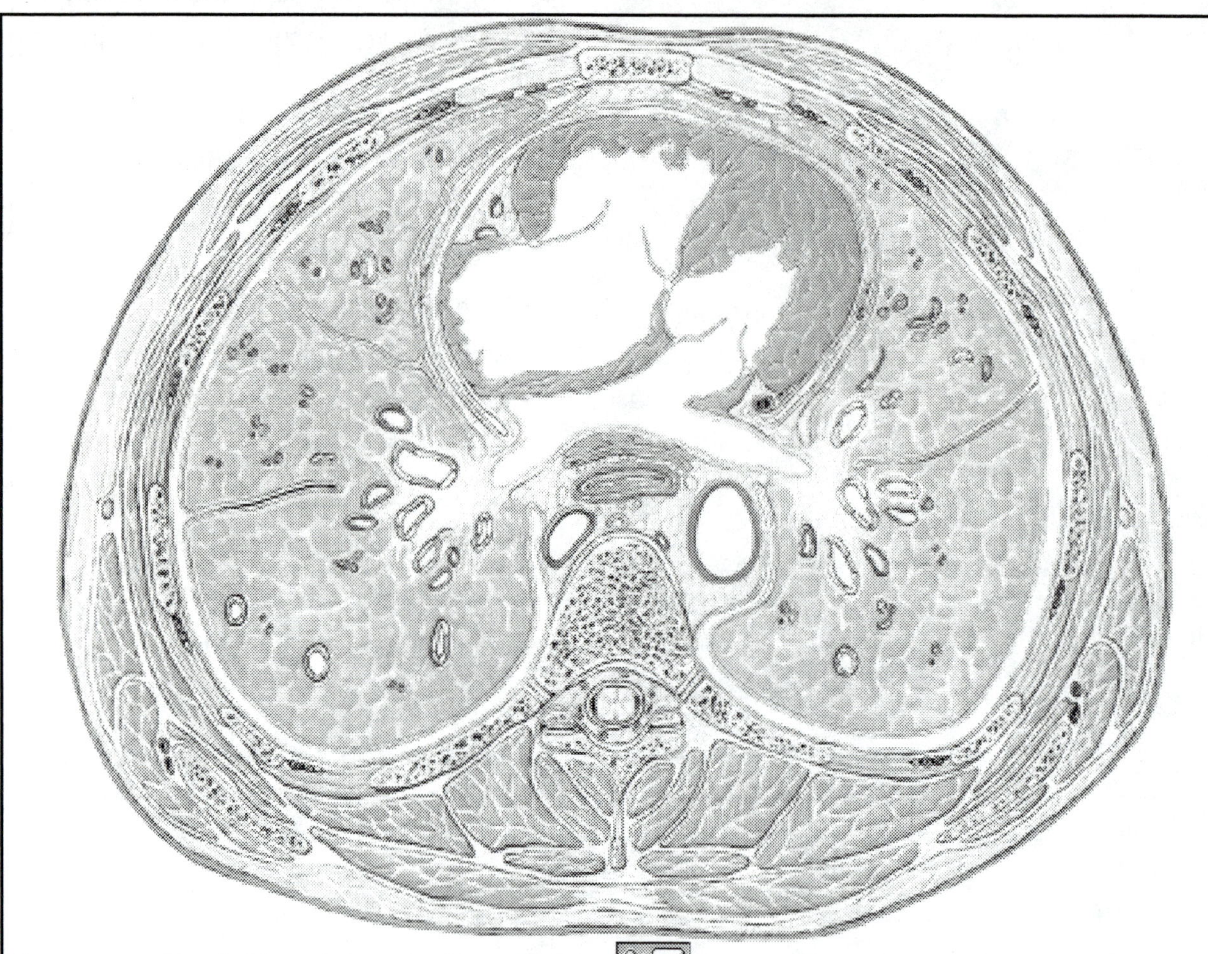

If you are working in Mac, click on the **Show All Pins** button and scroll to **All Systems**. Identify:

- **Horizontal fissure of right lung**
- **Esophagus**
- **Inferior lobe of left lung**
- **Inferior lobe of right lung**
- **Left superior pulmonary vein**
- **Superior lobe of left lung**
- **Superior lobe of right lung**
- **Middle lobe of right lung**
- **Oblique fissure of left lung**
- **Oblique fissure of right lung**
- **Parietal pleura**
- **Pleural cavity**
- **Thoracic duct**
- **Visceral pleura**

When you are finished, **close** the window.

Each lobe of the lungs is divided into ______________________________________________.

## Cadaver Images of the Lower Respiratory System

Under the **File** menu, drag down to **Open** (then to **Content** if you are working in Windows). Click on **Atlas Anatomy**. **System** and **Respiratory** should still be selected. Choose **Thoracic Viscera (Ant)**. If the list is not in alphabetical order, the image is three quarters of the way down the list.

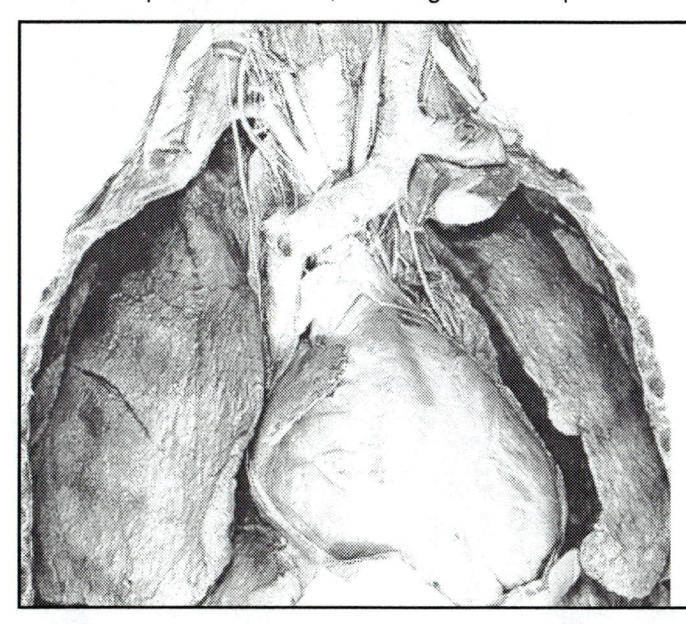

Identify:

- **Horizontal fissure of right lung**
- **Inferior lobe of right lung**
- **Superior lobe of left lung**
- **Middle lobe of right lung**
- **Oblique fissure of left lung**
- **Oblique fissure of right lung**
- **Superior lobe of left lung**
- **Superior lobe of right lung**
- **Trachea**

When you are finished, **close** the window.

Under the **File** menu, drag down to **Open** (then to **Content** if you are working in Windows). Click on **Atlas Anatomy**. **System** and **Respiratory** should still be selected. Choose **Posterior Mediastinum (Ant)**. If the list is not in alphabetical order, the image is close to the bottom of the list.

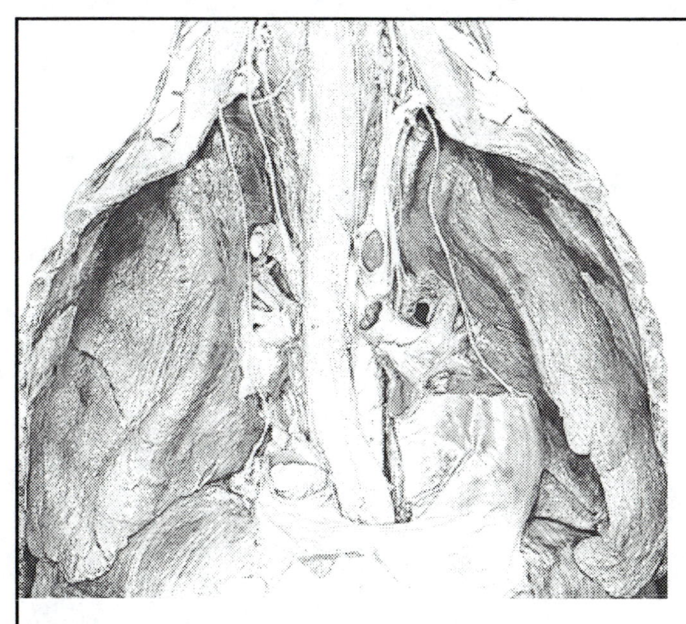

If you are working in Mac, click on the **Show All Pins** button and scroll to **All Systems**.
Identify:

- **Horizontal fissure of right lung**
- **Inferior vena cava**
- **Left inferior pulmonary vein**
- **Middle lobe of right lung**
- **Left phrenic nerve**
- **Oblique fissure of left lung**
- **Superior lobe of left lung**
- **Left pulmonary artery**
- **Right inferior pulmonary vein**
- **Right phrenic nerve**
- **Right superior pulmonary vein**
- **Secondary bronchus**

When you are finished, **close** the window.

## Objectives: Lower Respiratory System

Students will learn the anatomy of the trachea, bronchi, and lungs.

**Lungs**

- Apex of the lungs
- Hilum of the lungs
- Cardiac notch of the left lung
- Horizontal fissure of the right lung
- Lobes of the lungs
- Oblique fissures of the lungs
- Parietal pleura
- Pleural cavity
- Visceral pleura
- Pulmonary arteries
- Pulmonary veins
- Pulmonary trunk

**Trachea & Bronchi**

- Carina
- Primary bronchi
- Secondary bronchi
- Tertiary bronchi
- Bronchopulmonary segment
- Tracheal ring cartilage
- Bronchial cartilage

**Other**

- Phrenic nerve
- Diaphragm
- Esophagus
- Descending thoracic aorta
- Lymph nodes
- Thoracic duct

# Upper Digestive System

## Opening A.D.A.M Interactive Anatomy

Open A.D.A.M. Interactive Anatomy according to the directions in the Tutorial.

## 3D Anatomy of the Teeth

Under the **File** menu, drag down to **Open** (then to **Content** if you are working in Windows). Click on **3D Anatomy**, then on **3D Skull**.

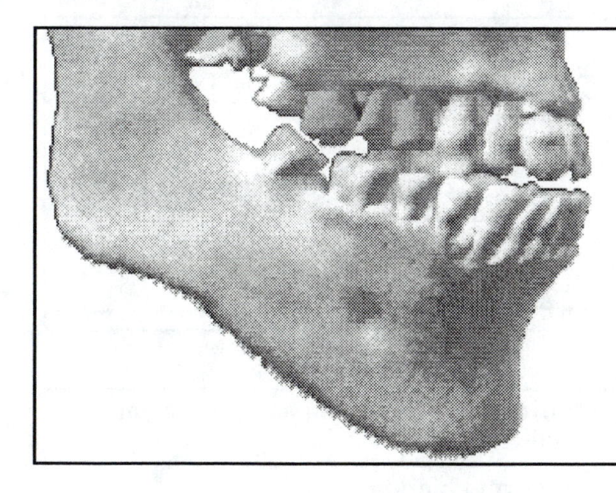

Identify the following by clicking on the **3D Anatomy Structure List**:

- **Incisors**
- **Canines**
- **Premolars**
- **Molars**

When you are finished, close **3D Anatomy**.

How many incisors are there in the adult? ________

How many canines are there in the adult? ________

How many premolars are there in the adult? ________

How many molars are there in the adult? ________

## Anterior Dissection of the Upper Digestive System

Under the **File** menu, drag down to **Open** (then to **Content** if you are working in Windows). Choose **Dissectible Anatomy**.

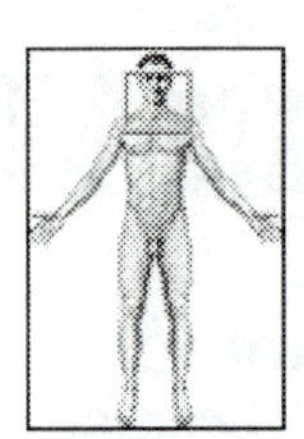

Choose either male or female. Select **Anterior.** Click **Open**.

Enlarge the window and position the image over the mouth and neck.

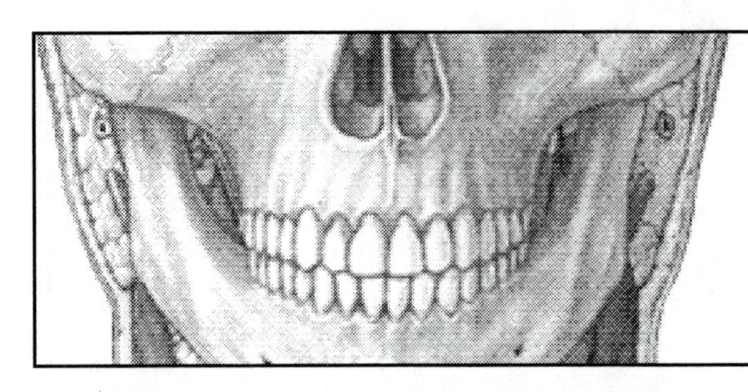

Click on the **Depth Bar** until you reach **layer 48**. Identify:

- **Incisors**
- **Canines**
- **Premolars**
- **Molars**
- **Parotid glands**

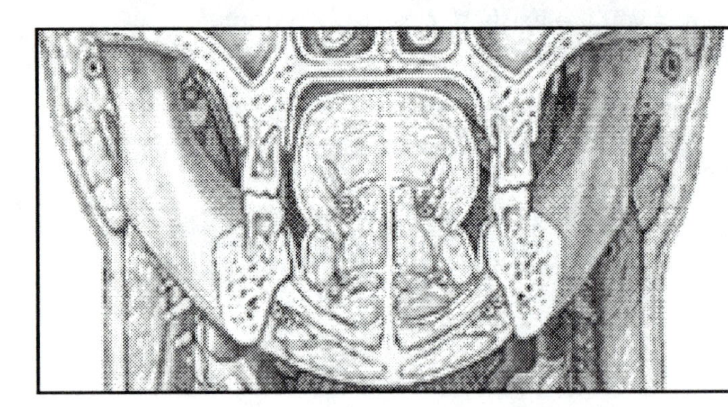

Click on the **Depth Bar** to **layer 58**. Identify:

- **Parotid glands**
- **Sublingual glands**
- **Submandibular glands**
- **Molars**

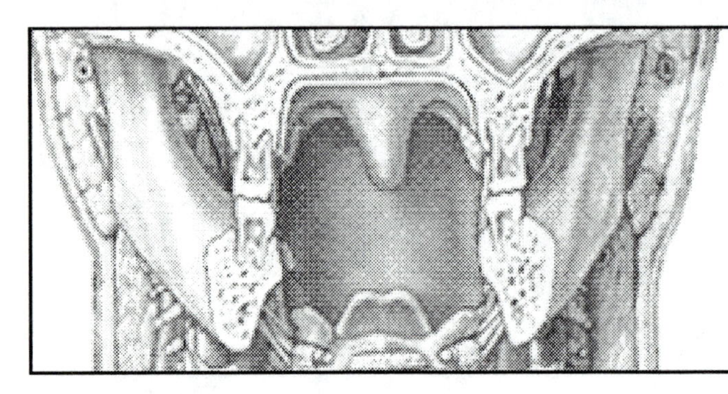

Click once on the **Depth Bar** to **layer 59**. Identify:

- **Parotid glands**
- **Mucosa of uvula and soft palate**

Scroll down on the **Depth Bar** to **layer 256**, then click once to **layer 257**. No graphic has been provided for this layer. Identify:

- **Esophagus**

## Medial Dissection of the Upper Digestive System

Click on the **View** button and choose **Medial**.

Click on the **Normal** button. Windows — Mac

Drag the **Depth Bar** to **layer 0.**

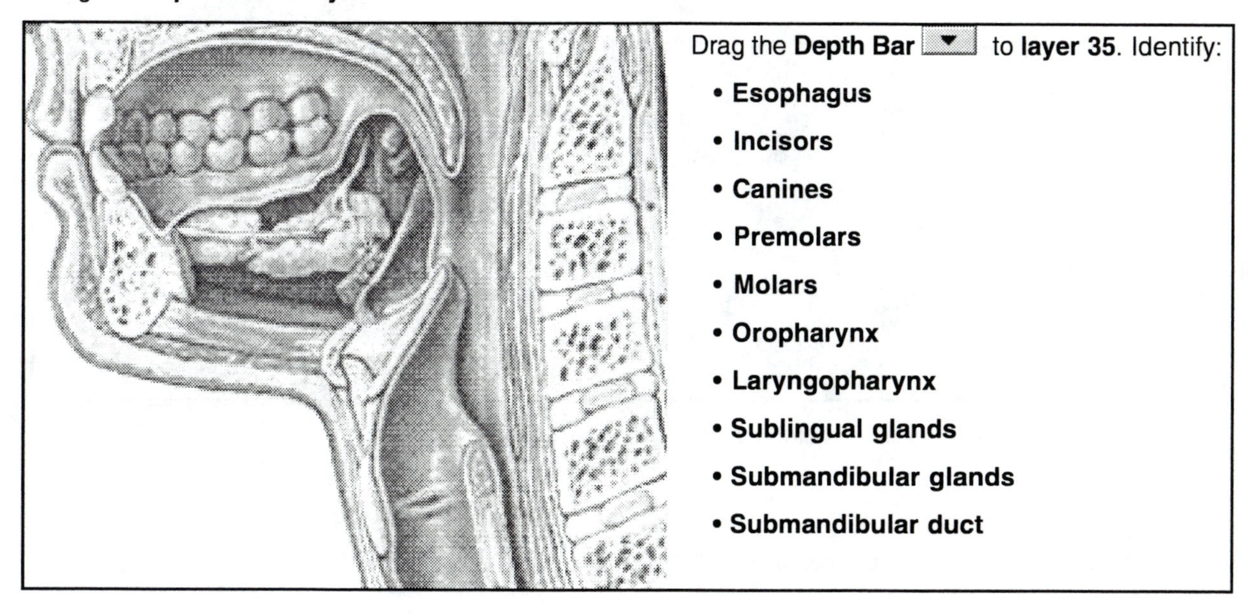

Drag the **Depth Bar** to **layer 35**. Identify:

- **Esophagus**
- **Incisors**
- **Canines**
- **Premolars**
- **Molars**
- **Oropharynx**
- **Laryngopharynx**
- **Sublingual glands**
- **Submandibular glands**
- **Submandibular duct**

## Lateral Dissection

Click on the **View** button and choose **Lateral**.

Click on the **Normal** button. Windows— —Mac

Drag the **Depth Bar** to **layer 0.**

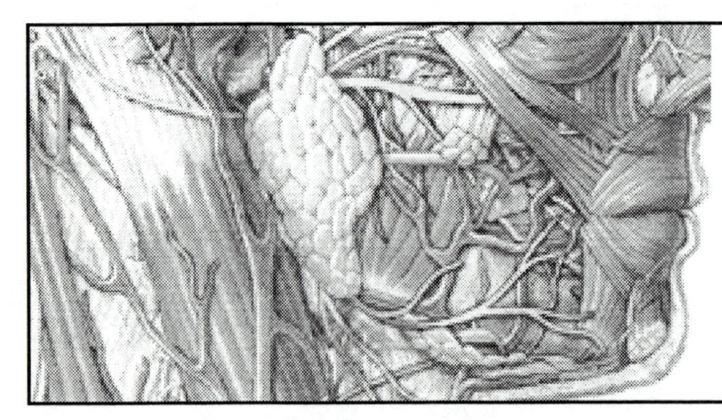

Drag the **Depth Bar** to **layer 8**. Identify:

- **Parotid gland**
- **Parotid duct**
- **Submandibular gland**

Which salivary gland is generally targeted by the mumps virus?

---

The parotid salivary gland is innervated by which cranial nerve?

---

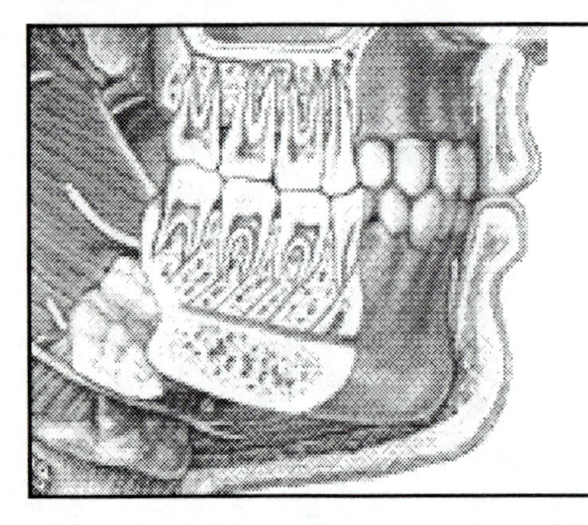

Scroll on the **Depth Bar** to **layer 188**. Identify:

- **Incisors**
- **Canines**
- **Premolars**
- **Molars**
- **Submandibular gland**

What cranial nerve innervates the sublingual and submandibular salivary gland?

---

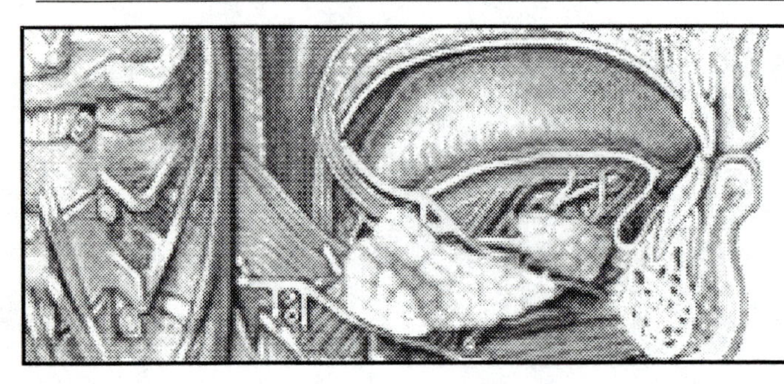

Click on the **Depth Bar** until you reach **layer 231**. Identify:

- **Incisors**
- **Sublingual gland**
- **Submandibular gland**

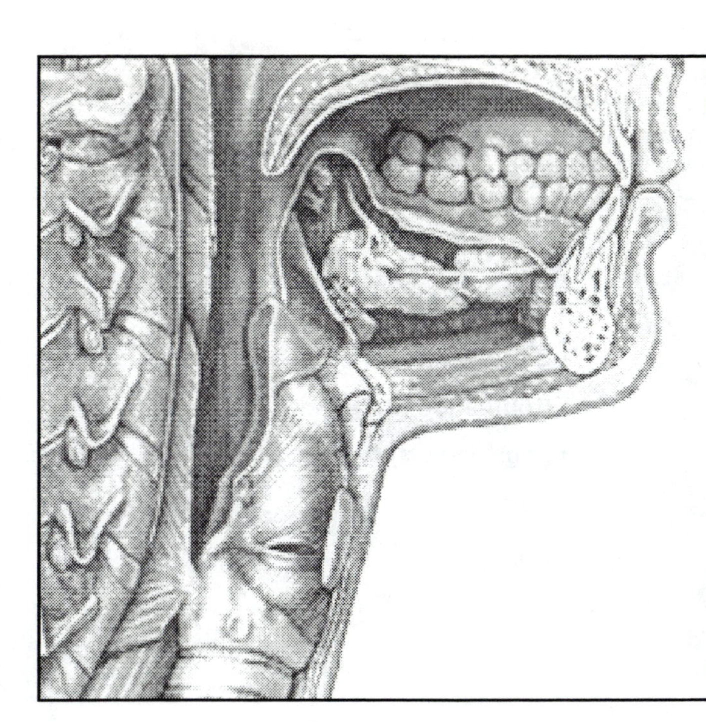

Click on the **Depth Bar** [▼] until you reach **layer 266**. Identify:

- **Esophagus**
- **Sublingual gland**
- **Submandibular gland**
- **Submandibular duct**
- **Incisors**
- **Canines**
- **Premolars**
- **Molars**
- **Oropharynx**
- **Laryngopharynx**

When you are finished, **close** the window.

## Atlas Anatomy of the Glands of the Upper Digestive System

Under the **File** menu, drag down to **Open** (then to **Content** if you are working in Windows). Click on **Atlas Anatomy**. Select **Region** and **Head & Neck**. Choose **Glands of Head and Neck (Lat)**. If the list is not in alphabetical order, the image is about half way down the list.

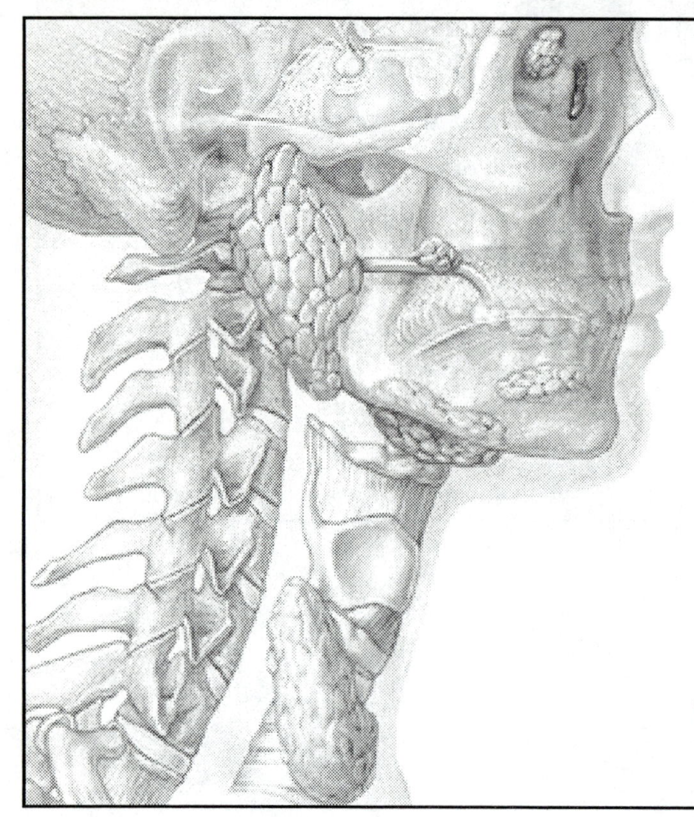

Identify :

- **Parotid duct**
- **Parotid gland**
- **Sublingual gland**
- **Submandibular gland**
- (Right lobe of) **thyroid gland**
- **Lacrimal gland**

When you are finished, **close** the window.

## Atlas Anatomy of the Teeth

Under the **File** menu, drag down to **Open** (then to **Content** if you are working in Windows). Click on **Atlas Anatomy**. **Region** and **Head & Neck** should still be selected. Choose **Mandible (Sup)**. If the list is not in alphabetical order, the image is close to the top of the list.

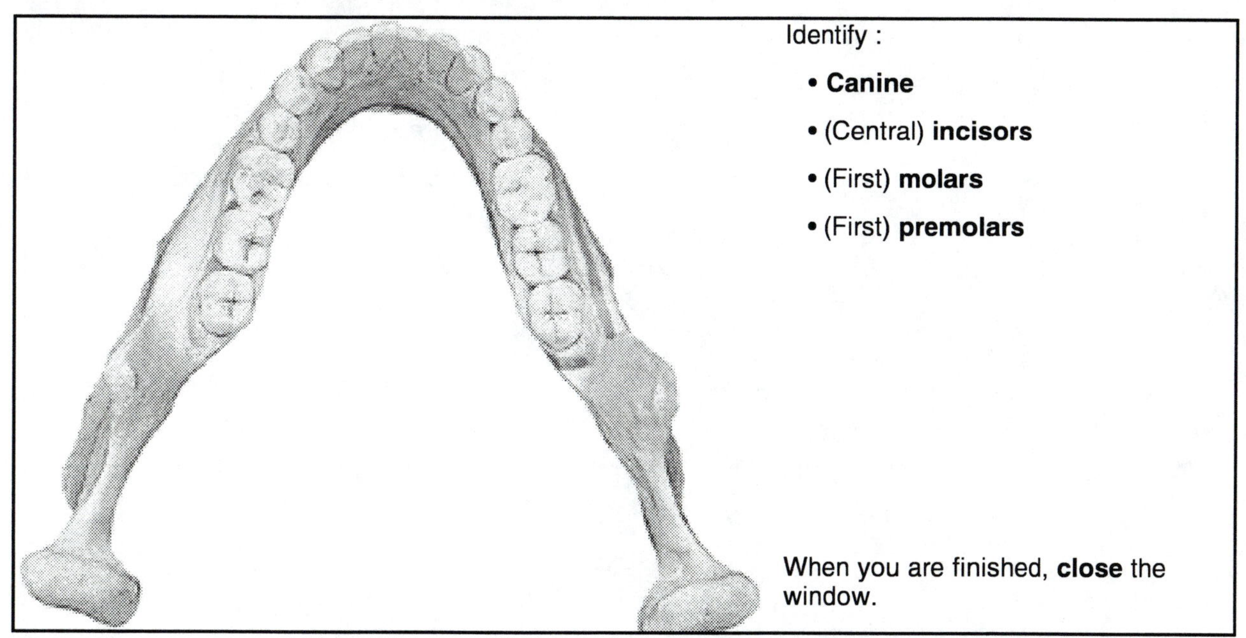

Identify :

- **Canine**
- (Central) **incisors**
- (First) **molars**
- (First) **premolars**

When you are finished, **close** the window.

Under the **File** menu, drag down to **Open** (then to **Content** if you are working in Windows). Click on **Atlas Anatomy**. **Region** and **Head & Neck** should still be selected. Choose **Muscle Atts-Infratemporal Fossa**. If the list is not in alphabetical order, the image is close to the top of the list.

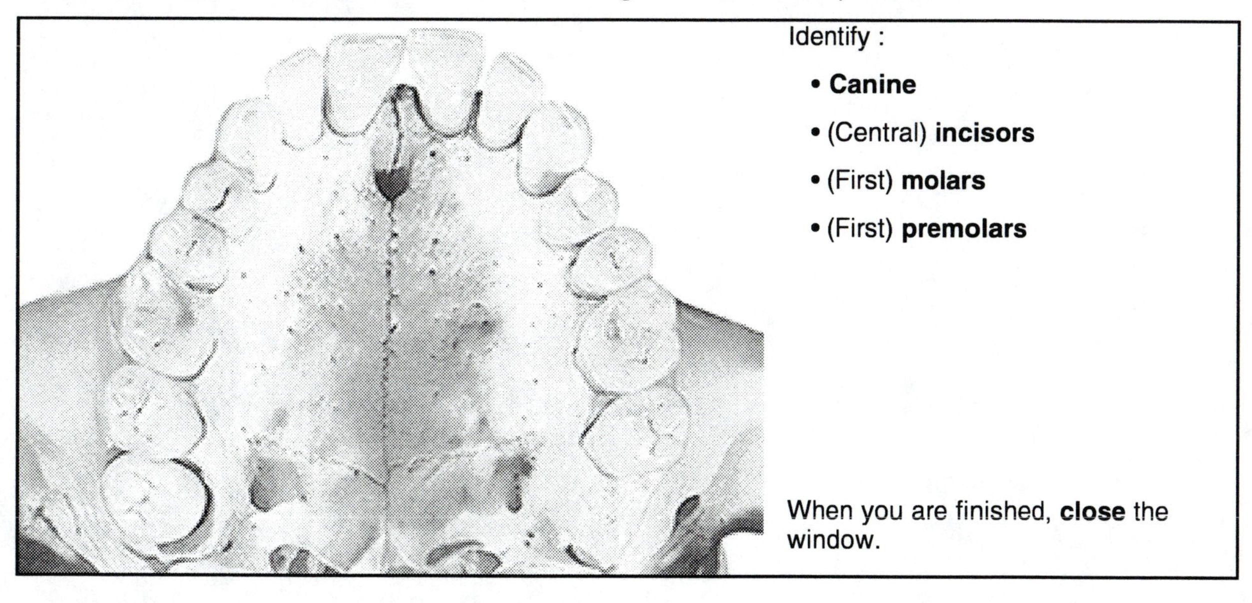

Identify :

- **Canine**
- (Central) **incisors**
- (First) **molars**
- (First) **premolars**

When you are finished, **close** the window.

## Cadaver Image of the Upper Digestive System

Under the **File** menu, drag down to **Open** (then to **Content** if you are working in Windows). Click on **Atlas Anatomy**. **Region** and **Head & Neck** should still be selected. Choose **Facial Nerve (Lat) 2**. If the list is not in alphabetical order, the image is about half way down the list.

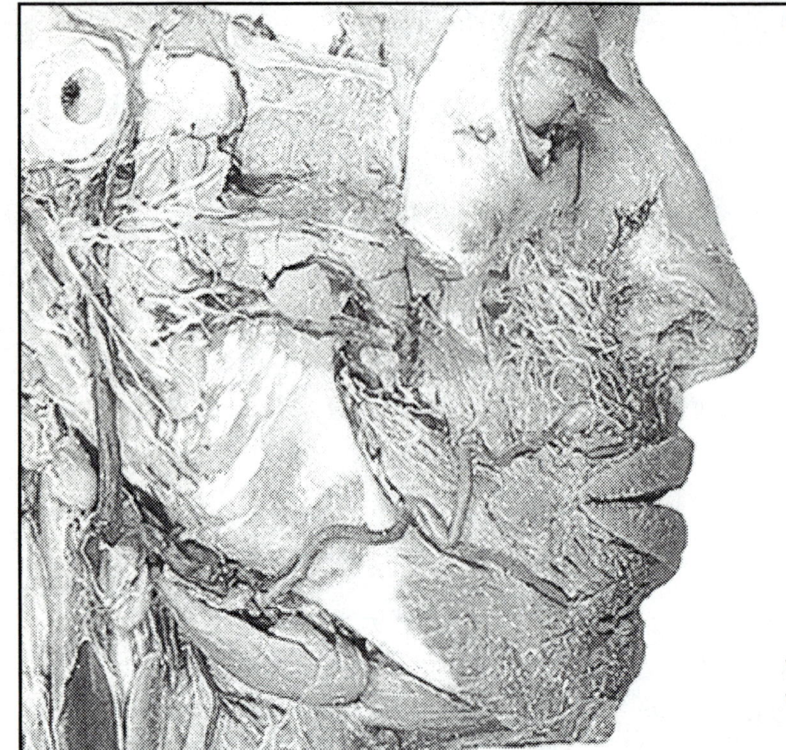

Identify :

- **Parotid duct**
- **Parotid gland**
- **Submandibular gland**

When you are finished, **close** the window.

## Objectives: Upper Digestive System

Student will study the anatomy of the upper digestive system.

**Teeth**

- Incisors
- Canines
- Premolars
- Molars

**Glands**

- Parotid gland
- Parotid duct
- Submandibular gland
- Submandibular duct
- Sublingual gland
- Thyroid gland
- Lacrimal gland

**Other**

- Uvula
- Soft palate
- Esophagus
- Oropharynx
- Laryngopharynx

# Liver

## Opening A.D.A.M Interactive Anatomy

Open A.D.A.M. Interactive Anatomy according to the directions in the Tutorial.

## Anterior Dissection of the Liver

Choose **Dissectible Anatomy**.

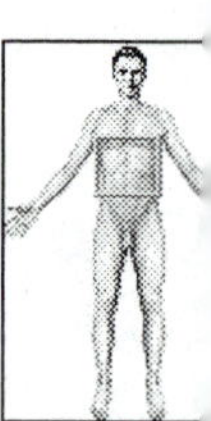

Choose **male**. Select **Anterior.** Click **Open**.

Enlarge the window and position the navigation rectangle over the upper abdomen.

Drag down on the **Depth Bar** to **layer 173**, then click several times on the **Depth Bar** ▼ until you **layer 194.** No graphic has been provided for this layer. Identify:

- **Parietal peritoneum**

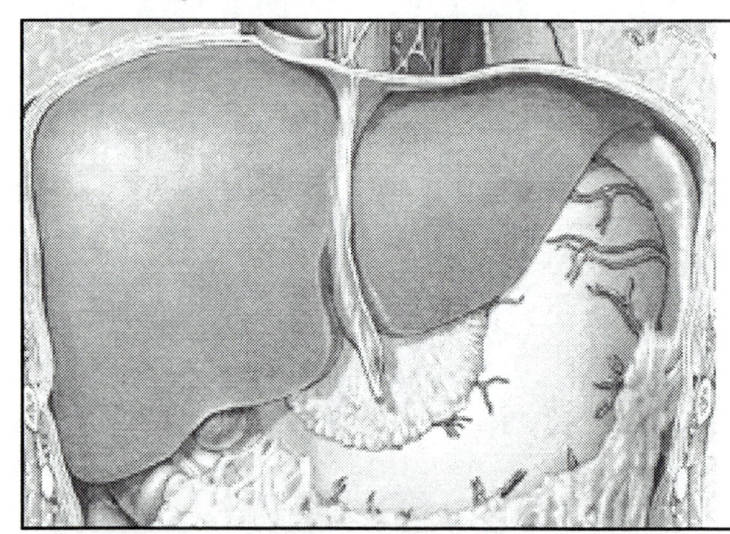

Click once on the **Depth Bar** ▼ **layer 195**. Identify:

- **Right lobe of liver**
- **Left lobe of liver**
- **Falciform ligament**
- **Round ligament of the liver**
- **Gallbladder**

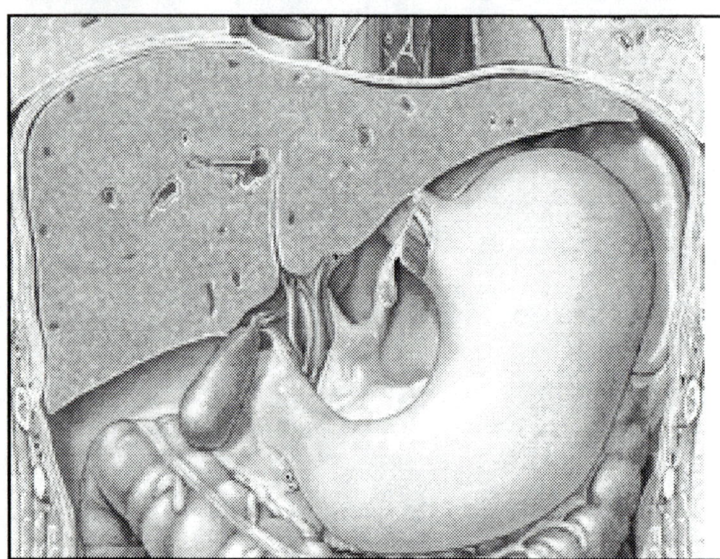

Click several times on the **Depth Bar** until you reach **layer 202**. Identify:

- **Right lobe of liver**
- **Left lobe of liver**
- **Body of gallbladder**
- **Common hepatic duct**
- **Right hepatic duct**
- **Hepatic artery** (Proper hepatic artery)
- **Hepatic portal vein** (Portal vein)
- **Branches of portal vein** (within

What two blood vessels bring blood to the liver?

_______________________________________________

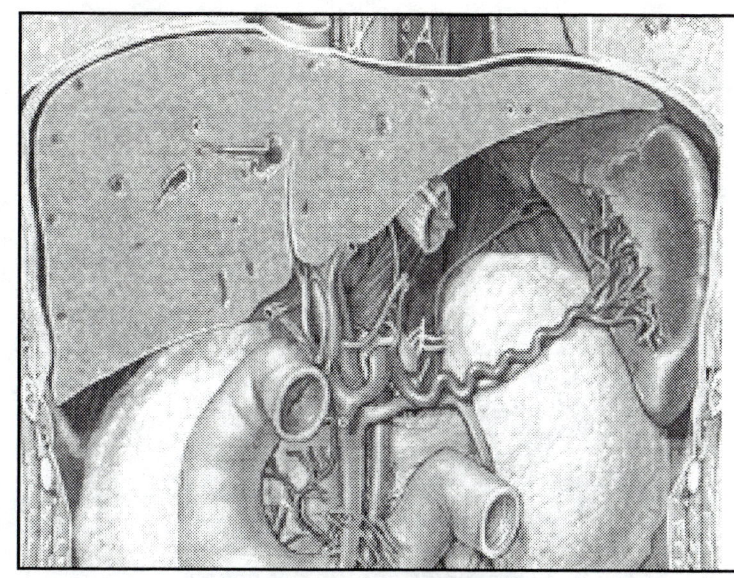

Click several times on the **Depth Bar** ▼ until you reach **layer 216**. Identify:

- **Right lobe of liver**
- **Left lobe of liver**
- **Cystic duct**
- **Common hepatic duct**
- **Right hepatic duct**
- **Hepatic artery** (Proper hepatic artery)
- **Hepatic portal vein** (Portal vein)
- **Branches of portal vein** (within liver)
- **Common bile duct** (see layer 215)

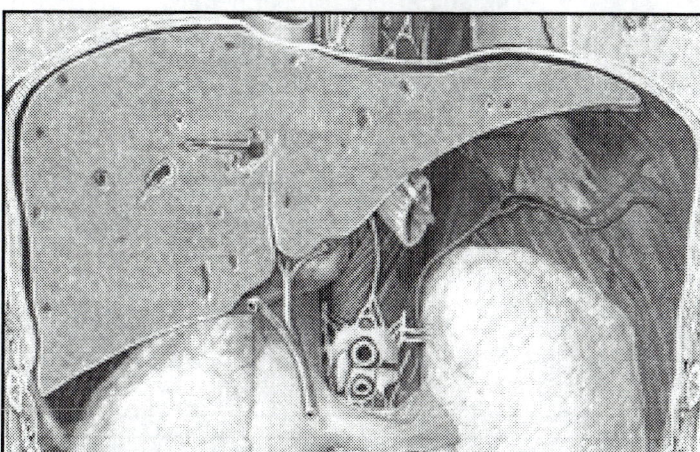

Click several times on the **Depth Bar** ▼ until you reach **layer 221**. Identify:

- **Caudate lobe of liver**
- **Cystic duct**
- **Common hepatic duct**
- **Right hepatic duct**
- **Left hepatic duct**
- **Common bile duct**
- **Inferior vena cava**

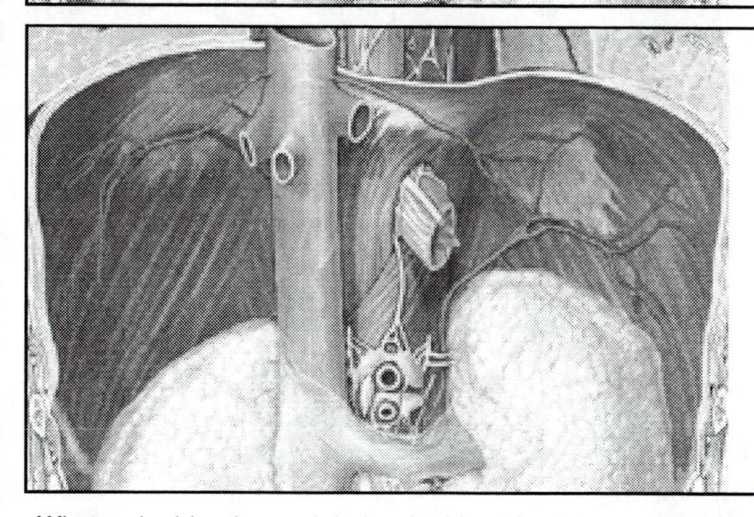

Click on the **Depth Bar** ▼ to **layer 228**. Identify:

- **Diaphragm**
- **Inferior vena cava**
- **Hepatic vein** (coming off vena cava)

What major blood vessel drains the blood from the liver?

____________________________________________________________

 

## Medial Dissection of the Liver

Click on the **View** button and choose **Medial**.

Click on the **Normal** button. Windows — Mac

Drag the **Depth Bar** to **layer 0.** No graphic has been provided for this layer. Identify:

- **Left lobe of liver**
- **Caudate lobe of liver**
- **Vein of hepatic portal venous system** (within liver)

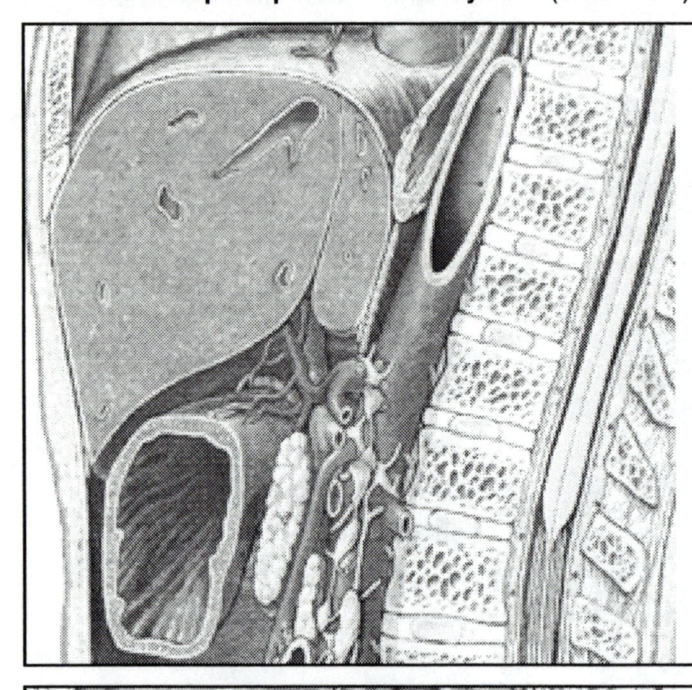

Click several times on the **Depth Bar** until you reach **layer 46**. Identify:

- **Descending thoracic aorta**
- **Abdominal aorta**
- **Hepatic artery** (Proper hepatic artery)
- **Branch of hepatic portal vein** within liver (Portal vein)

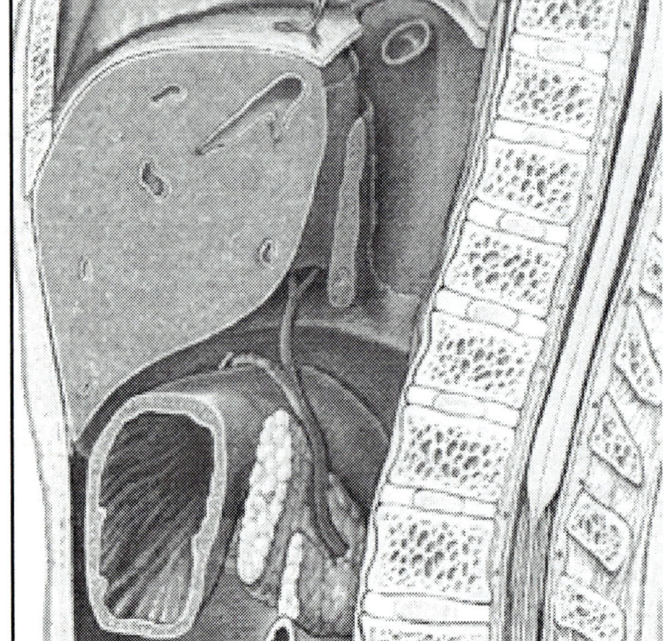

Click on the **Depth Bar** until you reach **layer 74**. Identify:

- **Right lobe of liver**
- **Left lobe of liver**
- **Cystic duct**
- **Common hepatic duct**
- **Common bile duct**
- **Left hepatic duct**
- **Right hepatic duct**
- **Caudate lobe of liver**
- **Quadrate lobe of liver**

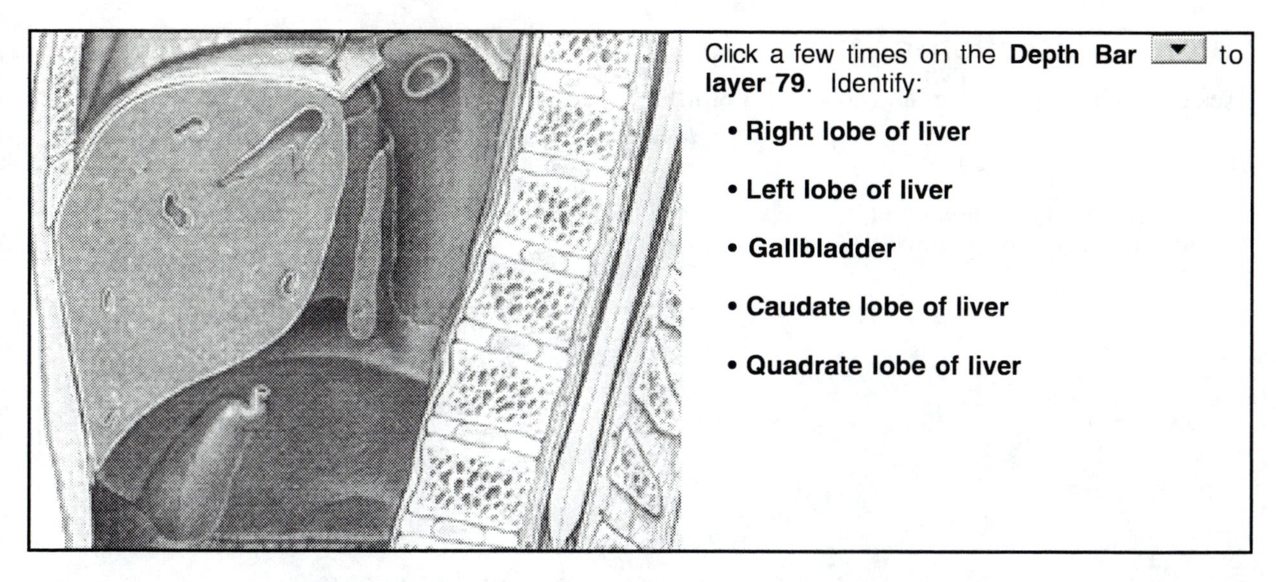

Click a few times on the **Depth Bar** ▼ to **layer 79**. Identify:

- **Right lobe of liver**
- **Left lobe of liver**
- **Gallbladder**
- **Caudate lobe of liver**
- **Quadrate lobe of liver**

How would you classify the position of the gallbladder with respect to the liver?

______________________________________________

List the four lobes of the liver:

______________________ ______________________

______________________ ______________________

## Lateral Dissection of the Liver

Click on the **View** button and choose **Lateral**.

Click on the **Normal** button. **Windows**—— ——**Mac**

Drag the **Depth Bar** to **layer 144**, then click several times on the **Depth Bar** until you reach **layer 151.** No graphic has been provided for this layer. Identify:

- **Liver**

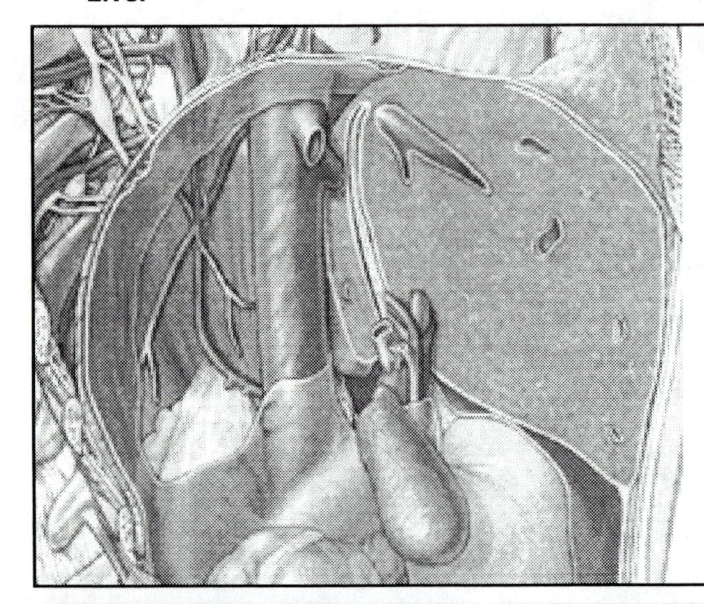

Click on the **Depth Bar** to **layer 156**. Identify:

- **Left lobe of liver**
- **Caudate lobe of liver**
- **Gallbladder (fundus & body)**
- **Inferior vena cava**

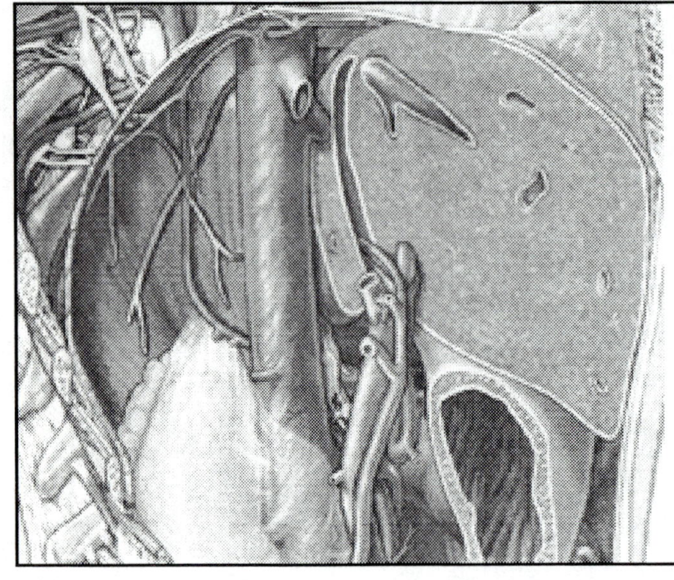

Click on the **Depth Bar** until you reach **layer 164**. Identify:

- **Left lobe of liver**
- **Common bile duct**
- **Common hepatic duct**
- **Hepatic artery** (Proper hepatic artery)
- **Hepatic portal vein** (Portal vein)
- Branches of **portal vein** (within liver)
- **Inferior vena cava**

## Atlas Anatomy of the Liver

Under the **File** menu, drag down to **Open** (then to **Content** if you are working in Windows). Click on **Atlas Anatomy**. Select **Region** and **Abdomen**. Choose **Liver (Ant)**. If the list is not in alphabetical order, the image is three fourths of the way down the list.

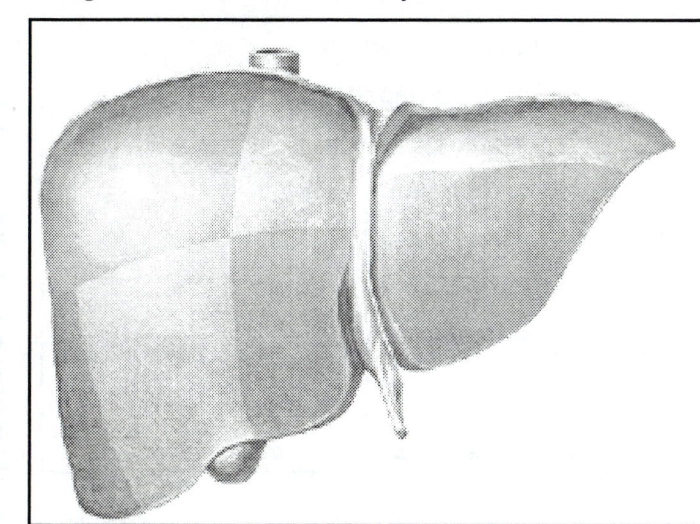

Identify:

- **Coronary ligament of liver**
- **Falciform ligament**
- **Fundus of gallbladder**
- **Inferior vena cava**
- **Left lobe of liver**
- **Right lobe of liver**
- **Round ligament of liver**

What attaches the liver to the diaphragm?

---

What attaches the liver to the anterior wall of the abdominal cavity?

---

When finished, **close** the window.

Under the **File** menu, drag down to **Open** (then to **Content** if you are working in Windows). Click on **Atlas Anatomy**. **Region** and **Abdomen** should still be selected. Choose **Liver (Inf)**. If the list is not in alphabetical order, the image is near the bottom of the list.

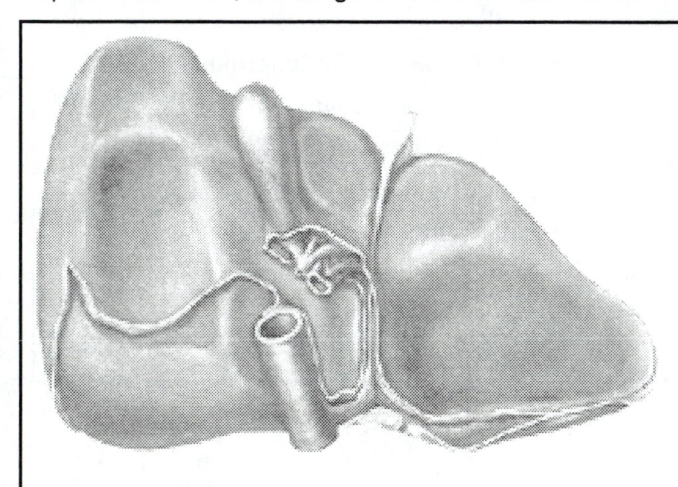

Identify:

- **Bile duct (Common bile duct)**
- **Body of gallbladder**
- **Fundus of gallbladder**
- **Falciform ligament**
- **Hepatic vein**
- **Inferior vena cava**
- **Round ligament of liver**
- **Portal vein**
- **Proper hepatic artery**
- **Quadrate lobe of liver**

When finished, **close** the window.

Under the **File** menu, drag down to **Open** (then to **Content** if you are working in Windows). Click on **Atlas Anatomy**. **Region** and **Abdomen** should still be selected. Choose **Liver (Post)**. If the list is not in alphabetical order, the image is near the bottom of the list.

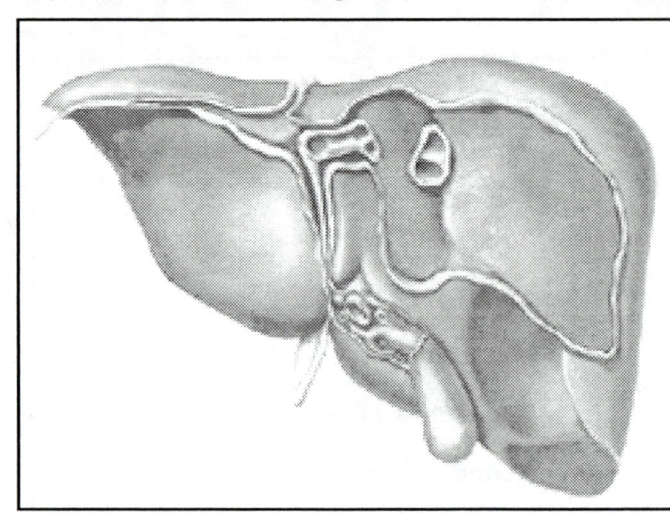

Identify:

- **Body of gallbladder**
- **Common hepatic artery**
- **Coronary ligament of liver**
- **Hepatic vein**
- **Cystic duct**
- **Round ligament of liver**
- **Portal vein**

When finished, **close** the window.

Under the **File** menu, drag down to **Open** (then to **Content** if you are working in Windows). Click on **Atlas Anatomy**. **Region** and **Abdomen** should still be selected. Choose **Pancreatic & Bile Ducts 1**. If the list is not in alphabetical order, the image is three fourths of the way down the list.

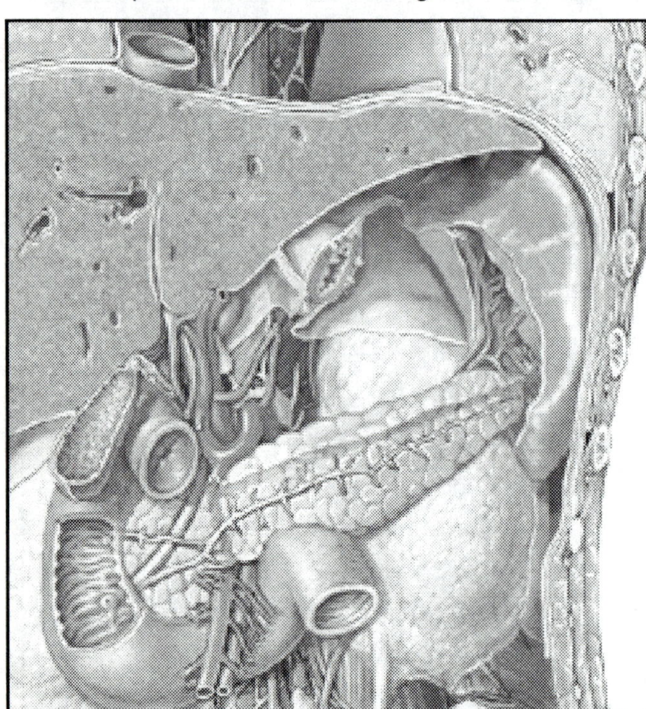

Identify:

- **Bile duct (Common bile duct)**
- **Body of pancreas**
- **Common hepatic duct**
- **Duodenal papilla** (entrance of bile duct into duodenum)
- **Cystic duct**
- **Duodenum**
- **Gallbladder**
- **Hepatopancreatic ampulla**
- **Inferior vena cava**
- **Left hepatic duct**
- **Left lobe of liver**
- **Main pancreatic duct**
- **Proper hepatic artery**
- **Right hepatic duct**
- **Right lobe of liver**

When finished, **close** the window.

Under the **File** menu, drag down to **Open** (then to **Content** if you are working in Windows). Click on **Atlas Anatomy**. **Region** and **Abdomen** should still be selected. Choose **Hepatic Portal Vein**. If the list is not in alphabetical order, the image is three fourths of the way down the list.

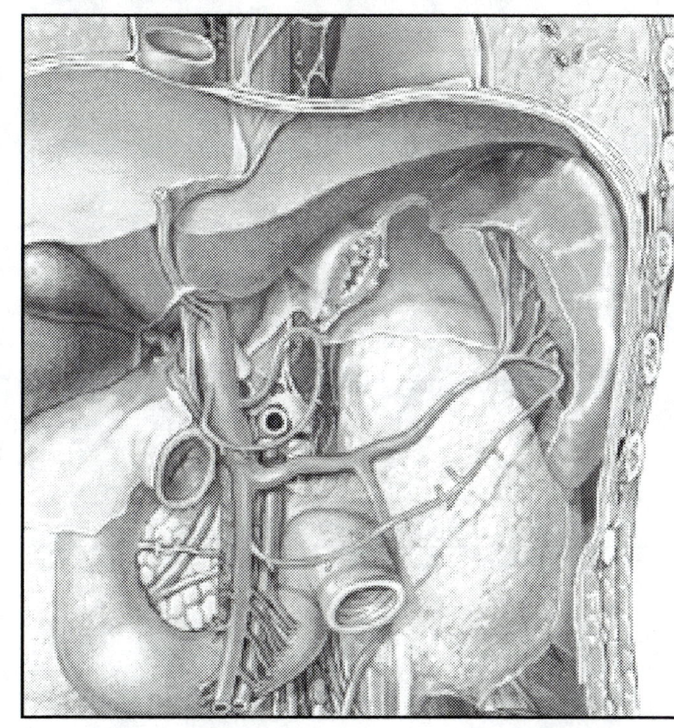

Identify:

- **Common hepatic duct**
- **Cystic duct**
- **Falciform ligament**
- **Gallbladder**
- **Left lobe of liver**
- **Portal vein**
- **Right lobe of liver**

This image clearly displays where the hepatic portal vein comes from. List several organs that drain into the hepatic portal vein.

______________________________

______________________________

______________________________

When finished, **close** the window.

 

## Cadaver Image of the Liver

Under the **File** menu, drag down to **Open** (then to **Content** if you are working in Windows). Click on **Atlas Anatomy**. Select **Region** and **Abdomen**. Choose **Abdominal Viscera**. If the list is not in alphabetical order, the image is near the top of the list.

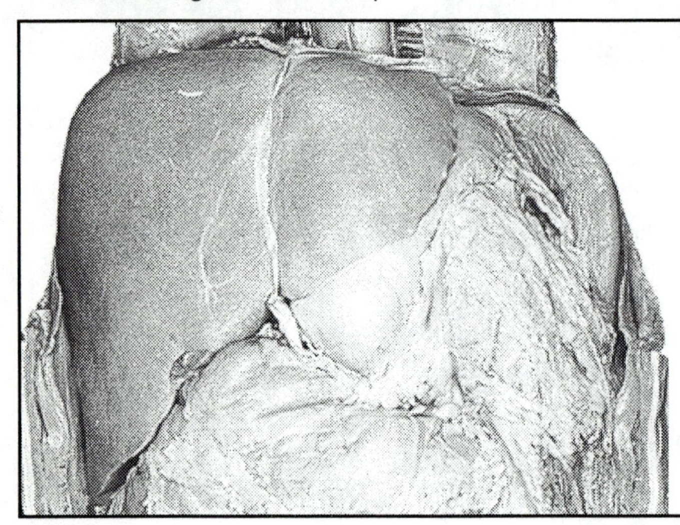

Identify the following by clicking on the heads of the pins.

- **Diaphragm**
- **Falciform ligament**
- **Gallbladder**
- **Left lobe of liver**
- **Round ligament of liver**
- **Right lobe of liver**

When finished, **close** the window.

## Objectives: Liver

Students will learn the anatomy of the liver, gallbladder, and associated structures.

**Lobes of the Liver**

- Right lobe
- Left lobe
- Caudate lobe
- Quadrate lobe

**Ligaments**

- Falciform ligament
- Round ligament
- Coronary ligament

**Ducts**

- Main pancreatic duct
- Hepatopancreatic ampulla
- Duodenal papilla
- Cystic duct
- Common hepatic duct
- Left hepatic duct
- Right hepatic duct
- Common bile duct

**Blood vessels**

- Hepatic vein
- Hepatic artery
- Hepatic portal vein
- Inferior vena cava
- Descending thoracic aorta
- Abdominal aorta

**Other**

- Body of gallbladder
- Fundus of gallbladder
- Diaphragm
- Body of pancreas
- Duodenum

# Abdominal Digestive System

## Opening A.D.A.M Interactive Anatomy

Open A.D.A.M. Interactive Anatomy according to the directions in the Tutorial.

## Anterior Dissection of the Abdominal Digestive System

Choose **Dissectible Anatomy**.

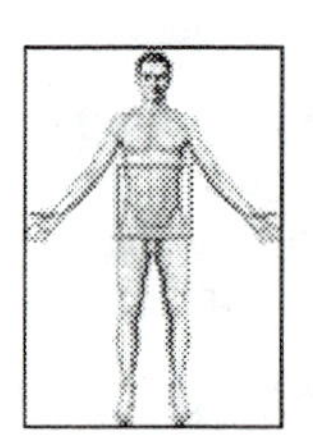

Choose either male or female. Select **Anterior.** Click **Open**.

Enlarge the window and position the navigation rectangle over the abdomen.

Drag down on the **Depth Bar** until you reach **layer 194**. No graphic has been provided for this layer. Identify:

- **Parietal peritoneum**

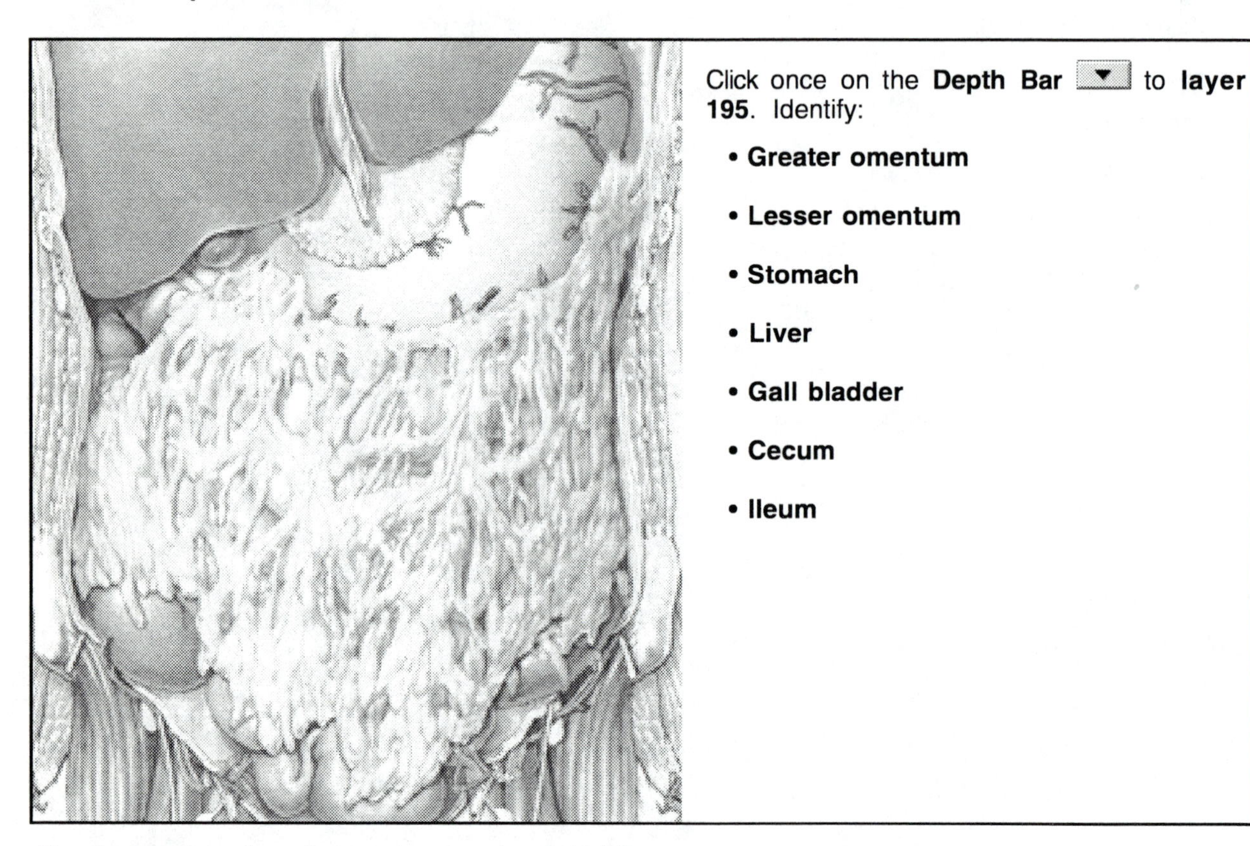

Click once on the **Depth Bar** to **layer 195**. Identify:

- **Greater omentum**
- **Lesser omentum**
- **Stomach**
- **Liver**
- **Gall bladder**
- **Cecum**
- **Ileum**

To which organs does the greater omentum attach?

______________________________________________

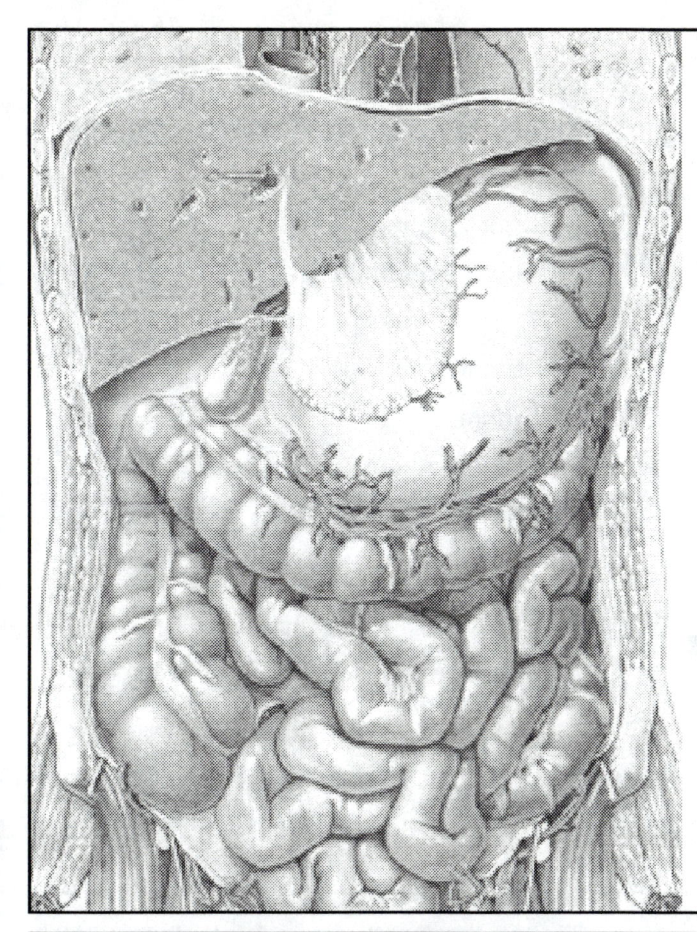

Click a few times on the **Depth Bar** ▼ to **layer 199**. Identify:

- **Esophagus**
- **Lesser omentum**
- **Right lobe of liver**
- **Left lobe of liver**
- **Gallbladder**
- **Body of stomach**
- **Fundus of stomach**
- **Pylorus of stomach**
- **Duodenum**
- **Transverse colon**
- **Ascending colon**
- **Cecum**
- **Sigmoid colon**
- **Ileum**
- **Jejunum**
- **Taenia coli** (Tenia Coli)
- **Hepatic flexure of colon**

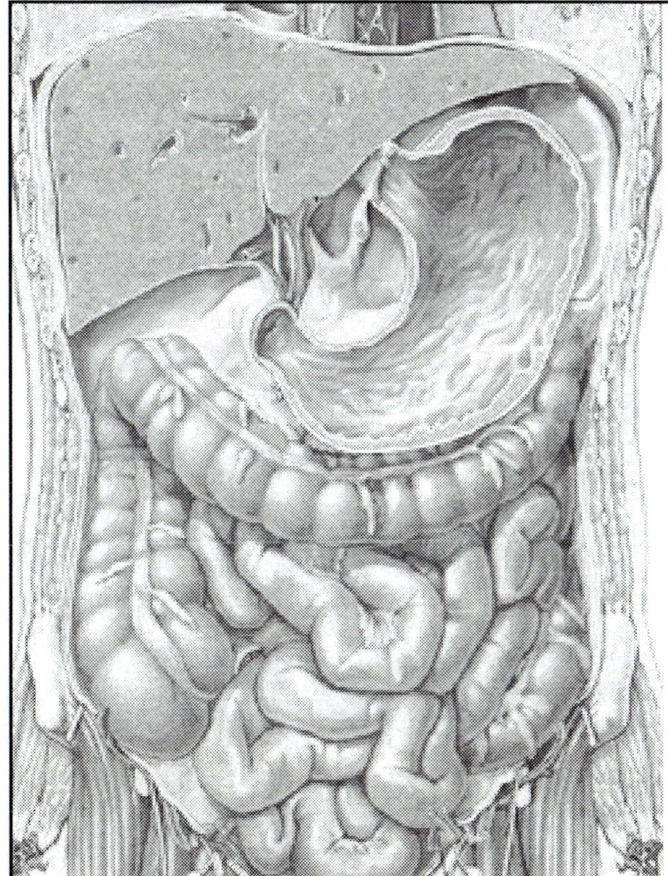

Click on the **Depth Bar** ▼ until you reach **layer 205**. Identify:

- **Esophagus** (above & below diaphragm)
- **Cardiac orifice of stomach**
- **Right lobe of liver**
- **Left lobe of liver**
- **Rugae of stomach**
- **Pylorus of stomach**
- **Pyloric sphincter**
- **Transverse colon**
- **Ascending colon**
- **Cecum**
- **Sigmoid colon**
- **Ileum**
- **Jejunum**
- **Taenia coli** (Tenia Coli)
- **Hepatic flexure of colon**
- **Splenic flexure of colon**
- **Mesentery**

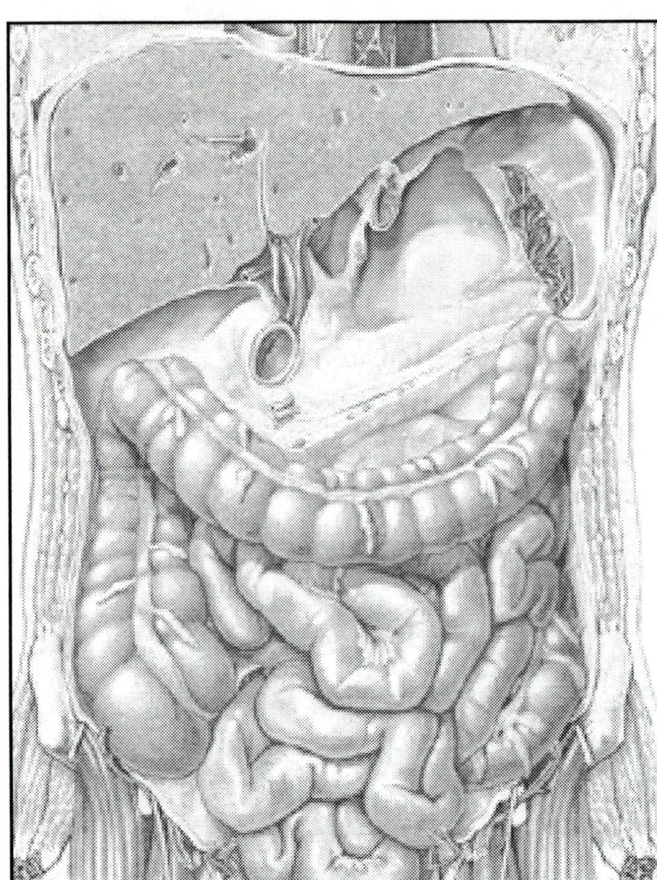

Click a few times on the **Depth Bar** ▼ until you reach **layer 207**. Identify:

- **Esophagus** (above & below diaphragm)
- **Peritoneum**
- **Right lobe of liver**
- **Pyloric sphincter**
- **Root of transverse mesocolon**
- **Ascending colon**
- **Cecum**
- **Sigmoid colon**
- **Ileum**
- **Jejunum**
- **Taenia coli** (Tenia Coli)
- **Hepatic flexure of colon**
- **Splenic flexure of colon**
- **Mesentery**
- **Spleen**

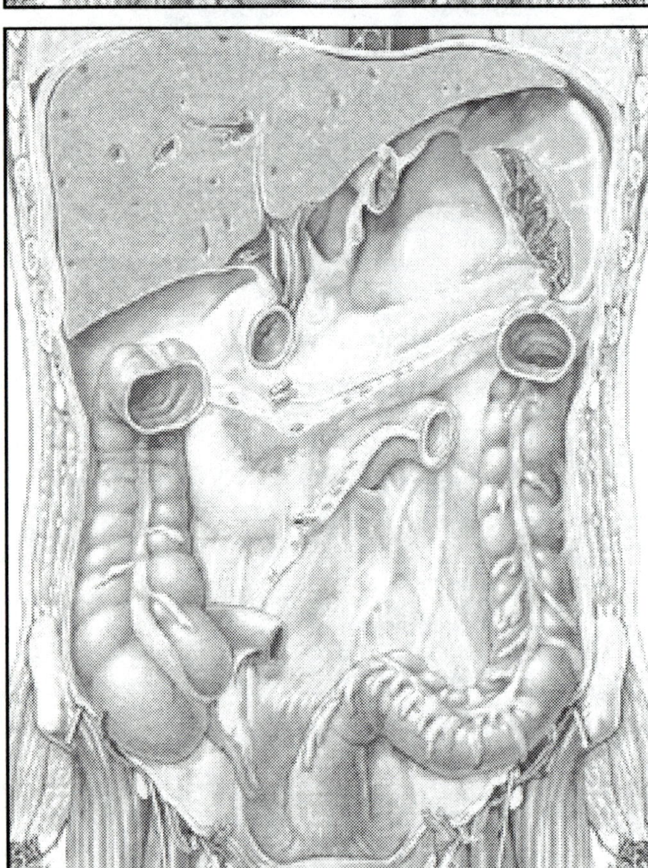

Click on the **Depth Bar** ▼ to **layer 209**. Identify:

- **Peritoneum**
- **Right lobe of liver**
- **Descending colon**
- **Transverse colon**
- **Ascending colon**
- **Cecum**
- **Sigmoid colon**
- **Ileum**
- **Taenia coli** (Tenia Coli)
- **Hepatic flexure of colon**
- **Splenic flexure of colon**
- **Root of mesentery**
- **Rectum**
- **Vermiform appendix**
- **Duodenum**

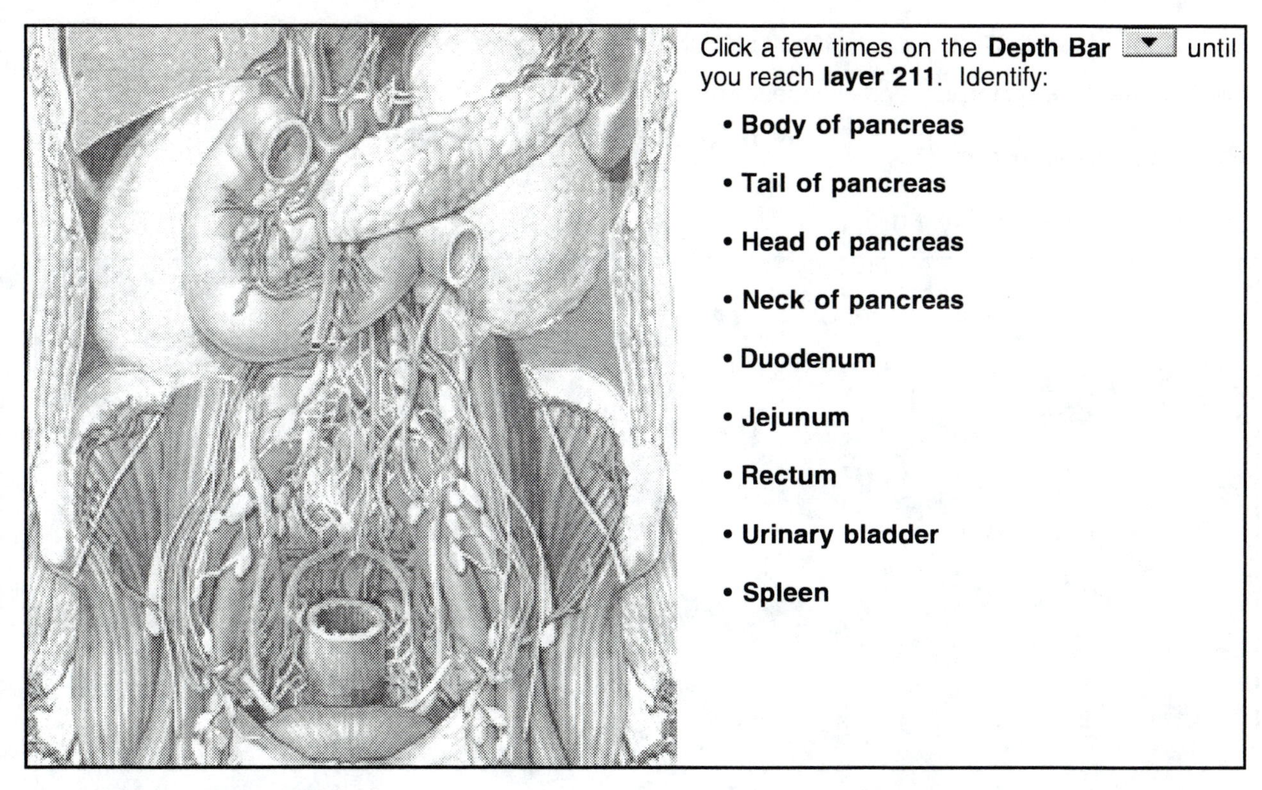

Click a few times on the **Depth Bar** ▼ until you reach **layer 211**. Identify:

- **Body of pancreas**
- **Tail of pancreas**
- **Head of pancreas**
- **Neck of pancreas**
- **Duodenum**
- **Jejunum**
- **Rectum**
- **Urinary bladder**
- **Spleen**

List three retroperitoneal organs:

______________________________

______________________________

______________________________

List three intraperitoneal organs:

______________________________

______________________________

______________________________

 

## Medial Dissection of the Abdominal Digestive System

Click on the **View** button and choose **Medial**.

Click on the **Normal** button. Windows— —Mac

Drag the **Depth Bar** to **layer 0.**

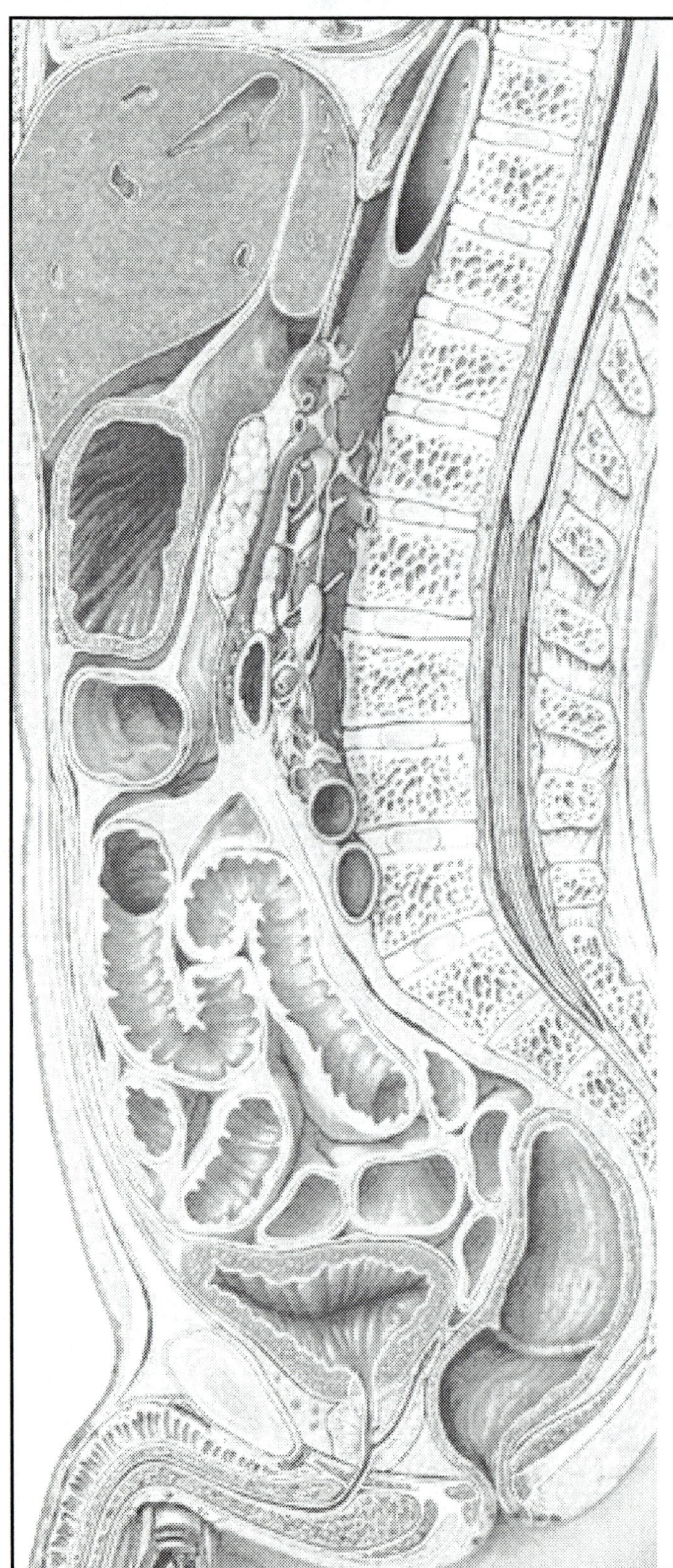

You will need to scroll to see all of the structures. Identify:

- **Esophagus**
- **Lesser omentum**
- **Liver**
- **Stomach**
- **Duodenum**
- **Lesser omentum**
- **Pylorus of stomach**
- **Pancreas (neck)**
- **Jejunum**
- **Transverse colon**
- **Ileum**
- **Mesentery** (highlight it to see the whole thing)
- **Urinary bladder**
- **Transverse mesocolon**
- **Peritoneum** (try to follow it all the way around)
- **Rectum**

List one antiperitoneal organ:

______________________________

Drag the **Depth Bar** to **layer 127**. No graphic has been provided for this layer. Identify the **internal anal sphincter**.

When you are finished, **close** the window.

# Atlas Anatomy of the Digestive System

Under the **File** menu, drag down to **Open** (then to **Content** if you are working in Windows). Click on **Atlas Anatomy**. Select **System** and **Digestive**. Choose **Large Intestine in Situ**. If the list is not in alphabetical order, the image is about half of the way down the list.

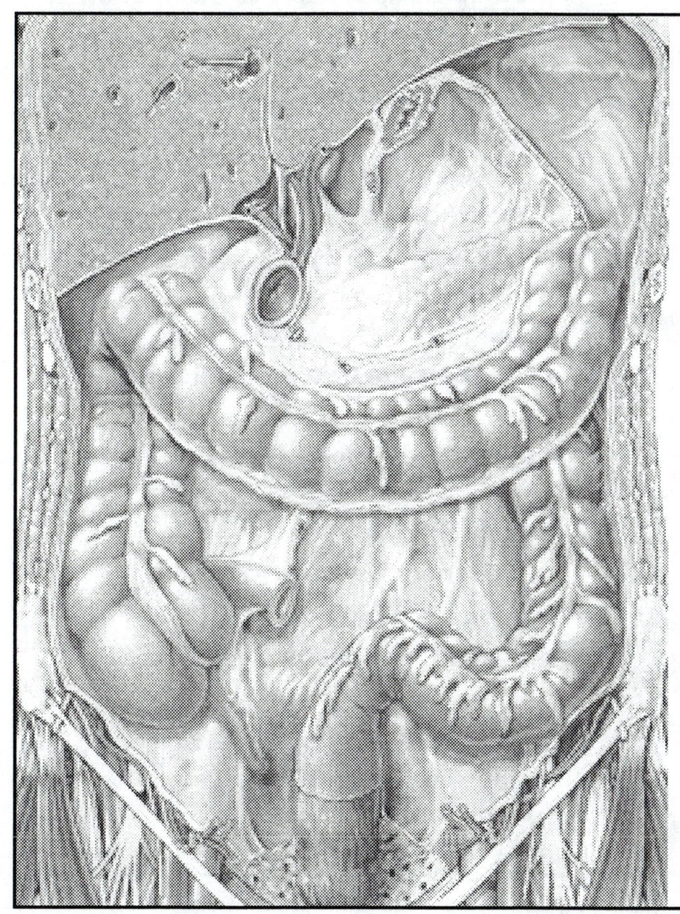

If you are working in Mac, click on the **Show All Pins** button and scroll to **All Systems**. Identify:

- **Ascending colon**
- **Cecum**
- **Descending colon**
- **Duodenum**
- **Esophagus**
- **Ileocecal junction**
- **Ileum**
- **Left (Splenic) flexure of colon**
- **Mesentery**
- **Pancreas**
- **Parietal peritoneum**
- **Peritoneal sac**
- **Rectum**
- Right lobe of **liver**
- **Right (Hepatic) flexure of colon**
- **Sigmoid colon**
- **Sigmoid mesocolon**
- **Tenia coli (Taenia coli)**
- **Transverse colon**
- **Transverse mesocolon**

List the parts of the large intestine in order.

________________________________________

________________________________________

________________________________________

When you are finished, **close** the window.

Under the **File** menu, drag down to **Open** (then to **Content** if you are working in Windows). Click on **Atlas Anatomy**. **System** and **Digestive** should still be selected. Choose **Inferior Mesenteric Artery 1**. If the list is not in alphabetical order, the image is almost half of the way down the list.

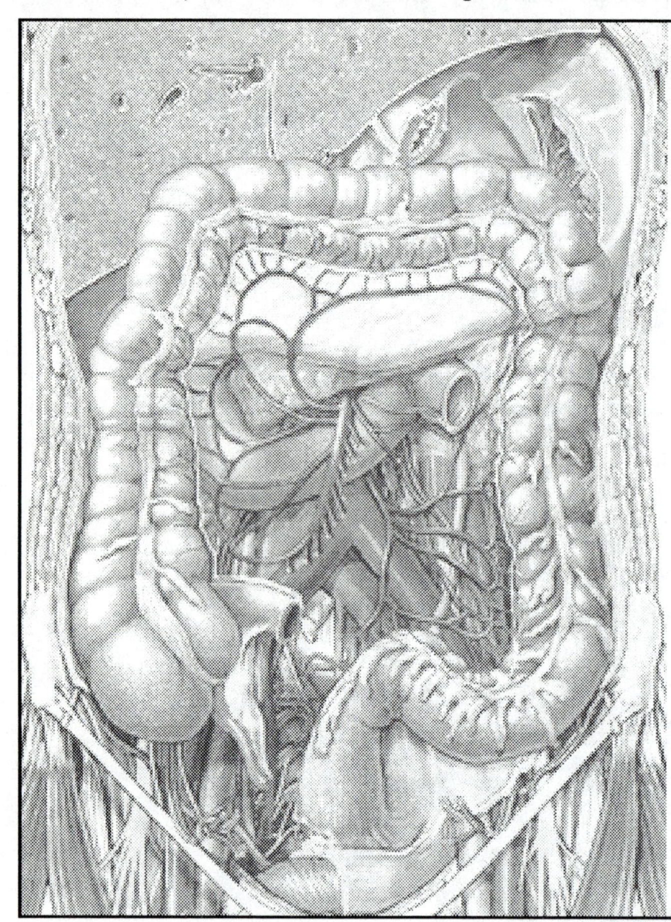

If you are working in Mac, click on the **Show All Pins** button and scroll to **All Systems**. Identify:

- **Ascending colon**
- **Cecum**
- **Descending colon**
- **Duodenum**
- **Esophagus**
- **Ileocecal junction**
- **Ileum**
- **Left (Splenic) flexure of colon**
- **Jejunum**
- **Omental (Epiploic) appendage**
- **Pancreas**
- **Rectum**
- **Right (hepatic) flexure of colon**
- **Sigmoid colon**
- **Sigmoid mesocolon**
- **Tenia coli (Taenia coli)**
- **Transverse colon**
- **Transverse mesocolon**
- **Vermiform appendix**

When you are finished, **close** the window.

What is the function of the transverse and sigmoid mesocolon?

______________________________________________________________________

Under the **File** menu, drag down to **Open** (then to **Content** if you are working in Windows). Click on **Atlas Anatomy**. **System** and **Digestive** should still be selected. Choose **Superior Mesenteric Artery 1**. If the list is not in alphabetical order, the image is about half of the way down the list.

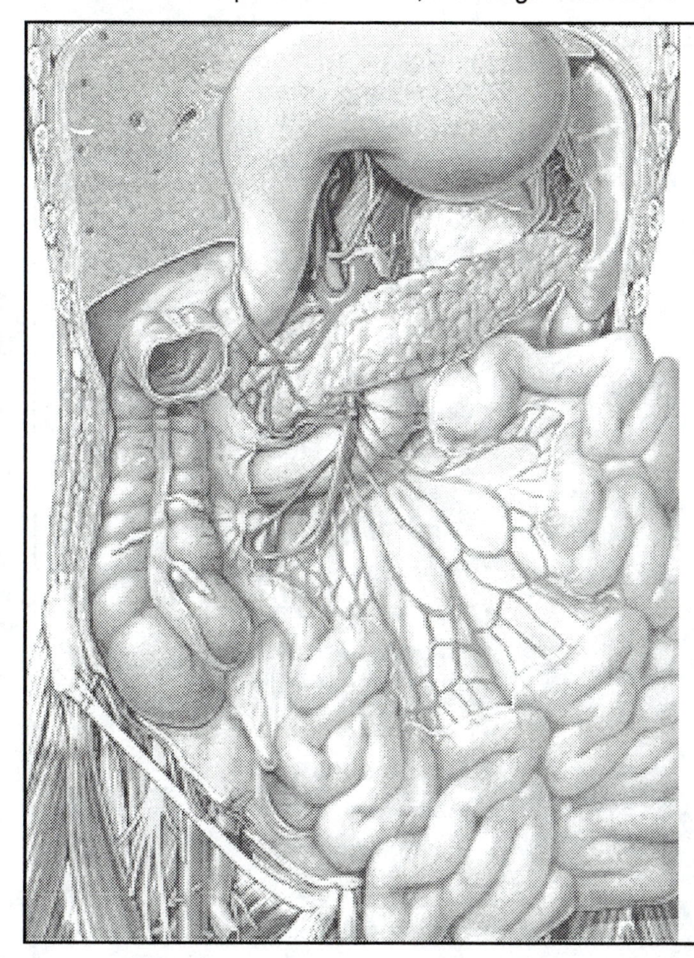

If you are working in Mac, click on the **Show All Pins** button and scroll to **All Systems**. Identify:

- **Ascending colon**
- **Body of pancreas**
- **Body of stomach**
- **Duodenum**
- **Cecum**
- **Ileocecal junction**
- **Ileum**
- **Mesentery**
- **Jejunum**
- **Mesoappendix**
- **Neck of pancreas**
- **Tail of pancreas**
- **Pylorus of stomach**
- **Spleen**
- **Right (hepatic) flexure of colon**
- **Superior mesenteric artery**
- **Transverse colon**
- Left lobe of **liver**
- **Tenia coli (Taenia coli)**
- **Vermiform appendix**

When you are finished, **close** the window.

What is the function of the mesentery?

______________________________________________________________________

Under the **File** menu, drag down to **Open** (then to **Content** if you are working in Windows). Click on **Atlas Anatomy**. **System** and **Digestive** should still be selected. Choose **Stomach & Spleen (Ant) 1**. If the list is not in alphabetical order, the image is almost half of the way down the list. If you are working in Mac, click on the **Show All Pins**  button and scroll to **All Systems**.

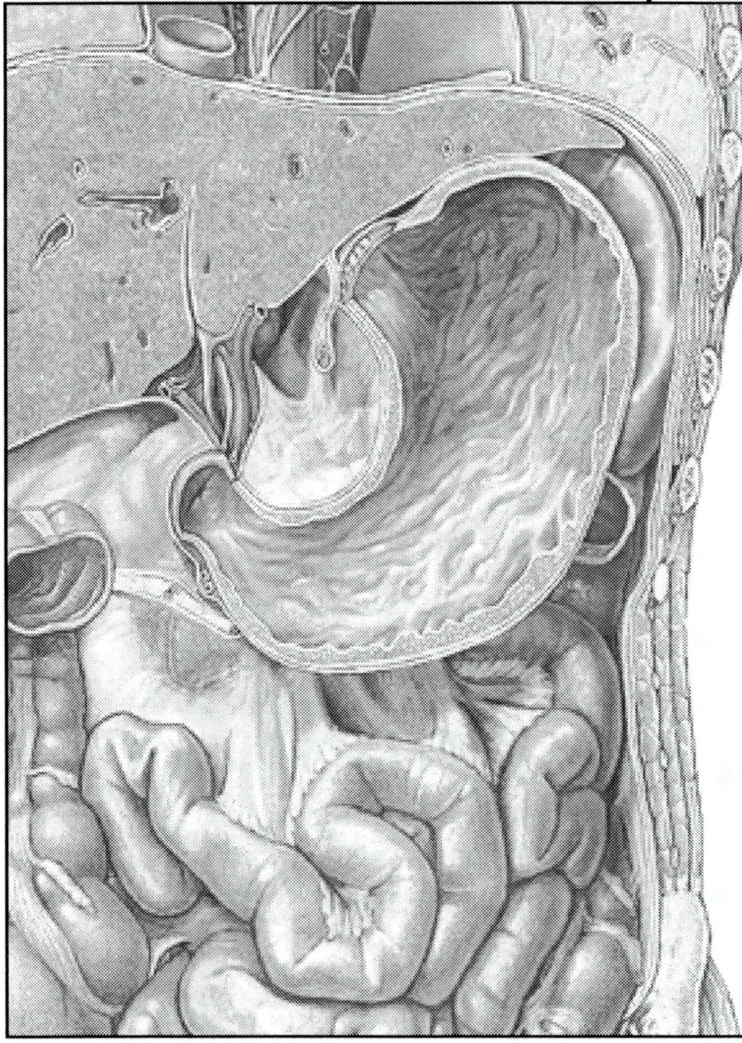

Identify:

- **Ascending colon**
- **Cardiac part of stomach**
- **Fundus of stomach**
- **Duodenum**
- **Cecum**
- **Gastric rugae**
- **Ileum**
- **Greater curvature of stomach**
- **Jejunum**
- **Lesser curvature of stomach**
- **Lesser omentum**
- **Mesentery**
- **Pancreas**
- **Pylorus of stomach**
- **Spleen**
- **Right (Hepatic) flexure of colon**
- **Stomach**
- **Pyloric sphincter**
- **Visceral peritoneum**
- **Parietal pleura**
- **Peritoneal sac**
- **Tenia coli (Taenia coli)**
- **Transverse colon**
- **Transverse mesocolon**

What organs does the lesser omentum attach to?

______________________________________________

List four parts of the stomach.

______________________ ______________________

______________________ ______________________

When you are finished, **close** the window.

Under the **File** menu, drag down to **Open** (then to **Content** if you are working in Windows). Click on **Atlas Anatomy**. **System** and **Digestive** should still be selected. Choose **Female Peritoneal Cavity (Med)**. If the list is not in alphabetical order, the image is almost half of the way down the list.

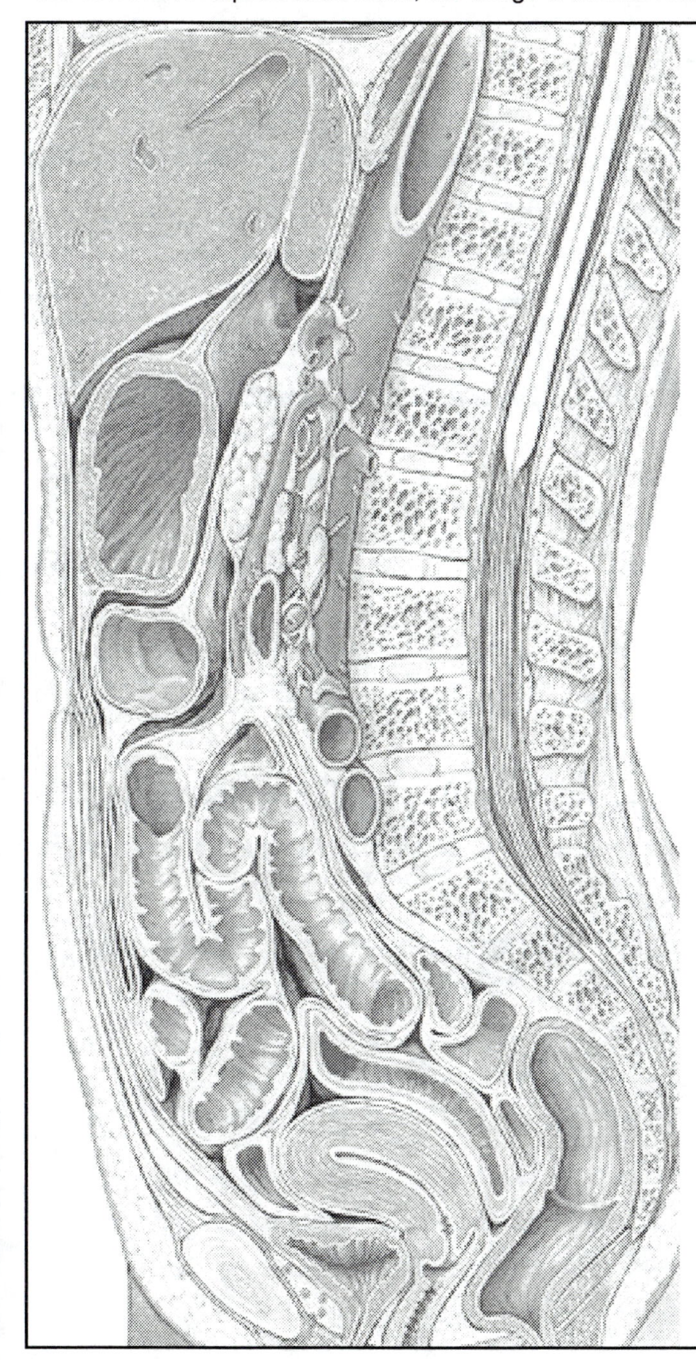

If you are working in Mac, click on the **Show All Pins** button and scroll to **All Systems**. Identify:

- **Anal canal**
- **Duodenum**
- **Esophagus**
- **Greater omentum**
- **Ileum**
- **Jejunum**
- **Lesser omentum**
- **Liver**
- **Mesentery**
- **Pancreas**
- **Parietal peritoneum**
- **Peritoneal sac**
- **Sigmoid colon**
- **Stomach**
- **Transverse colon**
- **Transverse mesocolon**
- **Uterus**

When you are finished, **close** the window.

Under the **File** menu, drag down to **Open** (then to **Content** if you are working in Windows). Click on **Atlas Anatomy**. **System** and **Digestive** should still be selected. Choose **Frontal Section of Anal Canal**. If the list is not in alphabetical order, the image is about three-fourths of the way down.

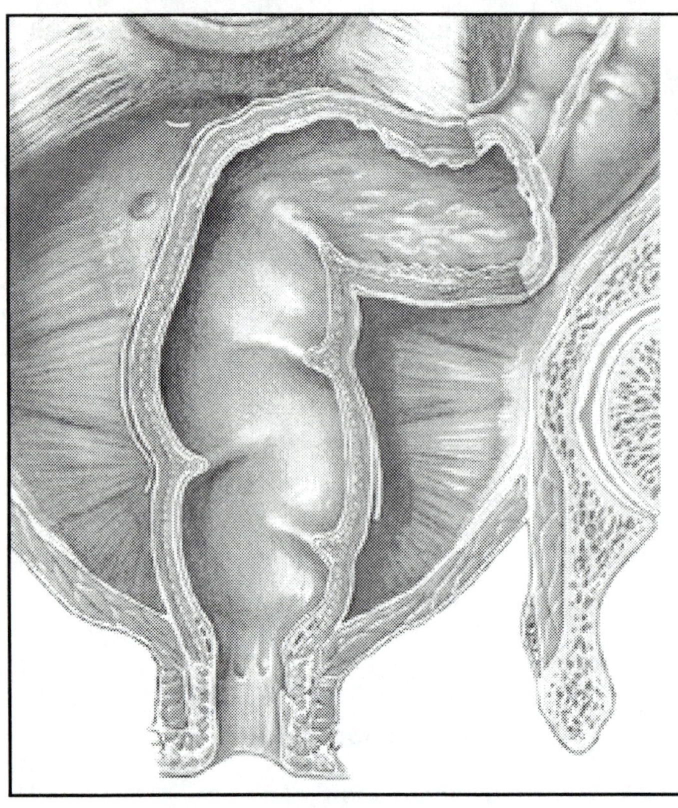

If you are working in Mac, click on the **Show All Pins** button and scroll to **All Systems**. Identify:

- **Anal canal**
- **Anus**
- **Circular muscle layer of intestine**
- **Rectum**
- **Sigmoid colon**
- **Tenia coli (Taenia coli)**

When you are finished, **close** the window.

Under the **File** menu, drag down to **Open** (then to **Content** if you are working in Windows). Click on **Atlas Anatomy**. **System** and **Digestive** should still be selected. Choose **Sagittal Section of Rectum**. If the list is not in alphabetical order, the image is about three-fourths of the way down the list.

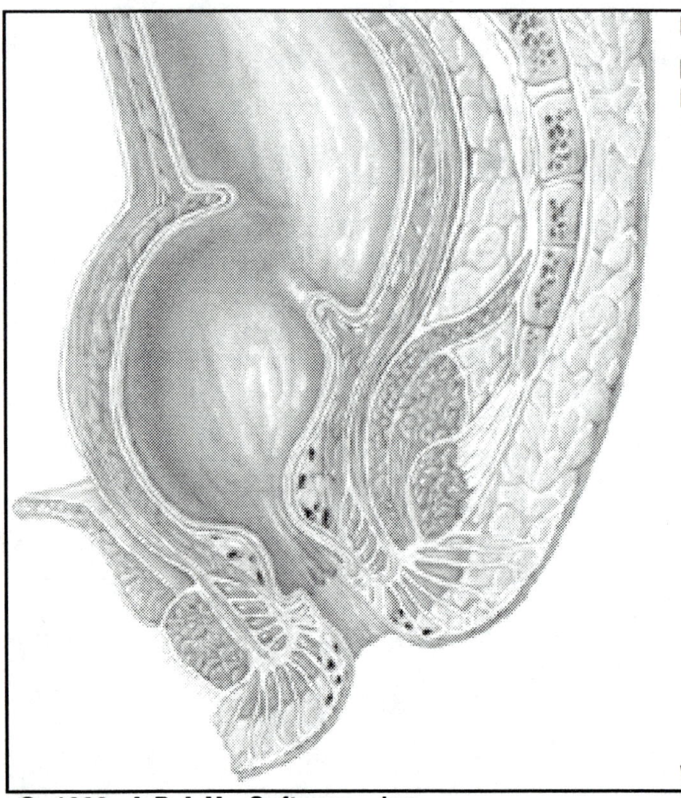

If you are working in Mac, click on the **Show All Pins** button and scroll to **All Systems**. Identify:

- **Anal canal**
- **Anus**
- **Coccyx**
- **External anal sphincter**
- **Internal anal sphincter**
- **Rectum**

When you are finished, **close** the window.

## X-Rays of the Digestive System

Under the **File** menu, drag down to **Open** (then to **Content** if you are working in Windows). Click on **Atlas Anatomy**. **System** and **Digestive** should still be selected. Choose **Barium in Large Bowel**. If the list is not in alphabetical order, the image is almost half of the way down the list.

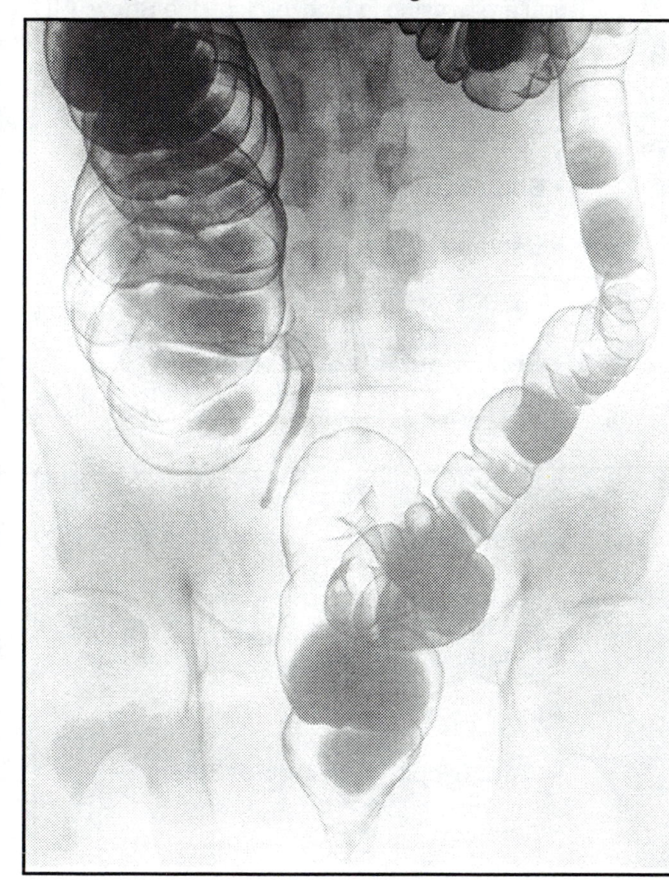

If you are working in Mac, click on the **Show All Pins** button and scroll to **All Systems**. Identify:

- **Anal canal**
- **Ascending colon**
- **Cecum**
- **Descending colon**
- **Rectum**
- **Right (Hepatic) flexure of colon**
- **Sigmoid colon**
- **Transverse colon**
- **Vermiform appendix**

When you are finished, **close** the window.

Under the **File** menu, drag down to **Open** (then to **Content** if you are working in Windows). Click on **Atlas Anatomy**. **System** and **Digestive** should still be selected. Choose **Barium in Small Bowel**. If the list is not in alphabetical order, the image is about half of the way down the list.

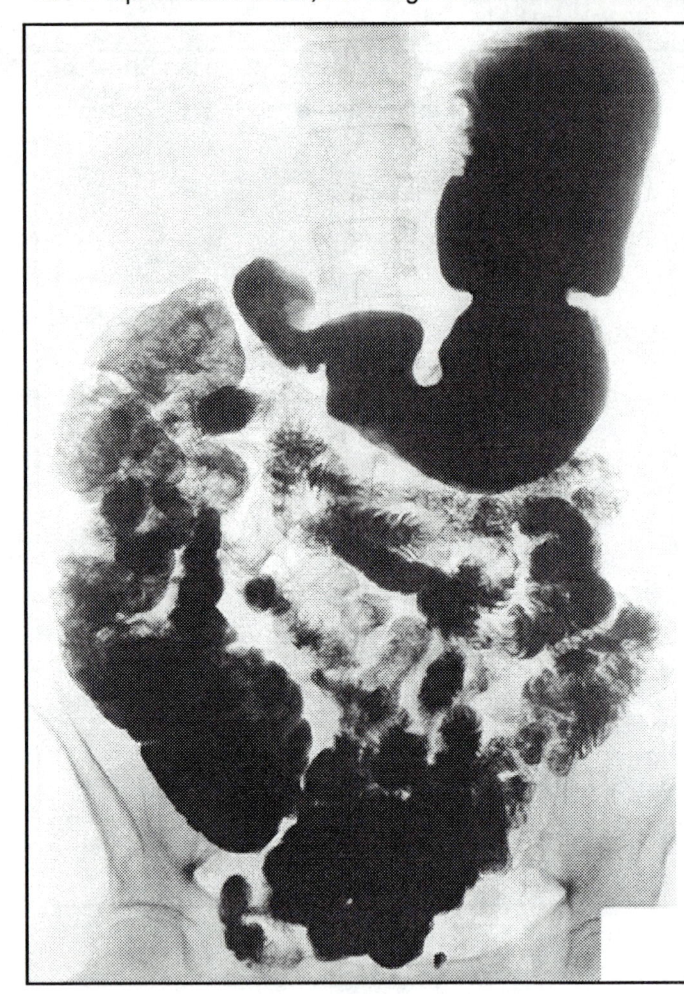

If you are working in Mac, click on the **Show All Pins** button and scroll to **All Systems**. Identify:

- **Ascending colon**
- **Body of stomach**
- **Cecum**
- **Fundus of stomach**
- **Greater curvature of stomach**
- **Ileocecal valve**
- **Ileum**
- **Jejunum**
- **Lesser curvature of stomach**
- **Pyloric sphincter**
- **Pylorus of stomach**
- **Right (Hepatic) flexure of colon**
- **Stomach**

When you are finished, **close** the window.

## Cross Sections of the Lower Digestive System

Under the **File** menu, drag down to **Open** (then to **Content** if you are working in Windows). Click on **Atlas Anatomy**. **System** and **Digestive** should still be selected. Choose **Abdomen at T12 Vertebra (Inf)**. If the list is not in alphabetical order, the image is less than half way down the list.

If you are working in Mac, click on the **Show All Pins** button and scroll to **All Systems**. Identify:

- **Left kidney**
- **Left (Splenic) flexure of colon**
- **Lesser omentum**
- **Liver**
- **Pancreas**
- **Parietal peritoneum**
- **Right kidney**
- **Spleen**
- **Stomach**
- **Visceral peritoneum**

When you are finished, **close** the window.

Under the **File** menu, drag down to **Open** (then to **Content** if you are working in Windows). Click on **Atlas Anatomy**. **System** and **Digestive** should still be selected. Choose **Abdomen at L5 Vertebra (Inf)**. If the list is not in alphabetical order, the image is less than half way down the list.

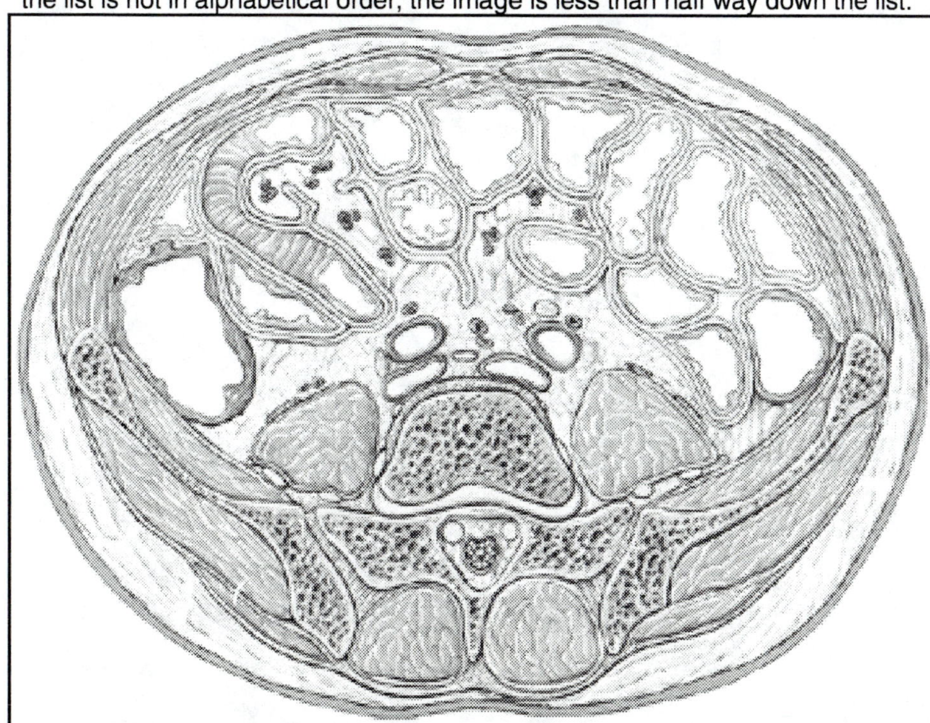

If you are working in Mac, click on the **Show All Pins** button and scroll to **All Systems**. Identify:

- **Ascending colon**
- **Descending colon**
- **Jejunum**
- **Peritoneal sac**
- **Parietal peritoneum**
- **Visceral peritoneum**

When you are finished, **close** the window.

Under the **File** menu, drag down to **Open** (then to **Content** if you are working in Windows). Click on **Atlas Anatomy**. **System** and **Digestive** should still be selected. Choose **Viscera of Female Pelvis (Sup)**. If the list is not in alphabetical order, the image is about three-fourths of the way down.

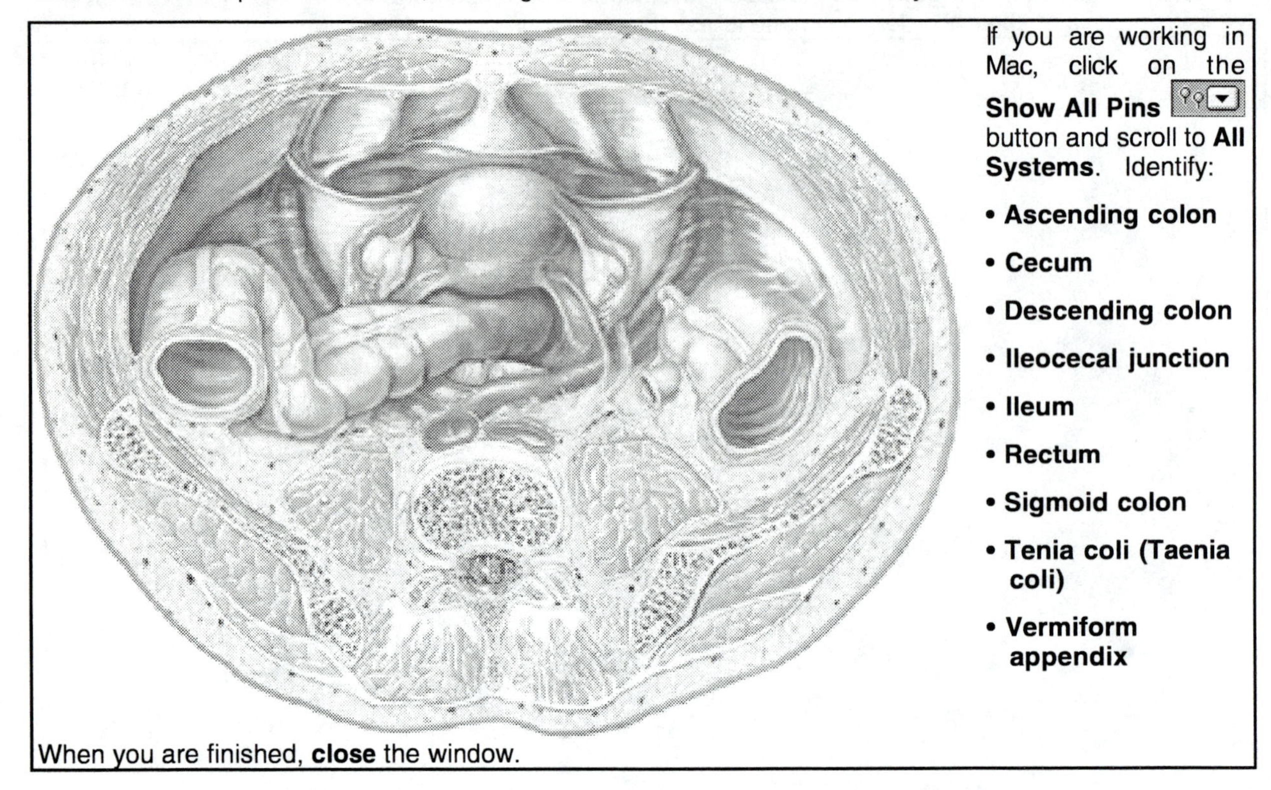

If you are working in Mac, click on the **Show All Pins** button and scroll to **All Systems**. Identify:

- **Ascending colon**
- **Cecum**
- **Descending colon**
- **Ileocecal junction**
- **Ileum**
- **Rectum**
- **Sigmoid colon**
- **Tenia coli (Taenia coli)**
- **Vermiform appendix**

When you are finished, **close** the window.

## Cadaver Images of the Lower Digestive System

Under the **File** menu, drag down to **Open** (then to **Content** if you are working in Windows). Click on **Atlas Anatomy**. **System** and **Digestive** should still be selected. Choose **Abdominal Viscera**. If the list is not in alphabetical order, the image is about three-fourths of the way down the list.

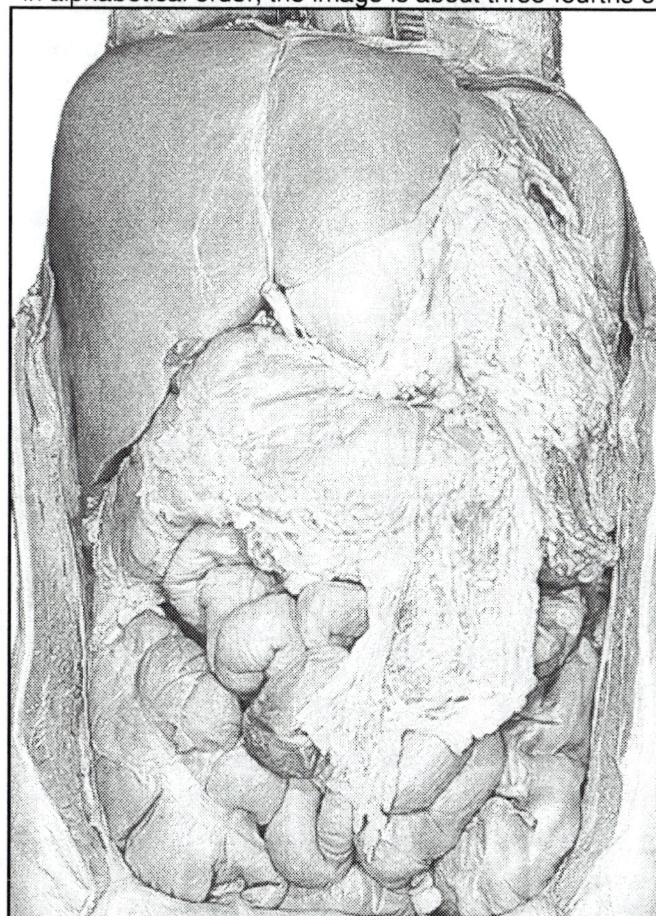

If you are working in Mac, click on the **Show All Pins** button and scroll to **All Systems**. Identify:

- **Body of stomach**
- **Cecum**
- **Fundus of stomach**
- **Esophagus**
- **Greater omentum**
- **Ileum**
- **Jejunum**
- **Liver** (Left lobe of)
- **Sigmoid colon**
- **Spleen**
- **Tenia coli (Taenia coli)**
- **Transverse colon**

When you are finished, **close** the window.

Under the **File** menu, drag down to **Open** (then to **Content** if you are working in Windows). Click on **Atlas Anatomy**. **System** and **Digestive** should still be selected. Choose **Superior Mesenteric Artery 2**. If the list is not in alphabetical order, the image is about three-fourths of the way down the list.

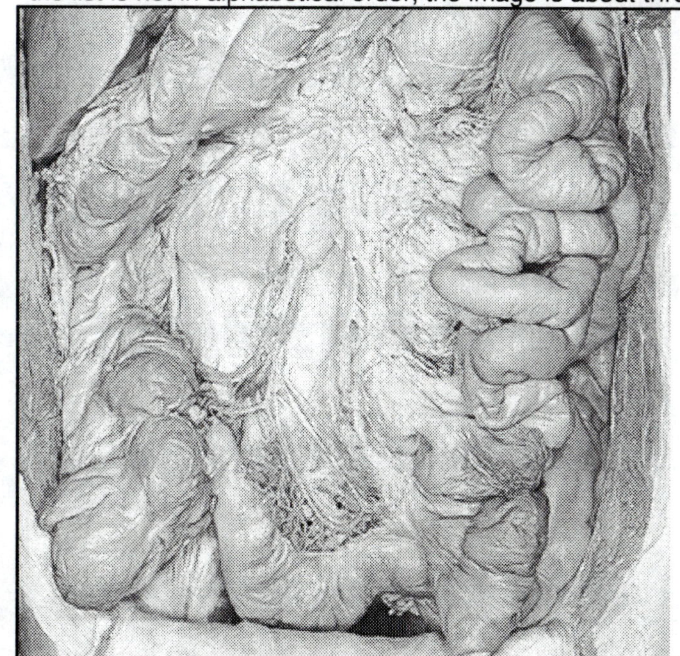

If you are working in Mac, click on the **Show All Pins** button and scroll to **All Systems**. Identify:

- **Ascending colon**
- **Descending colon**
- **Duodenum**
- **Gallbladder**
- **Mesentery**
- **Ileocecal junction**
- **Ileum**
- **Jejunum**
- **Right (Hepatic) flexure of colon**
- **Cecum**
- **Tenia coli (Taenia coli)**
- **Transverse colon**

When you are finished, **close** the window.

Under the **File** menu, drag down to **Open** (then to **Content** if you are working in Windows). Click on **Atlas Anatomy**. **System** and **Digestive** should still be selected. Choose **Inferior Mesenteric Artery 2**. If the list is not in alphabetical order, the image is about three-fourths of the way down the list.

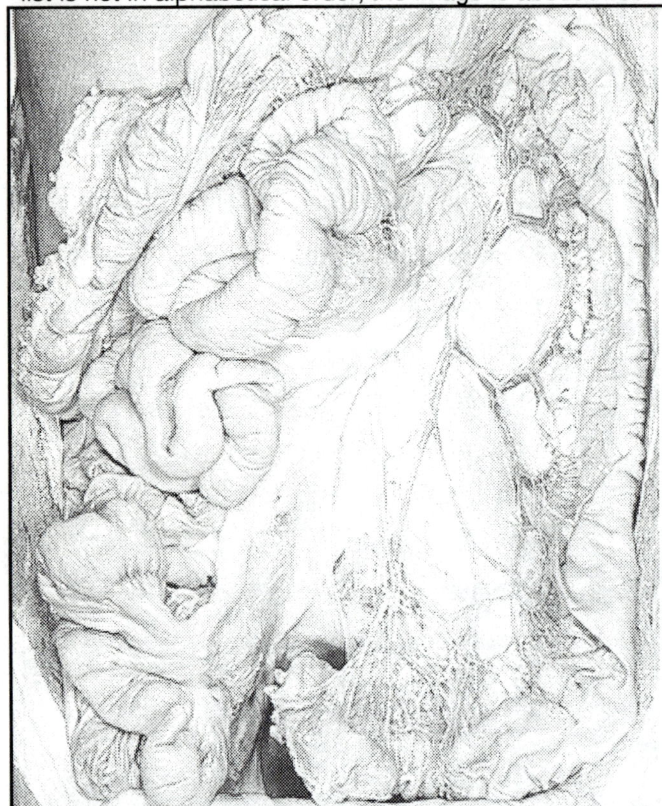

If you are working in Mac, click on the **Show All Pins** button and scroll to **All Systems**. Identify:

- **Ascending colon**
- **Descending colon**
- **Left (Splenic) flexure of colon**
- **Mesentery**
- **Greater omentum**
- **Ileum**
- **Jejunum**
- **Right (Hepatic) flexure of colon**
- **Sigmoid colon**
- **Spleen**
- **Tenia coli (Taenia coli)**
- **Transverse colon**

When you are finished, **close** the window.

Under the **File** menu, drag down to **Open** (then to **Content** if you are working in Windows). Click on **Atlas Anatomy**. **System** and **Digestive** should still be selected. Choose **Intestines & Mesentery**. If the list is not in alphabetical order, the image is about three-fourths of the way down.

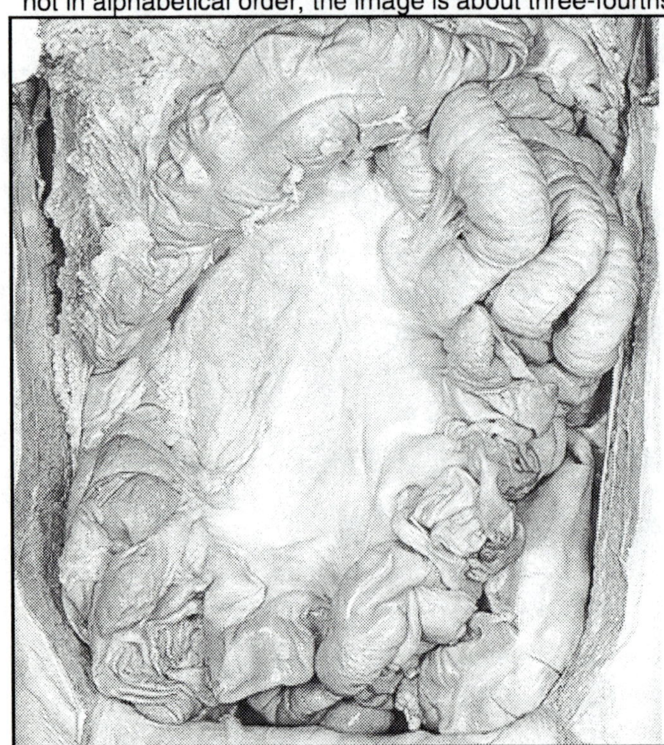

If you are working in Mac, click on the **Show All Pins** button and scroll to **All Systems**. Identify:

- **Ascending colon**
- **Descending colon**
- **Duodenum**
- **Left (Splenic) flexure of colon**
- **Mesentery**
- **Greater omentum**
- **Ileum**
- **Jejunum**
- **Right (Hepatic) flexure of colon**
- **Sigmoid colon**
- **Cecum**
- **Tenia coli (Taenia coli)**
- **Transverse colon**

When you are finished, **close** the window.

# Objectives: Abdominal Digestive System:

Student will study the anatomy of the digestive system in the abdominal cavity.

**Peritoneum**

- Greater omentum
- Lesser omentum
- Mesentery
- Peritoneum
- Visceral peritoneum
- Transverse mesocolon
- Sigmoid mesocolon
- Peritoneal sac
- Mesoappendix
- Parietal pleura

**Large Intestine**

- Ascending colon
- Transverse colon
- Descending colon
- Sigmoid colon
- Cecum
- Taenia coli
- Hepatic flexure of colon
- Splenic flexure of colon
- Ileocecal junction
- Omental appendage
- Rectum
- Vermiform appendix
- Anal canal
- Internal anal sphincter
- External anal sphincter
- Anus

**Small Intestine**

- Ileum
- Jejunum
- Duodenum
- Circular muscle layer of intestine

**Stomach**

- Cardiac orifice of the stomach
- Rugae of the stomach
- Pylorus of the stomach
- Body of the stomach
- Fundus of the stomach
- Greater curvature of the stomach
- Lesser curvature of the stomach
- Pyloric sphincter

**Other**

- Right lobe of the liver
- Left lobe of the liver
- Gallbladder
- Esophagus
- Spleen
- Urinary bladder
- Body of pancreas
- Neck of pancreas
- Head of pancreas
- Tail of pancreas
- Superior mesenteric artery
- Uterus
- Coccyx
- Kidneys

# Kidney

## Opening A.D.A.M Interactive Anatomy

Open A.D.A.M. Interactive Anatomy according to the directions in the Tutorial.

## Anterior Dissection of Kidney

Choose **Dissectible Anatomy**.

Choose either male or female. Select **Anterior.** Click **Open**.

Enlarge the window and position the image over the upper abdominal region.

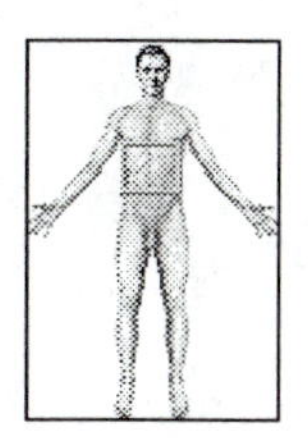

Drag down on the **Depth Bar** ▼ until you reach **layer 198**. Then click on the **Depth Bar** ▼, watching the structures disappear, to **layer 228**. No graphic has been provided for this layer. Identify:

- **Renal fascia**

Click once on the **Depth Bar** ▼ to **layer 229**. No graphic has been provided for this layer. Identify:

- **Perirenal fat**

Click once more on the **Depth Bar** ▼ to **layer 230**. Starting with **layer 198**, what organs had to be removed to view the kidneys?

______________________________________________

______________________________________________

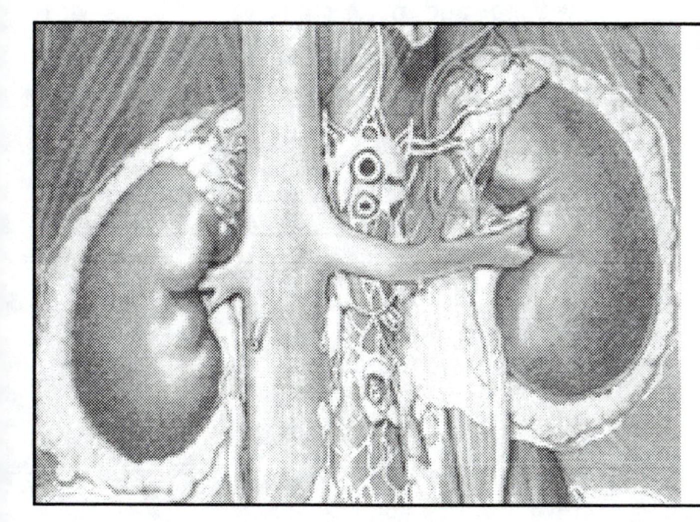

Remaining at **layer 230**, identify:

- **Right kidney**
- **Left kidney**
- **Right suprarenal gland**
- **Left suprarenal gland**
- **Right renal vein**
- **Left renal vein**
- **Inferior vena cava**
- **Right ureter**
- **Left ureter**
- **Perirenal fat**
- **Hilum**

What is the function of the **perirenal fat**?

______________________________________________

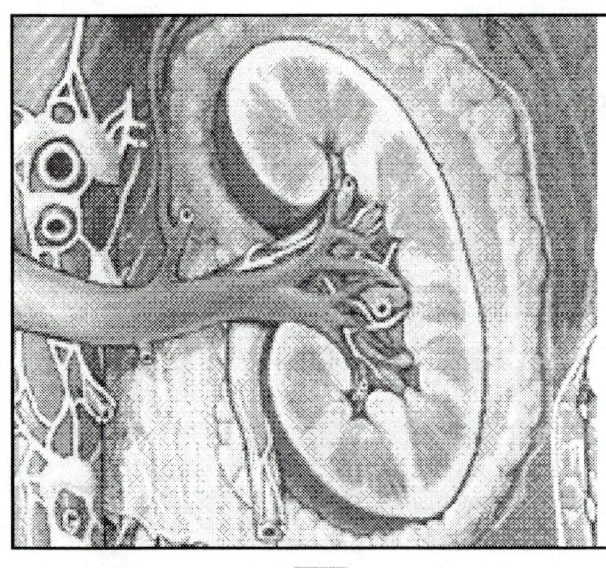

Click a few times on the **Depth Bar** ▼ to **layer 238**. Identify:

- **Left renal vein**
- **Renal cortex**
- **Renal pyramid** of renal medulla
- **Renal column**
- **Renal papilla**
- **Minor calyx**
- **Renal nerve plexus**
- **Renal pelvis**

Click on the **Depth Bar** ▼ to **layer 240**. No graphic has been provided for this layer. Identify:

- **Renal arteries**

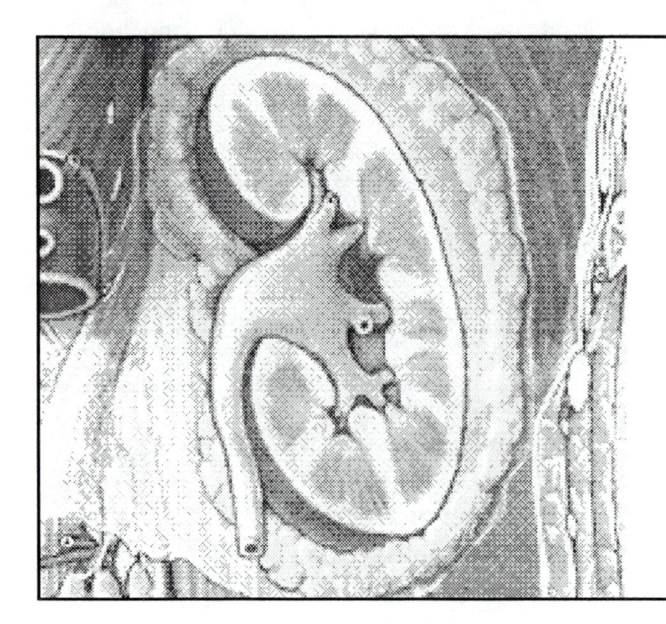

Click a few more times on the **Depth Bar** ▼ until you reach **layer 242**. Identify:

- **Renal cortex**
- **Renal column**
- **Renal pyramid** of renal medulla
- **Renal papilla**
- **Renal sinus**
- **Minor and major calyx**
- **Renal pelvis**
- **Ureter**

When you are finished, **close** the window.

## Atlas Anatomy of the Kidney

Under the **File** menu, drag down to **Open** (then to **Content** if you are working in Windows). Click on **Atlas Anatomy**. Select **System** and **Urinary**. Choose **Renal Arteries**. If the list is not in alphabetical order, the image is about one-third of the way down the list.

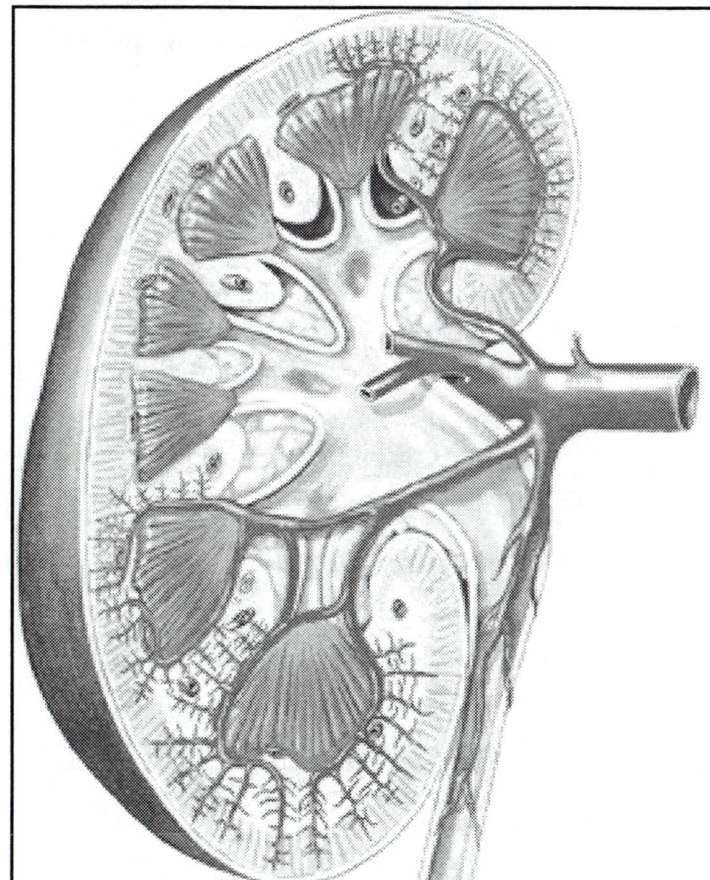

If you are working in Mac, click on the **Show All Pins** button and scroll to **All Systems**. Identify the following by clicking on the heads of the pins:

- **Arcuate artery**
- **Arcuate vein**
- **Interlobular artery**
- **Major calyx**
- **Minor calyx**
- **Renal artery**
- **Renal capsule**
- **Renal column**
- **Renal cortex**
- **Renal papilla**
- **Renal pelvis**
- **Renal pyramid of renal medulla**
- **Renal sinus**
- **Ureter**

When you are finished, **close** the window.

Under the **File** menu, drag down to **Open** (then to **Content** if you are working in Windows). Click on **Atlas Anatomy**. **System** and **Urinary** should still be selected. Choose **Diagram of Renal Glomerulus**. If the list is not in alphabetical order, the image is close to the top of the list.

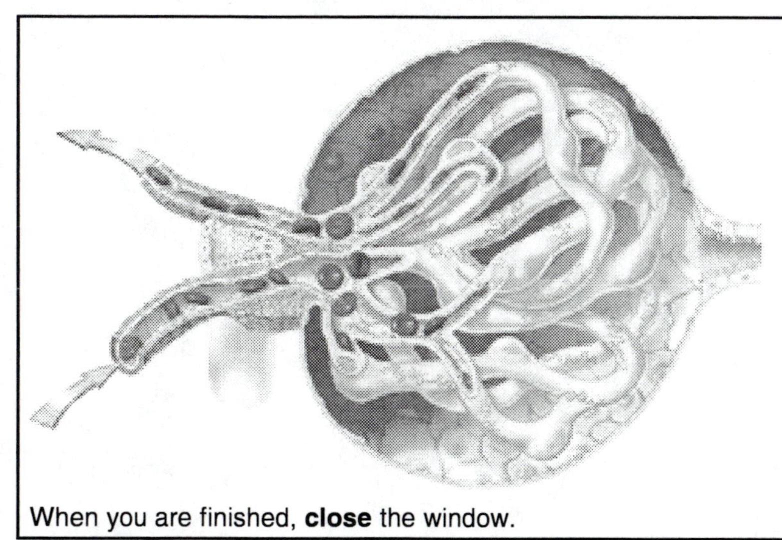

If you are working in Mac, click on the **Show All Pins** button and scroll to **All Systems**. Identify:

- **Afferent arteriole**
- **Distal convoluted tubule**
- **Efferent arteriole**
- **Erythrocyte**
- **Juxtaglomerular cell**
- **Macula densa cell**
- **Parietal layer of Bowman's capsule**
- **Podocyte**
- **Proximal convoluted tubule**
- **Urinary space**

When you are finished, **close** the window.

Under the **File** menu, drag down to **Open** (then to **Content** if you are working in Windows). Click on **Atlas Anatomy**. **System** and **Urinary** should still be selected. Choose **Diagram of Nephron**. If the list is not in alphabetical order, the image is close to the top of the list.

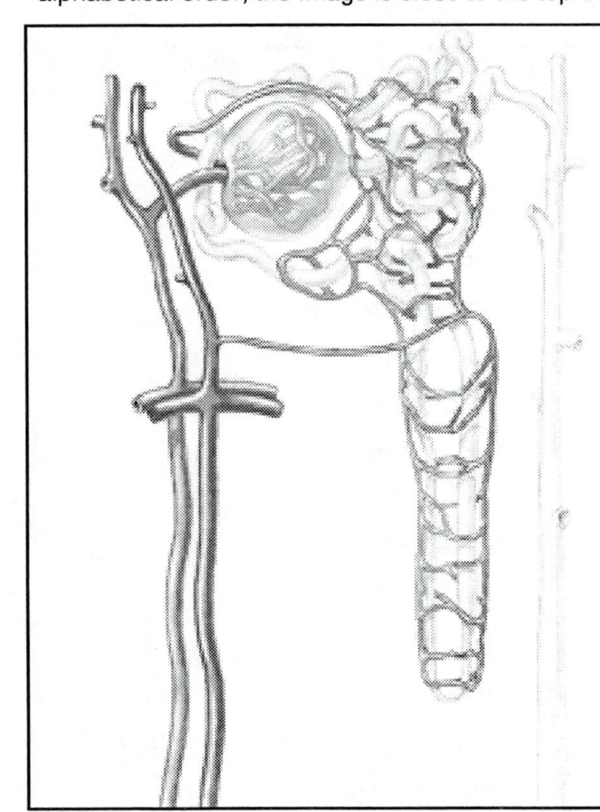

If you are working in Mac, click on the **Show All Pins** button and scroll to **All Systems**. Identify:

- **Afferent arteriole**
- **Ascending limb of loop of Henle**
- **Medullary collecting duct** (Collecting duct)
- **Cortical collecting duct** (Collecting tubule)
- **Descending limb of loop of Henle**
- **Distal convoluted tubule**
- **Efferent arteriole**
- **Glomerulus**
- **Peritubular capillary**
- **Proximal convoluted tubule**
- **Vasa recta**

When you are finished, **close** the window.

## Cadaver Images of the Kidney

Under the **File** menu, drag down to **Open** (then to **Content** if you are working in Windows). Click on **Atlas Anatomy**. **System** and **Urinary** should still be selected. Choose **Dissection of Kidneys (Ant)**. If the list is not in alphabetical order, the image is close to the top of the list.

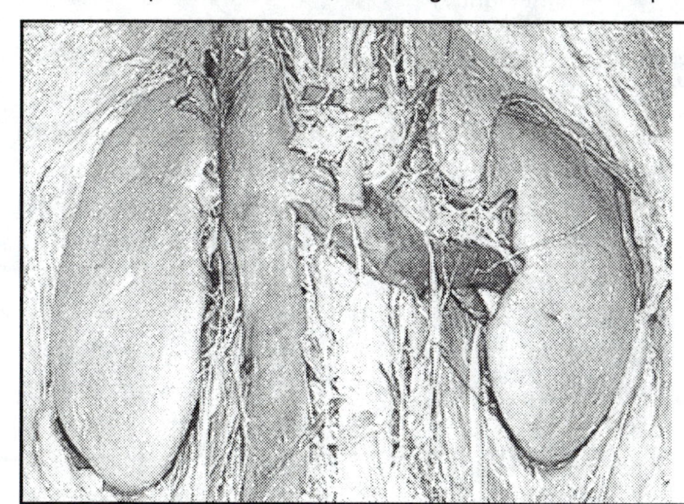

Identify:

- **Hilum of left kidney**
- **Hilum of right kidney**
- **Left kidney**
- **Left ureter**
- **Right kidney**
- **Right ureter**

When you are finished, **close** the window.

Under the **File** menu, drag down to **Open** (then to **Content** if you are working in Windows). Click on **Atlas Anatomy**. **System** and **Urinary** should still be selected. Choose **Posterior Abdominal Wall (Ant)**. If the list is not in alphabetical order, the image is about one-third of the way down the list.

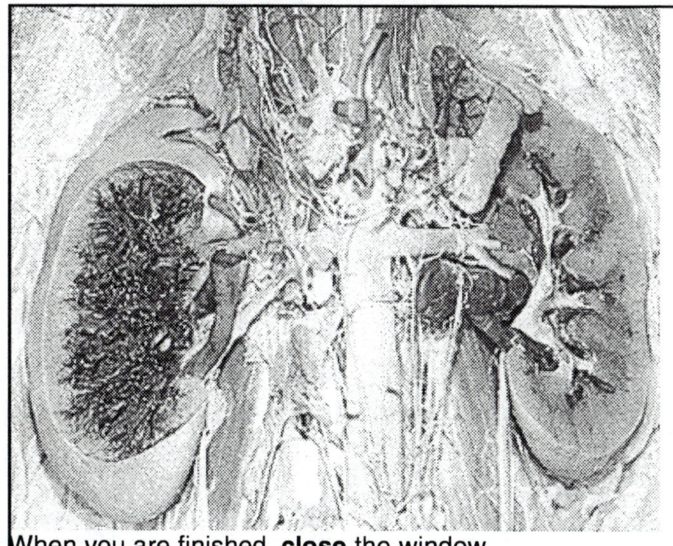

If you are working in Mac, click on the **Show All Pins** button and scroll to **All Systems**. Identify:

- **Left kidney**
- **Left ureter**
- **Major calyx**
- **Minor calyx**
- **Renal capsule**
- **Renal column in cortex**
- **Renal cortex**
- **Renal papilla**
- **Renal pelvis**
- **Renal pyramid of renal medulla**
- **Right kidney**
- **Right ureter**

When you are finished, **close** the window.

## Objectives: Kidney

To learn the anatomy of the kidney.

**Internal Kidney**

- Renal cortex
- Renal pyramid
- Renal column
- Renal sinus
- Renal papilla
- Renal pelvis
- Renal capsule
- Minor calyx
- Major calyx
- Descending loop of Henle
- Ascending loop of Henle
- Medullary collecting duct
- Cortical collecting duct
- Distal convoluted tubule
- Proximal convoluted tubule
- Glomerulus
- Juxtaglomerular cells
- Macula densa cells
- Parietal layer of Bowman's capsule
- Podocyte
- Urinary space
- Erythrocyte

**External Kidney**

- Right kidney
- Left kidney
- Right suprarenal gland
- Left suprarenal gland
- Ureter
- Renal fascia
- Renal nerve plexus

**Blood vessels**

- Right renal vein
- Left renal vein
- Inferior vena cava
- Renal artery
- Arcuate artery
- Arcuate vein
- Interlobular artery
- Afferent arteriole
- Efferent arteriole
- Peritubular capillary
- Vasa recta

# Lower Urinary System

## Opening A.D.A.M Interactive Anatomy

Open A.D.A.M. Interactive Anatomy according to the directions in the Tutorial.

## Anterior Dissection of Lower Urinary System

Choose **Dissectible Anatomy**.

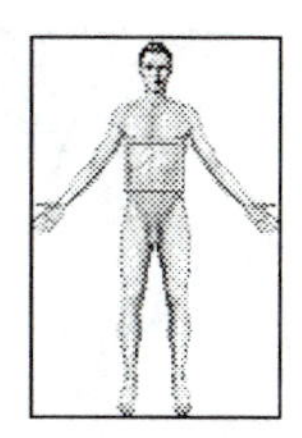

Choose **Male**. Select **Anterior.** Click **Open**.

Enlarge the window and position the image over the upper abdominal region.

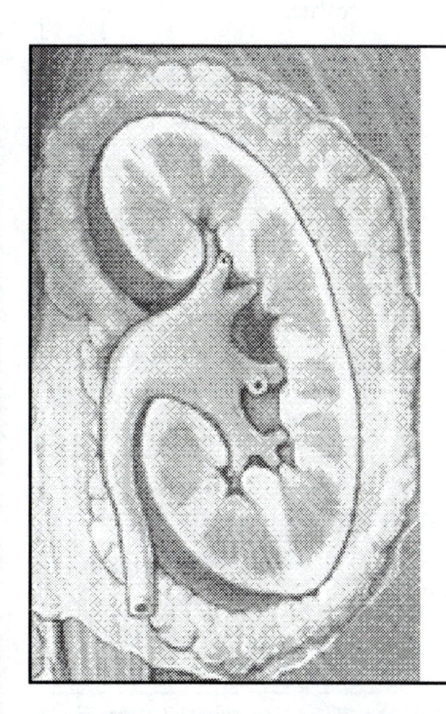

Drag down on the **Depth Bar** ▼ until you reach **layer 241**. Identify:

- **Minor calyx**
- **Major calyx**
- **Renal pelvis**
- **Ureter**

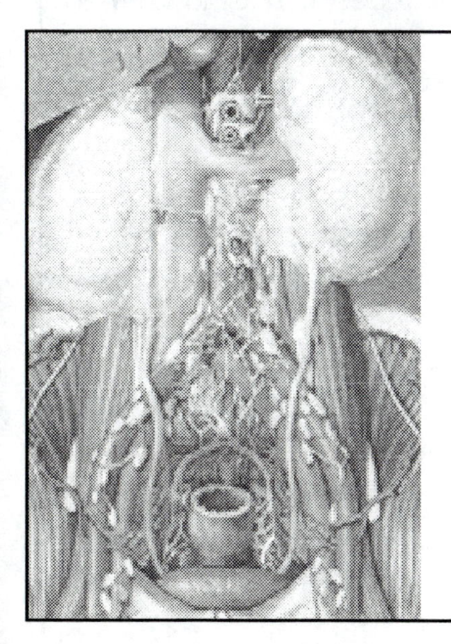

Click on the **Depth Bar** ▲ up to **layer 226**. Using the **Vertical Scroll Bar**, follow the **ureters** down to their insertion into the **urinary bladder**.

## Medial Dissection of Lower Urinary System

Click on the **View** button and choose **Medial**.

Click on the **Normal** button. **Windows**—**Mac**

Position the navigation rectangle over the pelvic region. Drag the **Depth Bar** to **layer 0.**

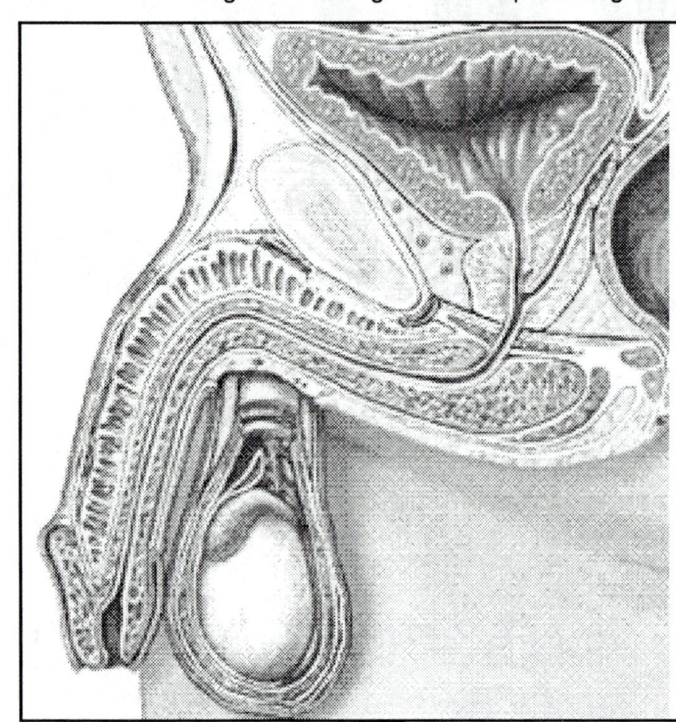

Use the **Depth Bar** to scroll to **layer 51**. Identify:

- **Urinary bladder**
- **Prostatic urethra**
- **Membranous urethra**
- **Spongy urethra**
- **Sphincter urethral muscle**
- **Ureter**

Which part of the male urethra is the longest?

_______________________________________

Which part of the male urethra is the shortest?

_______________________________________

Click on the **Gender Button** and choose **Female**.

Click on the **Normal** button. **Windows**—**Mac**

Drag the **Depth Bar** to **layer 0.**

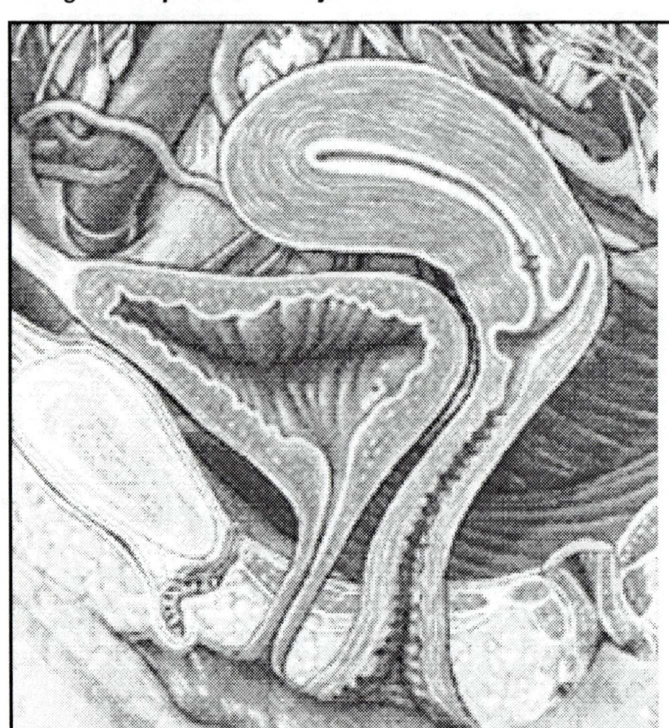

Use the **Depth Bar** to scroll to **layer 51**. Identify:

- **Urinary bladder**
- **Labium minus**
- **Vagina**
- **Ureter**
- **Sphincter urethral muscle**

Use the **Depth Bar** to scroll to **layer 56** and identify:

- **Urethra**

When you are finished, **close** the window.

## Atlas Anatomy of the Lower Urinary System

Under the **File** menu, drag down to **Open** (then to **Content** if you are working in Windows). Click on **Atlas Anatomy**. Select **System** and **Urinary**. Choose **Section of Male Bladder (Ant)**. If the list is not in alphabetical order, the image is half of the way down the list.

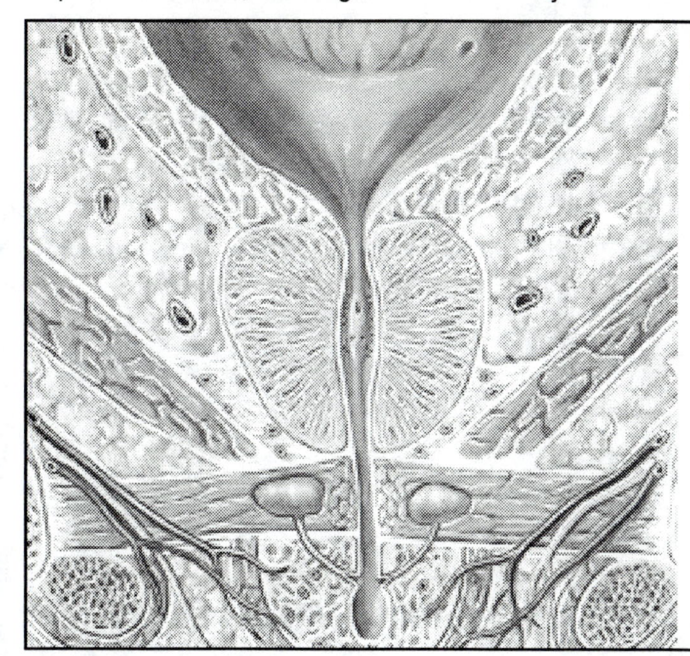

If you are working in Mac, click on the **Show All Pins** button and scroll to **All Systems**. Identify:

- **Orifice of left ureter**
- **Orifice of right ureter**
- **Prostate gland**
- **Prostatic part of urethra**
- **Sphincter urethra muscle**
- **Spongy part of urethra**
- **Trigone of urinary bladder**
- **Urinary bladder**

When you are finished, **close** the window.

Under the **File** menu, drag down to **Open** (then to **Content** if you are working in Windows). Click on **Atlas Anatomy**. **System** and **Urinary** should still be selected. Choose **Section of Female Bladder (Ant)**. If the list is not in alphabetical order, the image is half of the way down the list.

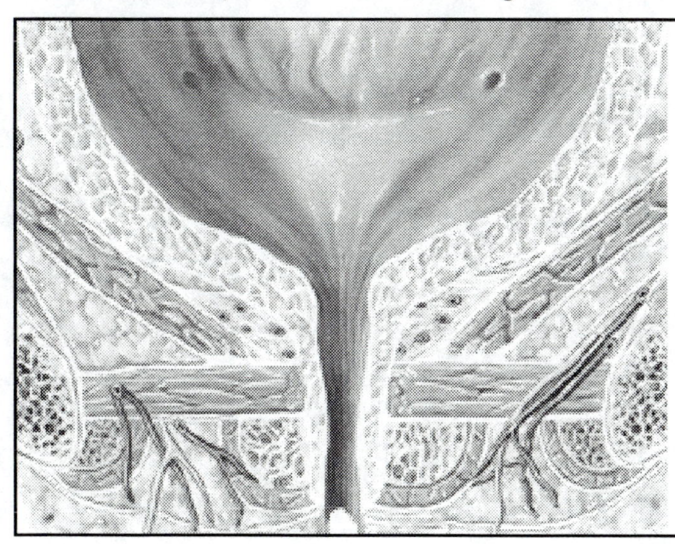

If you are working in Mac, click on the **Show All Pins** button and scroll to **All Systems**. Identify:

- **Orifice of left ureter**
- **Orifice of right ureter**
- **Membranous part of urethra**
- **Neck of bladder**
- **Sphincter urethra muscle**
- **Trigone of urinary bladder**
- **Urinary bladder**

When you are finished, **close** the window.

## Objectives: Lower Urinary System

Student will learn the anatomy of the ureters, bladder, and urethra.

- Minor calyx
- Major calyx
- Renal pelvis
- Ureter
- Orifice of left ureter
- Orifice of right ureter
- Urinary bladder
- Trigone of urinary bladder
- Neck of urinary bladder
- Sphincter urethral muscle
- Prostate gland
- Prostatic urethra
- Membranous urethra
- Spongy urethra
- Renal pelvis
- Urethra
- Vagina
- Labium minus

# Male Reproductive System

## Opening A.D.A.M Interactive Anatomy

Open A.D.A.M. Interactive Anatomy according to the directions in the Tutorial.

## 3D Anatomy of the Male Reproductive System

Under the **File** menu, drag down to **Open** (then to **Content** if you are working in Windows). Choose **3D Anatomy**. Choose **Male Reproductive**. Identify the following by clicking on the **3D Anatomy Structure List**:

- **Spermatic cord - Cut**
- **Suspensory ligament of penis**

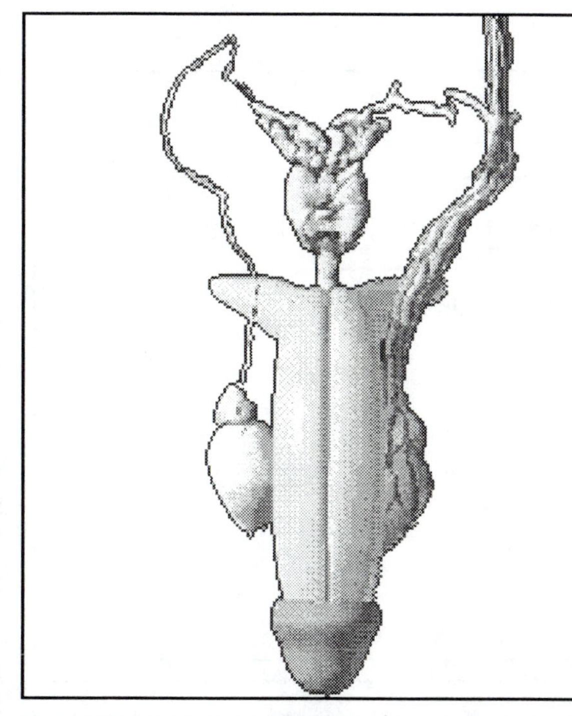

Identify:

- **Ampulla of ductus {vas} deferens**
- **Corpus cavernosum**
- **Testes**

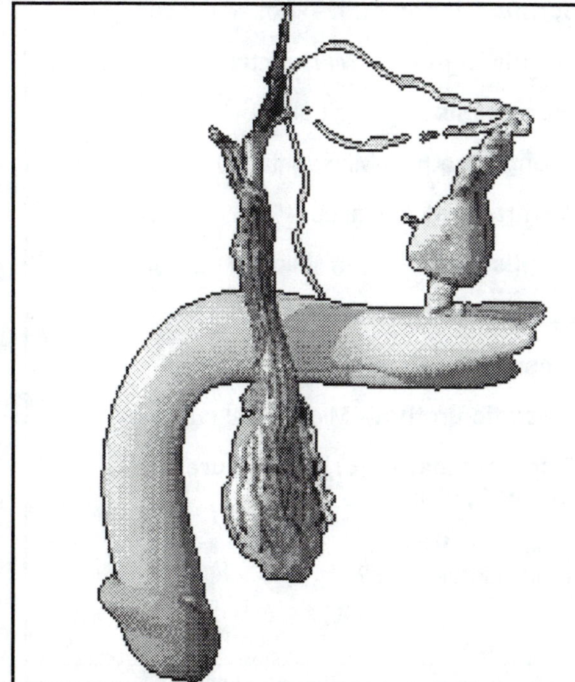

Identify:

- **Cremaster muscle**
- **Ductus {vas} deferens**
- **Epididymis**
- **Pampiniform venous plexus**
- **Testicular artery**
- **Testicular vein**

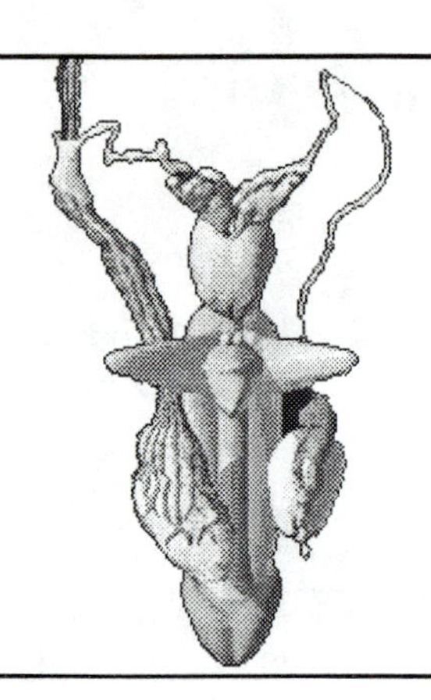

Identify:
- **Bulbourethral gland**
- **Prostate gland**
- **Seminal vesicle**

One function of the ______________________ gland is to neutralize the acidity of the urethra.

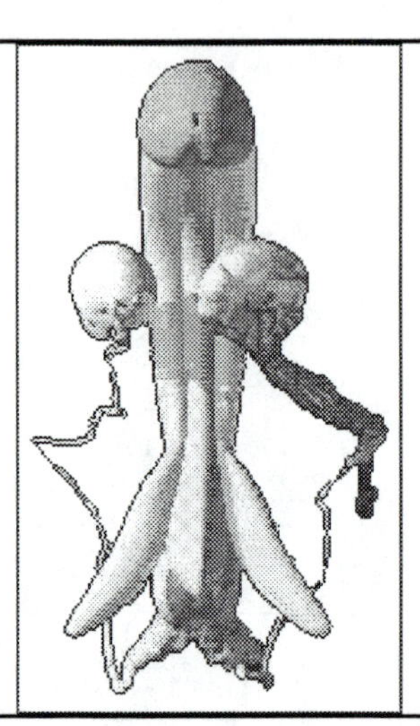

Identify:
- **External meatus {orifice} of urethra** (in uncircumcised male)
- **Glans penis**

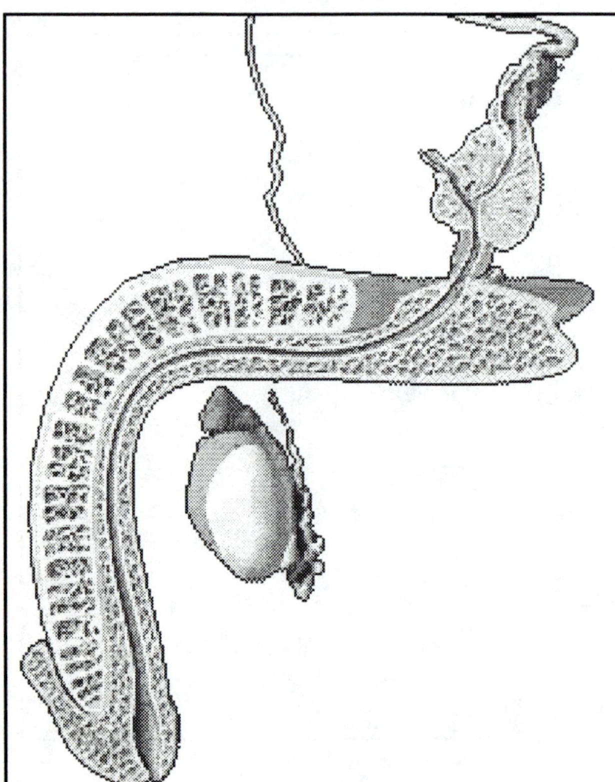

Identify:
- **Corpus spongiosum** - Midsagittal cut
- **Ejaculatory duct** - Midsagittal cut
- **Glans penis**- Midsagittal cut
- **Spongy urethra**- Midsagittal cut
- **Urethra** - Midsagittal cut
- **Corpus cavernosum**- Midsagittal cut
- **Corpus spongiosum**
- **Prostate gland**- Midsagittal cut
- **Prostatic urethra** - Midsagittal cut
- **External meatus {orifice} of urethra** - Midsagittal cut

The junction of what two structures marks the start of the ejaculatory duct?

______________________________________

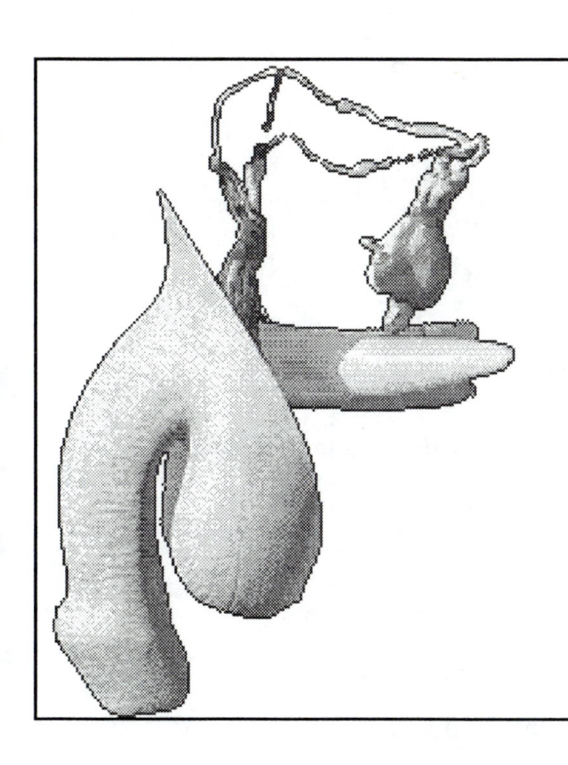

Identify:

- **Prepuce of penis** (removed during circumcision)
- **Scrotum**
- **Skin**

When you are finished, **close** the window.

## Atlas Anatomy - Surface Anatomy

Under the **File** menu, drag down to **Open** (then to **Content** if you are working in Windows). Click on **Atlas Anatomy**. Select **System** and **Reproductive**. Choose **Bony Landmarks of Male Pelvis**. If the list is not in alphabetical order, the image is close to the top of the list.

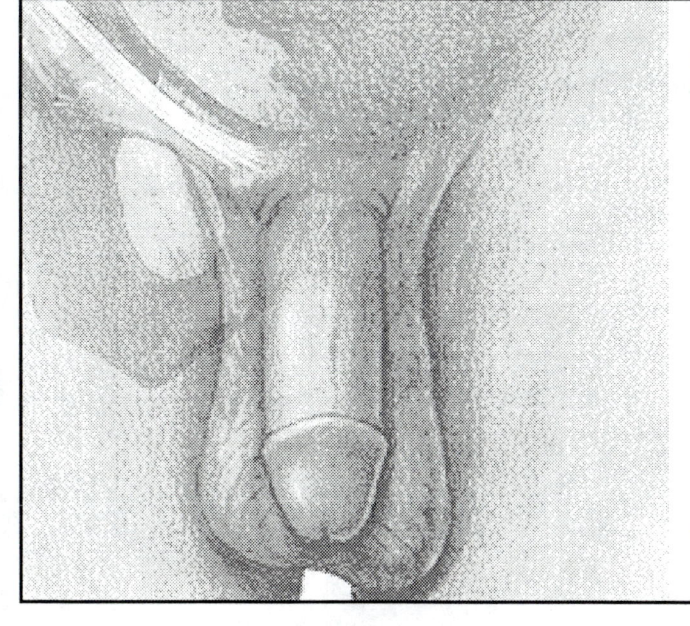

If you are working in Mac, click on the **Show All Pins** button and scroll to **All Systems**. Note this male is circumcised. Identify:

- **Inguinal ligament**
- **Body of penis**
- **Spermatic cord**
- **Glans penis**
- **External meatus of urethra**
- **Scrotum**

When you are finished, **close** the window.

## Anterior Dissection of the Male Reproductive System

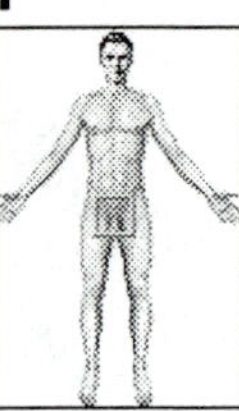

Choose **Dissectible Anatomy**.

Choose **male**. Select **Anterior.** Click **Open**.

Enlarge the window and position the image over the external genitalia.

Note that this male is uncircumcised. At **layer 0** note the **prepuce of penis** (foreskin).

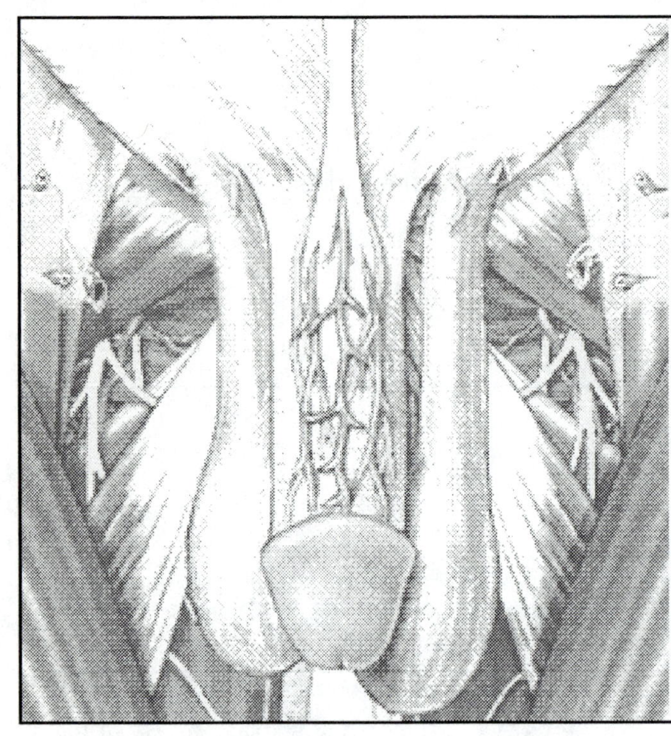

Click on the on the **Depth Bar** ▼ until you reach **layer 21**. Identify:

- **Suspensory ligament of the penis**
- **External spermatic fascia**
- **Penis**
- **Dorsal nerve of penis**
- **Dorsal artery of penis**
- **Vein of penis**
- **Glans penis**

Dissect down to **layer 27**. Identify the **cremaster muscle**. No graphic has been provided for this layer. What is the function of the cremaster muscle?

______________________________________________________________

______________________________________________________________

Scroll down to **layer 157**. No graphic has been provided for this layer. Identify:

- **Internal spermatic fascia**
- **Deep inguinal ring**

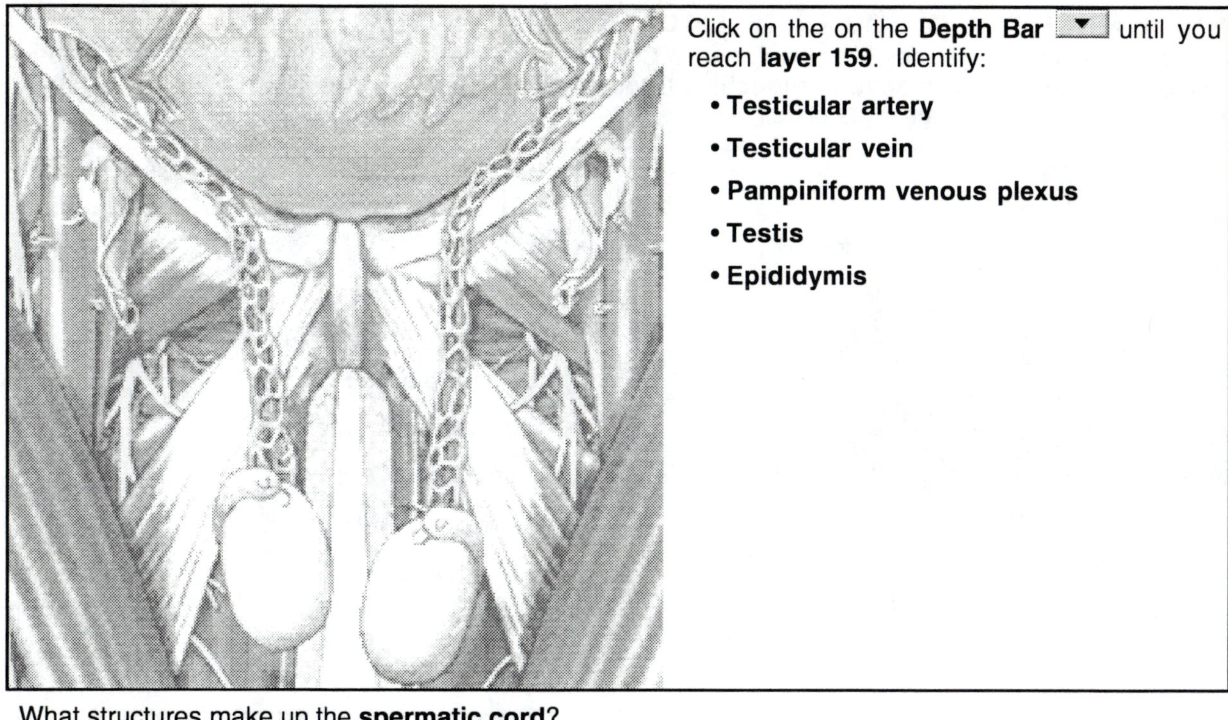

Click on the on the **Depth Bar** ▼ until you reach **layer 159**. Identify:

- **Testicular artery**
- **Testicular vein**
- **Pampiniform venous plexus**
- **Testis**
- **Epididymis**

What structures make up the **spermatic cord**?

________________________________________

________________________________________

________________________________________

________________________________________

Scroll down to **layer 180**. No graphic has been provided for this layer. Identify:

- **Ductus (vas) deferens**

When you are finished, **close** the window.

## Atlas Anatomy of the Male Reproductive System

Under the **File** menu, drag down to **Open** (then to **Content** if you are working in Windows). Click on **Atlas Anatomy**. Select **System** and **Reproductive**. Choose **Male Pelvic Organs (Ant)**. If the list is not in alphabetical order, the image is about half of the way down the list.

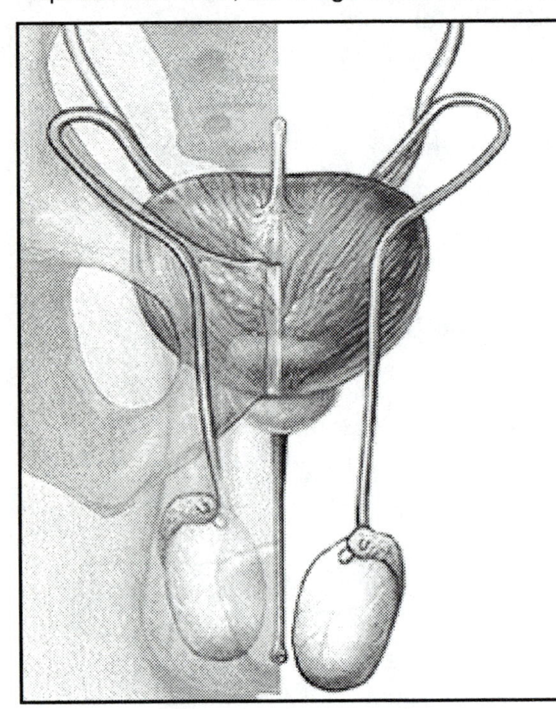

If you are working in Mac, click on the **Show All Pins** button and scroll to **All Systems**. Identify:

- **Ureter**
- **Median umbilical ligament**
- **Urinary bladder**
- **Ductus (Vas) deferens**
- **Seminal vesicle**
- **Prostate gland**
- **Epididymis**
- **Urethra**
- **Testis**
- **External meatus (orifice) of urethra**

When you are finished, **close** the window.

Spermatozoa mature in what structure? ______________________________

Which gland produces most of the volume of semen? ______________________________

Trace the path of the sperm from the testis to the urethra naming all structures (in order) through which the sperm would pass.

______________________________

______________________________

Under the **File** menu, drag down to **Open** (then to **Content** if you are working in Windows). Click on **Atlas Anatomy**. **System** and **Reproductive** should still be selected. Choose **Male Pelvic Organs (Lat)**. If the list is not in alphabetical order, the image is half of the way down the list.

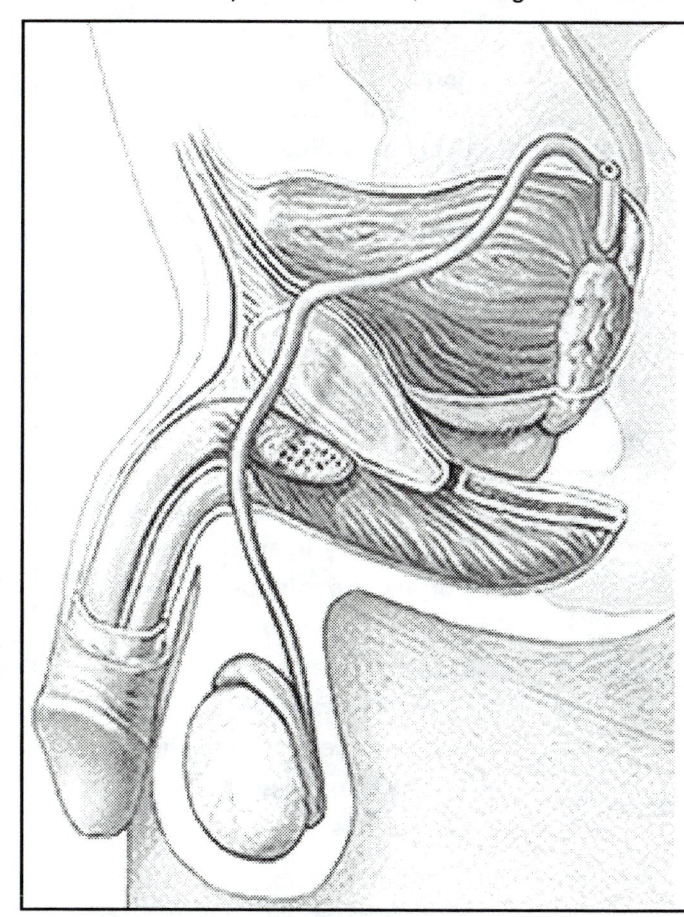

If you are working in Mac, click on the **Show All Pins** button and scroll to **All Systems**. Identify:

- **Testis**
- **Head of epididymis**
- **Body of epididymis**
- **Tail of epididymis**
- **Ductus deferens**
- **Corpus spongiosum**
- **Corpus cavernosum**
- **Prostate gland**
- **Seminal vesicle**
- **Left ureter**
- **Ampulla of ductus deferens**
- **Urinary bladder**
- **Right ureter**
- **Urogenital diaphragm**

When you are finished, **close** the window.

Under the **File** menu, drag down to **Open** (then to **Content** if you are working in Windows). Click on **Atlas Anatomy**. **System** and **Reproductive** should still be selected. Choose **Section of Male Bladder (Ant)**. If the list is not in alphabetical order, the image is one-third of the way down the list.

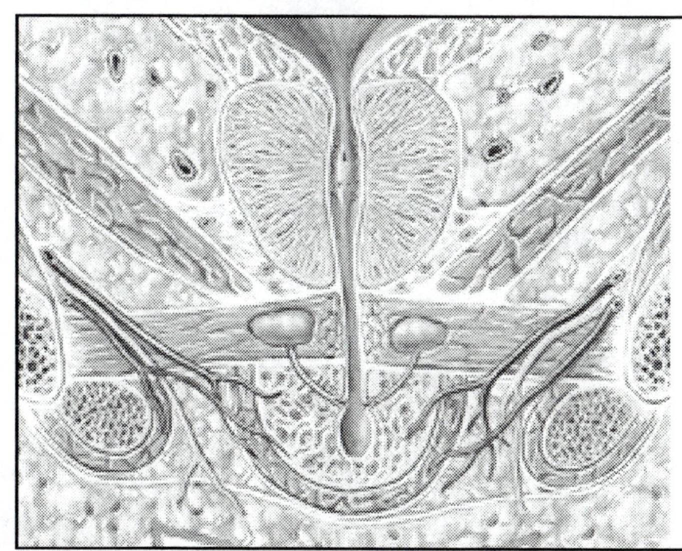

If you are working in Mac, click on the **Show All Pins** button and scroll to **All Systems**. Identify:

- **Ejaculatory duct**
- **Bulbourethral gland**
- **Corpus spongiosum**
- **Crus of penis**
- **Duct of bulbourethral gland**
- **Membranous urethra**
- **Prostate gland**
- **Spongy urethra**

When you are finished, **close** the window.

Under the **File** menu, drag down to **Open** (then to **Content** if you are working in Windows). Click on **Atlas Anatomy**. **System** and **Reproductive** should still be selected. Choose **Male Pelvic Organs (Med) 1**. If the list is not in alphabetical order, the image is half of the way down the list.

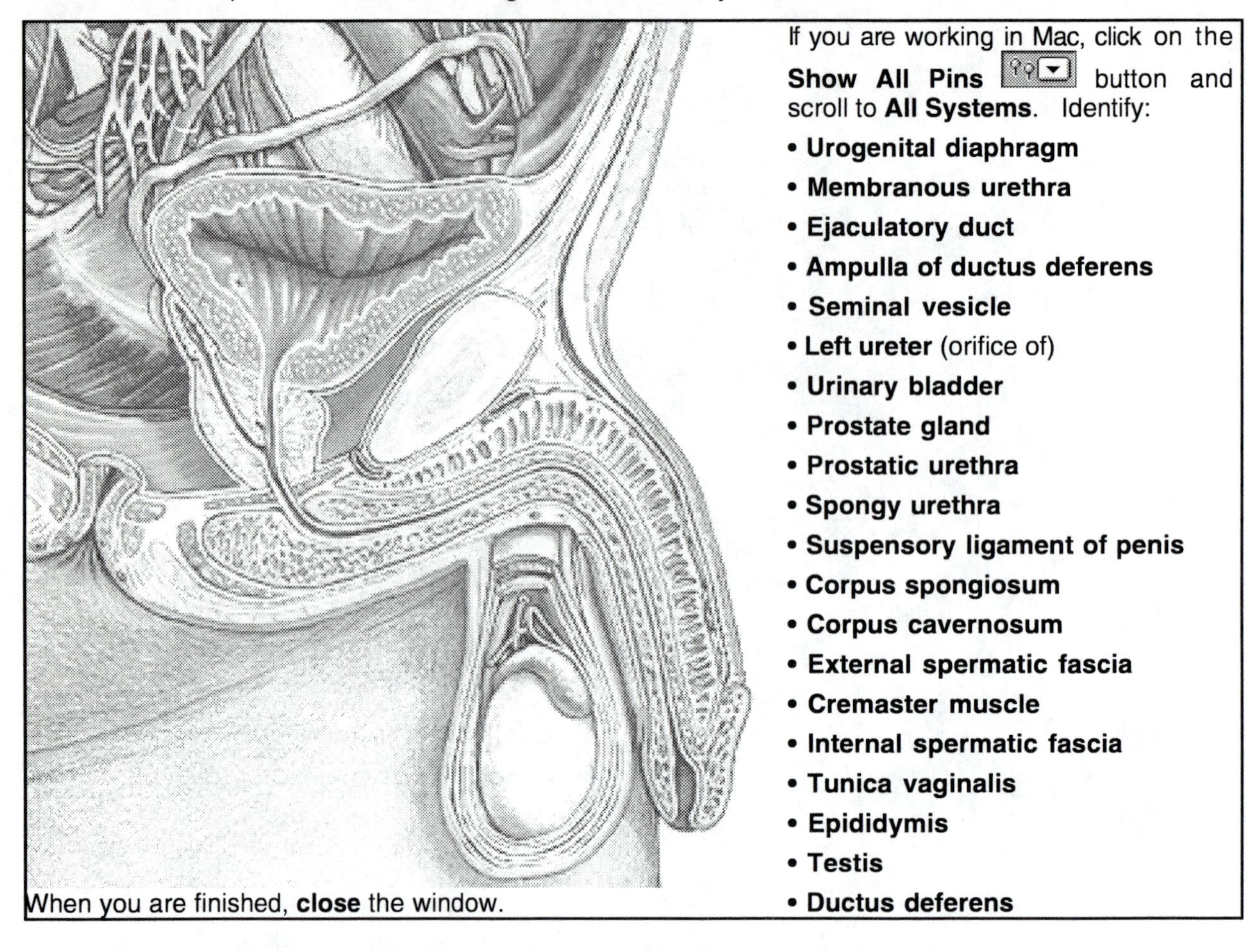

If you are working in Mac, click on the **Show All Pins** button and scroll to **All Systems**. Identify:

- **Urogenital diaphragm**
- **Membranous urethra**
- **Ejaculatory duct**
- **Ampulla of ductus deferens**
- **Seminal vesicle**
- **Left ureter** (orifice of)
- **Urinary bladder**
- **Prostate gland**
- **Prostatic urethra**
- **Spongy urethra**
- **Suspensory ligament of penis**
- **Corpus spongiosum**
- **Corpus cavernosum**
- **External spermatic fascia**
- **Cremaster muscle**
- **Internal spermatic fascia**
- **Tunica vaginalis**
- **Epididymis**
- **Testis**
- **Ductus deferens**

When you are finished, **close** the window.

What structure forms the crus of the penis?________________________________________

Under the **File** menu, drag down to **Open** (then to **Content** if you are working in Windows). Click on **Atlas Anatomy**. **System** and **Reproductive** should still be selected. Choose **Male Superficial Perineal Space 1**. If the list is not in alphabetical order, the image is close to the bottom of the list.

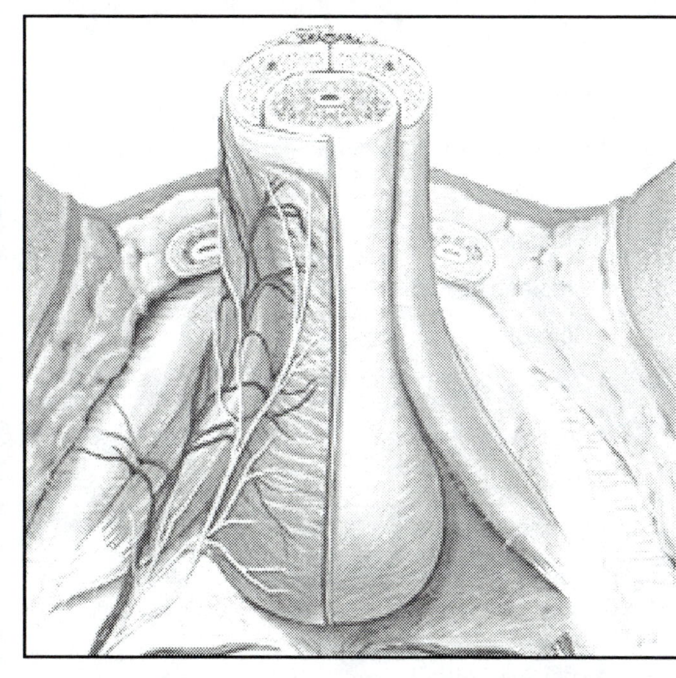

If you are working in Mac, click on the **Show All Pins** button and scroll to **All Systems**. Testes have been removed here. Identify:

- **Bulb of Penis**
- **Bulbospongiosus muscle**
- **Corpus cavernosum**
- **Corpus spongiosum**
- **Ductus deferens**
- **Ischiocavernosus muscle**
- **Spongy urethra**

When you are finished, **close** the window.

What structure forms the bulb of the penis? What structures form the erectile tissue of the penis?

---

Erection of the penis s under the control of what division of the autonomic nervous system?

---

Under the **File** menu, drag down to **Open** (then to **Content** if you are working in Windows). Click on **Atlas Anatomy**. **System** and **Reproductive** should still be selected. Choose **Perineal Muscles of Male Pelvis**. If the list is not in alphabetical order, the image is one-third of the way down the list.

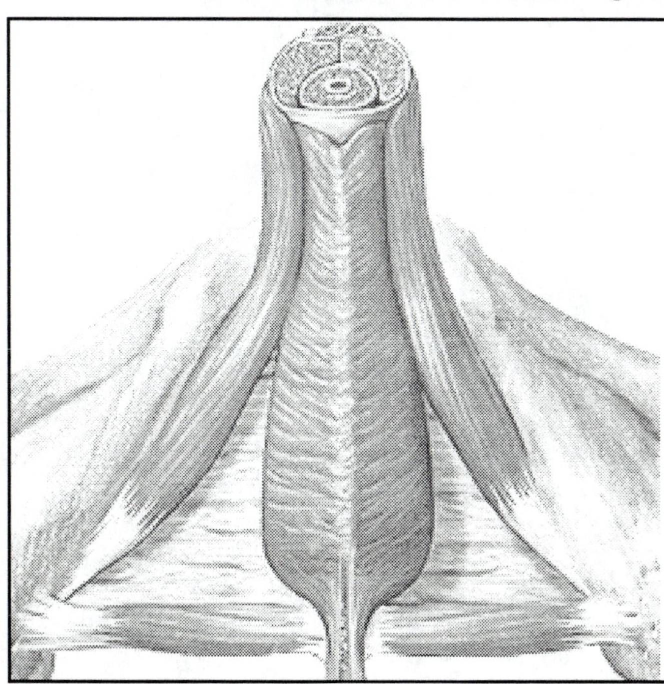

If you are working in Mac, click on the **Show All Pins** button and scroll to **All Systems**. Identify:

- **Bulbospongiosus muscle**
- **Corpus cavernosum**
- **Corpus spongiosum**
- **Ischiocavernosus muscle**
- **Urethra**

When you are finished, **close** the window.

## Objectives: Male Reproductive System

Student will learn the anatomy of the male reproductive system.

- Ampulla of ductus {vas} deferens
- Body of penis
- Bulbourethral gland
- Bulb of penis
- Corpus cavernosum
- Corpus spongiosum
- Crus of penis
- Cremaster muscle
- Dorsal artery of penis
- Dorsal nerve of penis
- Ductus {vas} deferens
- Ejaculatory duct
- Body of epididymis
- Head of epididymis
- Tail of epididymis
- External meatus {orifice} of urethra
- External spermatic fascia
- Internal spermatic fascia
- Inguinal ligament
- Glans penis
- Ischiocavernosus muscle
- Median umbilical ligament
- Membranous urethra
- Pampiniform venous plexus
- Prepuce of penis
- Prostate gland
- Prostatic urethra
- Scrotum
- Seminal vesicle
- Skin
- Spermatic cord
- Spongy urethra
- Suspensory ligament of penis
- Testes
- Testicular artery
- Testicular vein
- Tunica vaginalis
- Ureter
- Urethra
- Urinary bladder
- Urogenital diaphragm
- Vein of penis

# Female Reproductive System

## Opening A.D.A.M Interactive Anatomy

Open A.D.A.M. Interactive Anatomy according to the directions in the Tutorial.

## 3D Anatomy of the Female Reproductive System

Choose **3D Anatomy**. Click on **3D Female Reproductive System**. Identify the following by clicking on the **3D Anatomy Structure List**:

- **Broad ligament of uterus** - Cut
- **Cervix**
- **External os of cervical canal**

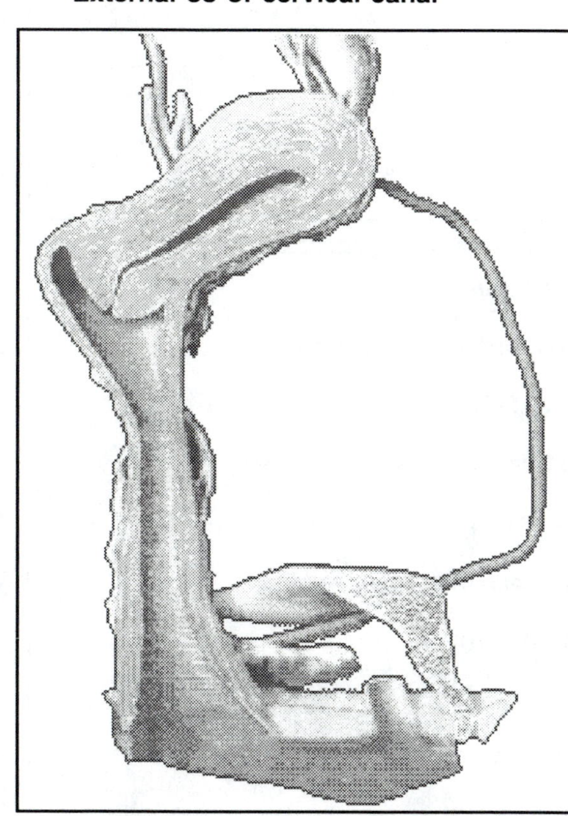

Identify:

- **Anterior wall of vagina** - Midsagittal cut
- **Body of uterus**
- **Cervical canal** - Midsagittal cut (2nd one)
- **Cervix** - Midsagittal cut
- **Endometrium** - Midsagittal cut
- **External orifice of urethra** - Midsagittal cut
- **Clitoris**- Midsagittal cut
- **External os of cervical canal**- Midsagittal cut
- **Fornix of vagina** - Midsagittal cut
- **Fundus of uterus**- Midsagittal cut
- **Myometrium** - Midsagittal cut
- **Orifice of vagina**- Midsagittal cut
- **Perimetrium** - Midsagittal cut
- **Posterior wall of vagina** - Midsagittal cut
- **Uterus** - Midsagittal cut
- **Vagina** - Midsagittal cut
- **Uterine cavity**- Midsagittal cut

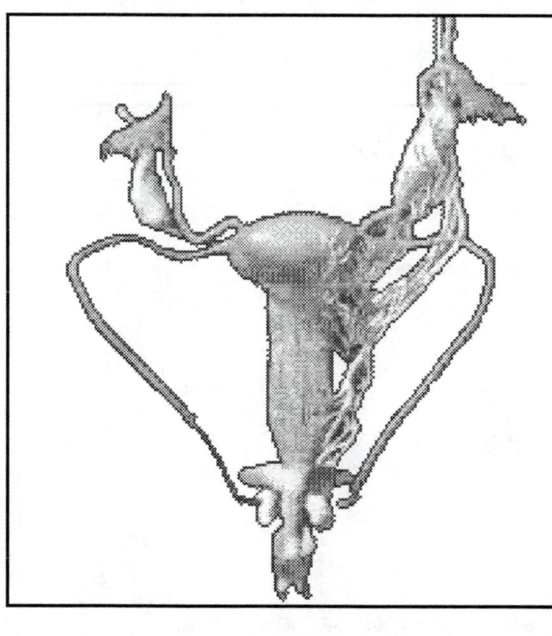

Identify:

- **Body of clitoris**
- **Crus of clitoris**- Midsagittal cut
- **Fundus of uterus**

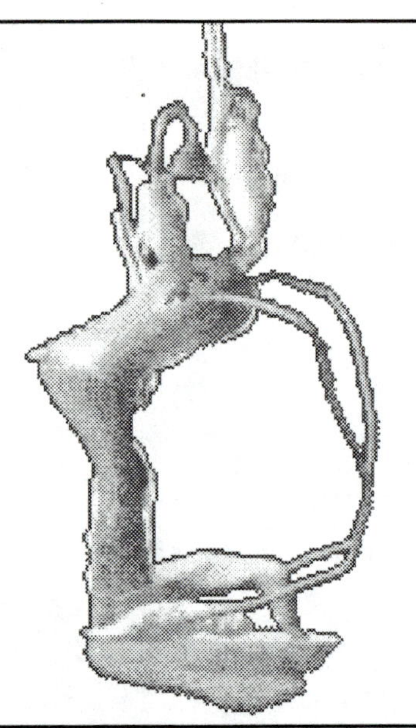

Identify:
- **Body of uterus**
- **Greater vestibular gland**
- **Uterus**
- **Vagina**

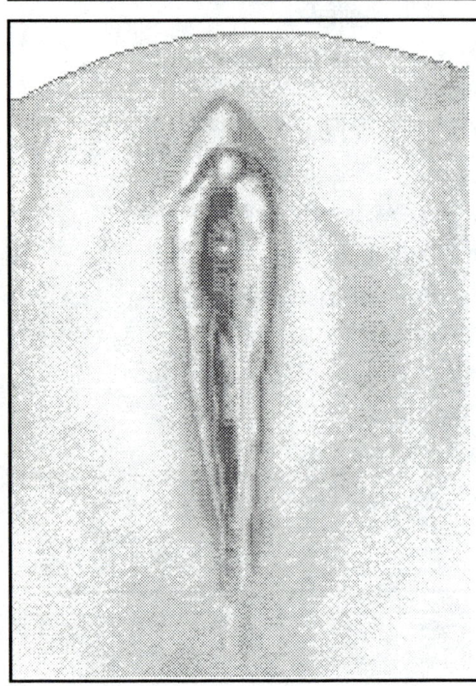

Identify:
- **Clitoris** - Skin
- **External orifice of urethra** - Skin
- **Hymen** - Skin
- **Labium majus**- Skin
- **Labium minus**- Skin
- **Mons pubis**- Midsagittal cut
- **Orifice of vagina**
- **Prepuce of clitoris**- Skin
- **Vestibule of vagina**- Skin
- **Vulva**- Skin

Identify:
- **Corpus albicans** - Cut
- **Corpus luteum**- Cut
- **Fimbria of uterine tube**
- **Ovary**
- **Ovary** - Cut
- **Uterine tube**

When you are finished, **close** the window.

# Medial Dissection of the Female Reproductive System

Choose **Dissectible Anatomy**.

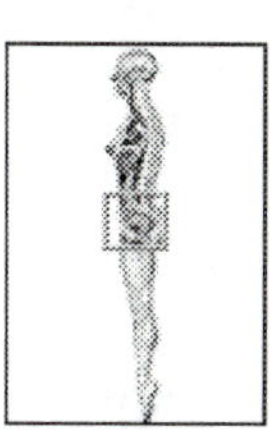

Choose **female**. Select **Medial.** Click **Open**.

Enlarge the window and position the navigation rectangle over the pelvic region.

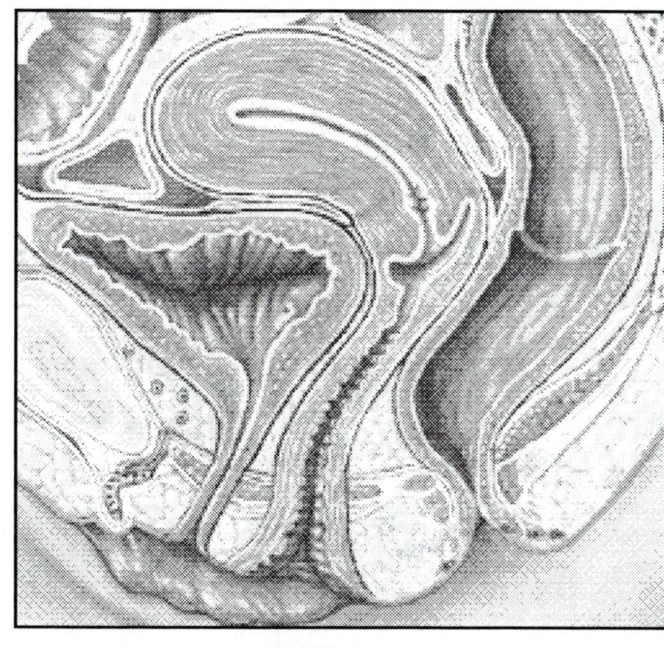

At **layer 0**, identify:

- **Labium minus**
- **Labium majus**
- **Clitoris**
- **Urinary bladder**
- **Vagina**
- **Fornix of vagina**
- **Cervix**
- **Body of uterus**
- **Fundus of uterus**
- **Rectum**
- **Pubic symphysis**

The vestibule is composed of what structures?

_______________________________________________

The folds in the lining of the vaginal wall are termed ____________________.

Which structure is the female equivalent of the male penis?

_______________________________________________

Use the **Depth Bar** [▼] to scroll to **layer 51**. No graphic has been provided for this layer. Identify:

- **Ovary**
- **Infundibulum**
- **Ampulla of the uterine tube**

Click on the **Depth Bar** [▼] a few times to **layer 56**. No graphic has been provided for this layer. Identify:

- **Urethra**

## Anterior Dissection of the Female Reproductive System

Click on the **View** button and choose **Anterior**.

Click on the **Normal** button. **Windows**—— ——**Mac**

Drag the **Depth Bar** to **layer 0.** No graphic has been provided for this layer. Identify:

- **Labium majus**

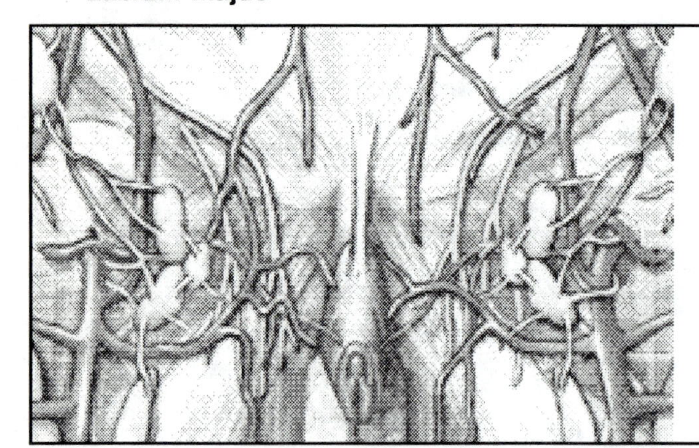

Click on the **Depth Bar** to **layer 1**. No graphic has been provided for this layer. Identify:

- **Labium minus**
- **Prepuce of clitoris**
- **Clitoris**

The prepuce of the clitoris is composed of what structures?

---

Click on the **Depth Bar** a few times to **layer 3**. Identify:

- **Suspensory ligament of clitoris**

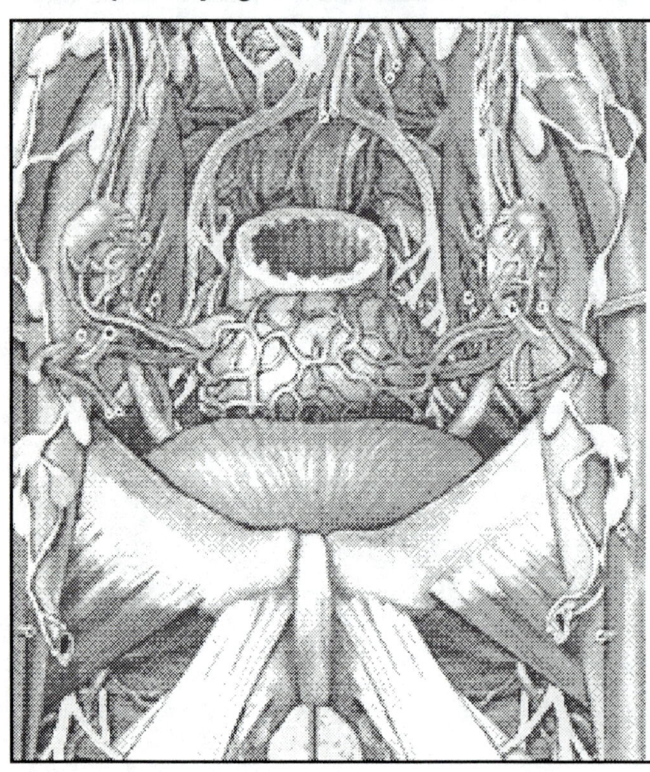

Use the **Depth Bar** to scroll to **layer 214**. Identify:

- **Pubis**
- **Pubic symphysis**
- **Ovary**
- **Infundibulum of uterine tube**
- **Fimbria of uterine tube**
- **Ampulla of uterine tube**
- **Ovarian artery**
- **Body of uterus**
- **Uterine venous plexus**
- **Uterine artery**
- **Fundus of uterus**
- **Urinary bladder**
- **Ureter** (Renal fascia covering the)
- **Rectum**

What cellular structures would you expect to dominate the cells lining the fimbriae?

---

The uterine tubes are lined with what type of epithelial tissue?

---

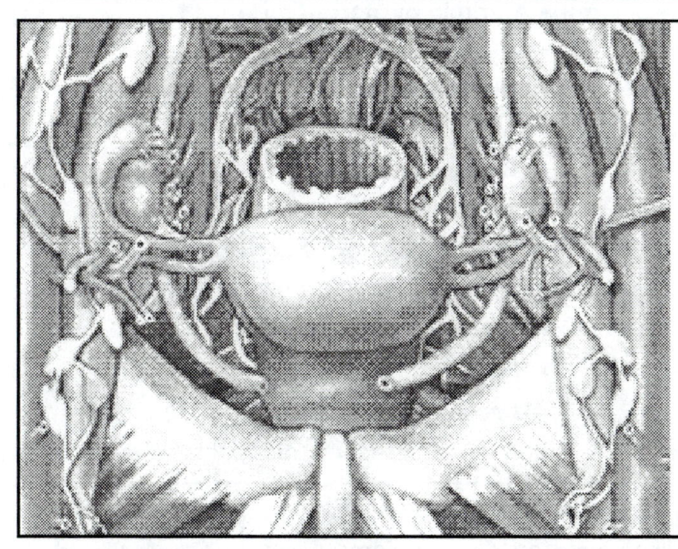

Use the **Depth Bar** [▼] to scroll to **layer 226**. Identify:

- **Ovary**
- **Fimbria of uterine tube**
- **Ampulla of uterine tube**
- **Isthmus of uterine tube**
- **Body of uterus**
- **Round ligament**
- **Fundus of uterus**
- **Cervix**
- **Vagina**

When you are finished, **close** the window.

The round ligament of the uterus would prevent movement in which direction? (anterior, posterior, medial, lateral)

---

## Atlas Anatomy of the Female Reproductive System

Under the **File** menu, drag down to **Open** (then to **Content** if you are working in Windows). Click on **Atlas Anatomy**. Select **System** and **Reproductive**. Choose **Female Pelvic Organs (Ant)**. If the list is not in alphabetical order, the image is about one-third of the way down the list.

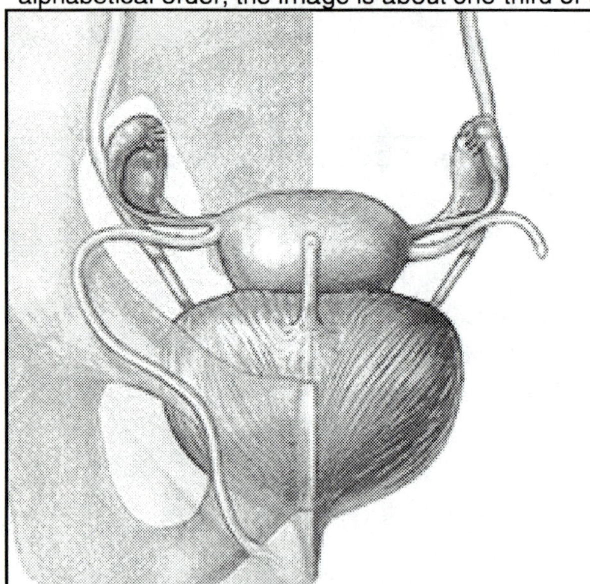

If you are working in Mac, click on the **Show All Pins** button and scroll to **All Systems**. Identify the following:

- **Ampulla of uterine tube**
- **Infundibulum of uterine tube**
- **Isthmus of uterine tube**
- **Median umbilical ligament**
- **Ovarian ligament**
- **Ovary**
- **Round ligament of uterus**
- **Urinary bladder**
- **Ureter**
- **Urethra**
- **Uterus**

When you are finished, **close** the window.

Under the **File** menu, drag down to **Open** (then to **Content** if you are working in Windows). Click on **Atlas Anatomy**. **System** and **Reproductive** should still be selected. Choose **Female Pelvic Organs (Lat)**. If the list is not in alphabetical order, the image is about one-third of the way down the list.

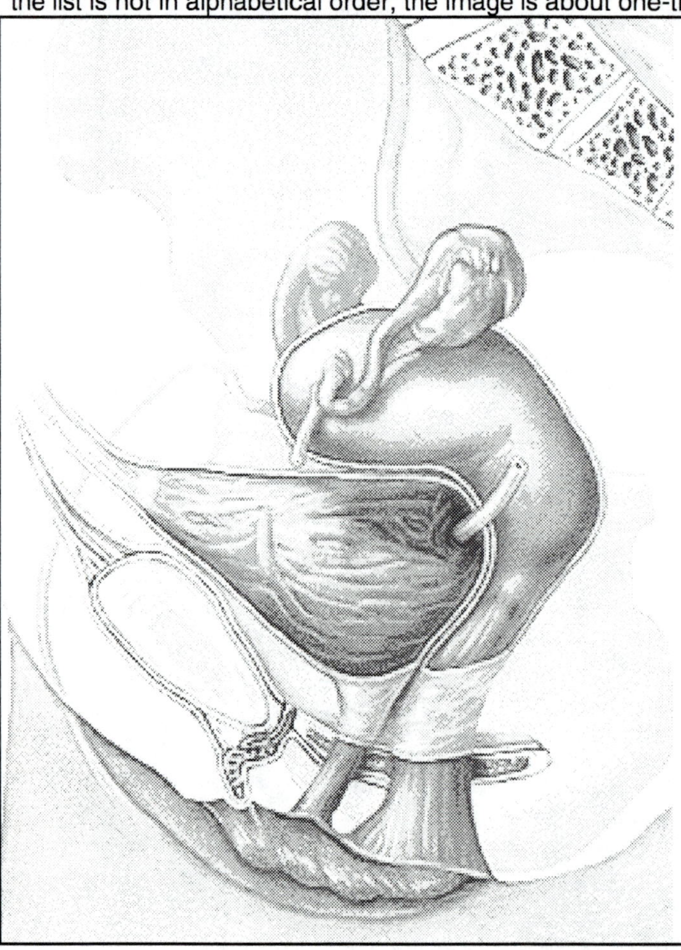

If you are working in Mac, click on the **Show All Pins** button and scroll to **All Systems**. Identify:

- **Ampulla of uterine tube**
- **Infundibulum of uterine tube**
- **Isthmus of uterine tube**
- **Median umbilical ligament**
- **Body of uterus**
- **Ovary**
- **Cervix**
- **Urinary bladder**
- **Ureter**
- **Urethra**
- **Vagina**
- **Clitoris**
- **Fundus of uterus**
- **Labium majus**
- **Labium minus**
- **Left ureter**
- **Urogenital diaphragm**

When you are finished, **close** the window.

Under the **File** menu, drag down to **Open** (then to **Content** if you are working in Windows). Click on **Atlas Anatomy**. **System** and **Reproductive** should still be selected. Choose **Female Superficial Perineal Space 1**. If the list is not in alphabetical order, the image is a little over one half of the way down the list.

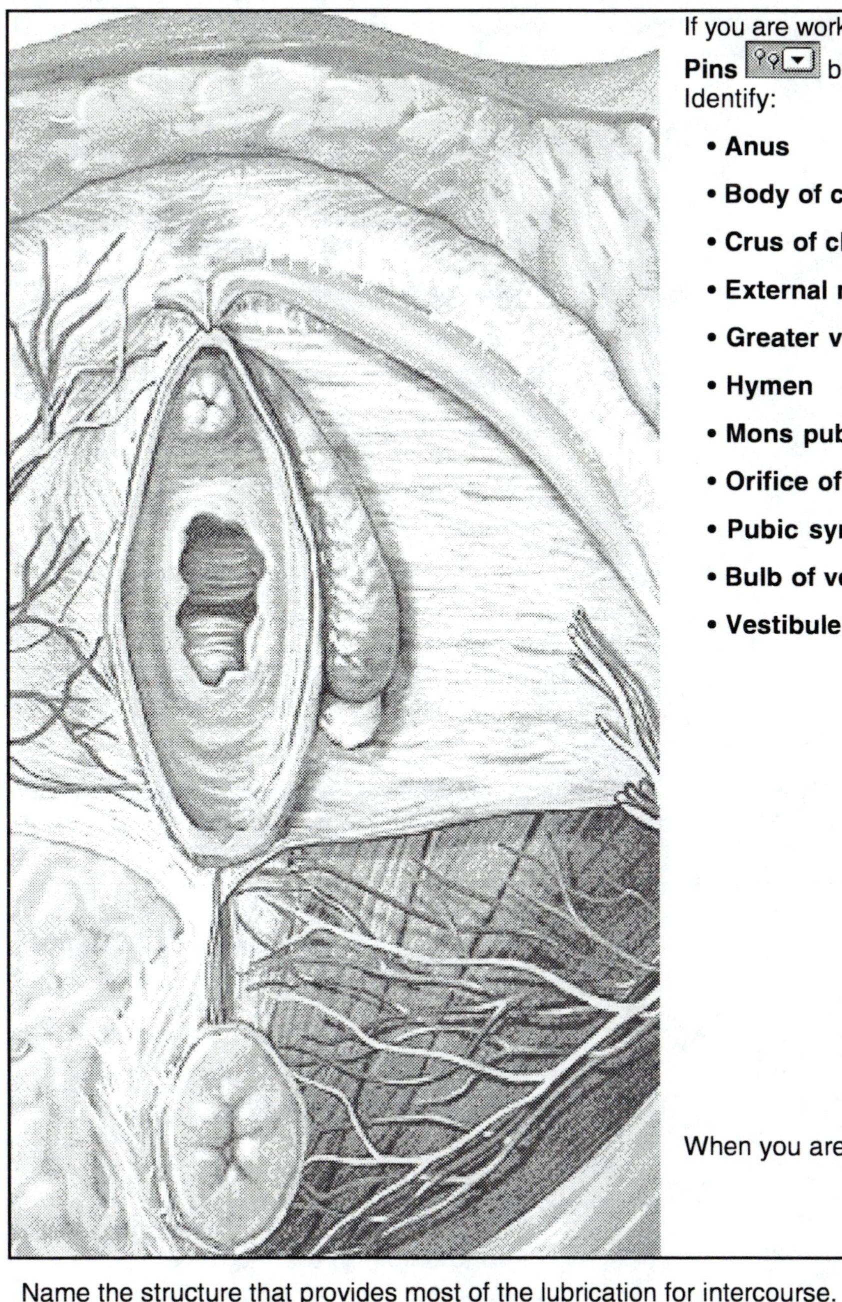

If you are working in Mac, click on the **Show All Pins** button and scroll to **All Systems**. Identify:

- **Anus**
- **Body of clitoris**
- **Crus of clitoris**
- **External meatus {orifice} of urethra**
- **Greater vestibular gland**
- **Hymen**
- **Mons pubis**
- **Orifice of vagina**
- **Pubic symphysis**
- **Bulb of vestibule**
- **Vestibule of vagina**

When you are finished, **close** the window.

Name the structure that provides most of the lubrication for intercourse.

_______________________________________________

The epithelial membrane that separates the vestibule and vagina prior to first intercourse is the

______________________.

The bulb of the vestibule is homologous to the ______________________ of the male.

Under the **File** menu, drag down to **Open** (then to **Content** if you are working in Windows). Click on **Atlas Anatomy**. **System** and **Reproductive** should still be selected. Choose **Section of Female Bladder (Ant)**. If the list is not in alphabetical order, the image is about one-third of the way down the list.

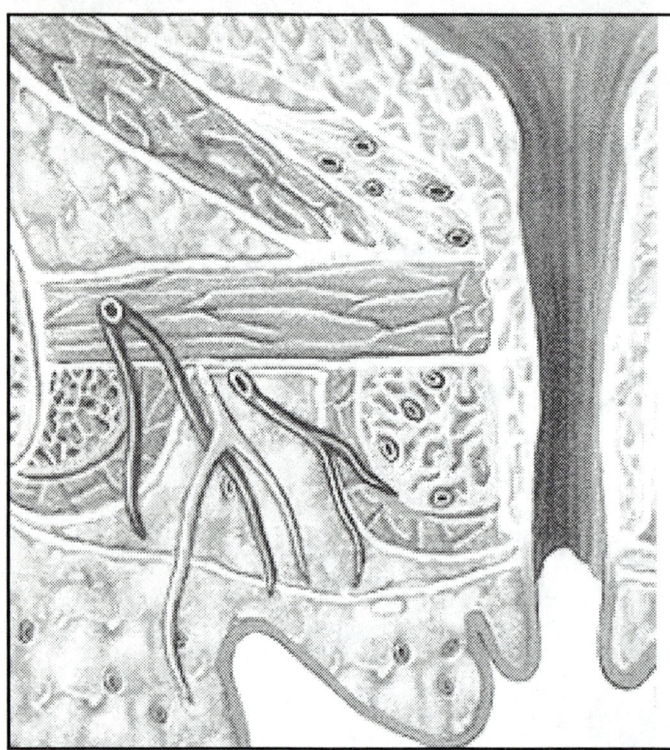

If you are working in Mac, click on the **Show All Pins** button and scroll to **All Systems**. Identify:

- **Bulbospongiosus muscle**
- **Crus of clitoris**
- **Ischiocavernosus muscle**
- **Labium majus**
- **Labium minus**
- **Urethra**
- **Bulb of vestibule**

When you are finished, **close** the window.

Under the **File** menu, drag down to **Open** (then to **Content** if you are working in Windows). Click on **Atlas Anatomy**. **System** and **Reproductive** should still be selected. Choose **Viscera of Female Pelvis (Sup)**. If the list is not in alphabetical order, the image is about one half of the way down the list.

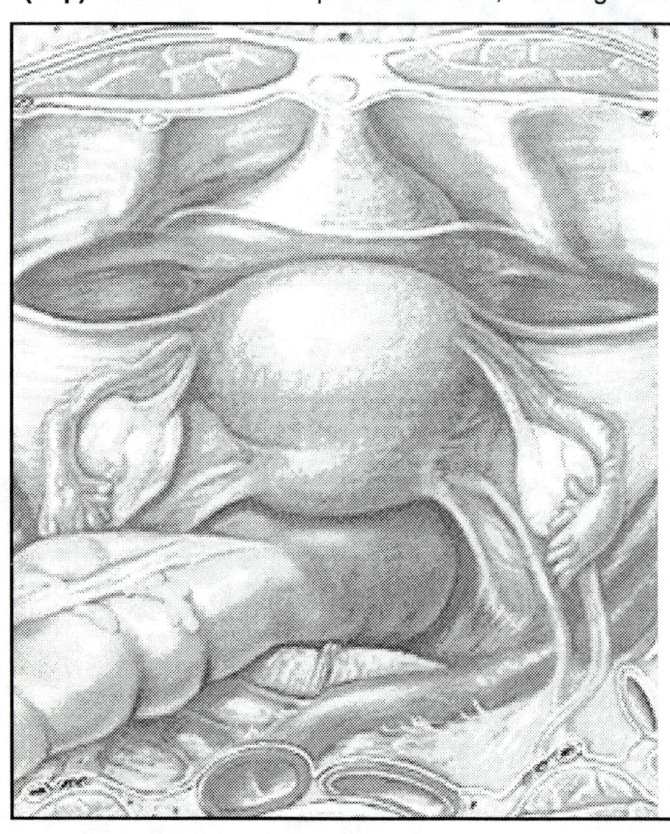

If you are working in Mac, click on the **Show All Pins** button and scroll to **All Systems**. Identify:

- **Ampulla of uterine tube**
- **Body of uterus**
- **Broad ligament of uterus**
- **Cervix**
- **Fimbria of uterine tube**
- **Fundus of uterus**
- **Infundibulum of uterine tube**
- **Isthmus of uterine tube**
- **Ovarian ligament**
- **Ovary**
- **Rectouterine pouch**
- **Rectum**
- **Round ligament of uterus**
- **Sigmoid colon**
- **Suspensory ligament of ovary**
- **Vesicouterine pouch**

When you are finished, **close** the window.

## Cadaver Images of the Female Reproductive System

Under the **File** menu, drag down to **Open** (then to **Content** if you are working in Windows). Click on **Atlas Anatomy**. **System** and **Reproductive** should still be selected. Choose **Contents of Female Pelvis (Sup)**. If the list is not in alphabetical order, the image is about one-third of the way down the list.

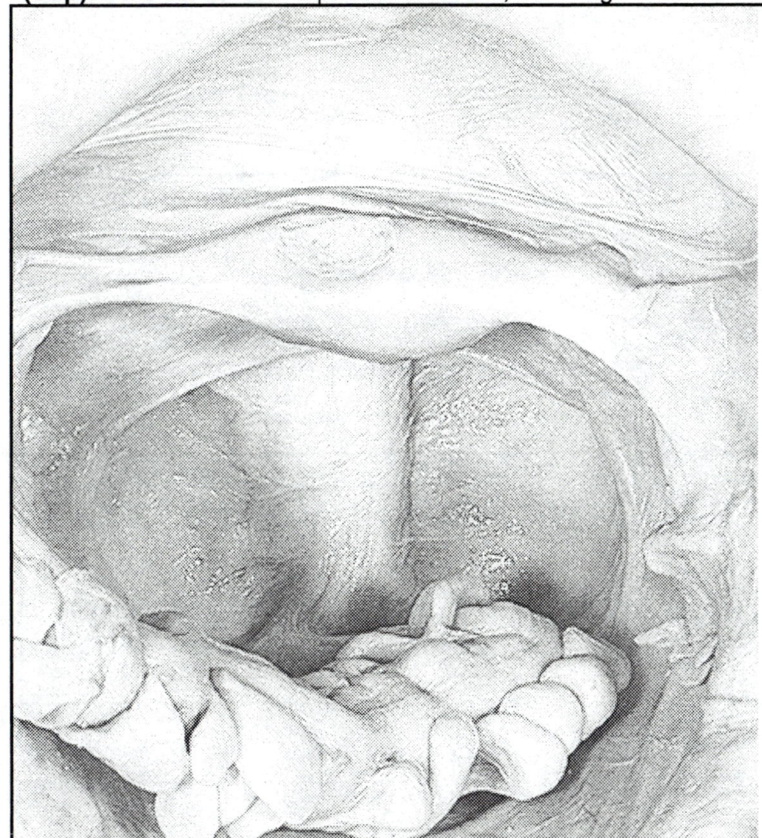

If you are working in Mac, click on the **Show All Pins** button and scroll to **All Systems**. Identify:

- **Body of uterus**
- **Broad ligament of uterus**
- **Ovary**
- **Rectouterine pouch**
- **Rectum**
- **Urinary bladder**
- **Sigmoid colon**
- **Uterine tube**
- **Vesicouterine pouch**

When you are finished, **close** the window.

Under the **File** menu, drag down to **Open** (then to **Content** if you are working in Windows). Click on **Atlas Anatomy**. **System** and **Reproductive** should still be selected. Choose **Female Pelvic Organs (Med) 2**. If the list is not in alphabetical order, the image is about one-third of the way down.

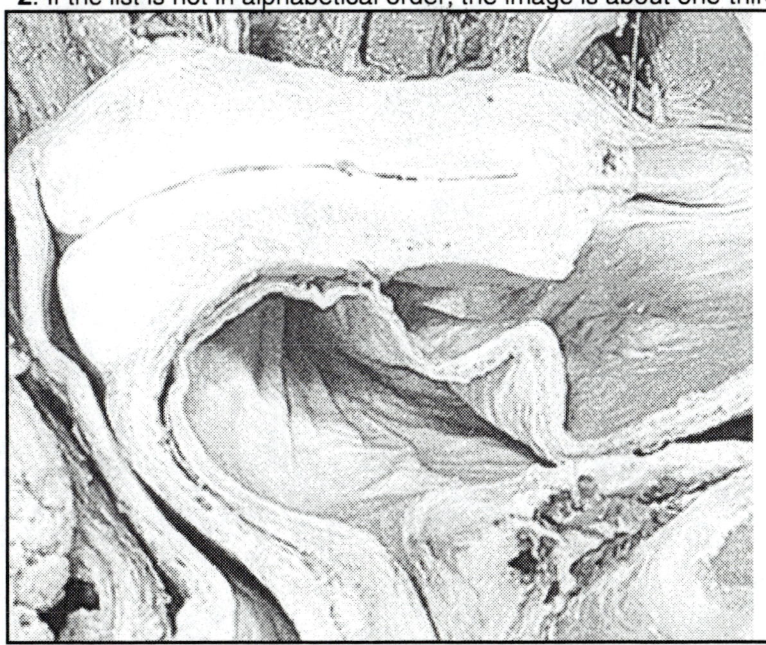

If you are working in Mac, click on the **Show All Pins** button and scroll to **All Systems**. Identify:

- **Body of uterus**
- **Cervical canal**
- **Cervix**
- **External os of cervical canal**
- **Fornix of vagina**
- **Fundus of uterus**
- **Pubic symphysis**
- **Rectum**
- **Urinary bladder**
- **Urethra**
- **Uterine cavity**
- **Vagina**

When you are finished, **close** the window.

# Anterior Dissection of the Breast

Choose **Dissectible Anatomy**.

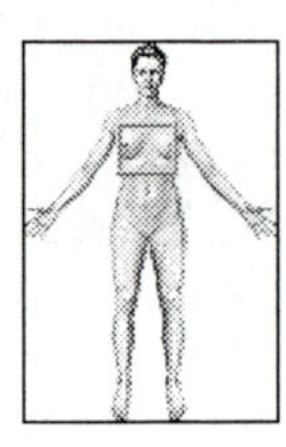

Choose **Female**. Select **Anterior.** Click **Open**.

Enlarge the window and position the image over the breasts.

Scroll up to **layer 0.** No graphic has been provided for this layer. Identify:

- **Nipple**
- **Areola**

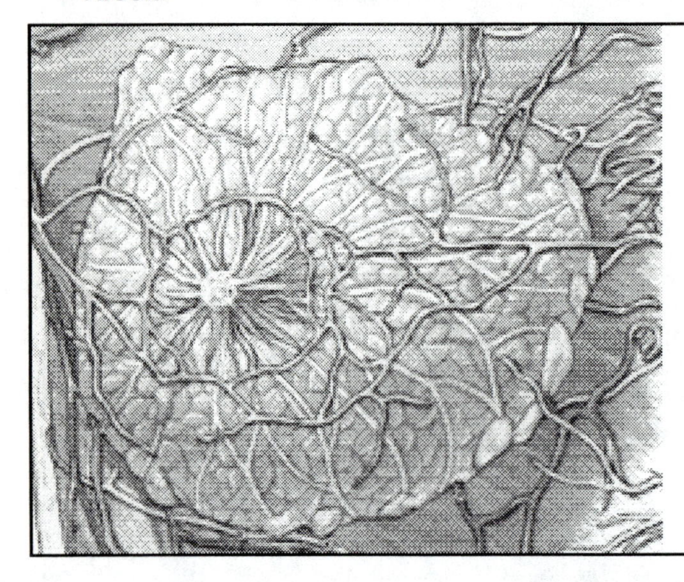

Click on the **Depth Bar** ▼ a few times to **layer 6**. Identify:

- **Adipose tissue**
- **Suspensory ligament of breast**
- **Lactiferous duct of breast**
- **Lobe of mammary gland**
- **Nipple**
- **Lymph vessel of breast**

The mammary glands are of what type? (merocrine, apocrine, holocrine). ____________________________

The mammary glands are specialized organs of what system? ________________________________________

## Lateral Dissection of the Breast

Click on the **View** button and choose **Lateral**.

Windows— —Mac

Click on the **Normal** button.

Drag the **Depth Bar** to **layer 0.**

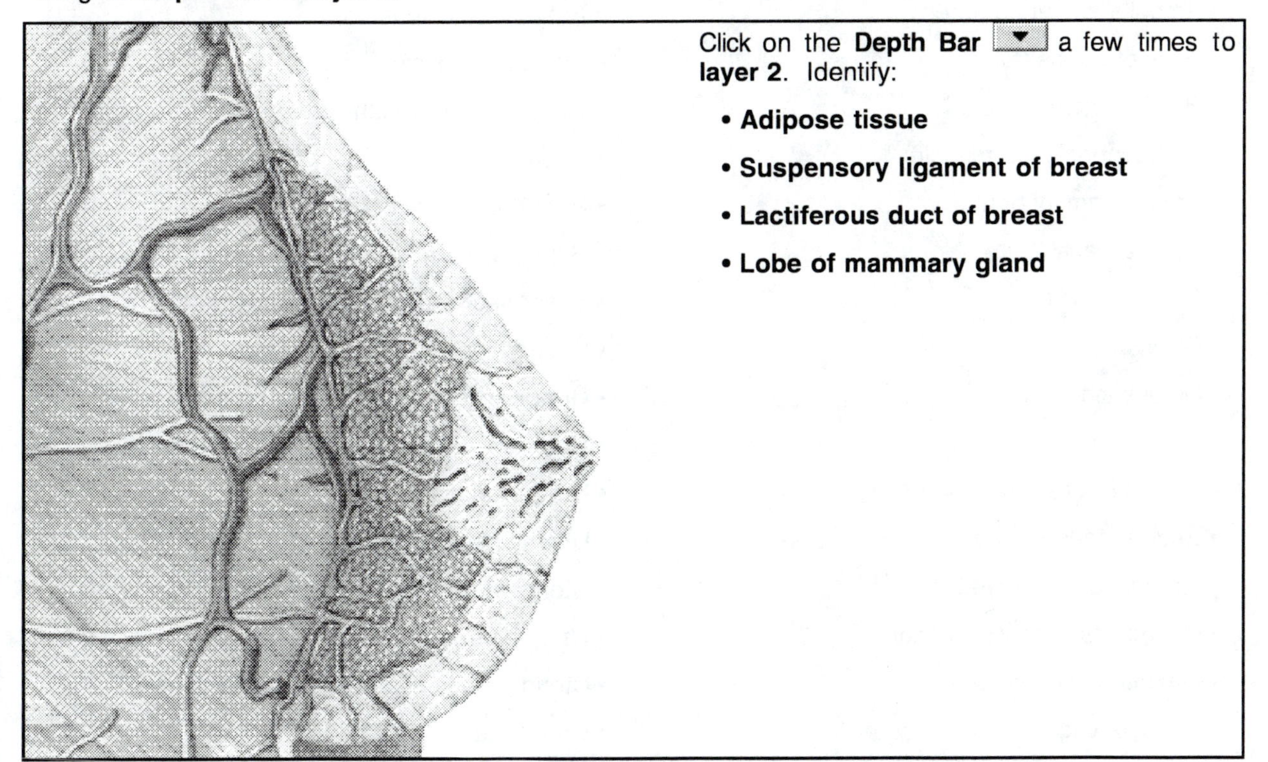

Click on the **Depth Bar** a few times to **layer 2**. Identify:

- **Adipose tissue**
- **Suspensory ligament of breast**
- **Lactiferous duct of breast**
- **Lobe of mammary gland**

# Objectives: Female Reproductive System

Student will learn the anatomy of the female reproductive system.

- Anterior wall of vagina
- Anus
- Ampulla of uterine tube
- Body of clitoris
- Body of uterus
- Bulbospongiosus muscle
- Broad ligament of uterus
- Cervical canal
- Cervix
- Clitoris
- Corpus albicans
- Corpus luteum
- Crus of clitoris
- Endometrium
- External orifice of urethra
- External os of cervical canal
- Fimbria of uterine tube
- Fornix of vagina
- Fundus of uterus
- Greater vestibular gland
- Hymen - Skin
- Ischiocavernosus muscle
- Isthmus of uterine tube
- Infundibulum of uterine tube
- Labium majus
- Labium minus
- Median umbilical ligament
- Mons pubis
- Myometrium
- Orifice of vagina
- Ovary
- Ovarian artery
- Ovarian ligament
- Perimetrium
- Posterior wall of vagina
- Prepuce of clitoris
- Pubic symphysis
- Rectouterine pouch
- Rectum
- Round ligament of uterus
- Sigmoid colon
- Suspensory ligament of clitoris
- Suspensory ligament of ovary
- Ureter
- Urethra
- Urogenital diaphragm
- Urinary bladder
- Uterine artery
- Uterine venous plexus
- Uterine cavity
- Uterine tube
- Uterus
- Vesicouterine pouch
- Vagina
- Vestibule (Bulb of the)
- Vestibule of vagina
- Vulva

**Breast**

- Adipose tissue
- Suspensory ligament of breast
- Lactiferous duct of breast
- Lobe of mammary gland
- Nipple
- Lymph vessel of breast